INTERPERSONAL PROCESS IN THERAPY

An Integrative Model

SIXTH EDITION

Edward Teyber

Faith Holmes McClure
California State University, San Bernardino

BROOKS/COLE
CENGAGE Learning™

Australia • Brazil • Japan • Korea • Mexico • Singapore • Spain • United Kingdom • United States

BROOKS/COLE
CENGAGE Learning™

Interpersonal Process In Therapy: An Integrative Model, Sixth Edition,
Edward Teyber, Faith Holmes McClure

Acquisitions Editor, Counseling: Seth Dobrin

Assistant Editor: Nicolas Albert

Editorial Assistant: Rachel McDonald

Media Editor: Dennis Fitzgerald

Marketing Manager: Trent Whatcott

Marketing Coordinator: Gurpeet S. Saran

Marketing Communications Manager: Tami Strang

Content Project Management: Pre-PressPMG

Design Director: Rob Hugel

Art Director: Caryl Gorska

Print Buyer: Linda Hsu

Rights Acquisitions Specialist, Text: Roberta Broyer

Rights Acquisitions Specialist, Images: John Hill

Production Service: Pre-PressPMG

Copy Editor: David Coen

Cover Designer: Carole Lawson, Lawson Graphics

Compositor: Pre-PressPMG

For product information and technology assistance, contact us at **Cengage Learning Customer & Sales Support, 1-800-354-9706**

For permission to use material from this text or product, submit all requests online at **www.cengage.com/permissions**

Further permissions questions can be e-mailed to **permissionrequest@cengage.com**

Library of Congress Control Number: 2010928004

ISBN-13: 978-0-495-60420-4

ISBN-10: 0-495-60420-8

Brooks/Cole
20 Davis Drive
Belmont, CA 94002-3098
USA

Cengage Learning is a leading provider of customized learning solutions with office locations around the globe, including Singapore, the United Kingdom, Australia, Mexico, Brazil, and Japan. Locate your local office at **www.cengage.com/global**

Cengage Learning products are represented in Canada by Nelson Education, Ltd.

To learn more about Brooks/Cole, visit **www.cengage.com/ brookscole**

Purchase any of our products at your local college store or at our preferred online store **www.cengagebrain.com**

Printed in the United States of America
1 2 3 4 5 6 7 14 13 12 11 10

Dedicated to those who are struggling to change;

To all of our children, who have strengthened our commitment to making a difference;

And to all of those who have come before and taught about the power of relationships to bring about change.

CONTENTS

PREFACE

WHO ARE WE AND WHAT IS OUR PURPOSE?

We are both therapists who continue to take great pleasure in helping clients change, and throughout the nearly 70 collective years we have been practicing, continue to find this work so deeply meaningful. We have spent our careers involved in every aspect of clinical work: seeing clients, supervising graduate student therapists on their initial client caseloads, directing treatment/training clinics, conducting psychotherapy process research, and carrying out large-scale, empirically validated intervention projects. As we grow older, we hold even more respect for the clients we see—often feeling genuine admiration for many who are coping and living as well as they are despite having grown up with so little help for the very big problems they were/are contending with. We find that the problems most clients bring to treatment are not simple or superficial but often complex, painful, and not so easily resolved. However, with helpful training and supervision, student therapists can learn much about how to help and make a meaningful difference in their clients' lives. We have been privileged to have taught and supervised so many rewarding graduate students, most of whom have gone on to become practitioners, teachers, and clinical supervisors. So, we love this work, feel honored to have the privilege to enter deeply into our clients lives, and believe strongly that therapists help clients change through the reparative relationships they can offer.

WHO IS THIS BOOK FOR?

Interpersonal Process in Therapy: An Integrative Model, 6/e, is a clinical training text written for graduate student therapists in practicum and internship courses who are seeing their first clients, and for students in upper division and prepracticum courses that provide an in-depth, applied introduction to counseling and therapy. Complex clinical concepts are introduced throughout each chapter and illustrated with numerous clinical vignettes and sample therapist–client dialogues that student therapists will find informative and compelling. Clinically authentic and personally engaging, this text will help developing therapists understand the therapeutic process and how change occurs.

We believe that the relationship between the therapist and the client is the foundation for therapeutic change. Across differing brands of treatment, and in short- or longer-term modalities, researchers find that the therapist's ability to establish a strong working alliance early in treatment may be the best predictor of treatment outcome. Also, psychotherapy researchers studying treatment outcome find only small differences at best between different theoretical approaches but, consistently over the decades, find large and robust differences between the effectiveness of different therapists working within the same theoretical orientation (that is, greater within-groups than between-groups differences). Thus, this text offers an integrative treatment approach that draws from different theoretical approaches and teaches student therapists how they can use the therapeutic relationship as a focus for understanding and intervening with their clients.

To be effective, therapists need to engage actively with the client and give them a sense that they are in a partnership with someone who is committed to their well-being. The process enacted between the therapist and client needs to honor the client's personal history and subjective worldview by acknowledging that the client's typical ways of managing intrapersonal and interpersonal conflicts are based on previous experiences, but that the client can begin to try new ways of relating with the therapist and that the therapist will be a "safe haven" and "secure base" as the client works through his or her difficulties. The data indicates that 40 to 60 percent of clients terminate therapy before sustainable benefits are realized. We believe this is largely because clients' expectations are unmet from the beginning— that is, the expectation that they will have a therapist who will be actively engaged with them, show a genuine interest, demonstrate understanding of what's going on for them and discern their key concerns, and offer something that is meaningful or helpful. Thus, developing therapists need help learning how to make sense of the complex interactions taking place in the therapeutic relationship, and to find effective ways to talk with clients about what may be transpiring between them. With a better understanding of what is going on between the therapist and the client, the therapist is able to intervene effectively by providing clients with a new and reparative relationship that disconfirms clients' pathogenic beliefs about themselves and counteracts faulty expectations of others. In this way, we are teaching student therapists how to intervene in the current interaction or interpersonal process that is occurring between the therapist and client to provide a corrective emotional experience. In sum, this text presents a comprehensive treatment approach that teaches student therapists how to use themselves, and the relationships they establish with their clients, as the most important way to help clients change.

WHAT IS THIS BOOK ABOUT?

Clinical training is stressful for many new therapists because they are so painfully uncertain of "what to do" and how to proceed with their clients. Although they take helpful courses on counseling theories, helping/micro skills, and psychopathology, student therapists need more specific help as they approach their first clients in session. Fully cognizant of their limited experience and knowledge, and often receiving contradictory input from different supervisors and practicum instructors, these trainees are often anxiously aware that they do not really know what to do or how to help their clients. Although bright and caring, many feel inadequate and worry about making "mistakes" or doing something wrong that would hurt their clients. To help with all of this, new therapists need a conceptual framework to help them understand where they are trying to go in treatment, and why, in order to help their clients change. One of the primary goals of this interpersonal process text is to replace the ambiguity that students often have about treatment with a well-developed framework for understanding how change occurs, and the role of the therapist–client relationship in the change process.

Specifically, this text aims to help new counselors understand therapeutic relationships by focusing on the current interaction or interpersonal process that is transpiring between the therapist and the client. It helps developing therapists learn how to use process comments and other "immediacy interventions," such as metacommunication, self-involving statements, and interpersonal feedback to intervene in the here-and-now, current interaction with clients. Written in a direct and conversational tone, this text provides an integrative treatment approach that highlights how new therapists can use the process dimension—talking sensitively yet forthrightly about "you and me" and what may be going on between the therapist and client, right now, to facilitate change. The model teaches readers to (1) identify significant cognitive patterns and relational themes in the client's behavior; (2) help the client recognize how these patterns function both for better and for worse in the client's life—that is, what they provide and what they cost; (3) change how these maladaptive relational patterns and outdated coping strategies that are causing problems with others are also coming into play during sessions in the real-life relationship between the client and therapist; and (4) generalize the in vivo relearning from this experience of change with the therapist to others beyond the therapy setting. This treatment model demonstrates a way of being with clients in session that is genuine, actively engaging but not directive, empathic, and highly collaborative. It encourages therapists to explore with clients what is going on between them so they can work together to change faulty patterns that are being enacted with the therapist and others, and provide an experiential relearning that best helps clients change.

Describing the course of treatment from the initial session through termination, this text provides practical intervention guidelines for working with the primary issues that emerge during different stages of treatment, and shows clearly how theory leads to practice. With clarity and immediacy, it highlights the challenging clinical situations that new therapists are facing, and captures the questions and concerns that are most salient for student therapists as they begin seeing clients. In sum, the interpersonal process model is an integrative approach that incorporates cognitive-behavioral, family systems, interpersonal-psychodynamic, and attachment theories, uses the therapeutic relationship as an organizing focus for treatment, and encourages student therapists to draw flexibly from varying theoretical perspectives and to develop their own personal styles.

WHAT'S NEW IN THIS SIXTH EDITION?

It has been both rewarding and challenging over the years to have the opportunity to return to this work and try to clarify better the interpersonal process model. The goal has always been the same—to make it simpler, clearer, and more practical for new therapists. The biggest change to the Sixth Edition is the addition of Dr. Faith H. McClure as coauthor. With her widely appreciated teaching and clinical talents, she has helped to infuse the book with inclusive multicultural sensitivities, expertise in attachment-informed psychotherapy, and clinical and research experience in trauma and resilience. Additional goals for the Sixth Edition were to develop further certain core concepts that run throughout the text, especially empathic understanding, working collaboratively, and rupture and repair in the working alliance. The Sixth Edition has thoroughly integrated the burgeoning literature linking attachment theory to clinical practice, in particular the concepts of safe haven/secure base, "mentalization" and reflective function, adult attachment styles and clinical presentation, and affect regulation. There is further attention to the transdiagnostic role of shame in many clients' symptoms and problems, and to its role in resistance and premature termination. Those who are familiar with previous editions of the text also will find in this new edition much more attention to clinical supervision, case formulation and treatment planning, and empirical support for the treatment guidelines offered. A significantly revised and expanded references section provides improved access both to classic and current clinical and research literature. Finally, it is challenging for new therapists to work "in the moment" with clients and try out the process-oriented interventions that can be so helpful yet intimidating in the beginning. To assist with these here-and-now interventions that bring intensity and a more authentic dialogue, we have provided numerous new sample therapist–client dialogues and case vignettes to illustrate effective and ineffective ways to intervene. We hope they provide helpful models as readers strive to become more effective therapists and to sort through their own identities as therapists and choose their own personal styles for how they wish to work with clients.

ANCILLARIES

THE STUDENT WORKBOOK

The Sixth Edition provides instructors and students with a much improved Student Workbook. The Student Workbook to accompany the text is an especially important aid to help readers integrate the personally challenging and evocative material presented in the text. It provides a study guide with key concepts and terms, and sample multiple-choice and short essay questions. The Student Workbook also includes extensive case studies of characters (clients) from well-known films (such as Conrad from *Ordinary People* and Dorothy from *The Wizard of Oz*) and plays (such as Willy Loman in Arthur Miller's *Death of a Salesman*). These comprehensive case studies follow the guidelines provided in the text for writing case conceptualizations and process notes utilizing the interpersonal process approach. The Student Workbook also includes self-assessment measures that student therapists can use to obtain feedback on their progress in developing the process-oriented competencies detailed in the text. Also, personal questionnaires to help new therapists identify and explore their own countertransference propensities are provided. Each chapter in the Student Workbook also provides a section—our favorite—of thought-provoking self-reflection questions to help students integrate and apply the material in the chapter to their own lives and professional development.

VIDEO DEMONSTRATIONS

Demonstration videos showing both authors applying interpersonal process concepts from the text in clinical situations are available online. Students and faculty can access the videos by visiting the companion website companion website for the text, found at www.cengage.com/counseling/teyber. The video clips focus on critical practice issues that are often especially challenging for new therapists.

INSTRUCTOR'S MANUAL AND TEST BANK

A suite of instructor's ancillaries is also available online at www.cengage.com/counseling/teyber. The instructor's resources include an updated Online Instructor's Manual and eBank Test Bank to help faculty prepare. The manual contains an instructor's lecture outline and in-class student exercises, and the test bank contains sample essay questions and multiple-choice questions for every chapter.

ACKNOWLEDGMENTS

I (Ed) began this clinical work that I love so much, and started percolating on this text as well, over 35 years ago as a trainee in the Student Counseling Center at Michigan State University. I learned much there that I still find helpful every day from good mentors such as Sam Plyler, Bill Mueller, and others. I have been fortunate to continue finding wise friends and colleagues over the years who keep teaching me about counseling relationships and helping clients change. Among many others, I would like to thank Lorna Smith Benjamin, Clara Hill, Jeff Hoffman, Jim Penrod, and especially, Margaret Dodds-Schumacher, who we lost in 2010 and miss greatly. Most importantly, I would like to thank Faith H. McClure for teaching me so much over the years—usually by her example rather than words. She has helped generously with previous editions of *Interpersonal Process in Therapy*, and I'm grateful that she has now joined me as coauthor and enriched this Sixth Edition with her deep understanding and compassion. She is a gifted therapist and supervisor, and a uniquely loving person who offers every child on the earth a trustworthy friend. I also wish to thank my family and three loving sisters for their lifetime of support—oldest sister Julie typed my dissertation for me in 1976 and, still typing, this manuscript in 2009! I also am deeply thankful for having two caring and creative sons, Ted and Reed—I am immensely proud of the person each has become. Finally, I wish to thank the clients I have worked with over the years. They have honored me by taking the risk of letting me become someone who mattered to them, and therein giving me the simple pleasure of being able to help.

I (Faith) have been privileged to have the support of a nurturing family, loyal friends, and affirming mentors. UCLA mentors Hector Myers, Vickie Mays, and Stanley Sue have always been generous with their time. My friends Bob Weathers and Laura Kamptner have provided both personal and intellectual fuel over the years. I am deeply grateful for the partnership I have with Ed Teyber—he is truly a master therapist and provides tremendous safety to those who have the pleasure of being in his circle. I am especially thankful to my sons, Andrew and Cameron, who have taught me so much about the power of relationships. I am also thankful to the many children and families who have given me the honor of letting me into their lives and allowing me to become a helper. My students, many of whom are now colleagues, have taught me as much as I have taught them—I have been fortunate to find this work.

Together, we would like to thank three delightful graduate students, Sarah Wolfarth, Stephanie Foy, and Kindra Edmondson, who have offered us skillful editing and helpful feedback with this revision. We appreciate how much they, and many other rewarding graduate students over the years, have given us. It will be a great pleasure to watch and share in all they are going to accomplish and offer to others in the years ahead. Additionally, we would like to thank the reviewers who made many helpful suggestions for this revision: Melissa D. Grady, University of North Carolina at Chapel Hill; Mona Bapat, Arizona State University; and Wendy Bianchini Morrison, Montana State University.

Finally, thanks to our patient editor, Seth Dobrin, and many good colleagues over the years at Brooks/Cole.

ABOUT THE AUTHORS

Edward Teyber is Professor of Psychology, director of the Psychology Clinic, and president of the Foundation Board of Directors at California State University, San Bernardino. He received his Ph.D. in clinical psychology from Michigan State University. Dr. Teyber is also the author of the popular book *Helping Children Cope with Divorce*, and coauthor with Dr. Faith H. McClure of *Casebook in Child and Adolescent Treatment: Cultural and Familial Contexts*. His work has focused on the effects of marital and family relations on child adjustment, child-rearing practices and attachment styles, and the counseling relationship. He is interested in supervision and training and enjoys maintaining a private practice

Faith Holmes McClure is Professor of Psychology at California State University, San Bernardino, and provides consultation to organizations serving the mental health needs of children. She grew up in South Africa during apartheid and received her Ph.D. in clinical psychology from the University of California, Los Angeles. Her clinical and research interests are in the areas of multicultural issues in mental health, and in at-risk children, families, and adults with a particular focus on factors that promote resilience in those who have experienced trauma. She is currently conducting an attachment-based, parent–child therapy intervention project with incarcerated mothers. She is coauthor with Dr. Edward Teyber of *Casebook in Child and Adolescent Treatment: Cultural and Familial Contexts*. Her publications are in the area of coping with stress and trauma.

INTRODUCTION AND OVERVIEW

CHAPTER ONE   The Interpersonal Process Approach

THE INTERPERSONAL PROCESS APPROACH

Claire, a first-year practicum student, was about to see her first client. She had long and eagerly awaited this event. Like many of her classmates, Claire had decided to become a therapist while working on her undergraduate degree. Counseling* had always been intrinsically interesting to her. For Claire, being a therapist meant far more than having a "good job"; it was the fulfillment of a dream. How meaningful, she thought, to make a living by helping people with the most important concerns in their lives.

At this moment, though, Claire felt the real test was at hand: Her first client would be arriving in a few minutes. Worries raced through her mind: What will we talk about for fifty minutes? How should I start? What if she doesn't show up? What if I do something wrong and she doesn't come back! Claire also worried if her client—a 45-year-old Hispanic woman—would have difficulty relating to her, a Caucasian woman in her mid-20s. Needless to say, Claire was very anxious. And even though she was painfully uncertain about how to proceed, she was still intent on finding a way to help this client with her problems. She had learned something about therapy in her undergraduate psychology classes and a good deal more from her volunteer experience with callers on the local crisis hotline. But even with these experiences and a supervisor to guide her, Claire was painfully aware of her novice status and the fact that she didn't know very much about actually doing therapy.

Claire's classmates shared her excitement about becoming a therapist. Many of them were older than Claire and far more experienced in life. Some had raised children; others had already had careers as teachers, nurses, and businesspeople. These new therapists were often coming from life roles in which they had already been successful and felt confident. A counseling career held new hopes for these classmates as well and also evoked anxiety about their ability to become effective therapists. Like Claire, they knew that realizing their hopes and plans for a rewarding new career would also depend on their ability to establish their credibility with their

---

*The terms *counseling* and *therapy* are used interchangeably throughout the text.

clients in order to work effectively with them. And with the arrival of their first clients, their ability to help their clients was about to be tested.

PERFORMANCE ANXIETIES DIMINISH NEW THERAPISTS' EFFECTIVENESS

It is indeed really hard to begin seeing one's first clients. Those initial sessions can be both exciting and intense for new therapists like Claire who often struggle painfully with performance anxieties and fears about their own inadequacy. We recall how worried we were, years ago, about "making mistakes" and as clinical supervisors often listen to bright and caring practicum students express concerns about somehow "hurting my client." If new therapists don't get help with these common and expectable concerns, however, they can get in the way and become a problem for the client and the treatment process. For example, if new therapists are too worried about making a mistake, or become overly preoccupied about what their supervisor might want them to say or do, they often become too quiet or passive—and the client doesn't feel heard, helped, or responded to. Similarly, when they are trying too hard to figure out what they are going to say or what they are supposed to do next, therapists cannot listen intently and actively engage with the client. In this way, performance anxieties can significantly impede the treatment process. These unwanted performance pressures are generated when new therapists frame what they don't know or can't do yet as a "deficit" or as evidence of their inadequacy, rather than more realistically framing it as merely their own inexperience. Because this book is about the relationship between the therapist and client, a useful starting point for learning how to use the therapeutic relationship to help clients change is to better understand these initial performance anxieties that routinely diminish therapists' effectiveness. Let's take a look at some of the typical concerns practicum students express as they begin seeing their first clients.

Jessica expresses concerns that are fairly realistic for a new therapist: *"My biggest anxiety about being a student therapist is that I really don't have any idea what I am doing—and feel like I've been thrown headfirst into the water! Don't get me wrong, I am definitely enjoying the chance to start seeing clients, but it's just a little scary taking other people's problems into my own hands and trying to help them without a clue of where to go next. I'm OK with exploring things, but when it comes to the "action stage" and actually trying to help somebody change, well, that frightens me most."*

Chanté brings a more personal concern to her initial sessions—the fear of being found out or exposed as an "imposter": *"My biggest anxiety stems from a fear of failing or being incompetent—a feeling that somehow I have managed to sneak through my educational career and land in this great training program without anyone figuring out that I'm really not cut out for this. Perhaps I am just as afraid myself of realizing that I may not be very good at this work. That would be devastating for me because the more I see my clients, the more I enjoy and appreciate every aspect of this. And even though I'm aware that I'm going to make mistakes as I go along, I am afraid of doing more harm than good sometimes. So, with every new case that gets passed out, the first thing I think about is all the reasons why I may not be the right therapist to help that person."*

Ginger connects how the anxiety that she experiences in her new role as a counselor links to a problematic role she played in her family. *"My biggest anxiety about becoming a therapist is feeling that I am inadequate. My instructors reassure me that this is a normal feeling, that many therapists experience this in their first year or two of training, and that we're not expected to be perfect. But it doesn't make any difference—it remains my biggest anxiety. I believe it's because I was always second best in my family of origin. No matter what I did, my sister was always smarter...more creative. I learned to feel really uncomfortable whenever I wasn't in complete command and didn't know just exactly what I was supposed to do. So, even though some part of me knows that I'm really not inadequate, it still churns my stomach when I am not good at something right away."*

Dennis relays how his performance anxieties keep him from grasping, in the moment, what his client is really saying or meaning. *"My anxiety level goes way up when I feel lost in the session. Sometimes I feel so lost in there, like I'm going in circles, and I just don't know what to do. And that just makes my anxiety even worse. Then I start wracking my brain about what to do next and I start missing even more issues. It just kills me when I review the videotape after the session and then, when it's too late, I "get" what they were really saying or think of what I could have said. But why couldn't I get it during the session, right then when it was happening...it's so frustrating!"*

As these promising student therapists are telling us, counselor training certainly is challenging. But new therapists need to be patient with themselves and appreciate that learning to be an effective therapist is a long-term developmental process (Rizq, 2009). It often takes three to five years to find one's own professional identity and feel comfortable in this work. Within a year or so, most new therapists do become more confident, start to feel like they can be themselves in the session, and begin to enjoy the hour as they find that they can indeed help their clients. Lorraine, a second-year student, captures this rewarding transition: *"Everything considered, I have enjoyed conducting therapy this year. Instead of being nervous in the room, like I felt last year, I feel at home now. I'm actually able to enjoy my clients instead of being caught up in my fear of failure. I'm not nearly as worried about making mistakes, and I view the misunderstandings or ruptures that occur as opportunities to develop a better relationship with my client. So, the best part of therapy this second year is that my performance anxiety has decreased exponentially. I spend a lot more time thinking about what's going on with the client and conducting therapy, and a lot less time thinking about myself and waiting to fail."*

What helps therapists like Lorraine resolve their initial performance anxieties so they can begin responding more effectively to their clients? We introduce three suggestions here, and will return to this important developmental issue throughout the chapters ahead. First, new therapists are encouraged to question the unrealistic performance expectations they often place on themselves, accepting that it is OK to make mistakes—all therapists do. Oftentimes, *this means they need to focus more on what they are learning than how they are performing.* This will progress readily for some, especially those who can be non-defensive, learn from their mistakes by being open to constructive feedback from supervisors, and move on. In contrast, this will be more difficult for other practicum students who may have grown up in families where they received too much criticism from their caregivers, were expected to be perfect and could not make mistakes, or too readily were judged. An important consideration to keep in mind is that we are human, are all going to make many "mistakes" with our clients, and will be most effective if we focus instead on how to recover from them. Join in this thought exercise:

Right now, think of the professional you regard as the single most effective therapist you have known—a particular instructor or supervisor, or perhaps a personal therapist you have worked with, that you trust and admire. Pause now, and reflect for a moment on the reality that this esteemed and seasoned therapist makes many mistakes in her or his practice every day. (If you don't think they do, ask them.) The therapist you are thinking about probably is very skilled, in part, because s/he can recover from mistakes (the inevitable misunderstandings that occur in every relationship or "ruptures in the working alliance") by being nondefensive. That is, s/he regularly checks out potential problems by asking clients about potential misunderstandings, is willing to look at his/her own contribution to whatever problem may be occurring, and talk through or sort out these misunderstandings that inevitably occur. For example:

THERAPIST: "You've become very quiet, Laura. I'm wondering if something I might have said, or something we're doing, doesn't feel quite right to you. Can I check in with you about that possibility? How are we doing together—what's it like for you to be talking with me right now?"

Second, new therapists will manage better whatever anxieties they may be experiencing if they can set their intention to de-center and *focus more on the client and what the client is really saying rather than on themselves and their own performance.* Too often, new therapists are self-absorbed, which can, at times, develop to a state of painful self-consciousness or obsessive self-awareness (for example, **Therapist:** "What should I say when she looks at me like that...what am I going to ask about next...I'm being way too quiet...I keep saying um hmm"). If new therapists can develop a reflective self-awareness that is more balanced—that is, process both what is going on for the client and consider what might be going on between client and therapist—they will be more effective. We are going to explore this "participant/observer" stance closely, a stance that involves being in the relationship experientially while simultaneously observing it and thinking about what might be going on between the therapist and client. This might feel elusive in the beginning, but it is an essential therapeutic skill that can be developed if we work at it.

Third, to help with these normative and expectable performance anxieties, new therapists need active support from their supervisors and instructors. In particular, they need to be reassured about the acceptable reality that most new therapists do not really know very much about the change process, what to do in their sessions, or how to help clients change. Student therapists also need practical guidelines and suggestions for how to proceed that they can apply, especially in their initial sessions. This support and preparation often includes real-life illustrations or exemplars such as watching videotapes of instructors conducting intakes and initial sessions, watching actual sessions of your supervisor with a client, and role-playing with instructors that model effective ways to respond and conduct initial sessions. These real-life role models are an especially effective way to learn in the beginning and can go a long way toward diminishing the student therapist's initial anxieties about seeing clients and knowing "what to do."

In sum, new therapists' anxieties about their ability to help are not to be dismissed as just neurotic insecurity or obsessive worrying. It is realistic to be concerned about one's performance in a new, complex, and ambiguous arena. It is simultaneously important that new therapists like Claire not lose sight of the significant personal strengths they already possess and bring to their first counseling experiences. Most students who are selected for clinical training already possess sensitivity, intelligence, and a genuine concern for others. Such personal assets, and all they have learned from their own life experiences, will prove helpful to their future clients. In this regard, treatment outcome studies repeatedly find that success rates in treatment have more to do with the personal characteristics and the skill of the therapist than the theoretical orientation or type of treatment (Luborsky, 1980; Seligman, 1995; Wampold, 2006).

Personal experience, common sense, good judgment, and intuition are useful, indeed. However, in order for a therapist to be effective with a wide range of people and problems, these valuable human qualities need to be wed to a conceptual framework. A conceptual framework provides a "compass" for the work you are going to do with the client. It is from this "compass" that you develop a case formulation for your client—we will discuss this further below. Suffice it to say that therapists working within every theoretical orientation become more effective when they understand more specifically just what really is wrong and the central problem for this particular client, and where they need to go in treatment, in order to help this client change. When therapists have this kind of case formulation, and know where they are going in treatment and why, they can be more consistently helpful. Without a conceptual framework as a guide, however, therapists do not have a focus for treatment and, in each successive session with the client, are just putting out the next fire that has come up. Let's turn to this key competency.

THERAPISTS ARE MORE EFFECTIVE WHEN THEY HAVE A CASE FORMULATION AND A TREATMENT FOCUS

The interpersonal-process approach is fundamentally grounded in the "core conditions" (see Imel & Wampold, 2008, for a review of the common factors literature). They are necessary for change and essential to every brand of successful therapy. For many clients, however, the core conditions of empathy, genuineness, and warmth are not enough, especially those who have more significant problems such as being violated or betrayed by a primary caregiver. For these and other clients with complex and challenging problems, something more is needed. Therapists will have much more impact when they are able to conceptualize or discern more precisely what this client's core problem really is, how it came about developmentally, and how it is being played out and causing symptoms and problems in his current life. The most widely used term for this key competency is a "case formulation" or "case conceptualization."

Having a case conceptualization to guide treatment can prevent therapists from getting "lost" in the different stories the client brings in each week—not knowing where to go or what to do next as the client keeps switching from topic to topic.

Clearly, the opportunity for more enduring and significant change is lost without a case conceptualization and the focus for treatment it provides. Without this treatment focus, clients might feel understood or cared about by their therapists, and may like them very much, but oftentimes little change occurs. In support, a burgeoning research literature highlights that greater effectiveness is associated with more focused interventions (Roth & Fonagy, 2005). Distinguishing a general sympathy, warmth, friendliness, or benevolence from a more specific accurate empathy or conceptual clarity, Allen et al. (2008) suggest the metaphor that "we should visualize psychotherapy as being less like lounging in a warm bath and more like swimming in a cool, crystal-clear lake. Warmth is easier; clarity is harder." Thus, while core conditions are necessary, they are insufficient—therapists also need to discern more clearly a focus for treatment that clarifies what's really wrong, how this came about, and what needs to be done to change it. In the chapters ahead, we will further elucidate how therapists can hone in and clarify more precisely the themes and patterns that link together the different problems and concerns the client is presenting.

Additionally, the process of learning how to contextualize or "make sense" of a client's problems involves exploring various theoretical frameworks. Before therapists can find their own professional identities and feel confident in their abilities, they need to integrate a theoretical framework that they can apply with diverse clients and is congruent with their own personal values and life experiences. Developing therapists cannot claim their own professional identity by simply taking on the same theoretical approach as their mentor or adopting what their instructors and supervisors do. To work most effectively in whatever theoretical orientation they eventually adopt, and to enjoy being a therapist, student therapists are encouraged to actively explore and try out different approaches for several years—eventually integrating and *choosing for themselves* how they are going to work with clients.

Continuing with this concern for the therapists' personal development, one purpose of this book is to provide a conceptual framework for understanding the therapeutic relationship, and learning how therapists can use the "interpersonal process" or current interaction with their clients to intervene and help clients change. Throughout this text, readers will be actively encouraged to modify this framework to better fit their own personalities and therapeutic styles. The interpersonal process approach is integrative and can be applied by therapists who choose to work within different theoretical orientations (Norcross et al. 2008). With support from supervisors and instructors, new therapists will be able to personalize this model, integrate it with other theoretical approaches, and modify it to make it their own.

In this approach the emphasis is on providing clients reparative experiences with the therapist that will allow them to create self-narratives that are more coherent, affirming, and flexible. Many clients who come to therapy have suffered significant developmental wounds—their experiences lacked safety, protection, boundaries, and predictability, and in many cases their attempts to differentiate and develop autonomy were met with threats of abandonment or guilt inducement (Cassidy & Shaver, 2008; Lieberman & Van Horn, 2008). As a result, many were left to cope in a variety of symptomatic ways in order to manage developmental challenges and unmet needs—such as adopting rigid coping styles and living their lives in ways that can be characterized as being pervasively compliant or pleasing, strident or rigidly controlling, invisible or without needs, detached or demanding, and so forth. Thus, therapists working within a relational framework, whether they eventually come to identify as cognitive-behavioral therapists, interpersonal/psychodynamic therapists, existential-humanistic therapists, family systems therapists, narrative therapists, or others, can use the interpersonal process approach to work more effectively.

Briefly, what will this approach entail? It entails helping therapists recognize the relational patterns and cognitive schemas that are shaping clients' problems. Tracking these themes, therapists will use "process comments" and other "immediacy interventions" to identify and change these problematic patterns—as they are occurring with the therapist and with others in their everyday lives. With this real-life or experiential re-learning with the therapist that expands clients' schemas, therapists are using the therapeutic relationship as a social learning laboratory, and then helping clients transfer this re-learning or generalize the change that has occurred in their relationship to others in the client's life. As we will see, an effective way for therapists working within different theoretical orientations to intervene is to assist clients to (1) identify and change these maladaptive patterns in the current interaction with the therapist; and (2) generalize this experiential re-learning to other relationships where the same patterns and problems are occurring.

THEORETICAL AND HISTORICAL CONTEXT

The therapeutic model presented here, termed the *interpersonal process approach,* is an integrative model that it is not based in any single theoretical orientation. It draws out the interpersonal elements or relational features from several different theories and synthesizes a common focus on the therapeutic relationship, which has emerged as the most consistent predictor of successful treatment outcome (Baldwin, Wampold & Imel, 2009; Horvath & Bedi, 2002; Norcross, 2002). More specifically, it clarifies how therapists can use their current interaction—the interpersonal process they are enacting with their clients—to help clients change. Theorists and researchers from different theoretical orientations all have contributed to our understanding of the therapeutic relationship and how therapists can use it to understand and guide what we do with our clients. In this regard, varying theories have focused on different aspects of clients' functioning and helped therapists respond to different aspects of people's problems. In particular, the Interpersonal, Cognitive, and Familial/Contextual domains have been emphasized in different treatment approaches. This text highlights the process dimension in each of these broad theoretical traditions, and links these concepts about the current interaction between the therapist and client to clinical practice These three domains are introduced below and provide a theoretical context for the clinical approach presented in the chapters ahead.

THE INTERPERSONAL DOMAIN

Let's begin with a historical perspective. The interpersonal dimension was originally highlighted by Harry Stack Sullivan and has been articulated further by Hans Strupp, Donald Kiessler, Lorna Smith Benjamin, Lorna Weissermann, Irvin Yalom, and many others. Sullivan (1968) first brought the interpersonal focus to psychotherapy in the 1940s, and he remains an enormously influential but insufficiently recognized figure. A maverick, Sullivan radically broke away from Freud's biologically based libido theory and was one of the first major theorists to argue that the basic premises of Freud's drive theory (for example, sexual and aggressive instincts) were inaccurate. Sullivan emphasized clients' current behavior and relationships with others rather than developmental fixations or how clients were stuck in the past. He argued that every major tenet of Freud's theory could be understood better through interpersonal or behavioral experiences. In this way, he and other relational theorists pressed for a more behavioral or reality-based focus on the child's actual experience with parents, or real-life interactions with caregivers, and stepped away from Freud's emphasis on fantasy and intra-psychic processes. Sullivan developed an elaborate developmental theory of psychopathology that emphasized what people do to avoid or manage anxiety in close relationships, and was especially interested in the hurtful or anxiety-rousing interactions that re-occurred in formative parent-child interactions. He viewed anxiety, and elaborate attempts to avoid or minimize it, as the central motivating force in human behavior. He described painful core anxiety as rooted in dreaded expectations of derogation and rejection by parents and others, and later by oneself. Following from this, Sullivan conceptualized personality as the collection of interpersonal strategies that the individual employs to avoid or minimize anxiety, ward off disapproval, and maintain self-esteem.

According to Sullivan, the child develops this personality, or "self-system," through repetitive interactions with her parents. More specifically, the child organizes a self around certain basic patterns of parent-child interactions. These repetitive interactions with parents, which may be anxiety-arousing or rewarding, are structured into the child's personality through a collection of complementary self-other relational configurations. For example, children may develop internal images of themselves as helpless or insignificant and expectations of parents and others as demanding or critical. Alternatively, more fortunate children who enjoyed better developmental/familial experiences may evolve images of themselves as love-worthy and expectations that others can be trustworthy.

Having learned these self-other relational patterns, people systematically behave in ways that avoid or minimize the experience of anxiety. For example, suppose that particular aspects of the child, such as feeling sad or crying, consistently result in parental responses of irritability or indifference, or even more problematically, ridicule or overt rejection. The child learns that these parts of herself, her vulnerability or emotional needs, are unacceptable and represent a "bad self." These unacceptable, anxiety-arousing aspects of the self are split off or disowned and the child develops interpersonal coping styles (for example, a pleasing/placating manner, a macho/dominating style, or a rigidly self-reliant and distant stance in all relationships) that preclude being subjected to anxiety-arousing rejection or ridicule again. *These coping styles are interpersonal defenses* (as opposed to Freud's intrapsychic defenses such as denial or projection) *that originally were necessary to protect the self in early parent-child relationships.* Unfortunately, these interpersonal defenses are rigidly applied and overgeneralized to other relationships in adulthood and become habitual behavior patterns as the child, now grown to an adult, inaccurately expects that new experiences with others will repeat the same hurtful, anxiety-arousing patterns of the past. Sullivan's early formulations have been greatly enhanced by attachment researchers (using the terms *internal working models* and *relational templates*) and by cognitive behavioral therapists (using terms such as *early maladaptive schemas, faulty expectations,* and *selective filters*). In a historical perspective, however, this was pioneering work.

Sullivan developed his rich, if sometimes elusive, interpersonal theory through his clinical work with seriously troubled patients who were often hospitalized. He was a gifted therapist and had great respect for the personal dignity of his clients. His sensitivity to clients, compassion, and genius for understanding the intricate workings of anxiety inspired nothing less than a new direction in psychotherapy. Sullivan influenced seminal thinkers in the interpersonal field, such as Carl Rogers, Eric Fromm, Frieda Fromm-Reichmann, Karen Horney, and Erik Erikson, and made far-reaching contributions to clinical theory and practice. He even prefigured family systems theory in his study of close relational systems. In fact, two of the founding pioneers in family therapy, Don Jackson and Murray Bowen, trained with Sullivan. Finally, he also set the stage for a host of contemporary approaches to short-term treatment that were shaped by this interpersonal approach and its more active and engaging therapeutic stance toward the client. In sum, many of the basic concepts in the interpersonal process approach can be traced back in some form to Sullivan and other interpersonal/relational theorists.

THE COGNITIVE DOMAIN

Approaches as seemingly diverse as **object relations theory,** Bowlby's **attachment theory,** which grew out of it, and **cognitive behavioral therapy** each help us understand and intervene with the faulty thinking that is central to clients' symptoms and problems. Let's begin with object relations theory—with its remote and alienating jargon, it is usually the most difficult theory for graduate student therapists to integrate. Following this overview, we will link the attachment theorists' core concept of "internal working models" and the "schema"-oriented approaches of contemporary cognitive therapists—which are concerned with the client's beliefs and expectations, and how they impact affect and interpersonal relationships.

Object Relations and Attachment Theory: Internal Working Models. In the present context, *objects* are people or, more precisely, internal representations of important caregivers. Object relations theory is about close interpersonal relationships—especially those between parents and young children, and how those important early relationships are internalized as enduring mental representations of parent-child relations or *internal working models* (similar in ways to some *schema* approaches). In particular, object relations theorists try to account for the meaning of attachment and the human need for safe, secure, and enduring relationships with one or a few significant others who can register and respond to our emotional needs. Sullivan's theory also centers on relationships, but the question that drove his work was very pragmatic: What do people do to avoid anxiety (that is, threat, danger, fear) in interpersonal relations? In contrast, the basic question for object relations theorists is more abstract and inferential. Object relations theorists are interested in understanding how formative interactions between parents and children become internalized by the child and, akin to cognitive schemas, serve as mental representations that shape or guide how children establish and carry out subsequent relationships with others. These representational models of self and other provide basic expectations or road maps for what will transpire in relationships (for example, Others are dependent and needy, I must provide and take care of them; Others are critical, competitive and controlling, and I need to protect myself by withdrawing or distancing). Although these early schemas for relationships will become more complex and change over time, object relations theorists argue that these formative parent-child interaction schemas provide the basic structure for developing a sense of self, organizing the interpersonal world, and shaping subsequent relational patterns (Cashdan, 1988; Ogden, 2002).

Extending these constructs, attachment theorists further clarify the important role of early parent-child relationships, and the security or anxiety that children find in their parents' responses to their attachment needs (see Cassidy & Shaver, 2008). When parents accurately register and respond to their children's emotional needs, children develop attachments that are labeled *secure*. In contrast, when parents do not accurately read the child's emotional cues, or are too preoccupied with their own concerns to hear or respond to their young child's distress or bids for engagement, the child will become insecurely attached. Attachment researchers describe three types of insecure attachment styles: two that are "organized" (Avoidant and Ambivalent attachment classifications) and a less common but more problematic "Disorganized" style.

When parents consistently ignore, dismiss, reject, or otherwise fail to console a distressed child, that child will be insecure and develop an *avoidant* attachment style. The child learns early on to cope with the parent's unresponsiveness by developing a psuedo or false form of "independence." This counter-dependency conveys that the child has absolutely no need of others, regardless of his or her actual experience or genuine emotional need. As an adult, this child is likely to have difficulty accessing and expressing his/her emotions and will run the risk of being isolated and emotionally disengaged from both him/herself and others. At the other end of the continuum from this Avoidant child who seemingly feels and needs nothing, other children develop an *anxious/ambivalent* attachment style. This occurs when children have caretakers who are intrusive, respond inconsistently, and have difficulty supporting the child's independence. Since the caretaker is sometimes responsive and sometimes unresponsive because they are self-absorbed or preoccupied, the child is uncertain if the parent will be there when they need them. Thus this Ambivalently attached child (and later, adult) is often clingy and needy, demonstrates significant worry and anxiety, and expresses contradictory or mixed messages (come close/go away) toward the parent and others. Relationships will be challenging for these children as they enter adolescence and adulthood—partners will experience them as dependent and needy, and reciprocity or mutuality will be lacking. Finally, children who develop the *disorganized* attachment style have no organized or characteristic attachment pattern or coping style to deal with their very troubled attachment figures. They have often experienced trauma or abuse, neglect, dissociative or other confusing or even frightening behavior from their parents. Main & Hesse (1999) articulate the maddening and unresolvable paradox that these children struggle with: the attachment figures who are their primary source of comfort are simultaneously the source of their fear. These Disorganized children therefore have difficulty sustaining consistent relationships and are at high risk for more serious psychopathology, such as Borderline Personality Disorder. Often, because they have experienced so much invalidation and frequently been traumatized, as adults they may engage in self-destructive behaviors (binge eating and purging, substance abuse, self-mutilation) and may also have dissociative episodes. They may, in the relationships they establish, have significant difficulty trusting yet be extremely vigilant for loyalty, commitment, and be demanding of friends and partners. In sum, each of these differing attachment histories is going to shape expectations and behavior in subsequent relationships, which, as we will see, includes clients' responses to the therapist and the therapeutic relationship.

Pulling this together, cognitive-behavioral therapists, attachment-oriented therapists, and object relations therapists all use these related concepts of schemas, internal working models, mental representations, and relational templates to conceptualize clients' problems and understand the problematic relational patterns that clients keep repeating with others in their lives (and with the therapist as well). For example, in order to identify these schemas or clarify faulty relational expectations, therapists working from an object relations, attachment, or cognitive behavioral framework often ask themselves (and their clients) questions like these:

1. What does the client tend to want from me or others? (For example, clients who repeatedly were ignored, dismissed, or even rejected might wish to be responded to emotionally, reached out to when they have a problem, or to be taken seriously when they express a concern.)
2. What does the client usually expect from others? (Different clients might expect others to diminish or compete with them, to take advantage and try to exploit them, or to admire and idealize them as special.)
3. What is the client's experience of self in relationship to others? (For example, they might think of themselves as being unimportant or unwanted, burdensome to others, or responsible for handling everything.)
4. What are the emotional reactions that keep recurring? (In relationships, the client may repeatedly find himself feeling insecure or worried, self-conscious or ashamed, or—for those who have enjoyed better developmental experiences—perhaps confident and appreciated.)
5. As a result of these core beliefs, what are the client's interpersonal strategies for coping with his relational problems? (Common strategies include seeking approval or trying to please others, complying and going along with what others want them to do, emotionally disengaging or physically withdrawing from others, or trying to dominate others through intimidation or control others via criticism and disapproval.)
6. Finally, what kind of reactions do these interpersonal styles tend to *elicit* from the therapist and others? (For example, when interacting together, others often may feel boredom, disinterest, or irritation; a press to rescue or take care of them in some way; or a helpless feeling that no matter how hard we try, whatever we do to help disappoints them and fails to meet their need.)

Internalizing the Attachment Relationship.   Object relations and attachment theorists believe that the child's primary motivation is to establish and maintain emotional ties to parental caregivers—a very different approach from that used by psychoanalytic and behavioral theorists (Eagle & Wolitzky, 2009; Mitchell, 1988). Parental caregivers are important in that these attachment figures are a source of comfort and protection. From their point of view, the greatest conflicts in life are threats to, or disruptions of, these basic attachment ties (separation anxieties or abandonment fears). In this regard, anxiety is a *signal* that emotional ties to caregivers are being threatened. If parents are "attuned" and consistently responsive to the child's distress, the child becomes secure in her attachment ties (parents don't need to do this perfectly—just "good enough"). What's so interesting here is that through the course of development, these secure children increasingly "internalize" their parents' emotional availability and responsiveness and come to hold the same constant or dependable loving feeling toward themselves that their parents originally held toward them (certainly, a beautiful developmental process to watch unfold in securely attached children). Said differently, cognitive development increasingly allows securely attached children to internally hold a mental representation of their emotionally responsive parents when the attachment figures are away and they can increasingly soothe themselves as their caregivers have done—facilitating the child's own capacity for affect regulation and independent functioning. Thus, as these children grow older and mature cognitively and emotionally, they become increasingly able to soothe themselves when distressed, function for increasingly longer periods without emotional refueling, and effectively elicit appropriate help or support when necessary. In this way, *object constancy* and more independent functioning develops—facilitating their ability to comfort themselves and become the source of their own self-esteem and secure identity as capable, love-worthy persons. Furthermore, they possess the cognitive schemas or internal working models necessary to establish new relationships with others that hold this same affirming affective valence (Farber & Geller, 1994). That is, these secure children are more likely as adults to be able to recreate their internal working models in subsequent real-life relationships—such as successfully choosing a similarly emotionally responsive and supportive partner and being able to enjoy a better marriage.

But what if the child cannot maintain secure emotional ties to the primary caregivers? In some families, parents do not respond well to the child's distress or attachment need because they are preoccupied, depressed, or even abusive. When emotional needs and attachment security are compromised by caregivers who are unreliable and respond to the child's cues inconsistently, or not at all, these children are trapped in an unsolvable dilemma. They cannot succeed in eliciting the attuned responsiveness they need from their parents, nor can they forsake their need for attachment. What do they do with the anguish this engenders? Object relations theory tries to account for the different **compromise solutions** and coping adaptations the child must make to this poignant and all-too-human dilemma. They may, for example, cognitively distort reality by denying that certain needs of their own, or reality-based problems with others, really do exist, and/or idealize others in unrealistic, problem-free ways.

> Alert and friendly, Bob entered treatment telling his therapist that he didn't have any real problems—things were "great," but his fiancée wasn't quite ready to go ahead with their wedding plans unless he talked with a therapist about his "short fuse" and a recent DUI. As the therapist explored these symptoms and tried to learn more

specifics about them, Bob gave very general explanations that minimized the clear significance of his alcohol-related problems and his temper outbursts—conveying that things really were OK by repeatedly using the phrase "No problem." Looking for a way to move their dialogue beyond the surface, the therapist suggested that, since Bob was thinking about getting married and starting a family of his own, maybe it would be helpful if they could talk a bit about his own childhood and help her get a feel for his parents' marriage and what it was like growing up in his family. Bob gradually unfolded a narrative that included a threatening and explosive father who yelled and hit his children when he drank, and a mother who provided little protection from this—or even acknowledgement that these frightening and humiliating interactions were occurring. Typical of an "avoidant" or "dismissive" attachment style, Bob described these experiences without much feeling, and dismissed these disturbing family interactions as "insignificant." They were "no big deal" because he had a "close and loving family" and "everybody came out great anyway."

Object relations and attachment theorists propose complex developmental theories to account for the cognitive and interpersonal strategies that children employ to protect against the painful attachment affects of separation anxiety, angry protest, and eventual despair evoked when they cannot maintain secure emotional ties. These theorists propose that all children mature through stages of development that progressively allow for a more integrated, stable view of self and others. However, if development is problematic, children remain "stuck" or regress under stress to less integrated stages. In particular, when parents are repeatedly unresponsive and don't register the child's distress or "hear" the child's concern, some children resort to **splitting defenses** to maintain ties to an idealized loving parent—akin to the polarized or **dichotomous thinking** that cognitive behaviorists describe. That is, the child internalizes the "bad" (threatening or rejecting) aspects of the parent in a sharply separate or split-off way from the "good" (loving or responsive) aspects of the parent. In object relations theory, in contrast to traditional psychoanalytic theories, symptoms and problems do not develop from unacceptable sexual and aggressive impulses that have been repressed, but from what has been painful, frustrating or threatening in children's real-life relationships with their caretakers. In healthier parent-child interactions that provide secure attachments, however, the young child can tolerate increasing levels of frustration and disappointment as she matures. Gradually, this healthier or more secure child will be able to increasingly integrate her ambivalent feelings toward the (inevitably) sometimes frustrating and sometimes responsive parent into a stable and affirming relational schema of self and other, which becomes the basis of her positive self-esteem and clear identity.

As previously noted, splitting defenses preserve the necessary image of an idealized, "all good," responsive parent with whom the child is internally connected or attached. However, the price is high in that reality is distorted, the self is fragmented, and the child becomes the one who is "bad." The frustrating or unresponsive parent is no longer "bad," allowing the child to view the external world and his attachment figure as safe, but the price is inner conflict (Fonagy, 2000). The child believes that, if only he were different or better in some way that the caregiver wanted, parental love would be forthcoming (Benjamin, 1993b). This situation is illustrated most poignantly by the physically abused child who, though repeatedly hit and afraid, nevertheless continues to idealize and defend the maltreating parent. For example, a child does this by believing that he is bad and maintaining that he actually deserves the punishment (for example, Seven-year-old: "My dad hit me because I wasn't being nice to my brother."). This self-blaming distortion enables the child to cope with his attachment dilemma (needing help and reassurance from the same person who is hurting and frightening him) and believe that he has some control over events and is not helpless. Early in their training, many new therapists will have clients in their caseloads who were physically mistreated or sexually violated. New therapists may be stunned in disbelief to observe that, commonly, these clients ultimately blame themselves for the mistreatment they suffered as children and unrealistically assume responsibility for what went wrong.

Why is this complex and abstract theory relevant to interpersonally oriented therapists? Realizing that the child must find some way to maintain ties with a good, loving parent helps therapists to appreciate why many clients persist in maladaptive behavior and self-defeating relationships. When clients relinquish symptoms, succeed in achieving a personal goal, or make healthier choices for themselves, subsequently many will feel anxious, guilty, or depressed. That is, when clients make progress in treatment and get better, new therapists understandably are excited. But sometimes they will also be dismayed as they watch the client sabotage her success by gaining back unwanted weight or missing the next session after an important breakthrough and deep sharing with the therapist. Thus, loyalty and allegiance to symptoms—maladaptive behaviors originally developed to manage the "bad" or painfully frustrating aspects of parents—are not maladaptive to insecurely attached children. Such loyalty preserves "object ties," or the connection to the "good" or loving aspects of the parent. Attachment fears of being left alone, helpless, or unwanted can be activated if clients disengage from the symptoms that represent these internalized "bad" objects (for example, if the client resolves an eating disorder or terminates a problematic relationship with a controlling/jealous partner). The goal of the interpersonal process approach is to help clients modify these early maladaptive schemas or internal working models by providing them with experiential or in vivo re-learning (that is, a "corrective emotional experience"). Through this real-life experience with the therapist, clients learn that, at least sometimes, some relationships can be different and do not have to follow the same familiar but problematic lines they have come to expect.

Although object relations theorists address these sensitive and personal dimensions of human experience in impersonal and alienating terms, we can use certain aspects of this theory to understand clients better, relate more empathically to them, and guide our therapeutic interventions. As noted above, we will see how anxiety and guilt are evoked for some clients as their attachment ties to internalized parental figures are threatened by the healthy changes they make in treatment. Later, we will return to the concept of internal working models (or relational templates) in order to understand and intervene with the cognitive component of clients' problems. Just as object relations and attachment theory help us understand how clients mentally represent themselves and their social world, cognitive behaviorists have been skillful in helping therapists identify the schemas that engender so many problems. In particular, let's explore cognitive approaches that clarify the central role of "Early Maladaptive Schemas" in understanding symptoms and problems.

Cognitive Behavioral Therapy: Early Maladaptive Schemas.   In his early pioneering work, Beck (1967) emphasized the importance of *schemas* in depression—referring to schemas as "a cognitive structure for screening, coding, and evaluating the stimuli that impinge on the organism." For Beck, schemas are the basic components of cognitive organization—the fixed or patterned ways we interpret, categorize, and evaluate our experiences. Put most simply, schemas can be thought of as broad organizing principles for making sense of one's life or experience. Beck emphasized that repetitive themes occur in clients' thinking through "schema bias"—a consistent and selective bias in organizing information and interpreting events that results in "the typical misconceptions, distorted attitudes, invalid premises, and unrealistic goals and expectations" that clients present (1967, p. 284). Cognitive behavioral therapy originally emphasized three aspects of cognition—automatic thoughts, cognitive distortions, and underlying assumptions. More recent theorists expand Beck's groundbreaking work and further emphasize the child's developmental experiences with caregivers in shaping affectively charged "Early Maladaptive Schemas" (Young, 2003).

In his rich and substantive approach, Jeffrey Young integrates cognitive behavioral therapy with object relations and psychodynamic approaches, and places more emphasis on the therapeutic relationship, affective experience, and early life experiences. Following Young and colleagues (1999), Early Maladaptive Schemas (EMSs) are defined as stable and enduring themes that develop from ongoing patterns of parent-child interaction during childhood that are significantly dysfunctional, and which are elaborated and extend on into adulthood. Young has identified 18 different schema types or core constructs that fundamentally shape an individual. He groups the 18 core schemas into five broad categories of unmet emotional needs called "schema domains." To illustrate, five related schema types are grouped in Domain 1, the **"Disconnection and Rejection"** domain. This domain clusters clients with five related but different schemas that share the broad theme that they are unable to form secure attachments with others and believe, in different ways, their needs for stability, safety, or love will not be met (Abandonment, Mistrust, Deprivation, Defectiveness/Shame, or Isolation schemas).

Continuing, a **"Subjugation" schema** is grouped in Domain 4—along with other schemas that share the common motif of "Other-Directedness." For example, clients with a Subjugation schema feel coerced and excessively surrender control to others. To avoid anger, retaliation, or abandonment from others, clients with Subjugation schemas suppress their own preferences or desires (that is, they comply and "go along") and suppress expression of their own emotions—especially anger. Other schema domains concern expectations about oneself and the environment that interfere with the ability to function independently, such as Dependence or Failure schemas, perhaps based on an overprotective or enmeshed family of origin that undermines the child's self confidence and initiative.

These schemas serve as templates for the processing or filtering of later experience, and are taken for granted by individuals as a priori truths. They feel familiar and are not questioned by the individual—they are just the givens of one's existence and form the core of a client's self-concept and conception of the environment. Early Maladaptive Schemas are self-perpetuating and, unfortunately, resistant to change because the individual distorts information to maintain the validity of these schemas. This occurs, for example, as the client magnifies information that confirms the schema and minimizes or negates information that is inconsistent with the schema (for example, the husband keys on the one complaint that his wife expressed toward him but does not really hear the genuine appreciation she feels for him that she also expressed). Clarifying further how EMSs selectively filter for corroborating experience and the need for "cognitive consistency," a client with a **Rejection/Shame schema** is prone to select a mate or spouse who repeatedly confirms his or her self-perception as inadequate and shame-worthy. Clearly, strong feelings accompany these EMSs, and intense anxiety (or other feelings) will be activated when life situations activate the client's EMS. For example, imagine a graduate student who has a **Failure schema**, which developed from repeated interactions with hyper-critical parents who were, in reality, virtually impossible to please. He is likely to worry excessively and experience significant anxiety or other distressing feelings as comprehensive exams, his dissertation defense, applications/interviews for internship sites, and other evaluation/performance situations that activate his Failure schema loom on the horizon.

Schema focused cognitive therapy helps us conceptualize clients—clarify what's really wrong for this particular client and needs to change in treatment, by paying close attention to:

1. events that trigger high levels of affect,
2. the client's most serious and enduring life problems, and
3. the early origins of the client's emotional distress.

The therapist intervenes with "empathic confrontation" and tries to find a balance between empathizing with the client's pain brought on by the schema and challenging the client with evidence that disconfirms the schema. The therapist is alert for EMSs and questions them whenever they come up—in the client's relationships with others and, especially, in the therapeutic relationship. The cardinal issue here, which will be emphasized throughout the interpersonal process approach, is that it is important to discuss schemas or misperceptions *in the moment*, when they are being played out or experienced in response to the therapist, and are accompanied by strong affect. For example, the client is hurt and ashamed with the therapist—feeling humiliated because she (mistakenly) believes that the therapist privately thinks she is "fat" and "disgusting"—just after she has disclosed another binge-eating episode. The opportunity for emotional relearning and expanding the client's schemas is far more powerful if the therapist questions the maladaptive schema *as it is occurring*:

THERAPIST: What do you think is going on for me—what could I be thinking and feeling inside—as you are telling me about this big disappointment?

CLIENT: Well, you have to say something nice and act understanding because that's what therapists are paid to do, but you probably really think that I'm just a fat disgusting pig—and that's all I'm ever going to be.

THERAPIST: Oh, there's such a big misunderstanding going on between us right now. No part of me is thinking anything like that—actually, I'm feeling sad right now as I see how much you have struggled with this—and how much pain this shame and self-hatred has caused you in your life. But it does seem like that Shame schema we've been talking about has really come into play—right now between us.

CLIENT: I think I've basically just hated myself all my life...but you're saying you really don't...even when I stuff myself like a pig...

Later in this chapter, we are going to frame this type of experiential or in vivo relearning as a "corrective emotional experience." But for now, let's continue with a different example. Aaron grew up with drug-involved parents who often were "away" and, when they were home and physically present, were often altered and emotionally absent. At age eight, Aaron was placed in a series of three foster homes by Child Protective Services until he was finally adopted and settled into a stable, predictable life with a supportive family. At the core of many issues that Aaron is going to be struggling with in his life will be an Abandonment Schema. Aaron is vulnerable or prone to feel anxious and depressed and struggle with other familiar symptoms that will disrupt his life, particularly at times when separation experiences are occurring, for example, when a significant other, such as his girlfriend, suggests that she wants to date others or break up, or when his daughter heads off to the first grade or college. Similarly, feelings of emptiness, loneliness, and despair ("Abandonment feelings") may be evoked toward the therapist at critical times such as at the end of an especially powerful session, when the therapist is preoccupied about a personal problem in his own life, just before the therapist leaves for vacation, when they discuss termination, and so forth. The key refrain here and in the chapters ahead is this: the real opportunity to help clients change is when the problem they are struggling with is occurring *between therapist and client*—not just being talked about in the abstract. This "immediacy" is anxiety-arousing for new therapists, for sure, but also the potent place where we have the chance to make a real difference in the client's life.

Working in the moment in this way, the key is to *address the schema when it is activated* or *in play between the therapist and client*. Experientially, this approach is far more meaningful to the client than talking about the same issue in the abstract—for example, relaying how this same theme played out with someone else, somewhere else, a few days ago. Drawing flexibly from a wide range of techniques, the therapist may address the EMS by reviewing contradictory evidence whenever it comes up in the session. The therapist can also intervene by helping clients write flashcards that describe the schema, carry the flashcard with them, and read it back to themselves when the schema is activated. This counters the EMS with a more rational response in the real-life moment when the schema is "triggered" and it feels to the client like the schema really is truth and reality, and the client is saying to herself something like:

"It's my fault, I did it wrong again, and they're going to blame me;"

"I'll always be left like this—I can't ever count on anybody;"

"They really don't want me, I don't belong here;" and so forth.

Additionally, experiential techniques of role play and rehearsal with the therapist, and holding imaginary dialogues with parent figures who originally instilled the schemas, and other techniques are employed to resolve the EMS. Most important, however, *the therapist tries to invalidate the client's schemas by providing a therapeutic relationship that counteracts or disconfirms what the client experienced in the past.* For example, within the appropriate limits and boundaries of the therapeutic relationship, the therapist tries to be responsive toward a client who was ignored, flexible with a client who grew up with rigidity, consistent and predictable with a client who grew up without discipline, and so forth. We will return to this concept later in this chapter and discuss "client response specificity"—how therapists can tailor their interventions flexibly to meet the differing needs of different clients. For now, we can see that theoretical approaches as disparate as object relations, attachment theory, and cognitive therapy can all teach us much about the schemas or internal working models that shape clients' lives. As we go on, we will explore further how these faulty beliefs and problematic expectations, which once made sense and accurately fit what really occurred in formative family relationships, are overgeneralized now and no longer serve clients well in most current relationships.

THE FAMILIAL/CONTEXTUAL DOMAIN

Children must adapt to their attachment figures, and they do so within a familial and cultural context. Family systems theory offers therapists an invaluable way of understanding their clients' strengths and problems—clarifying the familial rules, roles, myths, communication patterns, and boundary issues that defined their clients' development. Cultural norms and values are imparted through the arena of family interaction and, in this and many other ways, we develop our identity or "self" and become who we are in our families of origin. In particular, family process researchers have learned much about family interaction and relationships by adopting a multigenerational perspective.

In the late 1950s, Gregory Bateson, Virginia Satir, Murray Bowen, and other pioneers began studying communication patterns in families (Goldenberg & Goldenberg, 2008). They found that communication patterns in families followed definitive but often *unspoken rules* that determined who spoke to whom, about what, and when. These covert or unspoken rules also governed affective expression in the family, such that, for certain family members, being sad or disappointed, irritated or angry, or even excited and happy, may not be allowed. Years later, these unspoken family rules about communication or expressing certain feelings are still in play for many clients and cause problems in their current adult relationships. For example, adult clients often continue faulty communication patterns they learned years before in their families of origin. For example, they:

- address conflicts through a third party rather than speaking directly to the person involved (for example, Father: Tell your Mother that I'm sick and tired of..."),
- allow others to speak for them and define what they are thinking or feeling (Parent: "No, that's not what she really means..."), and

- never make "I" statements and communicate directly what they want, or set clear clear limits and define boundaries by saying "no" (for example, Mother: "You really don't want to do that, of course you don't.").

Early family researchers also illuminated how children in dysfunctional families are scripted into narrow roles, such as the responsible or good child; the problem or bad child; and other common roles that children are cast into from birth—including the family hero, rescuer, or invisible child. Following object relations theory, the most common configuration is the good child–bad child role split that occurs in so many dysfunctional families. Some therapists were the "good child" in their families of origin and had a sibling who fulfilled the "bad child" role and seemingly couldn't do anything right or was always viewed as the problem—independent of their actual behavior. Especially apt for therapists-in-training, many therapists, and others who enter the helping professions such as nurses or ministers, were "parentified." In this reversal of parent-child roles, they fulfilled a caretaking role for their parent. That is, they often were trying to meet the emotional needs of their parent, rather than having their own age-appropriate needs met, by serving as a confidant or "best friend" to their parent. This might have included trying to protect their parent or shoring up the emotional needs of an anxious, depressed, dependent, or otherwise vulnerable caregiver. For others who choose counseling careers, they may have been assigned to the role of mediator or "go-between" in family/parental conflicts.

Why is this family systems perspective important? Most clients are still fulfilling these limiting childhood roles, which deny many aspects of who they are and hold them back from becoming more of who and how they want to be in life. Thus, in the process of adapting in their families of origin, these children "internalized" these roles as self-schemas. These problematic familial roles have become their identity—who they are and what they do. And, even though their parents now may live far away or even be deceased, they continue to place their families' original expectations on themselves and re-create these problematic roles and relational patterns with others in their current lives. For example, many clients who have played the good-child role in their families are still perfectionistically demanding of themselves, feeling guilty about doing things for themselves, overly worried about the needs of others, afraid of their anger, and confused by their sad and empty feelings. Continuing in their role, the "bad child" may be defiant and act out personal conflicts externally. They may, for example, abuse substances, be promiscuous, or behave in other rebellious ways. In contrast, the "good child" is more likely to seek treatment and enter therapy, often presenting with vague complaints of guilt, anxiety, and an inability to make decisions or know what s/he wants that s/he can't understand.

There is great cultural variation in the balance of "separateness/relatedness" or degree of continuing responsibility that young adult offspring are expected to hold for their parents. However, within the parameters of these varying cultural contexts, many families also have unspoken rules about how, in late adolescence, offspring can leave home or "individuate" and further establish their own beliefs and values, work and careers, and marriage and families. For example, the oldest daughter may not be allowed to grow up and leave home successfully on her own. Perhaps she will seek emancipation through pregnancy, only to find herself even more dependent on her parents than before and forced to live at home again. A son may only be able to leave home through rejecting confrontations with parents, and then may live in another part of the country and have little or no continuing contact with his parents. In better-functioning families, young adults find support and guidance for launching their own lives, while continuing to maintain strong family ties and connections. Family rules, roles, and faulty communication patterns often serve to maintain family myths that, in turn, function to avoid anxiety-arousing issues in the family (Satir, 1967; 2000). For example, common family myths include:

- Dad doesn't have a drinking problem.
- We are a happy family and nobody is ever sad.
- Mom and Dad never fight, and are very happily married.

In sum, family myths and these other family characteristics are rule-bound *homeostatic mechanisms* that govern family relations and establish repetitive, predictable patterns of family interaction. They are homeostatic in the sense that they tend to maintain the stability of the family system—often making many changes uncomfortable or threatening.

Salvadore Minuchin (1985) and other family researchers have also explored the alliances, coalitions, and subgroups that make up the structure of family relations. In some families, for example, the maternal grandmother, mother, and eldest daughter are allied together, and the father is the outsider. This structural road map for reading family relations becomes even more illuminating when these family dynamics are examined in a three-generational perspective (Bowen, 1978; Boszormenyi-Nagy & Spark, 1973). It is fascinating—and sometimes disturbing—to draw "genograms" and see how these same family rules, roles, myths, and structural relationships can be reenacted across three or four generations in a highly patterned, rule-governed system (McGoldrick, Gearson, & Petry, 2008). Adding to the complexity of this tapestry, these family rules, roles, and communication patterns operate within a broader cultural context. For example, the oldest son in a traditional Chinese family or the oldest daughter in a traditional Latino family may be expected to assume significant, ongoing responsibilities for elders or other family members. It will be necessary for therapists to establish the extent to which these roles and behavior patterns are culturally sanctioned or determined.

To help, Sue et al. (1998) describe **cultural competence** in working with clients as including (1) the therapist's *awareness* of his/her assumptions, values, and biases; (2) the therapist's level of *knowledge* regarding the worldview of culturally different clients; and (3) the therapist's *skill* in developing interventions that are culturally sensitive. Sue and Zane (2009) suggest that one key consideration in working with clients is **"achieved credibility"**—the skills exhibited by the therapist that garner the client's sense of hope and confidence, which would include establishing a collaborative alliance, working relationship, or active engagement (Baldwin et al., 2007). Thus, knowledge of the client's culture is used in the service of developing credibility; for example, using language that resonates with the client's experience is associated with better treatment outcomes (Draguns, 2007). In this way, the client's cultural features or cultural group do not define the client but this cultural information alerts the therapist to potential credibility issues that can arise. Pedersen et al.'s (2008) informative book, *Inclusive Cultural Empathy*, provides student therapists with helpful exercises to increase self-awareness and sensitivity to a range of cultural issues.

The interpersonal process approach draws on these basic family systems and inclusive cultural concepts, and considers the effects of parental relations, familial experience, and contextual factors on clients' personality strengths and problems. Therapists want to help clients make realistic assessments of the resiliencies and vulnerabilities that coexisted in their families of origin. However, *for many clients and for many therapists, it is culturally taboo to speak critically of parents or talk about problems with someone outside of the family.* Alice Miller (1984) observes how, for some, this even breaks the Fourth Commandment: Honor thy parents. Herein, though, is one of the great strengths of family systems theory. The family systems therapist is trying to understand hurtful familial interactions, not to place blame or scapegoat any family member. The family therapist is concerned about the well-being of every family member, parents as well as children. In parallel, only as clients realize that their individual therapist wants to understand—rather than to blame their caregiver or make them bad in some way—do clients have the safety they need to explore threatening material and make significant gains in treatment. At times, however, therapists' own splitting defenses or dichotomous thinking leads them to blame or reject parental figures. For example,

INEFFECTIVE THERAPIST: I can't believe your mother did that to you, she's so mean!

Out of guilt and loyalty to their caregivers, it will be far more difficult for clients with this type of therapist to make progress in treatment. In contrast, it will be far more effective if therapists eschew blame and try only to understand and empathize with their clients' experience. For example,

EFFECTIVE THERAPIST: I'm sorry you were hurt so much when she did that.

There are some good things in almost every family, even in very troubled or abusive families, just as there are limitations and problems in the healthiest families. When clients continue to idealize parents and deny real problems that existed, as many do when they enter treatment, they are complying with binding family rules and protecting their parents at their own expense. On the other hand, however, clients cannot simply reject hurtful parents and emotionally cut off from them, or they will tend to reenact these same problematic relations with others. This means that therapists need to help clients find ways to keep parents (or the healthier, more benevolent aspects of them) alive inside themselves as "partial identifications." How can clients accomplish this?

In order to resolve problems and make progress in counseling, clients want to be able to come to terms with both the good news and the bad news in their families of origin. As already noted, however, clients cannot achieve this integration if the therapist also employs his own splitting defenses or dichotomous thinking. In other words, the therapist does not want to identify with or idealize the wounded child that exists within some clients and to reject the hurtful parent as "bad." Neither does the therapist want to support clients' continuing denial of familial problems and idealization of parents. Instead, the therapist's appropriate role is to try to understand what actually occurred in clients' development and help them realistically come to terms with both the good news and the bad in their experience.

Family systems concepts and other contextual approaches inform the interpersonal process approach. In particular, they will help us identify the problematic interaction patterns that clients are enacting with the therapist and others, and help both therapists and clients understand the familial context that shaped many current problems. With this awareness, clients are better able to stop reenacting problematic familial roles with others in their current lives but without breaking off important but conflicted familial relationships.

Let's pause here for a moment to appreciate two reasons why familial experience has such a profound, lifelong impact on the individual. The first is the sheer repetition of family transactional patterns. The same types of affect-laden interchanges are reenacted thousands of times in daily family life. To illustrate, suppose a parent has difficulty responding positively to her child's success experiences because of her own depression, narcissistic envy, or competitiveness. When the young child enthusiastically seeks this parent's approval for an accomplishment, the parent might ignore the child or change the topic, compare the child's achievement unfavorably with a sibling's greater accomplishment, turn away and look vaguely sad or hurt, or take the success away from the child by making it her own. The same type of parental response usually occurs when the child shows the parent a favorite drawing, makes a new friend, wins a race at school, or earns a star from her teacher. This transactional pattern becomes a powerful source of learning when it continues over a period of years and even decades. As an adult, this child is likely to feel conflicted about completing her educational degree, taking pleasure in a promotion earned at work, or even enjoying her marriage or good friends. In this way, the most significant and enduring problems in people's lives develop from these habitual response patterns. Contrary to popular belief and portrayals in Hollywood's movies, long-standing symptoms and problems are shaped far more by repetitive family transactional patterns (*strain* trauma) than by isolated traumatic events (*shock* trauma) (Wenar & Kerig, 2006). This represents a paradigm shift for many therapists-in-training, who now are learning to listen for patterns and themes in daily interaction, rather than keying on crisis events. In this way, enduring problems do not arise so much from traumatic or crisis events per se, but from the lack of validation or empathy from caregivers for the painful events that occurred.

Second, the impact of these repetitive transactions is magnified because of the intensity of the feelings involved. Parents are the pillars of the child's universe, and children depend on them with a life-and-death intensity. These repetitive transactional patterns have been reenacted in highly charged, affective relationships with the most important people in one's life. The child in the preceding example may well feel a desperate need somehow to win the parent's approval yet simultaneously feel increasing anxiety about trying to succeed or approaching the unresponsive parent. Thus, our sense of self in relation to others is learned in the family of origin and, in many ways, will carry over to adulthood.

In sum, family interaction patterns may be hurtful and frustrating for the individual, validating and encouraging, or most commonly, a mixture of the two. For better or worse, these repetitive patterns of family interaction, roles, and relationships are internalized and become the foundation of our sense of self and the social world. Of course, other factors are influential as well. Familial relationships are molded within a cultural context that can affirm or repudiate parental behavior.

Familial experience provides our first and most long-lasting model for what goes on in close relationships. It will figure significantly in the individual's choice of marital partner and career, in how adult offspring will, in turn, parent their own children, and in many of the other enduring problems and satisfactions of adult life. Although family-of-origin influences are profound, problematic rules, roles, and communication patterns can be relearned. Change is indeed possible, although often not easy, and it occurs in part through a relational process of *experiential relearning*, which we consider next.

CORE CONCEPTS

In the chapters ahead, we are going to apply three core concepts to understand clients and guide our interventions: the process dimension, a corrective emotional experience, and client response specificity. In this final section, these three orienting constructs are introduced and then illustrated with a case example.

THE PROCESS DIMENSION

The relationship between the therapist and the client is the foundation of the therapeutic enterprise and the therapist's most important means of effecting change. It shapes whether, and how much, clients will be able to change in treatment. In order to utilize the therapeutic relationship as a vehicle for change, however, therapists need to understand the complex interactions or "interpersonal process" that is taking place with their clients. Thus, the primary goal of this text is to help therapists understand what is going on between the therapist and client, right now or in the moment, in terms of the current interaction or *process*.

Let's clarify more specifically what we mean by the process dimension. The therapist-client relationship is complex and multifaceted as different levels of communication occur simultaneously. It is helpful, for example, to recognize the subtle but important distinction between the overtly spoken *content* of what is discussed and the *process* dimension of how the therapist and client interact (Kiesler, 1988). In order to work with the process dimension, the therapist is making a perceptual shift away from the overt content of what is discussed and beginning to track the relational process of how two people are interacting as well. This means that the process-oriented therapist is going to step beyond the usual social norms at times and talk more directly with the client about their current interaction or interpersonal process—that is, the way they might be interacting together or what may be going on between them right now. For example,

THERAPIST: Since I asked you about your canceling our session last week, and then coming late to our appointment today, you've become quiet. Maybe something about my question didn't feel quite right to you. Can I check in with you about that?

OR

THERAPIST: Right now, John, I'm having a little trouble keeping up with you. You're speaking fast, and sometimes it feels a little like we are almost jumping from topic to topic. Any thoughts here...what do you think might be going on between us?

In both of these queries, the therapist is not addressing what the client is saying (content) but how the therapist and client are interacting with each other (the interpersonal process). Let's look at one more example but this time, extend the therapist's initial process comment into a longer therapist-client dialogue and see how process comments like these can uncover important new issues to explore in treatment, resolve misunderstandings that arise between the therapist and client, and bring intensity to the therapeutic relationship by moving the dialogue beyond surface issues:

THERAPIST: I'm honored that you choose to share such an important part of yourself with me. What's it like for you to risk sharing this sensitive feeling with me?

CLIENT: Uh, I'm not really sure what you're asking me?

THERAPIST: You just said so much to me, and I'm wondering how that was for you. What do you think I might be thinking about you as you're telling me this?

CLIENT: Well, you probably have to be nice, you know, because you're a therapist so that's how you have to act. But maybe inside you're judging me or something.

THERAPIST: "Judging you." That sure wouldn't feel very good. Tell me more about how I might be judging you?

CLIENT: Well, come on, you can't possible respect me now that I've told you about this.

THERAPIST: I'm so glad we're talking about this misunderstanding, because no part of me is feeling judgmental toward you. Actually, I was appreciating your courage to bring this up and talk with me so honestly. But maybe you're telling me that others have often judged you?

CLIENT: Oh yeah, I know my mom loves me but she was always *so* judgemental…and now, whenever my husband says something that's the least bit critical, I just shut down inside and completely go away…

We are going to call these types of here-and-now, present-focused interventions **process comments**. Therapists working in differing theoretical traditions have variously termed these interventions, which focus on "you and me" or explore what is going on between the therapist and the client right now, as *process comments, meta-communication, immediacy interventions, interpersonal feedback, working in the moment, therapeutic impact disclosure, self-involving comments, mentalizing,* and other related terms. Because it may conflict with familial rules or cultural prescriptions for new therapists to speak directly in these ways, it may feel awkward at first for some. For others therapists, it may seem impolite or disrespectful to use process comments and speak forthrightly or inquire in open-ended ways about what may be going on between "you and me." However, as new therapists become more confident and less concerned about performance anxieties, and their supervisors help them understand better where they are trying to go in treatment with this particular client, they will be ready to begin trying out and exploring these process comments. New therapists soon will see how helpful it can be when they link the problem that the client is talking about with others to their current interaction in the therapist-client relationship—that is, highlighting how the client is utilizing the same problematic thought processes, faulty expectations, or ineffective coping strategies with the therapist, right now, that have been and are causing problems with others in their lives (Kasper, Hill & Kivlighan, 2008). Or, similarly, the therapist is *nondefensive* and willing to explore and sort through with the client potential misunderstandings, inaccurate perceptions, "mistakes," or other interpersonal conflicts that may be going on between them in their real-life relationship; that is, restoring "ruptures" (Safran et al., 2002). To help new therapists learn how to begin using these powerful but initially challenging interventions, each successive chapter will explore further how therapists can make these process comments effectively. That is, how therapists can metacommunicate and talk with clients about how they are interacting together, and utilize a wide variety of other immediacy interventions, in a tentative, empathic, and respectful manner. When we use process comments in this tentative, wondering-aloud way, clients will welcome them as a *collaborative invitation* for genuine understanding and honest communication—something that does not make them feel awkward or uncomfortable, and that *brings them closer to their key concerns or real problems* (Crits-Cristoph & Gibbons, 2002; Kiesler, 1996). Consider the following example.

During their first session together, the client tells the therapist that he resents his wife because she is "bossy" and "always telling me what to do." He explains that he has always had trouble making decisions on his own and, as a result, his wife has come to simply make decisions for him. Even though he felt that she was just trying to help him with his indecisiveness, he resents her "pushiness" and "know-it-all attitude." After describing the presenting problem in this way, the client asks the therapist, "What should I do?"

Let's think first about the process dimension in their interaction. Suppose the therapist complied with the client's request and said, "I think the next time your wife tells you what to do, you should..." If treatment with this particular client continues in this prescriptive vein, the therapist and client will quickly begin to reenact in their relationship the same type of conflict or scenario that originally led the client to seek treatment. That is, the therapist will be telling the client what to do, just as his wife has been doing. The client has certainly invited this advice, and will probably welcome the therapist's suggestions at first. In the long run, however, he will probably come to resent the therapist's directives just as much as he resents his wife's and will ultimately find the therapist's suggestions to be as unwanted and of as little help as hers.

As an alternative to this directive response, a more nondirective therapist might respond to the other side of the client's conflict and say, "I don't think I would really be helping you if I just told you what to do. I believe that clients grow and learn more when they find their own solutions to their problems." Frustrated by this nondirective response, which he only finds evasive, the client retorts, "But I told you I don't know what to do! It's hard for me to make decisions. Aren't you the one who's supposed to know what to do about these things?"

This nondirective response throws the client back on the other side of his conflict and leaves him stuck in his presenting problem—his own inability to make decisions. If this mode of interaction continues and comes to characterize their relationship, their process will reenact the other side of the client's problem. His inability to initiate, make his own decisions, and be responsible for his own actions will immobilize him in therapy, just as it has in other areas of his life.

One of the things we see here is that clients do not just talk with therapists about their problems in an abstract manner. Rather, in the interpersonal process they often "enact" or recreate with the therapist the same type of problematic interactions that originally led them to seek treatment. That is, clients also convey their problems in *how* they interact with the therapist (the process), bringing critical aspects of their problems with others into their current interaction (Hill et al., 2008). *This repetition or replaying of the client's problem is a regular and predictable phenomenon that will occur, at times, in most therapeutic relationships.* We will return to this construct and example later in the chapter and see how the therapist can use the process dimension to find more effective ways to respond. For now, let's turn to the second core concept, the corrective emotional experience.

THE CORRECTIVE EMOTIONAL EXPERIENCE

Like most far-reaching concepts in therapy, the **corrective emotional experience** has many grandfathers (and grandmothers). However, it was developed most fully in the 1940s by two psychoanalytically oriented therapists, Franz Alexander and Thomas French (1946), and has become the cornerstone of many short-term psychodynamic and interpersonally oriented therapies (Bernier & Dozier, 2002; Bridges, 2006). Although it was not appreciated by the psychoanalytic community of its day (it was disdained as being superficial), this more active, direct approach provided clients with a real-life experience of change in the here-and-now relationship with the therapist. Accurately anticipating the future of counseling and therapy, Alexander and French originally advocated the corrective emotional experience to shorten the length of treatment, put less emphasis on insight as the basic means of effecting change, and focused instead on a more experiential, behavioral, or in vivo relearning. Alexander actively encouraged clients to approach or engage in anxiety-arousing behavior that previously they had avoided, and even suggested giving homework assignments to clients between sessions. There was some dismay in traditional psychoanalytic circles when Horowitz (1974; 1976) published treatment outcomes from the Menninger Psychotherapy Research Project. He reported strong empirical support for enduring change in patients who received corrective emotional experiences in supportive psychotherapy—equivalent to patients receiving longer-term psychoanalysis.

Contemporary, short-term psychodynamic therapies (Binder, 2004; Crits-Cristoph & Barber, 1991; Sifneos, 1987; Strupp, 1993; and many others) describe the same direct, immediate experience of change in the real-life relationship between the therapist and client. The basic concept is that therapists, working within different theoretical orientations, can all help their clients change: they do so *by providing a new and more satisfying response to the clients' old relationship patterns than they have usually found with others. That is, working collaboratively with clients, the therapist helps clients identify maladaptive relational patterns or themes that commonly occur with others, and works together with clients to alter this problematic pattern, disconfirm the faulty expectations or schema, or change this familiar but unwanted interpersonal scenario in their real-life relationship or therapeutic interaction* (Andrews, 1991). Following Alexander and French, we are going to call this core concept a "corrective emotional experience" (some interpersonal and attachment-oriented therapists use the term "reparative relational experience," whereas behavior therapists often describe "exposure trials"). When therapy fails and clients terminate prematurely, or treatment bogs down and reaches an impasse, the therapist and client are often reenacting in their interpersonal process or the way they are relating aspects of the same conflict that the client has been struggling with in other relationships, although neither of them may be aware of this reenactment. Based on their schemas and expectations, clients soon come to hold the same misperception or faulty expectation toward the therapist, or respond with the same maladaptive relational patterns, that have been causing problems with others in their lives. For example, even though the therapist has not behaved accordingly, when clients become distressed or vulnerable, they soon may believe the following:

- They are being controlled by the therapist and have to do everything his way, just as they have done with others.
- They have to take care of the therapist and meet her needs, just as they have been doing with others in their life.
- They must please the therapist and win his approval, just as they keep striving to do with significant others.

When this phenomenon occurs, *the therapeutic process is metaphorically repeating the same type of conflicted interaction that clients have not been able to resolve in other relationships, and that they have often experienced in formative, attachment relationships.* As clients begin to play out with the therapist the same relational patterns that originally brought them to treatment, the therapist's goal is to respond in a new and more effective way that allows clients to resolve the conflict and change the pattern within their relationship. The benefit to clients from this in vivo experience of change is that fixed or narrow schemas will expand and become more flexible or realistic, and it will become easier to begin changing these problematic patterns with others, such as spouses, children, friends and co-workers. Although providing a corrective emotional experience may sound easy, it can be challenging to do—especially when all of this is so new to therapists-in-training. To help, Hill (2009) encourages therapists to be asking themselves the same process-oriented question throughout each session:

> Right now, am I co-creating a new and reparative relationship, or am I being drawn into a familiar but problematic interaction sequence that is reenacting for this client?

When the therapist and client are able to resolve these expectable reenactments, which occur in most therapeutic relationships, change has begun to occur. Clients learn that they no longer have to respond in their old ways (such as always having to be in control, be the responsible one, or have to take care of others) or always receive the same unwanted responses from others (for example, being ignored or dismissed, judged or criticized, envied or competed with). In this way, clients develop a wider range of expectations and begin to learn how they can respond in more flexible and adaptive ways—especially in the safety of their relationship with the therapist. As their behavior begins to change with the therapist, it is usually relatively easy to take the next step and help clients generalize this experiential relearning with the therapist and adopt similar, more adaptive responses with others outside the therapy setting.

We introduced this sequence earlier; let's return and examine it more closely. In order to provide a corrective emotional experience and help clients change, the therapist's intentions are to:

1. identify the faulty beliefs and expectations, maladaptive relational patterns, and ineffective coping styles that keep òccurring and causing problems in the client's life,
2. anticipate how these patterns or themes that are disrupting relationships with others may be expressed or come into play in the current interaction with the therapist—especially along the process dimension or way they are interacting together (for example, arguing, distancing, controlling),
3. provide new or corrective responses that help to resolve rather than repeat this familiar but problematic pattern in the relationship with the therapist, and
4. help clients generalize this new way of interacting to others in their everyday lives.

In this way, new therapists will find that their clients often change in a two-step sequence. First, clients have an experience of change when, this time, their old relational patterns or problematic expectations do not recur with the therapist in the way they interact. The second phase of treatment occurs as the therapist helps clients generalize this real-life or behavioral experience of change (as opposed to more intellectual modalities such as insight, cognitive reframing, interpretations, explanations, and so forth) to others in their lives. How does the therapist help clients transfer or link this experiential re-learning to others in their everyday lives? The therapist does this by:

1. clarifying when and how these same cognitive and interpersonal patterns are triggered and disrupt other relationships as well;
2. formulating and rehearsing (for example, role-playing) new, more adaptive responses that change how the client has been participating in most current interpersonal problems; and
3. helping clients anticipate future situations that are likely to activate or evoke their old problematic response patterns, so they can more actively choose how they wish to respond instead.

In this way, clients learn to respond in more flexible ways, rather than rigidly repeating old familiar response patterns which are not succeeding in many current relationships.

The basic idea here is that clients believe actions, not words. Long ago, Frieda Fromm-Reichmann (1960) captured this central tenet best by saying that the therapist needs to provide the client with an *experience* rather than an *explanation*. Interpretations, empathic understanding, cognitive restructuring, self-monitoring techniques, education, skill development, and other interventions are utilized in the interpersonal process approach and will be helpful with most clients. They are not the primary seat of action in this approach, however. This is an experiential learning model. Following Strupp (1980), clients change when they live through emotionally painful and long-ingrained relational experiences with the therapist, and the therapeutic relationship gives rise to new and better outcomes that are different from those anticipated and feared. That is, when the client re-experiences important aspects of her primary problem with the therapist, and the therapist's response does not fit the old schemas or expectations, the client has the real-life experience that relationships can be another way. When clients *experience* this new or reparative response, a response that differs from previous relationships and that does not fit the client's negative expectations or cognitive schemas, it is a powerful type of experiential re-learning that readily can be generalized to other relationships (Bandura, 1997). The key point here is that this behavioral experience of change with the therapist is far more compelling than words alone can provide.

We are going to be exploring many different ways that therapists can use this corrective emotional experience to facilitate change. In particular, the new or corrective response from the therapist helps clients change by creating *greater interpersonal safety* for the client. This is far more than the good feeling that comes from being taken seriously and treated with respect in the therapeutic relationship. Instead, it is a more significant sense of deep personal safety that results when the client is not hurt again by receiving certain interpersonal responses that are familiar and expected, yet unwanted or even dreaded. As we have noted, *this pivotal moment is not the end point in treatment but a window of opportunity for a range of important new behaviors to emerge.* For example, in the next few moments immediately following a reparative *experience* with the therapist, clients may feel safer and be empowered to:

- experience more fully certain painful or difficult feelings;
- resolve their ambivalence and make an important personal decision;
- risk trying out new ways of responding with the therapist or others in their lives;
- make significant self disclosures or bring up relevant new issues or concerns to explore;
- feel bolder and risk addressing unspoken misunderstandings with the therapist or bring up more directly a problem the client has been having with the therapist;
- feel better about themselves or forgive themselves for things they have felt unrealistically shameful or guilty about;
- grasp more fully how much this faulty coping strategy or self-defeating behavior has hurt others or cost them in their own lives;
- have insight or make meaningful links between current behavior and formative relationships where these schemas and patterns originally were learned.

In these ways, corrective emotional experiences with the therapist begin to expand maladaptive early schemas and repetitive relational patterns, and a wider range of responses becomes available to the client—often in the next few moments.

Generally, a single corrective emotional experience is not sufficient for sustainable change, but it is often the pivotal experience that initiates change. Based on the interpersonal safety that arises when clients find that they consistently or dependably receive this new, reparative type of response from the therapist, many clients become more willing to try out interventions from educational, behavioral, cognitive, and other theoretical approaches, and it facilitates clients' commitment to follow through on treatment goals. Thus, when clients experience a different type of relationship with the therapist that does not go down the familiar but problematic pathways they have come to expect from others, the therapist is seen as someone who can help—earning **achieved credibility** (Sue & Zane, 2009). Clients invest further in the therapeutic relationship and in the treatment process, and often become more willing to try out new coping strategies and ways of responding with others. In contrast, when clients elicit the same type of reactions from the therapist that they tend to find with others (for example, the therapist begins to feel frustrated, controlled, discouraged, or disappointed, as significant others in the client's life often feel), intervention techniques from every theoretical perspective will falter. In this way, the interaction between the therapist and the client provides a "meta" perspective for understanding what is occurring and how to intervene in the therapeutic relationship, and it readily can be integrated with other therapeutic modalities (for example, schema confirming vs. schema disconfirming interactions with the therapist).

Using the Process Dimension to Provide a Corrective Emotional Experience.

Linking our first two core constructs, the therapist needs to be able to work with the process dimension in order to provide this corrective emotional experience. Recall the previous example of the client who was complaining that his wife was "pushy" and always telling him what to do. When the client asked his therapist what he should do, one option would have been for the therapist to offer a process comment and make their current interaction overt as a topic for discussion:

THERAPIST: Right now, it seems that you're asking me to tell you what to do. But I'm wondering whether that will only bring up the same problem for us here in therapy that you are having at home with your wife. Let's see if you and I can figure out a way to do something different in our relationship. Rather than having me tell you what to do, let's try and work together to understand what's going on for you when you are feeling indecisive. Where do you think is the best place to begin?

CLIENT: I'm not sure.

THERAPIST: Take your time, and let's just see what comes to you.

CLIENT: (pause) I'll get criticized. No matter what I decide, she'll find something wrong with it.

THERAPIST: OK, that's a good place for us to start. You can't get it right in her eyes— you'll always do it wrong. Sounds like there's a lot of feeling in that for you, tell me more.

CLIENT: Well, I hate it because she wants to be the parent—you know, the one in control, and make me a child...

In this instance, the therapist's intention is to ensure that the same problematic but familiar pattern does not occur in their relationship. The therapist is trying not to take the bait and repeat the pattern by telling the client what to do. Instead, the therapist is trying to offer the client a new and different type of relationship—a collaborative partnership in which they can work together on the client's problems. If this collaborative effort continues to develop over the course of treatment, the client is having the real-life experience that he can make decisions and successfully initiate—at least in his relationship with the therapist. Then, with the therapist's

support for this new behavior, he will rehearse, role-play, and try out these new behaviors with his wife and others. He will have successes and failures in his attempts
to adopt this stronger way of relating to others and, as we will explore later, the
therapist will help him learn from both. However, the therapist's initial attempt to
offer the client a corrective response will need to be repeated in many different ways
for change to consolidate. If the therapist can maintain this type of collaborative
interaction throughout treatment (easier to say than to do with this client), this
new interpersonal process will be corrective, and it will facilitate other therapeutic
interventions as well. That is, educational inputs, behavioral alternatives, psychodynamic interpretations, and interpersonal feedback will all become more effective
with this corrective interpersonal process. This client's life has been ruled by his
Subjugation/Control schema, but now he is a collaborator in the treatment process
with his own voice and is not being told what to do again (Baldwin et al., 2009;
Miller & Rollnick, 2002). We certainly aren't around the block yet, but change
is under way.

Countertransference Issues and Therapists' Fears about Making Mistakes Diminishes Engagement.

As we have seen, the guiding principle in the interpersonal
process approach is to provide clients with an experience of change. With this in vivo
relearning, clients live out a new relationship with the therapist that disconfirms their
faulty expectations and expands their schemas for what they can have in life—for
what can take place for them in relationships. For example, clients learn that, at least
sometimes—first with the therapist, and then with some others in their lives, they find
that they can: ask for help and be responded to; say "no" and have others respect
their limits; give priority to their own needs sometimes rather than always submerging
them and going along with what others want; succeed without evoking competition or
envy; and so forth. Providing this type of corrective emotional experience is a personally engaging and highly rewarding way to work with clients but—make no mistake
about it—it demands much on the part of the therapist. In particular, countertransference issues, and our previous topic of new therapists' fears of making mistakes, make
it harder to provide a corrective emotional experience. Let's look further at both.

First, it requires personal involvement from therapists to work with the client in
this way. The therapist must be willing to genuinely engage with the client and risk
being personally affected by the client (Rogers, 1980). That is, in order to have the
emotional impact necessary to provide a corrective emotional experience and propel
change, the relationship must hold real meaning for both participants (Gelso &
Carter, 1994; Gelso & Hayes, 1998; Hovarth & Bedi, 2002). In this simple human
way, it is the relationship that heals—the relationship itself facilitates "collaborative,
purposive work" (Hatcher & Barends, 2006). If the therapist is merely an objective
technician, psychologically removed and safely distant from the client, the relationship will not hold real meaning for either person. It will be too insignificant to effect
change—even if the therapist does indeed respond in new, flexible ways that do not
follow along with problematic old patterns, beliefs, and expectations. This is the
critical component in the concept of the corrective emotional experience that, too
often over the years, has been lost or gone unappreciated.

On the other side of this issue, treatment also will falter if the therapist becomes
inappropriately close. That is, if therapists overidentify with the client or become

too invested in the client's choices or ability to change, perhaps in order to shore up their own feelings of adequacy as a new helper, therapy will not progress. In this situation, therapists often lose sight of the process they are enacting with the client and unwittingly begin to respond in problematic ways that reenact aspects of the client's maladaptive relational patterns. Thus, a prototypic situation for many new therapists is to find that they are experiencing the client's predicament or concerns as identical to their own. This occurs, for instance, when the therapist is thinking, "She's just like me!" The solution to this overidentification is a supportive yet honest supervisory relationship that helps therapists recognize their own counter-transference issues and see the numerous ways that their own experience really does differ significantly from the client's. There may be important similarities between them, yet, in reality, the client's problems—and the broader context that shapes their meaning—are never really the same as the therapist's. Once able to differentiate their own issues from those of the client, and able to discern the ways in which their situation or circumstances are perhaps similar but also different from the client's, therapists will often be able to see how the therapeutic process has been reenacting aspects of the client's conflict. The multicultural literature is instructive here in highlighting "zones of difference and similarity within and between cultures" (Pedersen et al., 2008, p. 63.) This understanding that each therapeutic relationship involves a range of similarities and differences often enables therapists to change the problematic way they have been reenacting with the client and establish a more productive, resolving interpersonal process.

Second, we have already seen that new therapists should expect to make many "mistakes" with their clients. For example, to become overinvolved or underinvolved with certain clients, to reenact the client's maladaptive relational patterns at times, and so on. Such mistakes are an inevitable part of the therapeutic process for beginning and experienced therapists alike. Few clients are fragile, however, and therapeutic relationships are often remarkably resilient. Unfortunately, most student therapists do not know this and, as emphasized above, concern about making mistakes remains one of their biggest anxieties. New therapists will do better therapy—and enjoy this work so much more—as they discover that *mistakes can be undone.* Mistakes can be worked through or resolved with clients, and mistakes can provide important therapeutic opportunities, when therapists are willing to work with the process dimension—that is, when therapist are willing to talk with clients about the misunderstandings or potential problems that may be occurring between them. For example:

THERAPIST: I'm wondering whether I might have misunderstood something you said. I'm asking about that because it seems as if you have become more distant from me in the last few minutes, and you've mentioned that other people often don't understand you. What do you see happening between us here?

CLIENT: Yeah, I guess I do feel a little bit distant. But I don't feel "misunderstood," I just don't really like it when you don't say very much—you're pretty quiet, you know.

THERAPIST: That's helpful to hear—I appreciate your honesty. So, let's change things— I'll start speaking up more and sharing more of my thoughts with you. And let's check back in on this later and see how it's going. How's that sound?

CLIENT: Thanks, that would be better for me.

THERAPIST: Good, I want to hear your concerns, and work together in ways that feel most helpful to you. Can you tell me a little more about what's it like for you when I'm quiet or don't say very much?

CLIENT: I get uncomfortable...I don't feel like I'm getting enough direction or help, and maybe then therapy isn't going to work for me and I'm just going to stay depressed. And I'm not really sure what a client is supposed to do. I've never been in therapy before, you know, so I worry about what I'm supposed to be talking about or what you expect me to be doing...

As their performance anxiety declines, new therapists will find that they are better able to identify and work with the process dimension. The very best way to realign the therapeutic relationship when it has gone awry is for therapists to talk through and resolve potential misunderstandings or mistakes. Following Hill (2009), student therapists are encouraged to be non-defensive and recognize their mistakes, apologize if necessary, and *talk through the event with the client*. This restores the therapeutic relationship, and it provides clients with an effective model for how to deal with problems in relationships. We will return to process comments and this all-important topic of "rupture and repair" in the next chapter. For now, we turn to the third core concept: Client Response Specificity.

CLIENT RESPONSE SPECIFICITY

We have already begun to see that new therapists need a theoretical framework to guide their therapeutic interventions. Let's look further at what a theory needs to provide if it is to be of help to therapists and their clients.

One feature of an effective clinical theory is that it has the flexibility and breadth to encompass the diversity of clients who seek treatment. For many student therapists, clinical training changes their fundamental worldviews in profound and enduring ways. One significant way this occurs is that new therapists recognize—to a far greater extent than they had appreciated before—that unifying patterns in personality and behavior exist. In parallel with this new organizing perspective, however, it also is clear that every client really is unique and different. Each client has been genetically endowed with a unique set of features and each has been raised with different values and beliefs in his or her family. Socialization is different for women and men, individuals growing up in varying cultures have vastly different political experiences and religious training, and economic class shapes opportunity and expectations. To be helpful, a clinical theory must be able to help therapists work effectively with all of these highly diverse clients—who also are entering treatment with different levels of motivation and at different points in the change process (Prochaska & Norcross, 2006). Moreover, each therapist is a different person. Like clients, therapists differ in age, gender, ethnicity, sexual orientation, and developmental background. Therapists also bring diverse values, subjective worldviews, and personal styles to their clinical work. How can any theory help such a diversity of therapists respond to the extraordinary range of human experience that clients present?

Our third core construct, *client response specificity*, will be one of our best tools. **Client response specificity** means that therapists need to tailor their responses to fit the specific needs of each individual client—one size does not fit all! There are no cookbook formulas or generic techniques for responding to the complex problems, diverse developmental experiences, and multicultural backgrounds that clients present. In this regard, we are going to see that the same therapeutic response or intervention that helps one client make progress in treatment will only serve to hinder another. For example, some clients respond well to warmth in the counselor, whereas others want objectivity (Bachelor, 1995). More specifically, a warm and expressive therapist can put off a distrustful or avoidant client; a businesslike therapist can fail to engage an anxious client in crisis. Clearly, this work is not easy, it's complex—and it will be simply maddening if you must have rules and can't tolerate ambiguity! The key point is that therapists need to have the flexibility to listen to the cues, assess clients' responses, and search for the best way to respond to this particular client (Lazarus, 1993). As we are going to see, however, *therapists often fail to hear and accommodate to clients' feedback about what they want and don't want from their them.*

As we consider client response specificity, the operative word is *flexibility*. This approach requires therapists to be flexible, so they can respond on the basis of each client's own personal history and ways of viewing the world. It is certainly helpful to be familiar with the experiences of particular groups (for example, to know that African Americans' history includes slavery and racial discrimination) and the characteristics that accompany certain diagnostic categories (for example, to know that bipolar clients in a manic phase are at risk to act out with aggression toward others, sexual promiscuity, or suicide). Additionally, however, with client response specificity, our abiding intention is to attempt to understand and respond to each client as a unique individual. Thus we will not seek treatment rules or intervention guidelines that apply to all clients, or even to specific diagnostic categories or groups, such as men, Latinos, or Christians. Rather, we will be trying to find case-specific recommendations in which we explore what it means to this unique individual—the client in front of us right now who we are talking with—to be depressed, lesbian, or biracial. Let's look more closely at the meaning in "client response specificity" and see how it can help us be more effective therapists.

Interpersonal process therapy is a highly "idiographic" approach; it emphasizes the personal experience or subjective worldview of each individual client. Diagnostic categories will tell us something about the client, but they are only the map and not the territory. Personality typologies also highlight potential directions that therapy might take, but the actual therapeutic process and issues covered will be client-specific. The two clients on your caseload who both have generalized anxiety disorder, or an avoidant or dependent personality disorder, or who are sexual assault survivors certainly will share some common ground, but the differences between the two—their uniqueness—is significant. Throughout each session, therefore, *therapists are encouraged to join the client in a process of mutual exploration to try to find the subjective meaning that each particular experience holds for this client.* For example,

THERAPIST: Let me check this out with you and make sure I'm understanding what you're really meaning here. When I hear you say that, it sounds to me more like you are feeling sadness, or maybe even grief, than depression. Am I getting that right—can you help me say it better?

From the first session through the last, the therapist is taking on the personal challenge to be *accurately* empathic or, in different terms, to have the cognitive and emotional flexibility to de-center and enter the client's **subjective worldview**. The meaning that a particular experience holds for the client often differs greatly from the meaning that this same experience holds for the therapist, or for other clients that the therapist might see. Therapists are encouraged to listen carefully to clients' language and choice of words, and to collaboratively explore clients' own metaphors. For example,

THERAPIST: Help me understand what you mean, or what you might be saying about our relationship, when you tell me that you just want to "sail right out of here?"

By continuously reaching for this specificity, and not taking for granted that we know what the client really meant, we are not belaboring the obvious. When someone cares about you enough to listen seriously, and works hard to understand just exactly what it is that you are trying to say, meaning is created.

With client response specificity, there is no standardized treatment protocol. Each individual client's developmental history, cultural context, and current life circumstances serve to guide the therapist's treatment plans and intervention strategy. The interpersonal process approach does not advocate a particular therapeutic stance toward clients in general, such as being directive or nondirective, problem-solving or exploratory, supportive or challenging, veiled or self-disclosing, and so forth. As we are going to explore, each of these modalities will be helpful with a particular client at times and ineffective with another. Thus, this approach asks something different and far more demanding of therapists:

- to conceptualize the specific relational experiences that this client needs in order to change, and
- to be flexible enough to modify their interventions and respond in the ways that this particular client utilizes most productively.

This recommendation may sound obvious or easy to do, but unfortunately researchers tell us that most therapists do not do this very well. Therapists continue to respond to clients in the same manner, based on their theoretical orientation— even when their preferred brand is not working! Psychotherapy process studies find that most therapists do not demonstrate *flexibility* in their technical approach to the client, or have the interpersonal range to modify their interventions, and *provide the responses that each client could utilize best* (Mash & Hunsley, 1993; Najavits & Strupp, 1994; Strupp, 1980). With client response specificity, our aim is to assess—strive to discern—how clients are responding to each of our interventions in an ongoing, moment-by-moment way, and flexibly adapting to respond in the ways they find most useful. To help with this, the counseling literature on **intentions** (Hill & O'Grady, 1985) suggests that therapists ask themselves the following questions:

Where am I right now and what do I want to accomplish? (For example, recognize and respond to the feeling or concern that the client is expressing right now.)

OK, what is the best way to do that? (This might include validating or providing an empathic reflection of the client's distress or concern.)

In this way, therapists are encouraged to think about their intentions—*what they are trying to accomplish with each response or turn of the conversation* (Hill, 2009; Ivey & Ivey, 1999). Therapists will be better prepared to respond to the specific needs of this particular client, at this particular moment, when they are asking themselves in an ongoing way,

What is needed right now, and how can I provide that?

Let's use an example to illustrate client response specificity more concretely, and then examine more closely how therapists can assess clients' positive or negative responses to the different interventions and varying responses they provide—so therapists can modify or tailor them if necessary, to work better for this client. Consider an often debated question: Should therapists self-disclose to their clients? Client response specificity emphasizes that self-disclosure (or any other intervention) *will hold very different meanings for different clients*. In fact, the same intervention or response may even have the opposite effect on two different clients with contrasting developmental histories and cultural contexts. For example, if a client's parent was distant or aloof, the therapist's judicious self-disclosure may be helpful for the client. In contrast, the same type of self-disclosure is likely to be anxiety-arousing for a client who grew up serving as the confidant or emotional caregiver of a depressed parent. Greater sharing with the therapist may help the first client learn that, contrary to her deeply held beliefs, she does matter and can be of interest to other people. In contrast, for the second client, the same type of self-disclosure may inadvertently impose the unwanted needs of others and set this client back in treatment as, in her mind, she experiences herself back in her old caretaking role again—this time with the therapist. This unwanted reenactment occurs because the therapeutic relationship is now paralleling the same problematic relational theme that this client struggled with while growing up. Considering client response specificity in this way, therapists working within every theoretical orientation will be more effective if they first consider how the client's cognitive schemas or maladaptive relational patterns are likely to shape the impact of their interventions on this particular client. This is a different way of working with clients than merely proceeding to respond in the way a client-centered/humanistic, psychodynamic or cognitive-behavioral therapist was trained to intervene.

Like our other two core constructs, client response specificity is a complex and multifaceted clinical concept that will take time to learn and apply with clients. Let's illustrate it further by posing another common question: Should therapists give opinions when clients ask for advice? As before, the therapist's response is informed more by each client's particular circumstances than by the brand of therapy standardly employed by the therapist. For example, offering advice may be counterproductive for a compliant, intimidated female client who is trying to become more independent and assertive with her dominating and controlling husband—who undermines her confidence and actively fosters her dependency on him. For such a client, it may be better for the therapist to suggest,

Well, let's start with your thoughts about this first, and then I'll add mine. What do you think might work best for you here?

In contrast, giving advice and directives may work well with a different client whose caregivers did not help him solve his problems, perhaps because the caregivers were too narcissistically self-absorbed or anxiously preoccupied with their own problems to be capable of or interested in helping him. To withhold suggestions or problem-solving advice from these clients may only reenact developmental deficits and impede their progress in treatment as, yet again in their experience, they do not get the help or guidance they need and want. Here again, the same response from the therapist will have very different effects on different clients. From this meta-perspective, techniques or interventions from any theoretical approach may help or hinder the client; it depends on whether the response serves to reenact or resolve maladaptive schemas and relational patterns for this particular client.

Finally, let's further highlight an important aspect of client response specificity—assessing clients' responses to therapists' interventions. Therapists can learn how to assess the effectiveness of their interventions by paying close attention to how the client utilizes or responds to what they have just done. For example, if the therapist observes that the client becomes anxious, irritated, or distant in response to the therapist's self-disclosure, the therapist is learning how to respond most effectively to this particular client. It's not that self-disclosure is ill advised for another client, or an ineffective way to respond in general, but it is not being helpful for this particular client. As noted earlier, when the therapist provides a corrective response, many clients will make progress right away by feeling safer and acting more boldly, bringing forth relevant new material, becoming more engaged with the therapist, and so forth. In contrast, when the therapist responds in a way that repeats problematic old patterns or confirms maladaptive schemas, clients will behaviorally inform the therapist of this by acting "weaker" in the next few minutes—for example, by being more distant, indecisive, compliant, and so forth (Hill, Thompson & Corbett, 1992; Weiss, 1993). Thus, if therapists track clients' immediate responses to the varying interventions they employ (for example, providing interpersonal feedback or making process comments, suggesting thought records or self-monitoring strategies as homework assignments, making interpretations or cognitive reframing, and so forth), clients will behaviorally show therapists how best to respond. However, as emphasized earlier, therapists then must be flexible enough to adjust their interventions when necessary and try out different ways of responding that this particular client might be able to utilize more effectively.

To sum up, working in this highly individualized way adds complexity to the therapeutic process and places more demands on the therapist. However, such an approach gives therapists the flexibility to respond to the specific needs and unique experiences of the diverse clients that seek help. Client response specificity is a core construct in the interpersonal process approach. Tools to assist with this process include guidelines for keeping process notes (see Appendix A) and for writing case conceptualizations (see Appendix B). These will help therapists identify developmental experiences that are likely to have caused faulty beliefs and relational difficulties, which need to be addressed along the process dimension in a corrective way, with specificity for this particular client. For now, the case example below illustrates these three core concepts.

TERESA: CASE ILLUSTRATION OF CORE CONCEPTS

The following vignette illustrates our three core concepts: the process dimension, corrective emotional experience, and client response specificity. In particular, the vignette highlights how the process dimension and the way the therapist and client interact together can contribute either to resolving or reenacting important aspects of the client's problem. Usually, therapists do not literally reenact with clients the same hurtful or problematic responses they have received from others (in this case, the problem is sexual abuse). Unintentionally, however, or without being aware of it, the way in which they interact often thematically evokes the same issues/feelings/concerns that the client is struggling with in other relationships (in this case, *compliance* and having to go along with the counselor and others). As you read this case example, think about what you would do similarly or differently if Teresa were your client.*

A first-year practicum student is talking for the first time with her 17-year-old client, Teresa, about Teresa's sexual contact with her alcoholic stepfather. Understandably, this new therapist is anxious. Her heart is breaking for this dear girl, and she wants to help Teresa so much, but she has never in her life spoken directly with anyone about their experience of being violated sexually. The *content* of what Teresa and this therapist are beginning to talk about is sexual molestation. Depending on the interpersonal process they enact, however, the effectiveness of this discussion is going to vary greatly. On the one hand, suppose the counselor is initiating this discussion and pressing Teresa for disclosure about what happened to her. The graduate student therapist is deeply concerned about Teresa's safety. However, her "need to know" is intensified by her concerns about her legal responsibilities as a mandated reporter, and her concern that her supervisor will want to know more details or facts about Teresa's molestation.

In response to the counselor's continuing press for disclosure, Teresa complies with the counselor's authority and reluctantly speaks about what happened. Useful information (that is, content) may be gained under these circumstances, but the opportunity for therapeutic progress is diminished or even lost because aspects of Teresa's problem are being reenacted with the counselor in the way they are interacting together (that is, process). How is this a problematic reenactment along the process dimension? Teresa is again being pressured to obey an adult, comply with authority, and do something she doesn't want to do. Of course, being pressed to talk about something she doesn't want to disclose in no way retraumatizes her as the original abuse did. However, their interpersonal process of demand/comply is awry, and painfully familiar to Teresa, and will evoke in Teresa similar types of feelings and concerns that the abuse initially engendered. That is, her helplessness will lead to depression and her compliance (having to go along) to feelings of shame. Because the therapeutic process is thematically or metaphorically evoking aspects of her original mistreatment, her disclosure is likely to hinder her progress in treatment and actually slow the process of reempowerment.

---

*The clinical examples in this text are based on actual cases; however, identifying information, including the gender of the client or therapist, has been altered to preserve confidentiality.

This unwanted situation readily can be further complicated if Teresa belongs to an ethnic group where family loyalty is highly prized, or if she is a member of a religious community where obedience to authority and hierarchical relationships are emphasized. In this cultural context, Teresa is being asked to violate rules sanctioned by her family's ethnic or religious group. Other family members and friends may strongly disapprove of the stepfather's behavior. Even so, they may not provide Teresa with the validation and support she needs because she took this information to someone outside the family (the therapist) or because she violated their religious proscriptions that emphasize obedience and forgiveness. Thus, although disclosure violates family rules and loyalties for most victims, this may take on additional significance and add more distress for Teresa if she were a Hispanic adolescent who belonged to a conservative religious community.

What should the counselor in this example do instead? Wait nondirectively for Teresa to volunteer this information—while she may continue to suffer ongoing abuse at home? Of course not. But by attending to the process dimension, the therapist may be able to begin providing Teresa a reparative or corrective experience while gathering the same information. That is, instead of pressing for disclosure, what if the therapist *honored* Teresa's "resistance" or cultural prescriptions. For example, instead of pressing for more disclosure and trying to get Teresa to talk about things she doesn't want to talk about, the therapist could meta-communicate and make a process comment by, compassionately or supportively, inquiring about Teresa's reluctance to speak:

THERAPIST: All of this is so difficult right now. What's the hardest thing about talking with me?

OR

THERAPIST: It seems hard for you to talk with me right now—maybe something doesn't feel safe. I'm wondering what might happen if you share this and let me try to help you. What might go wrong or how could things get worse?

OR

THERAPIST: Perhaps talking with me about this violates family rules, and you're concerned what your mother or stepfather might think. Would you feel OK talking about your family's rules and what's OK or not OK to talk about outside your family?

OR

THERAPIST: Let's work together and try to find a better way to talk about this. Can you tell me one thing that we might be able to change or do differently that might make this a little bit easier for you?

The purpose of process comments like these is to create a different and more reparative interpersonal process—one that helps Teresa feel less that she has to comply with the therapist—as she has with her stepfather and others. In response to these invitations, Teresa is likely to present a number of reality-based concerns that make it harder for her to disclose. For example, Teresa might reply that:

- her mother won't believe her;
- her stepfather will be sent away;

- she will be told at church that it is her fault;
- others will tell her that she is being selfish and should stop causing problems, be more forgiving, and not talk to anyone about this;
- she will be chastised for not keeping it in the family and trying to resolve it there;
- the counselor may not believe her or take her parents' side;
- the counselor may want to remove her from her home;
- the counselor or others will regard her as "dirty," "ruined," or shame-worthy in some way;
- the therapist may feel uncomfortable talking about this sensitive issue and will not want to explore it fully, and so forth.

If the therapist can help Teresa identify and resolve her concerns about disclosing, then she can find it safe and empowering to begin talking about what happened—which does not occur if the therapist simply directs her to disclose and say what happened. Although this process difference may seem subtle, its effect is powerful and will have a defining impact on the course and outcome of treatment. Teresa either will begin the empowerment process or confirm her problematic expectations that she must always go along with what others want, by the way she shares her trauma with the therapist. With the process comments and invitations suggested above, Teresa is able to *participate* in the decision process with an authority figure, have her concerns expressed and taken seriously, and, to the extent possible, accommodated as best they can. For example:

THERAPIST: Yes, Teresa, you don't have to do this "alone." Your aunt Nora brought you to see us today, and she is waiting in the lobby. If you like, we can invite her to sit in with us while we talk.

In a way that is new and different, Teresa finds that she can at least have some shared control over what she says, and to whom, and still remain supported. She is not "alone" or disempowered in her interaction with the therapist—as she had been while keeping this secret about her stepfather. This corrective interpersonal process with the therapist will help Teresa to begin acting in similarly empowered and self-affirming ways in some of her other relationships as well. In this and subsequent sessions, the therapist can help Teresa discern other people in her life with whom it is safe to be similarly assertive (her aunt Nora, her minister, and certain friends and family members), and those who will punish or thwart this and demand that she merely comply again (her stepfather, her grandmother, her soccer coach, and certain friends and family members).

But what if the therapist's efforts to create a different interpersonal process don't work so well and Teresa still does not want to speak or discuss the secret? The therapist is a mandated reporter and still will need to contact Child Protective Services. In some cases, the therapist's attempts to empower the client by giving her more choice or participation in the reporting/treatment process clearly will have helped, and other times it may seem as if it hasn't made a difference. Even if not, however, Teresa has found that the therapist is sincerely trying to find ways to include and empower her rather than merely demand that she comply again and do what the therapist wants. In this way, a small but significant difference is occurring in their relationship that will facilitate her recovery, as Teresa sees that the therapist's intentions are different than what she has come to expect from others.

In sum, *concepts and techniques from differing theoretical orientations all can be helpful, but they are not likely to be effective unless the interaction or process that transpires between the therapist and client is enacting a solution to the client's problems.* In the chapters ahead, a variety of therapeutic interventions will be utilized, ranging from providing empathy and a sense of being understood, to helping the client recognize thought processes or behavior patterns that are maladaptive, to role-playing new behaviors. Per client response specificity, however, the effectiveness of these interventions will depend on whether the client's maladaptive schemas and problematic interpersonal patterns are reenacted or resolved along the process dimension.

CLOSING

Student therapists reading this book need to have realistic expectations for themselves. New therapists who are seeing their first clients will find much to help them in their initial work with clients. However, more information is presented here than a new clinician can fully integrate and apply. The concepts presented in the interpersonal process approach are easy to say but challenging to do. New therapists are encouraged to incorporate these concepts into their practice gradually—at their own pace. Especially during their first year of seeing clients, some may begin to feel that the more they learn, the less they know. With more experience, however, second-year students will often begin to find that they are effectively employing many of these concepts with their clients. Typically, trainees are able to make these process-oriented concepts their own and utilize them easily and naturally—in whatever theoretical tradition they choose to work, in two or three years. Because it will take some time to learn how to work in this way, the best approach is to be patient with yourself and enjoy the learning.

In closing, new therapists are encouraged to *be themselves* with clients rather than trying to fulfill the role of a therapist. Perhaps Kahn says it best:

> When all is said and done, nothing in our work may be more important than our willingness to bring as much of ourselves as possible to the therapeutic session.... One of the great satisfactions of this work comes at the moment students realize that when they enter the consulting room, they don't need to don a therapist mask, a therapist voice, a therapist posture, and a therapist vocabulary. They can discard those accouterments because they have much, much more than that to give their clients (1997, p. 163).

Let's turn next to the first stage of therapy: establishing a working alliance.

SUGGESTED READINGS

1. Attachment theory and its core concept of "internal working models" helps therapists identify the patterns and themes that link together clients' symptoms and problems, and helps therapists find a treatment focus. Readers are encouraged to read the excerpt "Representational Models" in Chapter 1 of the Student Workbook that accompanies this text. We will be working with the

related concepts of cognitive schemas, relational templates, and internal working models throughout the chapters ahead.

2. Student therapists begin their clinical training with different levels of experience. Less experienced trainees may wish to read Clara Hill's (2009) integrative text, *Helping Skills: Facilitating Exploration, Insight and Action, 3/e.* This highly readable and practical text provides new therapists with the basic helping skills they need—especially in their initial sessions with clients. Student therapists with a little more experience may benefit from Irvin Yalom's (2003) book *The Gift of Therapy: An Open Letter to a New Generation of Therapists and Their Patients.* In the beginning, it may not be easy to differentiate the process dimension from the overt content of what clients are discussing. Yalom became masterful with process comments in his group psychotherapy work, and this book will help new therapists recognize and intervene with the process dimension. Finally, more experienced therapists who find that they can readily understand and utilize the concepts presented here will be informed by Lorna Smith Benjamin's illuminating text, *Interpersonal Reconstructive Therapy* (2003).

3. One useful book for trainees interested in the interpersonal process approach is Michael Kahn's (1997) *Between Therapist and Client: The New Relationship.* This clear and engaging book provides a historical overview of the major concepts and theorists of the therapist-client relationship. In particular, Kahn articulates the cardinal issue of *non-defensiveness* in the face of clients' attempts to "pull" or "hook" the therapist into their own relational reenactments. This elegant book deserves to be read and reread by beginning and experienced therapists alike.

4. In their rich and nuanced book, *Inclusive Cultural Empathy: Making Relationships Central in Counseling and Psychotherapy,* Pedersen et al. (2008) write about cultural issues in a relationship-centered context. They address a range of cultural issues and include examples and exercises that can be used to increase awareness and cultural sensitivity.

5. Chapters 3 and 5 of Salvadore Minuchin's (1974) *Families and Family Therapy* provide a succinct presentation of structural family relations and family developmental processes. This classic family therapy text will help therapists who work with individual clients understand the genesis and familial context of their clients' problems.

6. A highly readable application of object relations and attachment theory to psychotherapy can be found in Chapters 6–9 of John Bowlby's (1988) book *A Secure Base.* An excellent conceptual overview of object relations theory may be found in *Freud and Beyond,* by S. Mitchell and M. Black (1995).

7. The great novelists best bring to life the profound impact of childhood and familial experience on adult life. See, for example, John Steinbeck's (1952) *East of Eden* and Franz Kafka's (1966) *Letter to His Father.*

RESPONDING TO CLIENTS PART II

Chapter 2

ESTABLISHING A WORKING ALLIANCE

CONCEPTUAL OVERVIEW

Therapy is a profession based on trust. Clients enter treatment with a need: They are often in pain and asking for help with something they have not been able to resolve on their own. A trustworthy response is to respect clients' requests for help by responding nonjudgmentally—with compassion and empathic understanding—to their concerns. Yet to be most effective, the therapist also wants to respond in a way that helps clients achieve a greater sense of their own agency or voice. That is, the goal is not only to resolve specific situational problems but to do so in a way that leaves clients empowered with a greater sense of their own ability to cope with the situations and stressors that are likely to cause problems in the future. As noted by Erik Erikson (1968), crisis provides an opportunity for growth. A significant aim of the interpersonal process is to help clients resolve their presenting problems, and to do so in a way that also leaves them with a greater sense of their own personal power.

Clients cannot attain these twin goals of resolving problems and achieving a greater sense of self-efficacy in their lives in a hierarchical or one-up/one-down therapeutic relationship.

To become empowered, *clients need to share ownership of the change process* and be active participants who work collaboratively with the therapist, rather than being passively "cured" or told what to do. Clients do need help from a responsive and actively engaged ally. Thus, this chapter presents a model for a collaborative relationship or "**working alliance**" that accepts the client's need for understanding and guidance while equally encouraging the client's own initiative and responsibility. As highlighted by the work on Motivational Interviewing (see Miller & Rose, 2009) and the extensive research on the working alliance (Baldwin et al., 2007), a collaborative engagement between therapist and client is an essential component of therapy—contributing to the client's motivation and commitment to change and enhancing the overall effectiveness of treatment.

Although researchers have tried long and hard, they have not been able to find consistent empirical support for the long-term superiority of any one treatment approach over another (Lambert & Barley, 2002). When researchers do find differences in effectiveness between treatments, studies suggest this is suspect because the investigators already had an allegiance to the winning treatment—the researchers aren't cheating but their hopes and preferences affect the outcome in the form of researcher bias (Wampold, 2001). In contrast, researchers have found *strong evidence of great variability in the effectiveness of individual therapists within each treatment approach* (Teyber & McClure, 2000; Wampold, 2001). That is, greater differences in treatment effectiveness are found between therapists of the same theoretical orientation (within-group differences) than between groups of differing orientation (for example, interpersonal vs. cognitive behavioral orientations). Arguably, the strongest finding in the psychotherapy outcome literature is that the most common feature of effective therapists across different theoretical approaches is the therapist's ability to establish a strong *working alliance* early in treatment (Horvath & Bedi, 2002; Safran et al., 2009). Indeed, even when manual based treatments are used, therapist effects on outcome are large (Malik et al., 2003). This chapter explores how therapists can use empathic understanding to help establish a strong working alliance, and then use process comments to restore it when the misunderstandings and problems that inevitably occur rupture the alliance.

THE WORKING ALLIANCE IS A COLLABORATIVE RELATIONSHIP

At each successive stage of treatment, therapists have a different overarching goal to guide their interventions. In the first stage, the therapist's principal goal is to establish a working alliance with the client. A working alliance is established when clients perceive the therapist as a capable and trustworthy ally in their personal struggles—someone who is interested in, and capable of, helping them with their problems. To achieve this credibility with clients and become someone who matters to them, the therapist is able to successfully communicate that she:

- grasps their predicament and recognizes their distress;
- feels with them and is empathic to their pain;
- is an ally on their side who has their best interests at heart;
- has an abiding commitment to help them through this predicament.

The concept of the working alliance (also referred to as the therapeutic alliance and the collaborative alliance) originally was developed by the psychoanalyst Ralph Greenson (1965b, 1967). Bordin (1979, 1994) and others linked it to psychotherapy more broadly, and clarified that there are three separate but interrelated components in all therapeutic relationships: the real relationship; the transference component; and, most important, the working alliance (Gelso & Carter, 1994). Found in many other theoretical contexts, the working alliance is akin to Bowlby's *holding environment* in attachment terms, where the client's distress is emotionally "held" or contained in the safe relational envelope of the therapist's understanding (Karen, 1998). In existential therapy, Rollo May (1977) elucidates the therapist's *presence* with the client. He emphasizes how the therapist can be fully present with the client in an authentic manner throughout each session. Similarly, Borns and Auerbach (1996), Safran and Segal (1990), and others within a cognitive-behavioral framework have underlined the importance of empathy and the therapeutic alliance. Using the related construct of "mentalization," the psychoanalytic theorist Peter Fonagy argues that this "attuned" and validating response from the therapist helps clients change by increasing their ability to "reflect" on their own thoughts and emotions, consider their own perspective and others' point of view more flexibly, and thereby step beyond narrow schemas, resulting in better self-regulation of behavior and modulation of strong emotions (Allen, Fonagy & Bateman, 2008). Although present in some form in almost every theoretical orientation, the working alliance is most closely linked to Carl Rogers's core conditions of genuineness, warmth, and especially, *accurate empathy* (Rogers, 1980).

Researchers have defined the working alliance as a collaborative process whereby both client and therapist (1) agree on shared therapeutic goals; (2) collaborate on tasks designed to bring about successful outcomes; and (3) establish a collaborative relationship based on trust, acceptance, and competence (Gaston, 1990). Of these three components, Horvath (2006) reports that the relationship component is most important and accounts for most of the variance. The therapist's ability to establish a successful working alliance in the initial sessions has emerged as a strong predictor of effective treatment outcomes in both short-term and longer-term treatment (Kirsch & Tate, 2006; Hovarth and Greenberg, 1994; Henry et al., 1994). A robust ingredient common to all psychotherapies, the working alliance is correlated with successful outcomes across a wide range of clients' symptoms and problems, and across differing approaches and theoretical orientations (Beutler, 1997; Hintikka et al., 2006; Hovarth & Bedi, 2002).

COLLABORATION: AN ALTERNATIVE TO DIRECTIVE AND NONDIRECTIVE STYLES

One of the most important ways for therapists to establish a strong working alliance with clients is to work together collaboratively—as partners. In the initial sessions, the therapist's primary aim is to articulate clear expectations for working in this collaborative manner and, more important, to enact behaviorally these spoken expectations by giving clients the *experience* of partnership. If a collaborative relationship is maintained throughout treatment, this interpersonal process will go a long way toward our twin goals of helping clients resolve their presenting problems and achieve a greater sense of their own self-efficacy. Thinking of the working alliance as a collaborative partnership, therapy is not something therapists "do" to clients, it is a shared interaction that requires the participation of both parties in order to succeed (Aron, 1996). Using in-session process measures to assess their Motivational Interviewing approach, Miller et al. (1993) contrasted two counseling styles. Counselors who acted as expert, focused on persuading clients to change, and confronted reluctance to change actually increased clients' resistance to change, compared to collaborative counselors who focused instead on understanding client perspectives through empathy and on evoking the clients' own concerns.

How does a beginning therapist go about fostering this collaborative process? At the outset, the therapist can ask clients what they know about counseling, explore their expectations of the therapist and the treatment process, and invite their thoughts and suggestions about how the therapist and client could best work together. When clients have been in treatment before, it is important to ask about what was helpful and what wasn't. This is especially important if clients previously dropped out of treatment prematurely, felt they didn't change, or experienced problems or difficulties with the therapist that were not discussed or resolved. When clients are asked why they dropped out or what they didn't like about treatment, they often give one of these two types of responses:

CLIENT: She was a nice person and I liked going to see her in the beginning. But she didn't say very much—she was pretty quiet and I never really knew what she was thinking. I guess I needed more because, after awhile, I realized that not much was changing for me. I did feel kind of bad about stopping though.

OR

CLIENT: The therapist didn't really listen to me very well. Sometimes it almost seemed like he wanted to start telling me what to do before he really even heard what was wrong.

Let's look at how both of these directive and non-directive styles often fail.

A widely held misconception among clients is that the therapist is a doctor who will prescribe their route to mental health. Often, the therapist is perceived as the sole agent responsible for change—via advice, explanations, interpretations, or simply telling the client what to do. It is the therapist's job to help the client understand the therapy process and actively engage the client so they are active partners or collaborators. The research indicates that clients' successful outcomes are facilitated by working with therapists who promote rapport, instill hope, encourage "change talk," are more active (not directive), and engage in a collaborative, focused relationship (Beck et al., 2009; Miller & Rose, 2009; Moyers et al., 2007). Engaging the client can be done by assessing together:

- the issues and concerns that are most important—what's most painful or distressing right now;
- what the client and others have done in the past that has been helpful and what has not been helpful;
- shared treatment goals—brainstorming and sorting through together what's really wrong, what they would like to be able to change, and how they can best work toward those goals.

Therapists may also explain that, even though the client has not been able to change on his own, the therapist will be an active partner and the client's willingness to work with the therapist to understand problems and explore solutions will be essential. Unless the therapist works with this process dimension, a hierarchical doctor-patient or teacher-student relationship is likely to develop; this undermines the client's self-efficacy and fosters dependency. An important note here, especially for clients who come from a family or culture that is hierarchically structured, actively inviting partnership and collaboration becomes even more important. For example,

THERAPIST: "Delia, tell me your thoughts and concerns if you chose this way...and share with me what your thoughts are should you choose to go another way with this issue."

A one-up/one-down relationship brings with it a number of problems. As we would expect from client response specificity, a doctor-patient relationship will work well with some clients in the short run. Based on early maladaptive schemas and what they learned growing up in their families, many clients who enter treatment believe they have to be *compliant* in close relationships. Regardless of what the therapist says about collaborative relationships, these clients still believe that the therapist, like others in their lives, actually wants them to just "go along" and follow her lead. Such clients will passively follow the therapist early in treatment and, commonly, even go so far as to actively seek direction from the therapist. However, this compliance will ultimately evoke shame over being controlled and anger toward the therapist, just as it does in other relationships. Although most clients will not recognize or understand what is transpiring, this faulty interpersonal process will prevent them from utilizing the therapist's help, making progress on their problems, or even remaining in treatment. Even if clients don't have compliance issues, clients remain dependent on the therapist if this hierarchical mode is enacted. They will not be able to gain a greater sense of their own personal power as long as they hold the dysfunctional belief that the source of adequacy or responsibility for success resides with the therapist, rather than in themselves. Let's explore this further.

So, if the interpersonal process approach isn't directive, is it non-directive instead? No. Many clients bring cognitive schemas to the treatment setting that do not encompass collaborative relationships or mutuality, and these clients may insist that the therapist assume a directive or leadership role in the beginning. For example, if clients bring to therapy an authoritarian upbringing and hierarchical schemas for relationships, then it would be fruitless to insist that they go along with a more egalitarian process. They have not experienced this in past, and may have been punished for acting stronger, speaking up and having their own mind, or being more independent. Thus the therapist can be flexible, meet the client on his terms, and accept the client's request to provide more direction or advice—sometimes this is all the client can do in the beginning. At the same time, however, the therapist can start the change process by giving this client overtly spoken permission to initiate, and by enthusiastically joining this client whenever he does speak up or initiate. For example,

THERAPIST: That sounds like an important issue for us to work with, Sue, I like what you're bringing up. What's the best way for us to go further with this—what comes to mind right now as you bring this up?

The therapist can also find others ways to actively engage the client in a more collaborative treatment process. For example,

THERAPIST: Tell me about your strengths and successes, John. When you've had this problem in the past, what have you done, or how have others responded, that was helpful?

In particular, the therapist can also change this unwanted, hierarchical relational pattern that is causing problems with others in their everyday lives as well, *by using process comments to begin talking with the client about how they are interacting together.* For example, if the client seems to be rejecting the advice and direction he has just elicited from the therapist—as often occurs—the therapist can begin to talk with the client about their current interaction or interpersonal process:

THERAPIST: I'm a little confused here, John, let's try to figure this out together. You keep asking me what to do, but, whenever I suggest something, you say "Yes, but..." I'm wondering if one part of you feels like you are supposed to go along with what I say, but another healthier part of you doesn't want to be told what to do. What do you think might be going on between us here?

Whereas a directive stance may work in the short run, a purely nondirective approach will often sputter right from the start. Understandably, most clients feel frustrated when their requests for help or direction are merely reflected back (the "hot potato" game). A negative cycle may ensue: The client becomes increasingly frustrated and seeks some direction from the therapist, who may further eschew this role and talk about inner direction and finding one's own answers. This, in turn, further frustrates the client, who may not feel that she has the answer to anything at that moment, and only sees the therapist's attempts to be nondirective as evasive.

Many clients drop out of therapy prematurely because the therapist was too quiet—hesitant to say what she was observing, share what she was thinking, ask further about what seemed important or didn't make sense to her, and so forth. As we saw in Chapter 1, therapists' concerns about making a mistake often inhibit them in this way and prevent them from becoming more actively involved and responsive—as most clients need and want. Therapists need to be an engaging presence in the relationship—so clients find they have a helpful collaborator with them in the room. If therapists want to adopt an active stance toward the client, and not be passive or inhibited in the therapeutic relationship, how can they do so without taking over and being directive? To illustrate this active but not directive stance, therapists can respond in these ways:

- provide feedback about the relational or cognitive patterns they observe (for example, As I listen to you, it often sounds as if you are working hard to take care of everybody else. That leaves me wondering who meets your needs?);
- help clients consider alternative frames of reference and consider situations from new perspectives that expand their schemas (for example, I can see why you didn't like that—what do you think her intentions might have been when she said that to you?);
- offer empathic understanding of the client's feelings and validate their experiences (for example, I see why you felt so "unimportant" there—your feelings were being minimized. It's painful to feel as though you are overreacting—when your needs actually are being dismissed as insignificant);
- provide interpersonal feedback (for example, James, you're speaking to me in a really loud voice right now. I'm wondering if others have ever given you this feedback before...and how do others usually respond when you talk this way?);
- use process comments to make the current interaction overt and utilize the therapeutic relationship as a social learning laboratory (for example, You just

asked me a question, Tina, and then interrupted me as I was answering. That's happened a couple of times today, and I'm wondering what might be going on inside for you when that happens. Any ideas?);

- check out the client's reactions toward the therapist (e.g., clients may drop out prematurely when they disclose significant information and feel vulnerable, and the therapist has not acknowledged the significance of what has happened or explored what this may have evoked).

Stacy, a 28-year-old mother of two daughters, revealed that her children had been repeatedly molested by her father. She had the courage to report him and have him prosecuted. However, in the process, she had lost most of her family's support for "airing" family secrets. The therapist, recognizing that Stacy had yet again revealed her family's secret to another, inquired about how this could effect their relationship:

THERAPIST: Stacy, we have had an especially good session today. If it is hard for you to come back to see me next week, what do you think that might be about?

CLIENT: Well, I'm afraid that you will be wondering where I was when my kids were hurt, that they were hurt repeatedly before I knew about it.

THERAPIST: Thank you for telling me that. I feel compassion for you and see how hard you are trying to respond to their needs right now...

Throughout the chapters ahead, we will further illustrate this balanced therapeutic stance that is active, responsive and engaging—yet cannot accurately be characterized as directive or non-directive. It doesn't get clarified in most counseling theories courses.

⁎ New therapists, who lose clients in their first practicum because of their own inactivity or lack of responsiveness, often report feeling afraid of "making a mistake" or feeling too worried or preoccupied about "what to do." More subtly, new therapists' self consciousness or inhibition may be prompted by their own concerns about acting more strongly and expressing their own thoughts and observations—in other words, having their own voice in the hour. Others hold themselves back because they feel apprehension about becoming someone important to the client and accepting the responsibility of trying to have an impact on the client's life. For these and other reasons that we have already begun to explore, new therapists often are painfully concerned about doing something wrong and as a result become passive or emotionally distant and lose the opportunity to have an impact. This is unfortunate because clients benefit from active engagement with their therapists. Further, clients are less likely to drop out if their therapist engages in "role induction" and provides education about the therapy process (Reis & Brown, 2006). For now, it is important to recognize clients do need direction at times—especially when they are in crisis (James & Guilland, 2000). However, the therapist can provide guidelines for how to proceed without falling back on a hierarchical or one-up/one-down relationship. How can this best be accomplished? Beginning with the initial interview, *the therapist can make an overt bid to establish a collaborative relationship in which the therapist and client work together to resolve the client's problems.* For example, this attitude may be conveyed by the following overture.

THERAPIST: Let's work together and see if we can figure out what's been going wrong. Tell me your thoughts about what the problem has been, and I'll follow along and join in.

CLIENT: I'm not sure where to begin with all of this.

THERAPIST: It sounds like there's been a lot going on. Out of all the things that have been happening, what's been the hardest thing for you?

CLIENT: Well, I guess that would be my wife and I. We're not getting along very well anymore, and I guess I'm really scared about that.

THERAPIST: Scared. Tell me more about that feeling...

The therapist is trying to communicate, in words and in actions, that she will be an active and responsive ally. The therapist needs to actively create a collaborative partnership where clients are encouraged to shape the treatment agenda by articulating the concerns that are most important to them and contribute their own ideas about the problems. The therapist shows interest in the client's perspective and input by saying things like:

"What was your intention when you responded that way?"

"What do you think she was thinking when she said that?"

"What do you understand her to mean by that?"

When actively invited to participate, clients will begin to experience that their input is listened to, valued, and incorporated—essential as so many clients come in feeling disempowered and disconnected. Creating a working partnership is a treatment goal that establishes a new, middle ground of shared control in the therapist-client relationship. This balanced therapeutic stance is active, responsive, and engaging—but still collaborative. There is a spacious middle ground here between the polarized positions of prescriptively taking charge versus passively following the client's lead—although it doesn't get clarified in many counseling theories courses.

For many clients, this collaborative partnership, in itself, provides a corrective emotional experience. This collaborative interpersonal process is a new way of interacting that many clients have not experienced in other close relationships. This will challenge or expand the client's early maladaptive schemas, and provide a practical role model to help them see how they can begin to respond differently and more effectively in their relationships with others. *Regardless of the therapist's theoretical orientation or techniques, one of the most important determinants of treatment outcome is whether the therapist and client can continue this process of mutual collaboration.* Let's look further at ways to work collaboratively so therapists can begin facilitating a strong working alliance right away—even in the initial session.

COLLABORATION BEGINS WITH THE INITIAL INTERVIEW

One of the most important concerns for new therapists is whether their clients will return after the first session and remain in treatment. Many clients do indeed drop out after the first sessions (Garfield, 1994; Reis & Brown, 2006), and such experiences painfully exacerbate new therapists' concerns about their own adequacy and performance. One of the most helpful guidelines for conducting a successful first session is for therapists to initiate a collaborative relationship in their initial contact with the client. As we will illustrate, the therapist structures the session by providing the client with guidelines and direction for what is going to occur in the interview. However, *it is essential that the treatment focus reflect the client's own wishes and that the treatment goals be experienced by the client as her own.* The therapist's intention is to attend to the issues that the client views as most important. In other words, the therapist is trying to help clients articulate their own goals and agenda, and then collaboratively join clients in addressing their concerns. Sure, there will be further ideas about what's wrong, but case formulations and ideas about what most needs to change must remain closely connected to the client's agenda—and her subjective experience of what's wrong—or we will lose partnership with the client through this empathic failure. This process is best accomplished by cultivating an attitude and presence that is respectful, warm, open, and alert to all that the client is presenting (Ackerman & Hilsenroth, 2003). With this in mind, therapists can help clients begin by offering open-ended invitations to talk about the issues or concerns that matter most to them.

THERAPIST: What's the difficulty that brings you to treatment? Help me understand what's wrong.

CLIENT: There's so much going on, it's not easy to explain.

THERAPIST: Things are complicated. Maybe you can start by telling me about the concerns you feel are most pressing or important right now.

This type of inquiry communicates several important things to the client. It ends the opening phase of social interaction that occurs as the therapist and client are introduced and walk to the interview room. The client realistically knows that most people in life do not really want to hear about their problems in depth, so it is important to define this new relationship by saying, in effect, this is the right place—and I'm the right person, to talk about what's really wrong, what you are feeling, and what you need. It also tells clients that the therapist is someone who is willing to talk directly about their personal problems *as they experience or perceive them*—and is ready to respond to their need for help by listening. However, it does so in a way that still gives clients the freedom to choose where they want to start and leaves them in charge of how much and what they want to disclose. From the outset, clients are sharing control of the interview by choosing what they want to talk about, yet the therapist is an active participant who has offered some direction for where they are heading. This type of collaborative interpersonal process does not occur if the therapist gives the client a more specific or directive cue such as:

THERAPIST: When we talked on the phone, you said you were having trouble with your boss. What is the problem there?

OR

THERAPIST: You've been having anxiety symptoms. How long has this been going on, and how severe have they been?

The difference between these two opening bids may seem insignificant—but they are not. It is important to communicate that clients should talk about what they want to talk about and not feel that they have to follow the therapist's agenda. From the start, that is, we want the client to take an active role in directing the course of treatment, while still feeling that the therapist is participating as a supportive ally. Enacting this interpersonal process in the initial session is more important than the content of what is discussed, and it will fundamentally shape the course and outcome of treatment.

In most cases, clients will readily accept the therapist's open-ended invitation and begin to share their concerns. The therapist can then follow the client's lead and begin to learn more about this person and his difficulties. Therapy is under way when this occurs. Before we go on to the next step, however, we need to examine two exceptions in which the client does not accept the therapist's offer to begin: when another therapist has conducted an initial screening interview and when the client has conflicts over initiating.

Previous Screening Interview.   If another therapist has previously conducted an intake interview, the client may not be so ready to begin.

CLIENT: *(impatiently)* I've already been through all of this in the intake with Dr. Smith. Do I have to go over it all again just for you?

OR

CLIENT: (hesitantly) I don't know how much you already know about me. What has Dr. Smith told you about me?

Our initial goal is to develop a working alliance between the therapist and the client. In these two examples, the previous intake therapist is a third party who, psychologically, is still in the room with them. This is especially problematic for those clients who have come from families in which a third person was repeatedly *triangulated* into the marital relationship when too much conflict, or closeness, evoked anxiety (Bowen, 1978). For many clients, that is, when parents or other family members became uncomfortable with the level of intimacy or conflict between them, a third family member would be drawn in to disrupt the dyad (Haley, 1996; Minuchin, 1984). To keep therapy from reenacting this widespread but dysfunctional family pattern, the therapist and client want to begin their own relationship as a stable dyad. As the family therapists teach us, it is important to maintain a dyadic therapeutic relationship that does not allow others to disrupt the therapeutic coalition—as so often occurred for many clients in the past.

THERAPIST: I know that you have already spoken with Dr. Smith, and I have learned a little bit about you from his intake notes. But just you and I are going to work together from now on, and I'd like to hear about you in your own words. It may be a little repetitious for you, but this way we can begin together at the same point.

Third-party interference or triangulation in the therapist-client dyad may also occur in the initial therapy sessions because of the unseen but felt presence of the clinical supervisor. As we will see, this can occur from either the therapist's concerns or from the client's. Through insecurity, compliance issues, or other factors, new therapists may triangulate their supervisors into the therapeutic dyad. For example, the therapist might tell the client, "I'll have to ask my supervisor about that." More commonly, new therapists may silently invoke their supervisors—by wondering what they would say or do at a particular point in the session or how they would evaluate the therapist's performance at that moment. This self-critical monitoring does not facilitate treatment—and it takes the fun out of seeing clients! It diminishes both the therapists' self-confidence and their ability to be emotionally present with the client. In particular, it disempowers therapists by inhibiting them from being themselves in the session—from finding their own words, acting on their own ideas and utilizing their own perceptions, listening to their own intuitions, and developing their own therapeutic styles. Thus, new therapists are encouraged to trust themselves enough to simply be a person who is sharing the life story of another. Supervisory feedback and guidance, though necessary, usually will be more productive when it is processed outside of the therapy hour. Again, in accord with our theme: The therapeutic process is awry when the dyadic relationship between the therapist and the client is broken by third-party interference.

Although we don't want the therapist to triangulate the supervisor into the counseling dyad, clients may be curious or concerned about what the supervisor thinks of them or the progress they are making. This is a different situation and requires a therapeutic response. This is especially likely when a videotape or audiotape is being used during sessions. It can be informative for the therapist to explore clients' concerns about the recording equipment or thoughts about the supervisor—such inquiries often reveal the cognitive schemas or transference distortions they bring to relationships. For example, if the therapist asks clients what they imagine the supervisor might be thinking about them, clients often reveal key concerns:

- I'm worried that he might be talking about me to other people.
- He probably thinks I'm a weak person because I'm so anxious about this, you know. He doesn't respect me because I'm always so worried about everything all the time.
- She's probably bored with me—like everybody else always is.
- I think she's probably critical of me—you know, judging me.

By the client revealing these problematic schemas or expectations in this way, the therapist has the opportunity to highlight, address, and resolve them in treatment—and keep them from leading the client prematurely out of treatment. For example:

THERAPIST: No, I don't think she's feeling "critical" or judging you at all. Actually, I think she probably feels compassion for what you've been coping with all of these years, and respects your willingness to come here every week and work so hard on your problems—as you've been doing.

In addition, the therapist has the opportunity to begin making connections and working on how these faulty schemas or problematic expectations may be contributing to problems with the therapist and others in their current lives. For example:

THERAPIST: You feel he's judging you critically. Well, that sure wouldn't feel very good. I'm wondering if there are others in your life right now who might be judging you, or maybe you've been concerned sometimes that I might be feeling critical of you as well? Can we check this out together for a minute?

Finally, some clients may feel uncomfortable about working with a trainee or student therapist.

CLIENT: So, does your supervisor tell you what to say to me, or do they let you say what you think?

THERAPIST: I consult with a supervisor, but I have my own mind and will say to you what I think and believe.

Finally, many clients grew up "parentified"—taking care of the emotional needs of their caregiver at the expense of their own. Continuing to live out this familial role, these clients often will try to protect or take care of the student therapist as well. In order to make the therapist feel competent or look good to the supervisor, they may try extra hard to improve in treatment, talk about positive changes they are making and minimize areas that are not improving, express their appreciation for how helpful or understanding the therapist is being, and so forth. This problematic but common reenactment, if not addressed and resolved in the therapeutic relationship, will prevent clients from changing this care-taking role with others in their everyday lives. In all of these ways, clients make progress in treatment when they can talk through with the therapist such concerns and ensure that the therapeutic relationship does not subtly reenact long-standing but unwanted relational patterns. In sum, whenever therapists sense the psychological presence of a third party in the room, they are encouraged to make this presence overt, explore the impact this third-party may be having on the therapeutic relationship, and reestablish the therapeutic relationship as a dyad.

Conflicts Over Initiating.   There is a second circumstance in which the client does not respond to the therapist's initial request to begin. The problems that some clients bring to treatment involve conflicts over initiating. This type of client cannot begin at the therapist's request. By asking these clients to begin talking about whatever is most pressing or important for them, the therapist has inadvertently presented them with their central or presenting problem. It is difficult—if not impossible—for this client to decide what to talk about, take the first step and initiate any activity, or to share responsibility for the course or direction of treatment. Recalling client response specificity, a therapist—with the very best of intentions—who nondirectively waits for the client to lead places an impossible demand on this client.

On the other hand, a directive therapist who takes charge and begins the session by telling this client where to begin or what to talk about is often replaying the same unwanted scenario the client has found with others. Whether communicated in a genuinely friendly tone, or impatiently after waiting awkwardly for a while to get started, the therapist—like others in their lives—is again telling them what to do. Thus, treatment stalls right from the start if the client has problems with initiative and the therapist responds in either a directive or non-directive manner. What can the therapist do instead with this type of client? One of the best ways to find this more effective middle ground between directive and nondirective extremes, and to establish a more collaborative relationship, is to make a process comment. That is, an effective intervention is simply to wonder aloud or ask about what may be occurring between them at that moment.

THERAPIST: It seems like it's being hard for you to get started. Maybe we can begin together right there. Is it often hard for you to begin, or is there something about me or this situation in particular that is difficult for you?

By first identifying this as a problem, and then encouraging the client to explore it, the therapist has offered the client a focus and assisted the client in moving forward. However, the therapist has provided this focus without taking over and telling the client what to do, which would only reenact the familiar but unwanted relational pattern. The therapist's open-ended inquiry is supportive, in that it responds to the client's immediate concern. Yet the client can take this issue of initiating wherever he wants, and is sharing responsibility for the direction of treatment. This type of response provides a new opportunity for clients to explore their problems in a supportive context. Our first goal is met as the client *experiences* a collaborative interaction with the therapist, rather than merely having a conversation about the need to work together—which usually doesn't take the client very far. In contrast, such a new, collaborative experience from the outset engenders hope about the therapist, the therapeutic relationship, and the possibility of change.

It is worth noting, in keeping with client response specificity, that for some cultural groups, it will be more challenging to engage initially since family structure or culture endorse a more hierarchical teacher-student (therapist-guided) interaction. That is, age, economic class, gender, culture, religion, and other issues of diversity are important to consider when engaging in therapy. It may be seen as disrespectful to presume to lead with one's elders, an educated person, or an authority figure such as a therapist for some. Therapists can best respond to such concerns by acknowledging differences and inviting clients to express or explore the concerns together. While it is always important to ask clients, "What is it like for you to be here?" this question can be especially important for clients who come from different cultural backgrounds, especially backgrounds where family information is closely guarded and talking about family issues with others will often feel disloyal (see Johnson, 2009; Sue & Sue, 2008; Vasquez, 2007).

To illustrate, one 20-year-old Korean client came to treatment—as it turned out, to get help for her 12-year-old brother. Her parents had been divorced for two years but no one in the family actually knew about the divorce because they went to family events "as a family." However, her father had a separate residence and her brother was acting out both at home and at school. It took her most of the session to state her real reason for coming to counseling:

CLIENT: There are problems in my family but no one knows about them.

THERAPIST: It is hard to talk about your family to someone you don't know.

CLIENT: Yes...

THERAPIST: What are you most concerned will happen if you talk to me?

CLIENT: I'll be telling what has been a family secret.

THERAPIST: What would be most helpful about talking?

CLIENT: I won't be alone with the secret anymore.

THERAPIST: Yes, family secrets can be a heavy burden, a big load to carry alone. You can decide in a little while if and with whom you want to share this secret. Is there anything about your family that feels safe, that doesn't feel like a secret, that you feel OK talking to me about?

CLIENT: Yes, can I just describe them? Like how old they are and where they were born?

THERAPIST: Absolutely. That sounds like a great place to start. I'd like that.

As the client began talking about her family, safety in the relationship developed. This security helped her realize that she did in fact have another potentially safe relationship in her family, an aunt. She was then able to use the therapist's help to reach out to her aunt and get this family member to work with her to help her brother.

EMPATHIC UNDERSTANDING: THE FOUNDATION FOR A WORKING ALLIANCE

The therapist begins the session by giving the client an open-ended invitation to talk about whatever feels most important to her. Responding to this bid, the client begins to share her concerns with the therapist and clarifies the background and context of her problems. The therapist's intention now is to try and find the subjective meaning that each successive vignette holds for the client and grasp what is most significant to the client from the client's point of view. In this instance, the therapist is striving to have the cognitive and affective flexibility to de-center, enter into the client's subjective experience, and capture the core meaning that this particular issue holds for the client (McClure & Teyber, 2003). The client will feel deeply understood when:

1. the therapist repeatedly captures and reflects the most basic feeling, key issue, or core meaning in what the client just said; and
2. the therapist can identify a common theme or pattern that links the client's varying concerns into a more coherent narrative—helping the client better make sense of this particular issue in the broader context of his/her life.

These themes are often informed by early developmental experiences with early caregivers and attachment figures and have come to shape the client's worldview. As the therapist and client together explore these themes and developmental experiences, the accurate empathy that ensues earns credibility with the client. Perhaps more than any other component, such empathic understanding is the basis for establishing the working alliance: Clients become more engaged with the therapist, invest further in the treatment process, and risk exploring their problems more fully (Angus & Kagan, 2007). Let's explore this pivotal sequence more closely.

As used here, *empathic understanding* connotes a genuine feeling of warmth and concern for the client—and that their distress matters to the therapist. It is not a technique but rather a respectful attitude and nonjudgmental stance toward the client. Different than being friendly or nice, it behaviorally shows clients that the therapist "gets it" or "sees" them in a way that others generally do not. When therapists can grasp their dilemma and articulate or express this understanding, they are providing the secure base to "contain" the client's distress that the attachment researchers describe (Bowlby, 1988). Although this is not sufficient to resolve clients' problems, such empathic understanding will often ease clients' initial distress and, for some, their presenting symptoms may even abate. A secure attachment configuration is established, and the therapist's attuned responsiveness often begins to relieve the entire family of attachment affects:

• Anxiety may diminish from this reassuring emotional contact.
• Depression may lessen; the experience of being seen and accepted, rather than dismissed or judged, engenders hope in the client.
• Anger from frustration over not being seen or feeling invalidated by others may be assuaged by the therapist's affirmation.
• Shame over having emotional needs revealed by entering therapy and asking for help may be lessened by the therapist's respect.

Empathic understanding is the basis for establishing a successful working alliance, and a key concept for change in the interpersonal process approach. However, researchers consistently find that some therapists are significantly more effective than others in providing the therapeutic skill of empathic understanding (Catley et al., 2006; Gaume et al., 2008). To illustrate some of these studies, Lafferty, Beutler, and Crago (1991) summed results across groups of more and less effective therapists from 11 different studies based on the extent of clients' symptom change. They found that clients of less effective therapists *felt less understood* by their therapists, whereas more effective therapists were seen as more empathic by their clients. Similarly, Rollnick and Miller (2002) report that the same manualized treatment program for Motivational Interviewing had very different treatment outcomes—based on therapists who were rated as more or less empathic. Readers can find practical suggestions for ways to build their own empathy and listening skills in the Suggested Readings section at the end of this chapter.

Empathic understanding is highly challenging for therapists to provide consistently or reliably. Again, we are not talking about being nice or friendly—those are good things but not at all what we are getting at here. Empathic understanding goes beyond the surface and reaches the client's hidden, needy, or unacceptable self. It is a highly discriminating response that requires the therapist to discern closely the key concern or central meaning in what the client just said. Here, the therapist accurately (1) discerns the client's thoughts and feelings, and (2) effectively communicates her understanding of this to the client, engendering in the client the feeling, "My therapist really gets me!" Many clients cannot enter their own experience deeply—cannot tolerate exploring, experiencing, and then sharing painful feelings and memories, without the safety and security that this deep understanding provides.

Yolanda was drifting through community college without purpose or direction, and was working in a "nowhere" job. She had as beautiful a face as a girl could have but weighed nearly 300 pounds. Sometimes a cutter during stressful phases in her life, she would "scrape" herself on her arms. Developmentally, Yolanda had been molested by her godfather as a young girl. When she told her mother about what was occurring, her mother said she "didn't want to hear this." She made six-year-old Yolanda feel guilty about voicing this problem—telling her that it would "cause a lot of problems in the family" if she talked any more or to anyone else about it. Confused why her mother wouldn't help her, and feeling painfully alone and "bad" about herself, Yolanda complied and continued to suffer her violation silently. Two years later, when now her mother's boyfriend began "touching" her, Yolanda knew it was useless to try and speak up.

Striving to make a connection with this endearing but vulnerable young person, the therapist soon felt that she had heard enough to grasp Yolanda's dilemma and tried to offer her empathic understanding:

THERAPIST: Yolanda, you have kept so many painful experiences to yourself—you have been so alone with them. I hear you talk about how much you hate how you look, especially your weight. Is there anything about being heavy that makes sense to you now?

CLIENT: What do you mean?

THERAPIST: What happens when you begin to lose weight?

CLIENT: Guys begin to make remarks, pay attention; I hate that. They're just gross...

THERAPIST: So, when you are heavy, what happens to that attention from guys?

CLIENT: It stops.

THERAPIST: So I'm wondering, does being heavy make sense to you?

CLIENT: (crying) Ya, I know, it protects me, I guess I always knew but never quite talked it. After my godfather, I got chubby, then my mother's boyfriend, I got really...(cries harder), Ya, I guess I gotta deal with this first before I can lose the weight...I did try to tell my mom the first time but she just...(trails off)

THERAPIST: Yes, Yolanda, you have been alone with the hurt and sadness...

Therapists wish to treat clients with respect, restrain their rush to judgment, and *look instead for the life story or context that helps clients make sense of their symptoms and problems.* As we have just seen, therapists' ability to help clients like Yolanda will largely depend on their ability to use empathic understanding to establish a strong working alliance. The four sections that follow examine different aspects of this key concept.

WAYS IN WHICH CLIENTS DO NOT FEEL UNDERSTOOD

Most clients are concerned that others do not really listen to them, take them seriously, or understand what they are saying. Clients often describe themselves as feeling alone, unseen, different, unimportant, or dismissed. Many clients feel this way because their subjective experience was not *validated* in their family of origin. While growing up, most clients with significant and enduring problems repeatedly received messages from caregivers that denied their feelings and invalidated their experience:

* You shouldn't feel that way.
* Why would a silly thing like that make you mad?
* You can't possibly be hungry now.
* I'm cold. Put your sweater on.

- How can you be tired? You've hardly done anything.
- You shouldn't be upset at your mother. She loves you very much.
- You don't really want to do that.
- We don't talk about those things.
- What's the matter with you—how could you feel that way!

Sometimes, family members changed the topic or simply did not respond when the client expressed a certain feeling or concern. As Carl Rogers, Virginia Satir, Irvin Yalom, and Robert Stolorow model so effectively, one of the most effective ways therapists can help their clients change is to affirm their subjective experience. The pioneering family therapist R. D. Laing goes so far as to suggest that people stop feeling "crazy" when their subjective experience is validated (Laing & Esterson, 1970).

Consistent invalidation in the client's family of origin, or "misattunement" in attachment terms, has profound, long-lasting consequences. In its severe forms, some authors describe it as "soul murder" because clients lose themselves—their sense of self or their own voice—when they lose the validity of their own experience (Schatzman, 1973). Whereas disconfirmation has occurred to some extent in most clients' families, in alcoholic, eating disordered, highly authoritarian, and abusive families, children have experienced pervasive invalidation of some of their most important experiences in their lives (for example, "Of course he didn't do that. Don't you ever say anything like that again!"; or poignantly, simply changing the topic or continuing as if nothing was said). One of the most serious consequences of such systematic invalidation is inefficacy or disempowerment. When reality-based feelings and perceptions are repeatedly denied, children become incapable of setting limits, saying "no," and refusing to go along with what does not feel right to them. More significantly, consistent denial of their experience leaves clients unsure of what has actually happened to them and of the subjective meaning that events hold. That is, they lose the ability to "have their own mind"—to trust their own intuition, to listen to their own gut feeling and know what they know. They no longer trust their own perceptions of what may be occurring or making them uncomfortable. Typically, they won't register that something someone said or did bothered them until well after the problematic interaction is over, if at all. Denying the validity of their own experience, these clients characteristically say to themselves, "Oh, nothing really happened"; "It wasn't that bad"; or "It doesn't really matter," when it actually was significant to them. Routinely, they cannot find words to communicate their own experience or point of view with any clarity or specificity and, even if they can, do not expect others to understand or be interested in what they say (Linehan, 1997).

When their subjective experience has been denied repeatedly, clients live feeling confused, vulnerable, and anxious—not knowing what they are feeling, what they like or value, or what they want to do. In many dysfunctional families, such invalidation is a pervasive, everyday experience that continues throughout childhood and adolescence. In place of clear feelings and confident perceptions, a vague, painful feeling of internal dissonance results. Fortunately, this undifferentiated feeling state can be replaced with emotional clarity and a stronger sense of voice or personal identity if the therapist consistently listens to clients with respect, captures what is most central or key in what they just said, and validates their experience. Therapists will be more able to fully appreciate clients' actual needs when they relinquish the need for the session to proceed in a particular way and, instead, enter the client's subjective worldview and experience and, together with the client, explore and clarify the themes presented. Further, helping clients understand the themes presented within the context of their developmental life experiences will make the current concerns and issues make sense. Clients feel affirmed and empowered when they are listened to intently, taken seriously, and have their concerns reflected back with understanding. Although such simple empathy and validation may sound like common or ordinary human responses, they are not. Many clients have not had their most important feelings and perceptions validated in their most significant relationships—and in turn, they do not expect to be seen or understood by the therapist either (Linehan, 1997; 1993). Thus, the therapist is trying to do this repeatedly throughout each session. For example, in the following questions and comments, *the therapist is not agreeing or disagreeing with the client, but trying to capture the emotional meaning or distill the key issue in what the client has just said.*

- It didn't seem fair to you.
- I'm wondering if you were frightened when he did that?
- It's been too much for you, more than you can stand.
- It felt great to be so effective and in charge!
- It was disappointing; you wanted more than that.
- Here again, are you having to take care of everyone else?
- Yes, I really do hear the quandary you are struggling with there.

Stepping back from unrealistic performance demands, the therapist does not have to be an exceptionally perceptive person to understand the client's experience, and new therapists are reassured that, of course, they will not be accurate all the time. For example, when discussing parenting, Winnicott (1965) has a wonderfully reassuring phrase to diminish our performance anxieties: children need only "good enough mothering," just as clients, too, need their therapists only to be "good enough." What clients need is that the therapist convey their sincere striving to better understand them and "get" what this experience means to them. The therapist's empathy and genuine concern and the effort to foster a working alliance means more to the client than being "right." Thus, when therapists feel that they do not quite understand what the client is saying, they should not feign understanding by saying "Yes," "OK," or "I know what you mean" when they really don't. Instead, they can acknowledge the uncertainty and ask the client to try to restate something in a different way. This will convey that the therapist is not just going through the motions but is trying hard to clarify so he can really understand what she means and they can get on track together. This furthers a partnership and becomes an opportunity to get more connected to the client—by inviting this clarification, *the therapist is intervening with a more collaborative approach to empathy.* Here is an example of how this might sound with a client:

THERAPIST: That sounded important, Susie, but I didn't understand it as well as I wanted. Can you say that again, or put it differently?"

OR

THERAPIST: As I listen, it sounds as if you feel so discouraged that you just want to give up. Am I saying that right—or can you help me say it better?"

In both of these examples, the therapist is giving the client an invitation for a dialogue and trying to *jointly clarify the client's meaning with this back and forth interaction.* This collaborative approach to empathy takes the performance pressure of having to be "right" off the therapist, diminishes unwanted concerns about making mistakes or being wrong, and furthers the working alliance.

Carl Rogers and other pioneers in client-centered therapy originally highlighted the cardinal role of accurate empathy in the change process (Truax & Carkhuff, 1967). Unfortunately, empathy came to be regarded as if it were a stable or enduring personality characteristic of the therapist. Instead of thinking about empathy in this way as a relatively fixed personality trait, researchers have found that effective empathy comes from attempts to *collaboratively* understand the client's experience within an emerging, shared frame of reference (Barkham and Shapiro, 1986). That is, measuring across client-centered, cognitive, and dynamic therapists, *the therapeutic mode of mutual exploration—characterized by active negotiation between therapist and client—is crucial to the client's experience of being understood* (Allen, Fonagy & Bateman, 2008; Hardy & Shapiro, 1987; Hobson, 1984). For example,

THERAPIST: What did he do when you took the risk to say that?

CLIENT: He just went on talking, as if I hadn't said anything at all.

THERAPIST: I'm sorry he wasn't able to listen better—where did that leave you inside?

CLIENT: I don't know...*(pause)*...I hated it.

THERAPIST: Uh-huh, you "hated" it. I can sure see why...like you were invisible or didn't count?

CLIENT: Yeah, something like that.

THERAPIST: Hmm, it sounds like "invisible" doesn't quite capture it. What would be a better word?

CLIENT: I'm not sure...*(pause)* maybe just erased. Yeah, like my whole existence was being erased by him!

THERAPIST: "Erased." That captures it much better—you say that with a lot of feeling. Tell me more about being erased—help me understand what's that like for you.

CLIENT: Oh, I hate that feeling more than anything, and it happens over and over again...

THERAPIST: I can see why this is so significant...to have him erase you, given the cultural issues you mentioned before...

CLIENT: Yes, being an African-American woman, you know...I've worked so hard to be where I am, to earn my place...be visible......have dignity...

In this collaborative exploration, the therapist and client work together as partners to clarify the specific meaning this experience held for her. In this way, *accurate empathy is less a personality trait or characteristic of the therapist than it is an interpersonal process characterized by mutual exploration and collaboration.*

Further, understanding the developmental history (for example, the extent to which the client was neglected, rejected, or aggrandized and made a hero or star, or expected always to succeed, and so on) facilitates this process. It is further enhanced by attending to the client's cultural milieu, such as history of discrimination, whether based on social class, gender, race, or other factors. Thus, active engagement with clients and awareness of the attitudes, values, and biases we bring to treatment as we build alliances with clients of varying backgrounds is crucial (Arrodondo, 2007; Faoud & Constantine, 2007).

To sum up, therapists intervene by validating their clients' experience, grasping the core messages, and affirming the central meaning in what they relay. As emphasized above, being accurately empathic is sometimes misunderstood as merely being "nice," friendly, or supportive. These are benevolent, well-intended responses, yet they do not do much to help most clients change. Instead, with accurate empathy, we are looking for something more specific and far more potent—the ability to repeatedly identify what is central—discern what is most important, in what the client just said to us. To go beyond the cliché and appreciate how important accurate empathy and validation is, join in this thought exercise and recall for a moment what others have said and done to help you during a significant crisis in your life. Almost universally, helpful responses include an accurate, empathic understanding of our experience and validation of our feelings—leaving us with both the feeling of being seen or understood, and with the sense of being accepted and not judged.

To reiterate, providing validation is especially important when working with minorities, gay men and lesbian women, physically challenged individuals, economically disadvantaged clients, and others who feel do not identify with a majority group. These clients will bring issues of oppression, prejudice, and injustice into the therapeutic process because their personal experiences have often been invalidated by the dominant culture. These clients, in particular, will not expect to be heard or understood by the therapist. The first step in working with all people is to enter their subjective worldview, listen empathically and hear what is important to them—from their point of view (Pedersen et al., 2008). Therapists respect the personal meaning that experiences hold for different people when they enter into and affirm the client's subjective experience. The most effective therapists, of every theoretical orientation, offer these basic human responses especially well. Pioneering clinicians as varied as Beck, Bowlby, Fonagy, Kohut, Rogers, Satir, and Yalom all share this deep respect for the client's subjective experience and emphasize the need for therapists to respond empathically by communicating their ability to grasp the client's subjective experience.

COMMUNICATE UNDERSTANDING OVERTLY AND SPECIFICALLY

To engage a client in a working alliance, the therapist listens to the client's experience, finds the feeling and meaning that each successive vignette holds for this client, and accurately reflects back what is most significant or key in the client's experience. This is not just a rote parroting of what the client has said. An effective reflection is more akin to an accurate interpretation or creative reframing.

It demonstrates behaviorally that the therapist understands the core message, registers the emotional meaning, or distills what is most important from the client's frame of reference. Rogers (1975) believed that communicating such understanding in a nonjudgmental way provides a deep acceptance that is a prerequisite for meaningful change. New therapists will find that they achieve credibility with their clients—that is, prove their competence or ability to help, and strengthen the working alliance when they can articulate or *demonstrate* their understanding in this tangible way.

According to one unfortunate stereotype, a therapist is someone who remains on the surface and says superficially, "I hear you," "I know what you mean," or "I understand." However, this type of global, undifferentiated response is not effective and, paradoxically, often furthers the client's sense of not being seen or heard. The therapist does not simply say, "I understand," but demonstrates that understanding by *articulating* the central meaning of what the client has said—that is, links what the client has said to what might be meaningful. Twenty-four-year-old Susan, with a history of child abuse (perpetrated by her maternal grandfather and minimized or denied by family members who continued with regular "family gatherings"), struggled with the approaching Thanksgiving get together.

CLIENT: I wish someone would just tell me what to do...give me guidelines...

THERAPIST: Do you have thoughts and feelings about what you would like to do but feel reluctant to act on your own wishes?

CLIENT: Yes...Because I'm not always sure what's real...I can be convinced that what I think and feel and know is wrong...

THERAPIST: From what you have told me about your family, I can see why this is a struggle...you have described how they repeatedly invalidated what you knew and experienced and felt...it makes sense that sorting through what is real and what you know is often difficult.

CLIENT: That's it!! I know what I feel, what has gone on...then, on the outside, it all seems so perfect, like this perfect, together family...yet my cousin was also molested. But then they go on as though all is OK...I start to feel as though I'm the crazy one, especially when they have these family gatherings and Grandpa is treated like this special person...

THERAPIST: I can see how confusing that would be...

CLIENT: Yes, like, what's real?

Thus, therapists show the client they can be helpful when they capture and express the specific meaning in the client's experience, rather than offering well-intended but vague reassurances. And by checking with the client to see if that understanding is accurate, the therapist is working collaboratively, which in turn strengthens the working alliance. For example

THERAPIST: Yeah, I know just what you mean—that's happened to me too.

VERSUS

THERAPIST: Let's check this out together and see if I'm hearing what you're really saying. As soon as you realize that others at work are taking you seriously—listening to your ideas and thinking about what you're suggesting, you become anxious or unsure—worried that you're going to be dismissed or put down. That's when you start to get quiet, go along with others' ideas and stop putting out your own, and end up just hating yourself when you leave that meeting. Am I getting that right—can you help me say it better?

Let's look more closely at how clinical trainees can start to put these ideas into practice. In the following illustration, we explore a brief case study of a client who initially had the unwanted experience of not feeling heard by her therapist, but subsequently felt seen and understood with a second therapist.

While growing up, Marsha did not feel heard or understood by her parents. Her father was distant—uncomfortable talking with his adolescent daughter, and believing his wife should "handle the children." Struggling with her own ongoing anxiety and dysthymia, her mother too often was critical, angrily demanding, and intrusive. In particular, whenever her mother felt or believed something, she expected her children to see it the same way—differing points of view were not permitted. For example, if Marsha felt something that her mother did not feel, her mother would chide her, "That's ridiculous. What's wrong with you?" By the time she was an adolescent, Marsha was painfully self-conscious and insecure. However, growing up with so much invalidation left her confused and uncertain about her own subjective experience, such that she didn't really understand why she felt so bad about herself all the time. Sadly, Marsha often thought to herself, "There's just something wrong with me." Marsha frequently cried alone in her room but without really understanding why she was crying or knowing what she was so distressed about.

Marsha always thought that everything would "get better" and be OK when she went away to college. To her dismay, though, she found herself becoming even more depressed during her first semester away. Frequently, she had to fight back tears over seemingly nothing, found it hard to reach out or talk to others, and started gaining weight. More insecure and confused about herself than ever, Marsha began seeing a counselor at the Student Counseling Center. Although she did not really have words for what was wrong, Marsha tried to be a good client and help her therapist understand her problems:

MARSHA: I'm not sure what's wrong with me. Maybe I'm just lonely…but I've always had this sort of empty feeling inside.

THERAPIST: Do you have any friends? What are your classmates like?

MARSHA: I guess I sort of have friends. My roommate in the dorm is nice.

THERAPIST: What do you do with your friends—go to the movies, shopping?

MARSHA: Yeah, I do those things. I belong to a foreign language club, too.

THERAPIST: Do you like your friends? You're new to the university; maybe you need some new friends here at school. There's lots of ways to meet new people here…

MARSHA: Well, I've had friends, but it's just hard being with people sometimes. So, maybe I'm lonely, but I don't know what's wrong with me. I just keep crying—like a big baby all the time.

THERAPIST: But you've just left your family and come to college. You must miss your family and feel lonely. It's natural to feel lonely when you move away from home.

Most of the other kids in the dorm feel that way, too. I know I sure did when I moved away to college.

MARSHA: Oh.

THERAPIST: This is not unusual at all. You're going to be just fine once you get through this transition.

MARSHA: I hope so. Maybe my family is different, though. In high school, I always thought that my family had more problems than my friends' did. And it seemed like my mom was always mad at me for doing something wrong.

THERAPIST: Yeah, but, like I said, it's normal for you to be kind of emotional at this time in your life.

MARSHA: It is? But I still feel different from everybody else—like there's something wrong with me...

THERAPIST: Sure, it's real natural for you to feel different from everybody else. Late adolescence, moving away to college—it's a tough time in life. You're going to be just fine. Do you have a boyfriend?

MARSHA: Yeah, sort of. I try to tell him what's wrong, but that never works. And then he gets frustrated with me because nothing he suggests helps. I don't know why I get so depressed a lot (pauses)...maybe my mother used to yell at me too much...or maybe I'm just "too sensitive"—that's what my family always said.

THERAPIST: Do you have trouble eating or sleeping?

MARSHA: No trouble eating, as anybody can see, I've gained 12 pounds and look like a pig. I wake up at night sometimes—worrying about things that don't really matter, trying not to start crying, wondering what's wrong with me...

THERAPIST: Are you eating alone? Maybe you should be eating with friends.

MARSHA: *(with resignation)* Maybe that would help.

Marsha could not say exactly why, but she did not like seeing the therapist and, after canceling and then missing a few sessions, she did not go back. However, as the semester went on, she became more depressed, continued to wonder what was wrong with her, and began to struggle in her classes as well. During advising for second semester, a concerned professor saw her distress, listened to her disappointing experience in her previous therapy, and encouraged her to go back and try again with another therapist. Reluctantly, Marsha agreed to try it when the professor suggested that she could go back, for just one session, and if she didn't like it again, just stop.

Therapy began more slowly this time. Marsha found herself feeling impatient and testy with the therapist and was reluctant to share much. She alternately acted blasé and then distressed but would never allow the therapist to remain joined with her on the same topic for very long. The therapist was not frustrated by her mixed message of come closer–go away, however, and was effective at communicating his continuing interest in her. After a few weeks, Marsha sensed that this therapist might be different than before and began disclosing her feelings again.

MARSHA: I'm just a mess—I don't know what I'm feeling—or why I keep crying all the time.

THERAPIST: There are a lot of difficult feelings there, and you can't sort them out yet. Let's just sit together for a minute, and see if one of them come to you.

MARSHA: OK, uh, well, I guess I feel sort of empty. And sad and lonely and frustrated… and who knows what else. But I've always had this empty feeling that I just hate.

THERAPIST: "Empty." OK, let's start there. Can you help me understand a little bit about what this empty feeling is like for you?

MARSHA: I don't know…I just feel empty inside—I always have.

THERAPIST: This empty feeling is important—it's been painful for a long time. Let's work together and try to get closer to it. Bring me in a little more on this emptiness—I don't want you to be alone with it anymore.

MARSHA: *(long pause)* I don't know what to say…how to begin.

THERAPIST: Sure, it's hard to get started. Maybe you can use just pick an adjective, or make some gesture, that will help me see this emptiness. Or maybe you've noticed when you start to feel it—who you are with or what's going on?

MARSHA: There's just an emptiness inside. The wind blows right through me. I'm always hungry—it never gets filled up.

THERAPIST: That's a good start for us. You just feel empty inside and nothing fills it up. Nothing takes care of this, nothing feeds or fills you up—the hole just stays there.

MARSHA: *(doesn't speak; nods and becomes teary; looks at therapist)*

THERAPIST: *(holding her gaze kindly)* I can see how much you are hurting right now as you share this with me. It's sad that things have happened to make you feel this empty and alone, but I'm honored that you are willing to risk sharing all of this with me.

MARSHA: *(cries harder)* There's just something wrong with me; there always has been.

THERAPIST: Uh-huh. It's sad for me to hear you say that—that there's always been "something wrong" with you. Tell me more about what's been wrong…and felt so wrong.

Marsha was heard and felt understood this time. Breaking the old relational pattern, she was not crying alone in her room anymore. In subsequent sessions, the therapist continued to listen well and kept responding to what Marsha thought was most important. There were ups and downs, of course, but this empathic understanding forged a working alliance between them that ultimately allowed Marsha to begin clarifying her own thoughts and feelings. As she increasingly found her own voice in this relationship, she started to become more assertive, felt more confident and relaxed with others, and her lifelong dysthymia began to lift.

The concept of understanding the client sounds simple enough—as in, sure, can we move on to the next topic now. After all, all the therapist has to do is listen carefully, grasp the central feelings or core message, and find a way to articulate or reflect this understanding to the client. In actual practice, however, demonstrating our understanding in this way is not so easy to do. Most of us have been strongly socialized to "hear" in a limited, superficial way that minimizes the emotional meaning and avoids the interpersonal messages embedded in clients' remarks.

This was why Marsha's first therapist could not hear what she was really saying and repeatedly steered away from the sensitive feelings and personal concerns she was trying to formulate and communicate. Reflecting her resiliency, Marsha kept trying to come back to her core message—something's terribly wrong. By keeping things on the surface and reassuring her, trying to talk her out of her feelings, and orienting her toward superficial problem solving before he really knew what was wrong, their interpersonal process reenacted her developmental history of not being seen or heard.

Most new therapists are aware that they possess a "third ear"—that they already have a highly developed ability to hear the key issue or core meaning in what people say. However, they often feel they must avoid acknowledging the true content of these underlying, and often nonverbal, messages because of their emotional content. Therapists can downplay or avoid the emotional message or basic meaning for many reasons:

- feeling awkward themselves or concerned that they might embarrass the client
- feeling incompetent or unsure of how to respond if they do go beyond the surface and respond to the bigger feeling or issue
- feeling reluctant to violate cultural norms or familial rules against forthright or more direct communication
- feeling a need to protect or take care of others by shielding them from their own pain or distress
- feeling afraid of potential boundary violations or feeling too close to the client if they share deeply with the client and risk personal involvement
- having their own unresolved issues activated by the client's underlying messages or overt communication

As a result, many therapists are adept at switching automatically to a more superficial, or everyday, social level of interaction to avoid the client's (and sometimes their own) vulnerability and pain. This also occurs when clients convey *embedded messages* about the therapist–client relationship or interaction together that therapists may not want to hear or approach. That is, *clients are often making covert statements about the therapeutic relationship or indirect references to what is going on between them—especially about problems or concerns they are experiencing right now in their current interaction with the therapist.* Let's provide an example of an ineffective and an effective response for therapists faced with embedded messages about problems in the therapist–client relationship:

CLIENT: Do therapists who work with clients like me ever see their own therapists for their own personal problems?

INEFFECTIVE THERAPIST: Yeah, they probably do, but why don't we get back to the real concerns that you came in for. You were just saying...

VERSUS

EFFECTIVE THERAPIST: Yes, most therapists seek help with their own issues at times. But I'm wondering if you may be having a question about me as well, or about something related to our work together, that we might talk about? If so, it would probably help our work together.

In contrast to her first therapist, Marsha's second therapist broke the social rules and responded directly to her core affective message, "I feel empty." As a result, Marsha felt understood and responded to. In a small yet significant way, this was a corrective emotional experience. The therapist was willing to join meaningfully with her in her experience—demonstrating his genuine interest and concern for what was really going on for her—which significant others had not been able to do as she was growing up. Disconfirming her problematic relational expectations, he did not move away from her feelings or try to talk her out of them; in sharp contrast, he genuinely welcomed them and actively encouraged her to enter them more fully. When the therapist is able to provide such corrective experiences consistently, trust builds as the client learns that his previous, early maladaptive schemas do not fit in this relationship—and perhaps this will be true for some other relationships as well. Specifically, behavior change begins first as the client becomes more confident in the expectation that the therapist will see and hear her. Then, with the therapist actively scaffolding, this new, expanded schema and more flexible way of relating is extended to others in their everyday lives. *New therapists are encouraged to try responding in this more direct, empathic way; use the ability they already possess to hear what is most important to the client; and while still being diplomatic, judiciously take the risk of saying what they see and hear.* Most student therapists are overly inhibited and hold themselves back too much. We can tentatively wonder aloud with the client and share what we are hearing and observing, or wanting to ask further about in this active way, without taking over, giving advice, and telling clients what to do.

THERAPIST: Gina, you just told me three stories about Robert and I wonder if you find anything in common in these stories: he wants you to stop working and let him take care of you; later, he flirts with another girl in your presence and tells you, you are "stupid" and overreacting; then, when you try to remove yourself from the argument, he blocks your way. Any thoughts about this?

CLIENT: I feel disrespected and pushed to get angry...

THERAPIST: Disrespected...that makes sense. You have asked him not to flirt and he is doing that and calling you "stupid" when you try to address it...

CLIENT: I hate it when he calls me stupid; it makes me feel like he is better than me.

THERAPIST: Better than you—can you say more about that?

CLIENT: Like he knows more, is smarter, has more to offer, more power, you know...

THERAPIST: More power...that's a huge word...

CLIENT: Yes, like when he won't let me leave when I get mad, I feel so powerless, I just want to scream and hit at him but I don't want to give him a reason to hit me or say, "See why I flirt with other girls, they are nicer to me than you."

THERAPIST: You feel unable to go someplace else to calm down, yet, it sounds as though speaking up for yourself has also been difficult and perhaps not felt safe?

CLIENT: Yes, he starts to yell, calls me names, it can feel a little scary, I don't want him to lose his temper and blame me for it...

THERAPIST: So you have felt disrespected, and scared, and yet he wants you to quit your job and become financially dependent on him. How does that sound to you?

CLIENT: Wow, you know, as we talk about it, I start to wonder why I keep going back to him, why I stay with him, because he'll eventually say, "Babe, I'm sorry I called you a bitch," but it happens again after a while…

THERAPIST: There's a lot going on here that has been hurtful and it has been hard in the past to leave…

CLIENT: Yes, it has been hard…it helps to talk about it without having to hide it. When I try to talk to my mom, she immediately says, "Leave that asshole," but doesn't let me think it through. Yet she's still with my dad, who uses drugs and has never worked…

THERAPIST: I'd like to hear all your thoughts…all you want to tell me. What has kept you with him and what might make you not want to be with him? Together, we can work out what *you* want and what's in *your* best interest.

CLIENT: I like that…

To summarize, therapy offers clients an opportunity to be understood more fully than they have been in other relationships. When this understanding is offered in the initial sessions, clients feel that they have been seen and are no longer invisible or alone, different or defective, dismissed or unimportant, and so forth. At that moment, the client begins to perceive the therapist as someone who is different from many others in her life, and possibly as someone who can help. In other words, hope is engendered when the therapist understands and articulates the personal meaning that each successive vignette holds for the client. With this goal in mind, let's look at guidelines to help therapists clarify the central meaning in what the client presents and learn how to be more accurately empathic.

IDENTIFY RECURRENT THEMES

Most clients will not share deeply or risk exploring their vulnerabilities with a therapist who they does not provide the security that comes from feeling that the therapist understands them well. One of the best ways to help therapists understand their clients better—be more accurately empathic and more specifically "get" their subjective experiences—is to identify *recurrent themes* in the narrative they relay (Daly & Mallinckrodt, 2009). Let's explore pattern recognition closely—it also helps us develop useful case conceptualizations.

The therapist's goals at the beginning of treatment are to:

- support clients' initiative and help them take ownership of the treatment process by encouraging them to lead and choose what they feel is most important to discuss
- actively join clients in exploring the concerns they bring up and help them tell their story
- identify the core message or register the central meaning in what clients are communicating
- demonstrate this understanding by accurately capturing and reflecting this empathic understanding to the client
- begin the process of assessing whether there are common themes in the problems or issues presented

Although these basic intervention goals may sound simple, they are not. What's so hard about it? *It requires the therapist to give up a great deal of control over the direction of therapy and over the timing and content of which issues are brought up for discussion.* The "letting go" of what the client discusses—not knowing what the client is going to say or do next—is anxiety-arousing for many. Most student therapists are interpersonally skilled in their everyday lives and readily able to subtly take control and quietly lead or direct many of their social interactions. As student therapists try to cope with the anxiety and ambiguity of doing therapy they often fall back on the familiar role of friendship—which they can provide easily and successfully. However, being a friend or "buddy" is not a good model for how to be with clients. That is, the same social skills that may facilitate friendships and other familiar relationships outside of therapy (that is, subtly leading or shaping where the conversation is going, filling awkward moments and facilitating transitions, keeping the conversation moving or staying on comfortable topics, having our emotional needs met in a reciprocal manner, and so forth) usually limit a client's progress. In treatment, the novice therapist is placed in the new and far more demanding position of responding to the diverse and unpredictable material that the client produces. A therapeutic stance that encourages the client's initiative and ownership of the treatment process fosters greater agency or self-efficacy in clients. The therapist needs to have the *flexibility* to be able to:

- relinquish control over what will occur next in the therapeutic relationship;
- tolerate the ambiguity of not knowing what clients will produce or where the current topic will lead; and
- make sense of the varied and unpredictable material that clients present.

No doubt, this is a tall order.

There are two important dimensions to working in this way. On a personal level, therapists need to "know themselves" (Bromberg, 2006). This self-awareness, also emphasized in the multiculultural literature (for example, Sue et al., 1998; Pedersen et al., 2008) will allow them to recognize more fully, when their reactivity is evoked, if it is something in their own life or if it is grounded in the client's experience. Supervisors can be very helpful in assisting therapists differentiate their own experience from that of the client, especially critical for new therapists. Secondly, tracking the client's experiences helps identify *recurrent themes* in the client's vignettes or stories that will help therapists make sense of the client's experience and better understand what is most important or central in the wide-ranging material the client presents. This empathic understanding, in turn, provides therapists with the only legitimate control they can have in therapeutic relationships—over their own responding—as opposed to ineffective attempts to subtly direct or control the material that clients produce. Additionally, the appropriate or legitimate internal controls that come from this understanding will allow therapists to *tolerate the ambiguity* inherent in this work. Thus, our challenge is less to shape or subtly lead where the clients are going, as occurs so commonly, than it is to follow wherever they lead, and be able to make sense of whatever thoughts or feelings they bring to us.

Patterns and themes help us understand. To illustrate, suppose the client tells the therapist about his reasons for coming to therapy. As the therapist follows the different recollections and descriptions the client chooses to relate, the therapist's goal is to discern an *integrating focus* for the wide diversity of material the client presents. The best way to do this is by identifying the patterns or unifying themes that recur throughout. Typically, these integrating themes occur in three inter-related domains:

1. Repetitive relational themes or interpersonal patterns
2. Pathogenic beliefs, automatic thoughts, or faulty expectations
3. Recurrent affective themes or central feelings

Identifying these unifying patterns or themes, that typically are present in most clients' narratives, helps therapists conceptualize their clients' problems and formu-late a treatment focus. Although the disparate material that clients present may ini-tially seem unrelated and disconnected—and can evoke in the therapist the unwanted feeling of being lost, confused, or even overwhelmed by the client, over time new therapists will improve in their ability to listen for and recognize the syn-thesizing themes that are present in all three domains. As expanded further below, developmental experiences, unmet needs, and unwanted, hurtful messages received during those experiences are often key to understanding the patterns and themes that clients present.

Therapists can help clients make sense of their long-standing problems by lis-tening for and highlighting integrating patterns in their thoughts, feelings, and be-havior. This linking helps to locate current symptoms in the broader context of their lives—which helps clients change by developing a more coherent life story or narrative for who they are and where they have come from. When developmental life experiences are taken into account and symptom presentation and repetitive re-lational patterns are assessed, therapists can then integrate the themes and share these tentatively, as hunches or possibilities, to help formulate a shared treatment focus (Curtis et al., 1988; Smith, 2006).

Sarah, age 28, was referred to treatment by her physician. She was the oldest of three children and her mother was chronically depressed and hypochondriacal. By the time Sarah was 10, she often made the family dinner and school lunches for herself and her siblings. She was "reinforced" for this by her father for being "such a great helper" and through her adult life gravitated toward jobs where the demands were high (where she excelled), and was appreciated, but the praise and financial compensation were rarely commensurate with the time and energy invest-ment she made. She also often chose partners who, while initially attentive, soon demanded extensive caretaking from her. By the time she was 28, she was having severe migraines and had developed high blood pressure. Her boss had just given her a "promotion"—a title change with additional responsibilities, no additional staff/assistance, and no increase in pay. The boss's statement had been, "You are my best worker. In time, we'll give you a raise and more help. We can't do it now, be patient. At least I can give you a better job title for now."

Sarah, consistent with her developmental story, felt conflicted—the praise was reminiscent of what she had experienced with her father and she felt she should *comply*; yet she felt resentful. She had just been asked to take on her boss's re-sponsibilities and had no help to complete the work and no financial compensa-tion for it. Further, her personal life was strained because she was asked to be the "primary caretaker" in that relationship and he resented all the hours she devoted to her work.

THERAPIST: It's sounding to me as if, here again, you have to be the responsible one who takes care of everything—and wind up feeling tired and resentful in the end. I'm wondering if this sounds like a familiar pattern in your life—does something like this go on a lot?

CLIENT: It doesn't sound like "something" in my life—that *is* my life! My work, my personal life, my childhood! That's been my role my whole life—take care of everybody—as if it's my fate or job to do. But if I don't stop doing this, it's going to kill me—I'm exhausted!

In this example, Sarah's developmental experiences provide an illuminating context for grasping the patterns and themes that are leading to the poignant life circumstance she currently is in.

How can new therapists begin to develop this essential skill of recognizing patterns and understanding presenting symptoms in a lifespan perspective? Writing process notes after each session is one way that student therapists can learn to attend better to identify patterns and recognize integrative themes in their clients' narratives and in the therapeutic relationship. Guidelines for doing this are provided in Appendix A. These can then be used to develop a treatment focus.

Repetitive Relational Themes.   Terminology varies, but the bedrock of interpersonal and contemporary psychodynamic treatment approaches is to identify the interpersonal scenarios that keep reoccurring for the client and working with the **repetitive relational patterns** that weave throughout clients' symptoms and problems. Therapists-in-training are working to develop their pattern recognition abilities—that is, to develop the eyes and ears to recognize the interpersonal themes that keep reoccurring throughout the different narratives the client relates. For example,

- Whenever I risk trusting someone, they let me down.
- They expect so much from me, but no matter how hard I try, it's never enough.
- I always have to give up what I want in order to be close. It's a lose-lose situation; go along and do what others want—or be alone.

Many of the patterns that emerge are rooted in early developmental experiences and include:

(a) Attachment. Insecure or disrupted attachments, feelings of anxiety or ambivalence, inconsistency or rejection from important others, and so on.
(b) Beliefs and schemas. The attitudes, expectations, and beliefs that developed in these formative caretaking relationships: "I'm unworthy," "I don't really matter," "I have to go along with whatever they want," and so on.
(c) Emotional reactions. Difficulty modulating them the feelings that resulted: "I hate everyone," "I'm the only one who matters," "I hate everyone," "I can't stop crying," and so on.
(d) Behavioral dysregulation. Screaming at others, acting on impulse, withdrawing on reflex, isolating, and so on.

As beginning therapists review the process notes they write after each session, they will often observe that the varying problems that each client talks about usually come back to only two or three basic themes or patterns that keep getting expressed in different ways.

Interpersonal therapists have always focused on clients' repetitive, maladaptive relationship patterns. Since the 1980s, however, this approach has been developed more systematically as a treatment focus in short-term or time-limited treatments (Levenson, 1995; Luborsky & Crits-Christoph, 1990; Sifneos, 1987; Strupp & Binder, 1984), and by integrative theorists such as Wachtel (2008). These treatments all focus on clients' maladaptive relational patterns, which different theoretical approaches variously label as the core conflictual relationship theme, cyclical maladaptive pattern, maladaptive transaction cycle, and other terms. In each approach, however, the therapist is trying to identify the central themes that keep repeating in the varying problems the client is experiencing with others—and often beginning to recreate with the therapist as well. The therapist begins the change process by *highlighting* these relational patterns—so clients can begin to recognize them and anticipate when they are coming into play, and exploring new and different ways to respond that could change the usual but unwanted scenario. The client's relational pattern, for example, might be to readily feel controlled by others, let down, or abandoned. Or, perhaps the client quickly feels criticized and put down, ignored or dismissed as unimportant, or even idealized by others. It is compelling for clients when the therapist can identify the same interpersonal themes occurring in three different spheres:

- in current interactions with others in their daily lives;
- in formative relationships with family members; and
- in the current, here-and-now interaction with the therapist.

Let's explore this basic sequence more closely—it offers therapists working within different theoretical orientations such an easy way to help many clients.

First, as clients are telling their stories and relaying vignettes from current or past relationships, the therapist tries to identify the repetitive relational patterns that are causing problems in their lives and to highlight or punctuate them. These observations are then shared in a **tentative manner**—so the client can easily modify, refine, or even reject them. Exploring them collaboratively, clients develop a focused awareness of just how these troubling patterns typically go, and begin observing on their own when these familiar but unwanted situations are occurring with others. For example (client to herself):

CLIENT: I'm starting to feel that he really doesn't want to be with me, and I'm trying too hard to figure out what he wants and how to please him. I need to change what I'm doing here.

At the same time, the therapist actively looks for opportunities to *link* the same maladaptive relational patterns that are occurring with others to their current interaction. The therapist does this via process comments: asking about or wondering aloud how this same, problematic pattern that they have been discussing with others could be coming in to play in their relationship as well. For example, suppose the client is acting in a controlling or dominating manner with the therapist;

interrupting or criticizing the therapist whenever she starts to speak; or acting help-less and speaking to the therapist in a child-like tone of voice. Especially if this has been a pattern that has been causing problems with others in the client's current life *the therapist needs to find a safe and supportive way to make this problematic pattern overt—to name it, and then help the client find a different and better way of relating—right now—in their current real-life relationship, than the client has been doing with others.* That is, the intervention goal is to help the client find new and more adaptive ways to respond in their current interaction together—so that the new behavior the client is trying to adopt with others in their everyday life is not just being talked about abstractly but is being enacted behaviorally right now. Below, to illustrate this experiential learning, the therapist makes a process comment and gives the client interpersonal feedback about how she is acting with the therapist right now:

THERAPIST: Michelle, I know people don't usually talk together this directly but, if I may, I'd like to share a reaction I'm having right now. Maybe there's something to it or maybe not—you tell me. You've been saying that people don't take you seri-ously, even though you are clearly such a bright and capable person. But as I listen to you right now, you are speaking to me in a child-like tone of voice—almost like a frightened or timid little girl. I'm concerned that if you are doing this with others in your life, many will just dismiss you—kind of like you have been describing. What do you think as I share this possibility?

CLIENT: Well, it's kind of embarrassing to hear, but I've heard this before so you're probably on to something. I don't know why I talk like a little girl sometimes—it's kind of automatic—I just seem to start doing it sometimes.

THERAPIST: I really respect your willingness to look at this. Let's be partners and work together on this here in our relationship. How would this be? Whenever you or I notice the "little girl voice," let's speak up and name it. And why don't you try talking to me in here in a stronger voice. I'd welcome that—I like the smart, strong part of you. I'm also thinking that if you can claim your own power in here with me, then we can help you start doing it out there with others. What do you think?

CLIENT: I'd like that. I do want to change this—that little girl voice does have something to do with why other people just shine me on and never take me seriously...

This is in vivo or experiential relearning—the therapeutic relationship is serv-ing as a social learning laboratory. As this relearning is occurring in the therapeu-tic interaction, the therapist also has a significant opportunity to help the client generalize or transfer this new, stronger way of being to others in her everyday life. That is, as they talk about and change the familiar but problematic way they are interacting together, the therapist and client can better explore when and how these same problematic patterns are being played out with others in their current relationships:

CLIENT: I think I slip into that "little girl voice" thing whenever I feel someone has control or power over me—like my boss, or my professors. I don't do it with my friends, but maybe I am doing that with my dad too...

As they explore and try out new ways of responding with others, and begin to succeed sometimes in changing these maladaptive behavior patterns with others, clients often become interested in doing a "cost/benefit analysis"—assessing what they get from this interpersonal pattern (for example, when I talk in the "little girl voice," no one will get mad at me or think I'm challenging them) and how they pay for it (for example, people in authority don't take me seriously or respect me). Oftentimes, clients become interested in trying to understand where they learned to act in this way—which helps clients understand their problems and be able to maintain the new behavior after treatment ends. This understanding also helps clients become more aware of the multiple arenas in which these developmental experiences impact them. For example, staying in a "little girl" role was one way to maintain connection in one's family of origin or one way to avoid verbal or physical abuse—yet this coping strategy is now hindering progress in the adult world and in intimate relationships. With this increased selfunderstanding, the client is now able to anticipate problematic interactions and relationships with more effective and flexible approaches. Recovery from faux pas is also enhanced. That is, this understanding empowers them, since they are now able to author a new and more accurate narrative for themselves that gives a more realistic account of their lives, their problems, and how they came about (Neimeyer, 2009). Thus, throughout this text, we will focus on how therapists can use process comments to help clients recognize maladaptive relational patterns with others (and the beliefs, affects, and behaviors that accompany these), and how aspects of them may be reoccurring with the therapist at times. We then will find ways to change these problematic patterns—first with the therapist and then with others in their lives. For now, let's go on to identify themes in the cognitive and affective domains—all three domains are equally important and inextricably entwined.

Pathogenic Beliefs.   Just as the therapist can better understand clients by identifying unifying themes in their interpersonal relations, cognitive patterns and thought processes provide an essential component for accurate understanding and effective intervention (Beck, 1995; Bjorgunsson & Hart, 2006; Lazarus, Lazarus & Fay, 1993). For example, clients' maladaptive relational patterns are closely associated with their problematic beliefs about themselves, inaccurate perceptions of others, and faulty expectations for what is going to occur in relationships and in the future. Ellis (1999) suggests that two of the most common dysfunctional beliefs concern being liked and loved (I must *always* be loved and approved of by the significant people in my life) and being competent (I must *always*, in all situations, demonstrate competence). Working in an integrative cognitive-psychodynamic approach, Weiss (1993) and Silberschatz (2005) highlight **"pathogenic beliefs"** that help to create and sustain the maladaptive relational patterns introduced above. For example, regarding clients who feel excessive or unrealistic guilt, the client's pathogenic belief might be:

- I am being selfish whenever I say no or do what I want.
- Others are hurt by my independence.
- I must not enjoy my own success, and I need to sabotage my own interests and goals, or I will be disloyal.

Similarly, a pathogenic belief for clients who are prone to feeling shame, or that others will ignore or reject them, might be:

- I am unimportant and do not matter.

- I am inadequate, and others will see me as lacking or weak.
- Others would ignore me if they knew what I needed, or resent my emotional needs as childish and demanding.

Long ago, George Kelly (1963) introduced his theory of "personal constructs," which became the forerunner for cognitive behaviorists today who try to discern clients' "core beliefs," such as those listed here, and the "early maladaptive schemas" we introduced in Chapter 1 (Young, 1999). In particular, Beck (1976) and Beck et al. (2003) have illuminated self-talk, automatic thoughts, and the dysfunctional interpretations that are central to clients' problems—and that reflect *core themes* about ourselves and our relationships. They write compellingly about the cognitive schemas people develop—that is, ways of viewing themselves, others, the world, and the future. In particular, these authors have identified a *cognitive triad* in which people come to view themselves as defective or unlovable; the world as unmanageable or overwhelming; and the future as bleak and hopeless. We began to see in the last chapter that, as a result of formative developmental experiences, clients begin to exercise a *selective bias* in processing information. As a result of this cognitive filter or selective attention process, *clients inaccurately synthesize the same repetitive themes from diverse interpersonal experiences.* They continuously construe or interpret events to arrive at the same faulty conclusions, such as disinterest or unfairness from others, which come to characterize their daily experiences—and their lives. Thus, this mind-set or fixed lens for viewing others creates patterns in the clients' cognition, as they keep slotting different experiences into the same narrow perspective (viewing others as wonderful or idealized, needy and dependent, disinterested and unresponsive, critical and judgmental, and so forth—even when they are not).

Similar to these cognitive schemas, we also saw that attachment theorists use the term "internal working models" to clarify how these overgeneralized beliefs and expectations are learned in repetitive interactions with caregivers, and how they may operate consciously or without awareness. As before, therapists try to

1. identify these core beliefs that provide integrating threads throughout the client's life and problems;
2. make them overt to clients or name them—especially as they are occurring in the current interaction with the therapist; and
3. help the client question the truthfulness or utility of these problematic beliefs as they come into play with others in current relationships.

As Beck and attachment oriented therapists emphasize, automatic thoughts and dysfunctional beliefs ("internal working models") are not so evident when clients are functioning well. It is when clients are upset or distressed (as Beck would say, in their *hot cognitions*) that pathogenic beliefs and faulty expectations will be revealed. At these crisis points, when the client is vulnerable, they "become" the client's reality and are experienced as the only way that relationships have ever been and ever will be! In other words, when clients are not distressed, they can imagine a wide range of interpersonal scenarios and outcomes, entertain a more realistic set of beliefs about themselves and others, and have flexible expectations about what may occur in relationships. In contrast, when distressed, clients' beliefs and expectations often become rigid and unidimensional. Thus, it is when clients are distressed or in crisis that therapists can best assess their orienting schemas, how they process information, and how they make sense of events and interactions. At these times, intervening and assisting them to develop their reflective capacities through cognitive reframing, and entering the client's experience or "mentalization," considering this familiar, problematic set of circumstances from a new perspective or different point of view is especially helpful (Allen & Fonagy, 2006; Bjorgunsson & Hart, 2006). In closing, therapists often learn more from asking, "How did you explain that to yourself?" than from asking, "How did you feel about that?"

Recurrent Affective Themes.   Therapists also can identify unifying themes in clients' emotional reactions. Recognizing and responding to these *recurrent feelings* can be nothing less than a gift that therapists offer to their clients. For example, the therapist might reflect:

THERAPIST: This is so hard for you. Here again, when _____ happens, you are always left feeling _____.

Often, a primary or **core affect** comes up again and again for the client in many different situations (Greenberg, 2002; Greenberg & Goldman, 2008). As the therapist listens to the client, an overriding feeling such as sorrow, bitterness, distrust, or shame may pervade the client's mood or characterize different experiences the client relates. For instance, the therapist might then reflect:

THERAPIST: As I listen to you, it sounds as if you have felt hopelessly *burdened* all of your life.

OR

THERAPIST: This doesn't quite sound like worry or anxiety that keeps coming up, it seems much bigger than that! Am I hearing more of a sickening feeling of *dread*?

When the therapist can identify and accurately name this "characterological" affect—the central or core feeling that some clients may experience as nothing less than the defining aspect of their existence—it has a profound impact. Clients feel that the therapist understands who they really are and is seeing them and what their life is like in a way that others have not done. A key in establishing the therapist's credibility and fostering the working alliance is the therapist's ability to identify a core affect and accurately reflect the far-reaching meaning it holds in the client's life.

Responding to the primary feelings and capturing the recurrent emotional themes that keep coming up in the client's life is one of the most significant interventions that therapists can provide. Unfortunately, new and experienced therapists often feel insecure about responding to strong feelings in their clients (Williams et al., 1997). Trainees need their supervisors to help them recognize and respond to these affective themes, especially by role playing or otherwise demonstrating what trainees might say or do when these primary feelings emerge. Because this topic is so important, and often receives too little attention in clinical training, specific guidelines for responding to clients' feelings will be provided in

Chapter 5. Completing the process notes provided in Appendix A will also help therapists identify recurrent patterns in the affective domain.

To sum up, if the therapist can identify the maladaptive interpersonal patterns, pathogenic beliefs, and primary feelings that recur throughout their lives, and link together the clients' experience and problems so that clients can better make sense of things, clients are reassured to know that the therapist understands them in a way that others do not. Providing this highly differentiated or specific empathy that "hits the nail on the head" is essential for clients. *This is not an isolated intervention or technique, but a consistent stance that characterizes the therapist's way of being with the client throughout each session over the course of treatment.*

IMMEDIACY INTERVENTIONS: WORKING IN THE MOMENT

> There are, in fact, no more important communications between one human being and another than those expressed emotionally, and no information more vital for constructing and reconstructing working models of the self and other than information about how each feels towards the other...it is the emotional communication between a patient and his therapist that play the crucial part. John Bowlby (1988, pp. 156)

USE PROCESS COMMENTS TO BUILD A WORKING ALLIANCE

Therapists help build a working alliance by intervening in the here-and-now and talking with clients about what may be going on between them, right now, in their current interaction. In this section, we look at several closely related interventions that therapists use to work in the moment and create *immediacy*. These **immediacy interventions** *are not for the faint of heart, however, they significantly increase the intensity of the therapist-client relationship. They are powerful interventions that allow therapists to engage clients in far more meaningful ways, and when used sensitively and compassionately they bring about a more genuine or authentic relationship.*

Process comments.   In essence, **process comments** make the interaction between the therapist and client overt and put the relationship "on the table" as a topic for discussion. Process comments are *not* confrontations, demands, intrusions, or judgments. Instead, they are simply observations, tentatively suggested, about what *may* be occurring between the therapist and client at that moment. Process comments are always more effective when they are offered *tentatively*—the therapist is often wondering aloud about what might be going on between them right now, and as an invitation for further dialogue and mutual sharing of perceptions. Researchers find that process comments offered tentatively ("It sounds like..."; "I'm wondering if..."; "Sometimes it seems like..."; "Maybe...") are more effective than direct challenges ("I think you...") (Jones & Gelso, 1988; Miller et al., 1993). Let's illustrate how therapists can use process comments, in this tentative and respectful way, as an invitation to discuss and explore together their relationship:

THERAPIST: As you're talking about him and what went on between you two, I'm wondering if you might be saying something about our relationship as well? What do you think—does that kind of frustration ever come up between us, too?

CLIENT: Well, sometimes I do feel frustrated with you, too, like when you're so quiet and I don't really know what you're thinking.

OR

THERAPIST: I'm wondering what it's like for you to be telling me about this sensitive issue. How is it for you to be sharing this with me?

CLIENT: I'm just dying inside—it's mortifying to talk about this. You're being nice—I know you have to act that way because you're a therapist—it's your job to act nice, but I just know you can't respect me anymore.

Clearly, process comments bring the therapeutic relationship to life—they make it far more meaningful for both the therapist and the client. As in both dialogues above, *process comments also reveal unspoken interpersonal conflicts*—significant misunderstandings or faulty expectations that are going on in the therapeutic relationship without the therapist's awareness (for example, the first client is frustrated because he wants more responsiveness from the therapist, and the second client is convinced that the therapist doesn't respect her anymore and is just acting nice because it's his job). Unless we use process comments to reveal these unspoken problems in the therapeutic relationship, oftentimes referred to as "ruptures" in the working alliance (Samstag, Muran & Safran, 2004), the therapist doesn't have the opportunity to address and resolve them—and that is what we are looking for in the interpersonal process approach.

Process comments are also useful when the interaction has become repetitive or treatment has lost its direction. For example:

THERAPIST: I'm not sure where we are going with this right now. Is this the best way for us to spend our time together today, or is there another issue that might be more important for you? What do you think?

CLIENT: Well, I don't really have a problem I want to talk about today. Actually, something pretty exciting happened this week and I'm feeling really good, so I guess I don't know what to talk about.

THERAPIST: We don't have to talk only about problems—I want to know about what's important in your life. You've got some good news—great! I'd love to hear about it.

These process comments bring *immediacy* to the therapeutic interchange as the therapist and client stop talking about others out there and begin talking instead about "you and me" (Binder & Strupp, 1997; Cashdan, 1988; Ivey & Ivey, 1994; Kiesler, 1996; Safran & Muran, 2000). Process comments also give therapists a way to identify or highlight the client's maladaptive relational patterns, and a way to intervene and change these reenactments *as they are occurring with the therapist.* For example,

THERAPIST: I'm wondering if something important might be going on between us right now. Can we talk together about what just happened here? I know people don't usually talk together this way, so directly, but maybe it could help us understand what's been going wrong with others as well.

CLIENT: Well, OK, I guess I don't like these homework assignments we're talking about. I just don't like people telling me what to do, and uh, I guess I can get kind of pissy with you, like I do with people at work.

THERAPIST: I'm glad you're telling me this. I don't want to tell you what to do in here in therapy, and I'm learning something important about what goes wrong with your boss, and maybe your wife, too. Let's see if we can work out this control thing better here in our relationship, and I have some ideas we could start talking about that might help make this better with others, too? How does that sound?

Interpersonal Feedback. In addition to activating relationships that have become stalled, and altering maladaptive relational patterns as they begin to replay with the therapist, process comments can be used to provide **interpersonal feedback.** Sometimes called "challenges," these immediacy interventions point out contradictions and help clients see discrepancies between what they are saying and doing. For example:

THERAPIST: What you are sharing with me right now is a very sad thing, John. Actually, it's kind of heartbreaking to hear. But you are saying it in such an offhanded manner—as if it doesn't really matter or mean much. Help me understand the two different messages I'm getting—such a painful story being told in such a half-interested, almost pleasant manner?

CLIENT: Well, no, I didn't really realize I was doing that. I do feel sad about what happened, but I sure wouldn't want anyone to think I was feeling sorry for myself—you know, being a complainer or anything like that.

OR

THERAPIST: You're telling me how furious you are about this, Janice, but your voice sounds so small or meek. Are you aware of this discrepancy between what you are saying and the way you are saying it?

CLIENT: Well, yeah, I guess so, but why do you bring it up?

THERAPIST: Because I think what you say matters. And if you talk to others like you are talking with me right now, people won't really listen to you or take you seriously.

CLIENT: (tearing). Well, that's sort of been my life story.

Metacommunication and Therapeutic Impact Disclosure. Across differing theoretical orientations, a variety of terms have been used to describe process comments and other related interventions that create immediacy. In particular, Donald Kiesler discusses **"metacommunication"** and *therapeutic impact disclosure*. This immediacy intervention uses interpersonal feedback from the *therapist to make overt how the client's maladaptive relational patterns are affecting the therapist at that moment* and how they are being played out in their current interaction (Kiesler & Van Denberg, 1993). For example:

THERAPIST: Sometimes it feels like you are jumping from topic to topic, James, and I have trouble keeping up with you. Maybe this is just me, but I often feel like I'm getting lost or losing the point you are trying to make. Could this be something that just goes on between you and me sometimes, or have others told you that they feel confused or have trouble following you as well?

Similarly, object relations theorists and communication theorists provide *meta-communicative feedback* as an intervention to register the unspoken emotional quality of a relationship (Cashdan, 1988). For example:

THERAPIST: I could be completely wrong about this, Barb, but I keep having a feeling that I would like to check out with you. Although you've never actually said anything like this, I sometimes have the feeling that if I disagree with you, you'll be angry and leave our relationship. You know, if I see something differently and disagree with you, you'll walk out the door at the end of the session and I won't see you again. What do you think about this feeling I'm having? Is this just me or is there something to it?

CLIENT: (sheepishly) Well, maybe, my boyfriend would sure agree with that...

Self-Involving Statements.   In the counseling literature, *self-involving statements present another type of immediacy intervention, and they may be especially useful for new therapists. Whereas **self-disclosing statements** refer to the therapist's own past or personal life experiences, self-involving statements express the counselor's current reactions to what the client has just said or done (McCarthy, 1982; Tyron & Kane, 1993). For example, the therapist might say:*

THERAPIST: Right now, I'm feeling _____ as you're telling me about this.

Contrast the first, self-disclosing response with the second, self-involving comment:

THERAPIST: I have a temper too, sometimes.

VERSUS

THERAPIST: As I listen to you speak about this, I find myself feeling angry, too.

As we would expect on the basis of client response specificity, self-disclosing comments may be useful at times. Too often, however, self-disclosing comments take the client's focus away from her own experience and shift it onto the therapist when therapists reveal personal information. More productively, in contrast, self-involving statements keep the focus on the client and reveal information about what is happening in the relationship or how something the client has said or done is affecting the therapist. For example:

THERAPIST: I can feel my stomach tighten right now as you are reading that letter from your father.

Sharing personal reactions like this to what clients have just said or done conveys personal involvement with clients and acknowledges emotional resonance. This type of self-involving statement, like all of these process-oriented, immediacy interventions that bring the therapeutic dialogue into the here-and-now, go a long way toward strengthening the working alliance. We have also seen that they give clients feedback about the impact they are having on the therapist—and others. It can be a gift when therapists use process comments to provide interpersonal feedback, and therapists can find constructive, noncritical ways to help clients see themselves from others' eyes and learn about the impact they are having on others (such as regularly making others feel bored, intimidated, impatient, overwhelmed, confused, and so forth). *With these types of process comments, an effective therapist is willing to take the risk to say to clients, sensitively and diplomatically and always with the intention to safeguard the client's self-esteem, what others may be feeling and thinking but don't want to risk saying:*

THERAPIST: Courtney, are you aware that you're speaking to me in a loud, harsh voice right now?

Although this may be hard for new therapists to believe, it really is not a big problem if the client misunderstands their good intentions and feels put down or criticized! How can that possibly be?! If this misunderstanding occurs—which it will sometimes based on the client's schemas—the therapist just takes the next step and talks through this misunderstanding. That is, the therapist can make it overt and address it with the client—which allows them to sort it through in a reparative way that does not occur in most other relationships. For example:

THERAPIST: You just interrupted me when I began to speak, and I commented that I thought that happens several times every session. That's a very direct thing for me to say, and maybe you disagree or didn't like that I said that. Can we talk about how it was for you to hear me say that?

CLIENT: It hurt my feelings, of course. I didn't like being told I had a big mouth like that.

THERAPIST: Thank you for being so honest with me. Let's sort this through together—I think we just had a big misunderstanding. I have never thought that you have a "big mouth," and I didn't say that to you, but I often do feel interrupted. I asked about it because if other people in your life feel interrupted like I do, it might have something to do with the loneliness and difficulty making friends that you've been talking about. So I took the risk to broach that delicate topic with you.

CLIENT: Well, I still don't like that you said that, but maybe I do interrupt people too often.

THERAPIST: OK, it still doesn't feel good that I said that, but maybe this is teaching both of us something important about what's going wrong for you. Sure, this is a very sensitive issue, but I'm hoping we can keep talking about it constructively so we can work this out better with others in your life—like with your wife and daughter.

CLIENT: I don't know why I interrupt people so much. I think I do it without even realizing it. I guess that's how we grew up talking in my family—I never could finish a sentence...nobody could...

Early in their training, it's hard for most new therapists to have the interpersonal confidence to intervene in these personal and forthright ways. It is challenging to take the risk of talking with clients about how they are interacting with you and others, and to make overt and talk about what's going on between you, right now, that might be related to their problems with others. Supervisors, practicum instructors, clinical researchers, and just about everyone else all agree: these present-centered, immediacy interventions that bring a different level of engagement and intensity to the therapeutic relationship are the most anxiety-arousing, challenging interventions for new therapists to adopt (Hill, 2009; Safran et al., 2002). To use them successfully, therapists need to do two things. First, they must balance the challenge of the metacommunication with being supportive and protective of clients' self-esteem (Kiesler, 1988). Second, therapists need to check in with the client and talk about the immediacy intervention they have just made:

THERAPIST: How was it for you when I said that? I know people don't usually speak so directly about their relationship or what's going on between them.

As always, therapists' self-awareness is critical in safeguarding them from not using these interventions intrusively but allows therapists to utilize these interventions respectfully and as they relate to client concerns rather than to the therapist's personal issues. Let's explore this further.

Depending on the therapist's ability to communicate in a respectful manner, any of these process-oriented interventions that focus on the current, here-and-now interaction can be used ineffectively or effectively. For example, contrast this ineffectively blunt response here with the more empathic collaborative and effective response that follows:

THERAPIST: I'm not getting much out of this. Why don't we try to talk about something more important?

VERSUS

THERAPIST: I'm concerned that I'm missing you right now and not grasping what really matters to you here. Can you help me understand this better—why you are choosing to tell me about this and why this is meaningful for you? Or, if it's not so important, maybe there is something else that you might prefer to move on to that might be more productive for us? What do you suggest?

Without question, working in the moment with process comments like these is a challenging way to intervene for most new therapists. Routinely, it is going to take a year or two before student therapists feel comfortable saying what they see and broaching tentatively with clients what may be going on between them. Before going further, let's explore three situations when process comments will not work

First, therapists do not want to jump in with process comments in a spontaneous or cavalier manner. We want to be thoughtful about when and how we use these powerful interventions, and the first thing we want to consider is the possibility that therapists' observations or reactions reflect more about their own *countertransference* issues than they do about the client or their current interaction. As a rule of thumb, experienced therapists usually want to wait until they have seen an interaction occur two or three times. New therapists may want to reflect and discuss it with a supervisor before tentatively venturing this observation with the client. Differentiating the client's concerns from the therapist's own personal issues is one of the single most important issues in psychotherapy. Because this self-awareness takes on even more significance for new therapists, we will return to it and explore it further in the chapters ahead.

Second, process comments create problems for clients when therapists promise something they cannot provide. That is, the process comments and other immediacy interventions introduced here encourage open and direct communication, which is also an invitation to share together in a more significant, meaningful relationship. However, if the therapist then responds in impersonal, distancing, or judgmental ways that do not honor or follow through with the honesty and personal engagement invited, clients are set back by this mixed message from the therapist. This double-binding situation will be an especially problematic reenactment for clients who were repeatedly disappointed by caretakers who gave promises of interest and involvement but failed to follow through. For example:

THERAPIST: You've become so quiet. I'm wondering if there is something going on here—you know, between us, that we should talk about?

CLIENT: Well, since you're asking about it, I'm feeling kind of discouraged about therapy. I don't know where we are going each week. For a while I thought things were getting better, but I don't have the feeling that things are changing anymore.

THERAPIST: *(in a defensive tone of voice)* Well, I don't really know what you expect me to do with all that. Maybe it would help if you just tried to work a little harder in here.

Let's redo this same conversation but, this time, the therapist follows through nondefensively with his invitation for straight talk about their relationship and work together:

THERAPIST: You've become so quiet. I'm wondering if there is something going on here—you know, between us, that we should talk about?

CLIENT: Well, since you're asking about it, I'm feeling kind of discouraged about therapy. I don't know where we are going each week. For a while I thought things were getting better, but I don't have the feeling that things are changing anymore.

THERAPIST: *(in a nondefensive, welcoming tone of voice)* Thanks for the honest feedback. I like the part of you that is willing to speak up like that—it's a strength of yours. I've been feeling that we're not moving forward either. It seems to me that we start out on issues that seem important at first, but they don't go very far and just sort of fall away. I've been hoping that we would clarify more of a focus or direction that would keep going further, but we haven't been able to do that yet. What do you think we could change or do differently in here to try to make this work better?

CLIENT: What you just said sounds like the story of my life. I'm always starting things—school, relationships, playing the drums—but I don't stay with any of them. Nothing stays alive for me—things never get finished...

Third, therapists want to be cautious when working with clients from a culture they lack familiarity with. They will need to take the time to learn what is acceptable and unacceptable within that culture's norm and to this individual as a member of that culture, both by gathering knowledge about that culture and tentatively exploring the range of acceptance behaviors, values, and beliefs that may impact intervention with this particular client (see APA, 2003; Constantine, 2007; Pedersen et al., 2008; Sue & Sue, 2008, for guidelines on working with diverse populations).

To sum up, process comments generally will be the most powerful interventions therapists make. When used effectively, that is, offered tentatively and as an invitation for further dialogue, they greatly facilitate the working alliance, help to repair it when it has been ruptured, reveal important new issues that are central to the client's problems, and provide the best way to alter problematic interpersonal patterns that are being reenacted with the therapist. As we have emphasized, however, these process-oriented interventions that create immediacy are new and very different for most beginning therapists. Trainees do not need to try out process comments until they feel ready to try communicating in this forthright way and are ready to meet the client in the authentic manner that process comments invite.

Use Process Comments to Repair Ruptures in the Working Alliance

Carl Rogers gets the credit—he's the person in our field who legitimized concern about the quality of the therapist-client relationship. His contribution and influence cannot be overstated: As Kahn observed, he "changed the game" (1997). Rogers's core conditions of genuineness, positive regard, and accurate empathy in particular, are at the heart of a strong working alliance. However, these core conditions are not easily achieved with certain clients. In particular, when clients present with a hostile, demanding, competitive, dominating, or critical interpersonal style toward others (including the therapist!), it is challenging indeed to maintain Rogers's core conditions. For decades, researchers have been reporting that even highly experienced therapists do not do a very good job of remaining empathic, or even neutral, with these "negative" clients who are difficult to be with because they are angry, provocative, controlling, and so forth toward the therapist (Binder and Strupp, 1997; Hill et al., 2003). As we explore below, process comments provide therapists an effective way to address the problems that occur with these challenging clients and, more widely, to resolve the misunderstandings or "ruptures" that occur with almost every client at times—and that too often lead to treatment failure. In the working alliance literature, this is the all-important work of "rupture and repair" (Safran & Muran, 2000).

To facilitate the working alliance, the therapist's role is to *sustain* an empathic, respectful, and interested attitude toward the client (Hovarth & Bedi, 2002; Strupp, 1995). Usually, however, this alliance-fostering therapeutic stance is not provided consistently as recurrent "ruptures" in the working alliance almost inevitably occur during the course of treatment (Safran & Muran, 2000; 1995). Alliance ruptures occur for three different reasons:

- the covert or overt hostility described above (referred to as "client negativity");
- reenactments or interpersonal scenarios that clients keep recreating with others that ensnare or embroil the therapist; and
- the simple human misunderstandings that occur in every meaningful relationship.

Illustrating "client negativity" first, irritability is a defining diagnostic feature for most clients who are depressed and for many who are anxious. Similarly, demandingness or provocativeness will often be expressed by clients with histrionic features, and, perhaps most challenging, with clients who have borderline and narcissistic disorders who can be overtly contemptuous of the therapist. Clients with obsessive-compulsive symptoms can be controlling and critical toward others in their lives—including the therapist, and so forth. Thus, even though most new therapists haven't really anticipated receiving this "negativity" as an everyday part of the career choice they have just selected, many clients at times are going to be critical, controlling, accusatory, dominating, provocative, demanding, intrusive, competitive, and so forth with the therapist—just as they are with others in their lives. For example,

CLIENT: *(irritated)* No, that's not right either. Do I have to explain this to you again?

OR

CLIENT: *(provocatively)* So, you're just a beginner. I see. And, uh, how many clients have they let you practice with so far?

Second, regarding maladaptive relational patterns, many clients are going to "pull" therapists into hostile, critical, competitive, controlling, rescuing, or other types of conflictual exchanges that reenact problematic scenarios that they grew up with and now are repeating with others in their current relationships. For example:

CLIENT: *(patronizing)* You know, my previous therapist was *so* smart and perceptive. He really understood me and so cared about me. Do you think you can do the same? Have you had enough training to be able to help me like he did?

Disturbingly, researchers consistently find that most experienced therapists do not deal in a therapeutic manner with these common and expectable types of client "negativity." Instead, they respond "in kind" toward the client with their own counter-criticism, judgmentalism, hostility, and so forth, or they withdraw (Kiesler & Watkins, 1989; Najavits & Strupp, 1994). For example, in a well-known series of studies known as Vanderbilt I and II, clients' problematic relational patterns routinely elicited counter-hostility and punitive control from seasoned therapists, and when these "complementary" responses occurred, it led to poor therapeutic outcomes. That is, *clients with angry, distrustful, and rigid interpersonal styles tended to evoke counter-therapeutic hostility and control—even in a carefully selected sample of highly trained and experienced therapists* (Binder & Strupp, 1997b; Henry, Schacht & Strupp, 1990; Strupp and Hadley, 1979). Process comments give us as a more effective way to respond to client negativity. For example:

THERAPIST: You've just said something about my shirt Suzie. You've made comments before about my haircut and about other aspects of how I look or how I dress. Do you critique others in this way? If you do, how do they typically respond?

Third, even with the many other clients who are "nice," simple misunderstandings and problems with treatment and/or the therapist occur at some point in most therapeutic relationships. And here is the big issue: unfortunately, *clients often do not bring up or voice their concerns to their therapists, and therapists as well often avoid or do not ask about them* (Hill et al., 1993; Rennie, 1992). Clearly, many therapists are uncomfortable approaching interpersonal conflict directly and are insufficiently trained in restoring disrupted relationships and resolving personal impasses between them (Watson & Greenberg, 2000). Unfortunately, unresolved misunderstandings between client and therapist undermine the working alliance and lead to poor treatment outcomes (Elliot et al., 1994; Johnson et al., 1995).

So, what's the solution to client negativity and everyday misunderstandings? Process comments provide an effective way for therapists to repair or restore ruptures in the working alliance. That is, rather than avoiding the interpersonal conflict as if it hadn't occurred, or responding in kind with counter criticism or anger, the most effective intervention is to address the problem by talking directly with the client about it as it occurring (Rhodes et al., 1994; Safran et al., 2002). That is, by using process comments and the meta-communicative feedback presented above, therapists can neutrally observe or tentatively wonder aloud about potential problems or misunderstandings that may be occurring between them. Let's consider some examples:

THERAPIST: I'm wondering if you might have experienced my last comment as critical or judgmental? Can we talk about that possibility?

OR

THERAPIST: Right now, it feels to me as though we're almost arguing with each other. I'm wondering if both of us are feeling misunderstood by the other? Let's work this out. What do you see going on between us?

OR

THERAPIST: I think I was feeling impatient there and pressing you to make a decision before you were ready. What was going on for you right then?

OR

THERAPIST: When you spoke over me just then, in that loud voice, I felt kind of squashed. Maybe something wasn't feeling very good to you, either. Let's have an honest conversation about what's going on for both of us here and see if we can sort this through.

Impasses result, and treatment cannot progress, unless therapists and clients can talk openly about problems in the relationship and sort through what went wrong between them (Petersen et al., 1998). Working out problems, disagreements, and the inevitable misunderstandings that arise between them provides an invaluable social laboratory where clients can learn how to address and resolve conflict with others in their lives. However, most trainees do not receive enough preparation or training for responding to conflict with their clients. Further, on a personal level, most new and experienced therapists find it far easier to be empathic or supportive with clients than to approach or address interpersonal conflict and try to work through problems directly. As with their clients, most therapists did not learn to do this in their families of origin. Oftentimes there was no way to restore the ruptures that inevitably occurred in the parent-child relationship as well. Many families also held strong but unspoken rules against discussing conflict openly and honestly—and for some it was nothing less than simply taboo. Working within every theoretical orientation, however, treatment success depends on the therapist's ability to restore the working alliance when it has been disrupted, and process comments give therapists the tools they need to do this successfully.

PERFORMANCE ANXIETIES MAKE IT HARDER TO ESTABLISH A WORKING ALLIANCE

Let's let therapists be human, too. They cannot always hear or fully assimilate what their clients have just told them or be accurately empathic. Understandably, all therapists will miss the feeling and meaning in the client's experience at times, but this takes on greater salience for therapists-in-training. Thus, this section returns to performance anxieties and new therapists' worries about "mistakes," which make therapists too self-conscious to be able to listen well and be present with their clients.

As we saw in Chapter 1, new therapists will not be as effective with clients when they are burdened by their own excessive performance demands—which can be amplified by reading about all the ideas and ways of responding presented in this chapter. We have already seen that beginning therapists often are trying too hard (to be helpful, to win approval from a supervisor, to prove their own adequacy to themselves, to be liked by the client, to avoid making a mistake and hurting the client or being criticized, and so forth). When therapists are trying too hard in one of these ways, it is almost impossible to de-center, enter the client's subjective world-view, and be emotionally present.

Too often, as well, novice therapists are anxiously preoccupied about where the interview should be going next, wondering how best to phrase what they are going to say next, trying to formulate alternatives or suggestions for the client, or worrying about what their supervisors would expect them to do at this point in the interview. Such self-critical monitoring can immobilize student therapists, block their own creativity, and keep them from enjoying this rewarding work. Moreover, it prevents therapists from stopping their own inner chatter, listening receptively to their clients, recognizing patterns and making connections, and understanding as fully as they could. Thus, new therapists are encouraged to slow down their own self-monitoring process and focus more on the moment—on the client rather than themselves and their own anxiety, on their current interaction with the client, and on grasping the meaning that this particular experience seems to hold for this client right now. That's how we begin to establish this all-important alliance.

Finally, they feel anxious about their helping abilities. New therapists often experience an unwanted, internal press to *do* something to make the client change. They frequently believe they are supposed to make the client think, feel, or act differently within the initial therapy session. As a result, they often change the topic abruptly (at least from the client's point of view) and pull clients away from their own concerns to what the therapist thinks is most important—losing any sense of collaboration or shared exploration. The most appropriate way for therapists to manage their realistic concerns about their performance is to discuss them openly with a trusted supervisor and to recognize that they are common and expectable. Therapists do not need to labor under the burden of always knowing the right thing to do or what is going on with the client. What is asked, instead, is only to foster a collaborative dialogue with the client in order to explore and learn together what may be occurring.

CARE AND UNDERSTANDING AS PRECONDITIONS OF CHANGE

The therapist does need to help the client change, of course, but change is most likely to occur if the client first experiences the therapist as someone who "sees" and understands him, and treats him with respect. As previously stated, in the initial sessions, the therapist's primary goal is to establish a working alliance by repeatedly providing clients with empathic understanding and to actively extend to the client and communicate his genuine interest in the client's safety and well-being. This understanding of the client's experience is done in a way that also communicates compassion and nonjudgmental concern. The ability to articulate the client's experience in an accurate and caring way is illustrated elegantly in a classic article, "Ghosts in the Nursery," by Selma Fraiberg and her colleagues (Fraiberg, Adelson & Shapiro, 1975). The therapist in this case study is working with a depressed young mother whom social services has judged to be at risk for physically abusing her infant daughter. Early in treatment, the therapist is disconcerted as she observes the mother holding her crying baby in her arms for five minutes without making any attempt to try to soothe her. The mother does not murmur comforting things in the baby's ear or rock her; she just looks away absently from the crying baby. Trying to formulate a case conceptualization and treatment plan, the therapist asks herself, "Why can't this mother hear her baby's cries?"

As the young mother's own neglectful and abusive history began to come out in treatment, the therapist realized that no one had ever heard or responded to the mother's own profound cries as a child. The therapist hypothesized that the mother "had closed the door on the weeping child within herself as surely as she had closed the door upon her own crying baby" (p. 392). This conceptual understanding led the therapist to a basic treatment plan: When this mother's own cries are heard, she will hear her child's cries.

The therapist set about trying to hear and articulate compassionately the mother's own childhood experience. When the mother was five years old, her own mother had died; when she was 11, her custodial aunt "went away." Responding to these profound losses and the mother's resultant feeling that "nobody wanted me," the therapist listened and put into words the feelings of the mother as a child:

> How hard this must have been.... This must have hurt deeply.... Of course you needed your mother. There was no one to turn to.... Yes. Sometimes grown-ups don't understand what all this means to a child. You must have needed to cry.... There was no one to hear you. (p. 396)

At different well-timed points in treatment, the therapist accurately captured the mother's experience in a way that gave her permission to feel her feelings. As a result, the mother's grief and anguish for herself as a cast-off and abused child began to emerge. The mother sobbed; the affirming therapist understood and comforted. In just a few more sessions, something remarkable happened. For the first time, when the baby cried, the mother gathered the baby in her arms, held her close, and crooned in her ear. The therapist's hypothesis had been correct: When the mother's own cries were heard, she could hear her baby's cries. The risk for abuse abated as this beginning attachment flourished.

This poignant case study illustrates how the therapist can use empathic understanding to resolve the client's symptoms and, in this case, stop the intergenerational transmission of abuse. A corrective emotional experience occurred when the therapist responded to the mother's pain with compassion and understanding—something she desperately needed as a child but did not receive. This case study illustrates how powerful it is when therapists can articulate their understanding of the client's experience, and do so in a way that also communicates their compassion for the client.

Using role models such as this, therapists-in-training are encouraged to explore their own ways of communicating that they are touched by the client's pain and are concerned about the client's well-being. Clients' inability to care about themselves is central to many of their problems, and most clients cannot care about themselves until they feel someone's caring for them (Gilligan, 1982). Therapists provide this care when they recognize what is important to the client, express their genuine concern about the client's distress, and communicate that the client is someone of worth and will be treated with dignity in this relationship. These are the therapist's goals in the initial stage of therapy.

CLOSING

The interpersonal process approach is integrative and draws on concepts and techniques from differing theoretical traditions. The basic concept in this chapter—establishing a collaborative alliance through empathic understanding, is grounded in Rogers's client-centered approach and, especially, accurate empathy.

The interpersonal process approach tries to resolve problems in a way that leaves clients with a greater sense of their own self-efficacy. This independence-fostering approach to therapy can best be achieved through a collaborative partnership. The client needs to be an active participant throughout each phase of treatment—not a good patient who waits to be cured or told what to do by the doctor. This process dimension of how the therapist and client work together is more important than the content of what they discuss or the theoretical orientation of the therapist.

In this chapter we also have explored the profound therapeutic impact of listening with presence and working collaboratively to discern what is most important to the client. Offering the deep understanding described here is a gift to clients—one of the most important interventions that therapists of every theoretical orientation can offer. Perhaps because they are so simple, these basic human responses are too easily overlooked. They are the foundation of every helping relationship, however, and the basis for establishing a working alliance.

SUGGESTED READINGS

1. Beginning therapists need to observe role models demonstrating the concepts they are reading about. Readers are strongly encouraged to watch two training DVDs that illustrate powerfully the concepts presented here. Readers can observe M. Hanna Levenson providing empathic understanding to a physically abused client, and M. Jeremy Safran use immediacy interventions with a bereaved client. These skillful therapists will help readers understand a relational approach to treatment in this DVD series produced by the American Psychological Association.

2. Two clinical studies explored the effects from therapists' use of immediacy in two complete cases of time-limited treatment—one with a high-functioning counseling center client and one with a client presenting with a significant

abuse history. These two cases offer in-depth examples of how therapists use immediacy in actual settings (Hill et al., 2008; Kasper, Hill & Kivlighan, 2008).

3. To help readers learn more about how to establish and maintain a strong working alliance with their clients, items from the *The Working Alliance Inventory* have been included in Chapter 2 of the Student Workbook that accompanies this text.

4. A. For helpful illustrations of how process comments facilitate group interaction, see the classic text by Irvin Yalom & Molyn Leszcz (2005), *The Theory and Practice of Group Psychotherapy*.
    B. Practical guidelines for identifying themes, using self-involving statements, and creating immediacy are provided in Chapter 11 of Egan, (2007) *The Skilled Helper*. Readers can learn much more about immediacy interventions in Chapter 12 of Hill (2009), *Helping Skills: Facilitating Exploration, Insight, and Action*, 3e.

5. Readers may wish to read the chapter "Therapist Variables" in *Handbook of Psychological Change: Psychotherapy Processes and Practices for the 21st Century* (Teyber & McClure, 2000). This chapter summarizes the research literature on the working alliance and characteristics of effective therapists across differing theoretical orientations.

6. Illuminating ideas about the effects of multigenerational family relations on individual personality are found in two classic articles on triangular coalitions: Haley's "Toward a Theory of Pathological Systems" and especially M. Bowen's (1966) "The Use of Family Theory in Clinical Practice." Both of these pioneering articles will help therapists apply family systems concepts to the practice of individual psychotherapy.

Honoring the Client's Resistance

Joan was nervous about seeing her first client, but the initial session seemed to have gone well. The client talked at length about his concerns and, to her relief, she found it was easy to talk with him. The client shared some difficult feelings and Joan felt that she understood what was going on for him. She was thinking they had begun a good therapeutic alliance; the client seemed so friendly and appreciative when he left. One week later, however, Joan received a brief telephone message from the client saying only that he was "unable to continue therapy at this time." Confused and dismayed, Joan sat alone in her office wondering what had gone wrong.

CONCEPTUAL OVERVIEW

Just when the therapist feels that something important is getting started, some clients put their feet on the brakes. The client cancels the second appointment, shows up 25 minutes late, or asks to reschedule for Sunday at 7:00 A.M. This *resistance* is puzzling and frustrating for the novice therapist: "Why didn't that client return? We had a great first session!" Although most clients will not be resistant in this particular way, other forms of **resistance** will occur regularly throughout therapy—even for motivated, responsible clients who are trying hard to change. Although most new therapists do not anticipate this as they begin their training, working with resistance is a normal or expectable component of the treatment process. Thus, the purpose of this chapter is to help new therapists learn how to recognize resistance and respond effectively to this frequently observed treatment issue.

All clients have both positive and negative feelings about entering therapy, although the positive feelings are more apparent at first for most clients. Clients seek therapy in order to gain relief from their distress. However, we must look further into the complexity of the client's feeling. Often, as clients seek help and genuinely try to change, they simultaneously resist or work against the very change they are trying to attain. Since Freud's early writings in the 1890s (Breuer & Freud, 1893, reprinted in 1955) and continuing with contemporary integrative treatments such as Motivational Interviewing (Miller & Rose, 2009), clinicians have been trying to understand and address this paradoxical, shadow side of the change process. For many clients, *shame* is associated with having an emotional problem they cannot solve on their own. In some cultures, revealing problems to people outside one's family meets with strong disapproval or is seen as *disloyal*. Other clients feel *guilt* about asking for help or doing something to meet their own needs. Still other clients, acting out of their cognitive schemas, have *anxiety* evoked by the expectation that the therapist, like significant others in their lives, will respond in familiar but unwanted/hurtful ways. Thus, if clients have a problem or need to ask for help, then shame, guilt, anxiety, and other difficult or threatening feelings are often evoked, especially the deeply held belief that the therapist, too, will ultimately respond in the same problematic ways that others have in the past. In addition to difficulties in entering treatment, we will see that *ambivalent* feelings are often triggered later in treatment when clients feel better, improve in treatment, and make successful changes in their lives (Miller & Rollnick, 2002). Thus therapists are trying to identify and work with the specific issues that make it difficult for this particular client to enter treatment in the beginning, and then to sustain behavior change later in treatment. In many instances, when the resistance is a coping strategy that once was adaptive for a client (it kept them from being abandoned, ridiculed, or in some way hurt, but is no longer needed), if the therapist can first find a way to _honor_ this (acknowledge its historic adaptive value—"it makes sense that asking for help was not safe because you were ridiculed when you showed vulnerability") before assisting the client see that the inflexible use of this approach is no longer functional (it is impeding the development of authentic, reciprocally nurturing relationships). Following the principle of client response specificity, specific concerns must be clarified for each individual client, but the following are common themes:

- If I let myself trust or depend on the therapist, he might criticize or judge me, try to control me, start to depend on me or need me, or in some unwanted way take advantage of me—just as others have done when I needed them.
- I cannot ask for help because I must be independent and in control all the time.
- I don't deserve to be helped by anyone, I don't really matter very much.
- I cannot need anything from others—if I'm not perfect I will bring shame to my family.
- Asking for help is admitting that there really is a problem, and that proves there really is something wrong with me.
- I am afraid of what I will see or what a perceptive therapist will learn about me if I stop and look inside myself.
- If I cannot handle this by myself, it means that I really am weak, just as they always said.
- I'm a therapist. I can't have problems—I'm supposed to have the answers.
- I'll start crying and won't be able to stop—it will be humiliating.
- If I ask for help, I'll lose my independence and become weak and dependent, just like the other needy people in my family.

In these and a thousand other ways, difficult or resistant feelings will be activated for many clients by the simple act of calling the therapist and coming in for the initial session (Young et al., 2001). Although reasons will vary greatly for each client, *clients' resistance and defenses are driven most frequently by shame*—as seen in many of the examples above. To help us gain some understanding for this resistance, and therein find a little more compassion for what it's about, pause now for a moment and think about your own life during a time of real crisis. Ask yourself, "What has it been like for me to have a personal or emotional problem and need to ask someone for help?" For many, sadly, it's not easy; for some, it's simply unacceptable.

This profound topic leads many places but, in particular, it takes us to the heart of the attachment story. What do we really mean when we say that a child is securely attached? It's this simple: *The securely attached child is secure in the expectation that, when they have a problem, their caregiver will consistently register or "hear" their distress and reliably orient to try and help them solve it.* In other words, they are not alone with their problems—they have a trustworthy partner or ally who genuinely wants to know when something is wrong and who dependably wants to help them with whatever problem they are having (Bowlby, 1988; Cassidy and Shaver, 2008). With resistance, in contrast, we are usually working with insecure attachment histories. They lacked both a safe haven and secure base—someone to whom they could come in distress and from whom they could find support to venture, explore, and discover their authentic self. That is, based on their real-life experience, many clients have learned realistically to hold the expectation that the therapist and others will not really be very interested in, or be capable of, responding to them when they are distressed or in need; or encourage their growth or independence when that conflicted with the other person's need. This belief, originally learned in repeated real-life experiences with primary caregivers, and then confirmed in subsequent important relationships, is going to come in play with the therapist at different points—and especially as treatment begins.

It is easy to respond to the approach side of clients' feelings—the fear or pain that motivates them to seek help and enter treatment. If you just scratch the surface, it is usually plain to see. One part of the client wants you to respond to it but, commonly, another part doesn't. This shadow side—the contradictory or ambivalent feelings that arise from the unwanted responses the client has received from others in the past for having a problem and needing help—acts as a countervailing force. If unaddressed, this resistance will keep some motivated and highly workable clients from being able to successfully engage in treatment, as in our opening vignette.

RELUCTANCE TO ADDRESS RESISTANCE

Imagine your client has missed, come very late for, or twice rescheduled the second appointment. Perhaps this has no significant psychological meaning at all. Cars do break down; traffic jams occur; children get sick; employees get called for work at the last minute. However, if this behavior reflects the client's ambivalence about some aspect of being in treatment, and not just reality-based constraints, the therapist needs to find an effective way to address this or the client is far more likely to terminate prematurely. Initially, therapists usually do not know whether the client's behavior is reality-based, psychologically motivated, or both. As we will see, however, by finding nonthreatening ways (that is, nonblaming) to explore this behavior with the client, therapists can greatly increase the chances of the client's continued participation. In these explorations, therapists' curious, nonjudgmental manner is as important as what they say. Otherwise, the client may misunderstand the therapist's observations as blame or criticism, which never help and only serve to exacerbate client resistance.

To be more specific, let's follow a two-step sequence for working with resistance. First, the therapist can enter supportively into the client's frame of reference, *affirm or validate the reality-based constraints they perceive*, and respond flexibly by doing whatever is reasonable to accommodate or help resolve the problem:

CLIENT: My child gets out of school at 2:30 and...

THERAPIST: Well, I can't meet at a different time on Wednesdays but I could meet on Thursdays at 4:30. Would that give you the shuttling time you need?

Second, only after taking seriously client's reality-based concerns, as they see them, can therapists begin to inquire about other conflicted feelings or psychological meanings that entering treatment may hold. The therapist can begin this joint exploration by wondering aloud, *in a tentative manner*, whether entering treatment may be evoking other more psychological concerns as well.

THERAPIST: I'm wondering how it is for you to call and come in to see me. What's it like for you to have this problem and come here to see if I can help?

OR

THERAPIST: What's the biggest fear or concern that coming here and talking to me brings up for you?

OR

THERAPIST: Help me understand how others in your life have responded when you needed help in the past? What have they tended to say and do?

If therapists approach resistance without following this two-step sequence, most clients will feel misunderstood or blamed, impeding their ability to engage in treatment (for example, a blaming therapist: "Why didn't you show up for our last session?" or, at the other extreme, an avoidant therapist: "Oh, you're late again, but I'm so glad you still came. Great, let's get to work, where should we begin?"). Let look now at both the therapist's and the client's shared reluctance to talk about potential resistance—and how this leads to "losing clients."

THE THERAPIST'S RELUCTANCE TO WORK WITH RESISTANCE

Resistance is not a welcome concept to most new therapists. To a greater or lesser degree, however, every client will be ambivalent, defensive, or resistant at times. One way to say it is that resisting clients are simultaneously struggling with a conflict between their genuine wish to change and need to maintain the status quo (Gabbard, 2007). This push-pull may occur at the beginning of treatment and it can continue to wax and wane throughout treatment. In their informative work on stages of change, Prochaska et al., (1992, 2006) report that most clients who terminate prematurely can be described as "precontemplators." Precontemplators do not "own" their problems, and usually enter treatment because of pressure from others. That is, they feel coerced to change by a spouse who threatens to leave, an employer threatening to dismiss them, judges threatening to punish them, or a parent threatening to withdraw support. Thus, resistance becomes a bigger issue when significant others press clients to seek help, and when courts or others mandate clients to attend treatment. Although these clients often can be helped by working collaboratively with them to find some aspect of the problem they genuinely wish to change, they are generally far more challenging for new therapists. With every client seeking treatment, however, therapists want to approach or address potential signs of client resistance—and not avoid or deny them. Researchers find that clients progress in treatment and have better outcomes when therapists recognize resistance and respond to it in ways that restore clients' active collaboration in the treatment process (see Beutler et al., 2002b; Orlinsky et al., 2006). Let's examine three reasons why many beginning—and experienced—therapists often find it difficult to approach and explore the resistance, ambivalence, and defense that is part of the treatment process.

✱ First, many new therapists—especially if they have never been in therapy themselves—are not so aware of the multiple meanings and conflicted feelings associated with the decision to enter treatment. These therapists are often surprised to find that, in the initial sessions, clients actually resist the help they are overtly seeking. Commonly, therapists are reluctant to explore and work with resistance because they fear it has to be addressed in a blaming or critical way that will anger clients or make them feel guilty. Or, they believe resistance has to be addressed in a manner that puts therapists in a superior position and denies the validity of the client's own experience—which we never want to do. This occurs, for example, when a therapist ineffectively says to a working, single mother of three:

THERAPIST: Hmm, I notice that you are five minutes late for our appointment today.

Ineffectively, this therapist is ignoring the reality-based constraints of this client's life-context. Resistance does not have to be associated with this unwanted type of hierarchical therapist-client relationship. Therapists can work with resistance and still be sensitive to the social context or reality-based constraints of the client's life. For example, clients who use public transportation will not be able to arrive on time consistently. Clients from cultures with varying views of time may not think that being 5 minutes late for a 50 minute appointment is of any significance. For some, it may even be impolite to come on time; arriving late allows the host extra time to prepare for the visit. Therapists can educate clients about the bounds and procedures of therapy and, as we will see, respond to resistance in collaborative, nonblaming ways that strengthen the client's commitment to therapy, enhance the therapeutic alliance, and empower rather than invalidate the client.

A second reason why therapists are reluctant to address their clients' resistance is more personal. Most new therapists have strong needs for their clients to like them. Understandably, student therapists want very much for their new clients to find them helpful and to keep coming to counseling. If the client does not show up or comes late, new therapists, in particular, often struggle with feelings of failure. If this becomes a pattern that occurs with other clients as well, new therapists often become painfully insecure about their ability to help, place even more performance demands on themselves, and become overly invested in pleasing or performing adequately for the next client. This is not a good spot for new therapists to find themselves in, indeed.

Third, therapists often do not inquire about signs of potential resistance in order to ward off unwanted criticism. Not wanting to hear what they may be doing wrong, it becomes especially hard for new therapists to make process comments and invite the (seemingly) bad news:

THERAPIST: What's it been like for you to talk with me today?"

OR

THERAPIST: Is there anything about our work together that doesn't feel quite right to you? If so, I would welcome hearing about it. You know, so we could make any changes we needed to make this work better for you.

OR

THERAPIST: What felt good about talking together today, and what didn't?

Like their clients, most therapists are not eager to approach issues that arouse their own anxiety—such as bringing up potential conflict with others and talking it through forthrightly. Thus, new therapists may be hesitant to invite clients to express any negative reactions they may be having toward the therapist's response, the concerns evoked for them by having to ask for help, or their ambivalent feelings about needing to be in treatment. Sure, it is difficult for most people to invite critical feedback or approach interpersonal conflict, but it is so very helpful when the therapist can be *nondefensive* enough to do so—many clients have not found this in any other relationship.

Exploring potential resistance or ruptures becomes far more important if the client makes an "embedded message" about the therapist or some aspect of the therapeutic relationship and alludes to some conflict with the therapist or difficulty about being in treatment:

CLIENT: Where I grew up, we don't really talk about problems with others outside the family.

THERAPIST: Thank you for telling me that. So, that must make it harder to come here and talk to someone like me. Let's talk some more about this and see what we can figure out together—I sure don't want this to keep us from being able to work together.

OR

CLIENT: My family really values self-reliance. You're not supposed to have any problems—ever. And if you do, you just handle them on your own. You'd never tell anyone about them or see a therapist.

THERAPIST: OK, so you're telling me that it's being hard to come here and do this—it violates a family value. I appreciate your willingness to risk doing it differently with me. How has it been so far for you to tell me about this problem when your family has always said that's verboten?

CLIENT: Kind of mixed. I like talking to you because you seem to understand me, but I guess I feel kind of guilty, too.

THERAPIST: I appreciate both sides of your feelings. I'm glad that you feel understood by me, and think that will help you. But I also appreciate that you want to honor your family and not be disloyal to them. I'm wondering if you can decide to do some things differently than your family does, you know, that work better for you, but still find other ways to continue to honor and respect them.

Continuing further, it becomes even more important for the therapist to address potential problems or misunderstandings in their relationship when the client implies there is some problem or personal concern with the therapist. For example:

CLIENT: So, how old are you anyway—have you been practicing long?

For many new therapists, *learning to address interpersonal conflict in this straightforward, nondefensive way is the single biggest challenge in their clinical training.* As with "client negativity" in the previous chapter, most new therapists did not expect that being forthright and addressing interpersonal conflict was part of the job description when they signed on for this career! The paradox is, *if therapists allow their own anxiety to keep them from asking clients about such potential signs of resistance, ambivalence, or rupture, their clients will be far more likely to act on these concerns and drop out of treatment prematurely.* That is, new therapists want to take a breath, gather themselves to respond nondefensively, and take the risk to meet this concern head-on. They do this effectively by simply inviting the client to discuss this concern more fully:

THERAPIST: I'm very happy to respond—but I'm wondering if you are asking because you have concerns about whether I have enough experience to help you or if there is something in particular that happened between us that brought that question on?

Thus, an effective approach is to make a process comment that suggests the possibility that something about entering treatment or talking to the therapist may be difficult for the client, and invites the therapist and client to explore this possibility together. (If the client is much older and is just focused on age, being forthright is often what is needed and responding to the question directly is appropriate. Often, however, the question is about competence and/or whether a good working relationship between the therapist and client can be established.) Before long, new therapists will find that they can do this in an affirming way that does not make the client feel accused or blamed, and that their nondefensiveness and willingness to engage so honestly enhances the working alliance.

THERAPIST: You seem discouraged today, and you've had some trouble getting here. I'm wondering if something isn't going right here. Any ideas?

CLIENT: I'm not sure what you mean.

THERAPIST: I'm wondering if something about our relationship—or being in treatment—isn't working very well right now. We seem to be missing each other in some way. Any thoughts about what could be going on between us, or what I could do differently to make this work better?

CLIENT: Well, maybe you could talk a little more or give me some more feedback. You're pretty quiet, you know, and I don't really know what you're thinking most of the time.

THERAPIST: I'm glad you're telling me this—thanks for your honesty. Sure, I'll be happy to start sharing more of my thoughts with you—that'll be easy to change. I'm wondering what's it like for you when you don't know what I'm thinking?

CLIENT: Well, maybe when you're quiet it's like your kind of judging me or something.

THERAPIST: "Judging you." That sure wouldn't feel very good. Let me clarify that no part of me has felt judgmental toward anything you've said, and if I ever was, it would certainly be my limitation or problem. But maybe you're also telling me that others have judged you too much in your life?

CLIENT: Oh yeah, my husband is *so* critical. And I guess my mother disapproved of just about everything I ever did...

In this way, taking the risk to explore the client's resistance will often "uncover" or reveal key concerns (that is, "judging") that (1) impede the client's participation in treatment, and (2) are central to his/her problems and clarify the treatment focus. Much more on this later.

THE CLIENT'S RELUCTANCE TO WORK WITH RESISTANCE

Unfortunately, the client usually shares the therapist's reluctance to address resistance. Clients often are unaware of their resistance and externalize it to others or outside events. For example, the client may say,

CLIENT: Yeah, I'm late for our sessions, but the traffic is always so bad.

Although there usually is some truth to this statement, if this is a consistent behavior pattern, it begs the need for further exploration. To acknowledge both the reality-based constraints that are possibly contributing, and the more personal or psychologically based resistance that is often in play as well, the therapist can ask whether leaving earlier for the appointment is possible or if an alternative session time needs to be negotiated. *In tandem*, the therapist also begins to explore any concerns or other thoughts the client might be having about treatment:

THERAPIST: Can you recall what you might have been thinking about our session last week as you walked out to your car or as you drove home afterwards?

OR

THERAPIST: I'm wondering if you had any thoughts about me, or concerns about our work together, during the week?

OR

THERAPIST: I'm wondering what people in your family, or friends at church, might think about you coming to counseling and talking about personal problems?

There is no criticism or blame here. And, even if the client is not ready to recognize or talk about whatever psychologically based resistance may be operating, the therapist has invited the client to consider the possibility of other psychological factors and laid the groundwork for further discussions. This is part of educating the client about the treatment process and changing the social rules that define their interaction. That is, the therapist is making it "normal" to talk about their relationship and explore what goes on between them—setting new expectations that are different from what goes on in most other relationships in the client's life. To illustrate,

THERAPIST: I know other people don't usually talk this directly, but one of the best ways I can help you is if we can talk together about the feelings and concerns you have about our relationship and the way we're working together.

In contrast, when clients see themselves resisting, they often feel frustrated and confused by their own contradictory behavior. For example, clients often exclaim with dismay:

• Why do I go to all the trouble and expense of coming here to see you, then I can't think of anything to say as soon as I walk in the door?
• Why would I forget our next appointment after we had such a great session last week? It doesn't make sense!
• Why do I keep asking you for advice, and then say, "Yes, but..." to whatever you suggest? What's wrong with me?

If it becomes clear to clients that they are resisting in some way and sabotaging their own efforts, they typically evaluate themselves harshly—as being bad or failing in some way—and assume that the therapist shares this critical judgment. Thus, when the therapist begins to inquire about resistance, most clients want to avoid the topic because they are afraid the therapist is going to *blame* them—for not trying hard enough, failing as a client, being unmotivated, and so on. Because it is never helpful for the client to feel blamed or criticized, the therapist tries to make this self-critical attitude overt and to help clients reframe their critical attitude toward their own resistance. How do we do this? First, *the therapist wants to help clients learn how to honor their resistance as an outdated coping strategy that originally served a self-preservative and adaptive function for them*—and actually was the best possible response they had available at an earlier time in their development. Second, the therapist helps clients realize that they no longer need this coping strategy in most current relationships, such as the present relationship with the therapist. Instead, they can work with the therapist to learn more flexible ways of responding, first with the therapist and then with others in their everyday lives. Reframing the client's resistance with this contextual or developmental understanding helps in many ways, but especially by inviting self-acceptance and understanding rather than self-criticism and shame. This is a complex and far-reaching intervention; we need to examine it more closely. Addressing the resistance or coping style with compassion and providing understanding for why it developed is critical—it provides clients with an opportunity to have compassion for themselves and thereby begin the process of change and growth.

Oftentimes, clients' resistance, and even the symptoms that bring them in for treatment, once served as a survival mechanism that was necessary, adaptive, and even creative. For example, anxiety signals danger—perhaps not a current threat but a past or developmental problem that really did exist. Depressive symptoms may reflect a client's attempt to cope with painfully unfulfilled longing for protection or love from a caregiver, or a way to cope with the exploitation or derision that she experienced and came to expect from others (for example, shut down, avoid and stop feeling). In other words, formative experiences with caretakers and current interactions with significant others have given clients very good reasons for not wanting to ask for help, share a vulnerable feeling, or risk trusting someone. If the therapist and client explore patterns in how significant others are responding now, and have responded in the past, the client's resistance will make sense and become understandable—it's not "irrational"—it makes sense developmentally.

To illustrate, suppose a client is having trouble entering therapy. The therapist is encouraged to use *immediacy interventions* and work in the moment to highlight maladaptive relational patterns:

- If you and I begin to work together on your problems, what could go wrong between us?
- What could I do that might hurt you, or make things worse, if you take the risk to let me try to help you?
- How could our religious and ethnic differences get in the way and be a problem for you and me—for our work together—here in therapy?
- As we talk about this, I'm wondering how others have responded in the past when you asked for help or needed someone? What did they do that helped, and what hurt?

Routinely, questions like these reveal the most meaningful issues in clients' lives. In particular, they highlight what the client is afraid of or doesn't want, yet expects, because that is how significant others have indeed responded in the past. In other words, these immediacy questions serve an *assessment function* and help shape the treatment focus. That is, they clarify the key issues in the client's life—making clear to therapist and client alike what's really wrong.

Of great value, such process-oriented questions help the therapist and the client identify the aversive consequences that followed when the client asked for help in the past. Routinely, clients will have very specific answers to these questions and describe in detail the interpersonal scenarios that typically transpired in the past. When they asked for help in their family, for example, they were ignored and felt powerless, were made fun of and felt ashamed, or were told that they were selfish or too demanding. They may have been invalidated and told that they didn't really need or want what they asked for. They may have learned that anything they received brought unwanted obligations as "strings were attached," or that anything they needed or asked for seemed to burden their caregiver—inducing guilt. In this way, therapists and clients together can clarify together the very good reasons why clients originally learned to respond in these ways that once made sense but have now outlived their usefulness and become maladaptive (Wei & Ku, 2007).

Marjorie, a 45-year-old schoolteacher, came in because of increasing depression that now involved long periods of self-isolation, crying bouts, and wondering how long she'd "keep up appearances." During the first session she described a long history of emotional deprivation and neglect: her mother was severely depressed and drank herself to sleep most afternoons. By the time Marjorie was ten years old, she generally made her own dinner, ate in front of the TV, and put herself to bed. Desperate for connection with others, Marjorie found herself helping many of the neighbors with their chores and errands and she became the "neighborhood helper." Some invited her to dinner. The reality, though, was that whenever someone was kind to her, Marjorie overextended herself—to use her own word, she "enslaved" herself in gratitude for their attention. By age 45, Marjorie was "depleted" by her "enslavement." As the session ended, the therapist, aware that Marjorie had seemed relieved to tell her story, wondered (given Marjorie's developmental experiences) how Marjorie would later feel about being responded and attended to so fully and whether this could evoke her anxiety and raise concerns for her about having to respond to yet another person.

THERAPIST: Marjorie, how was it for you to be here today?

CLIENT: You listened and I felt understood.

THERAPIST: Good. I'm glad you felt that way.

CLIENT: Yes, it was good to be able to talk to someone. I had isolated myself for a long time, afraid to get close to people again, afraid that if I did I would have to start taking care of others again...

THERAPIST: Marjorie, could there be any other concerns you might have that might make it harder for you to come back next week?

CLIENT: Hmm, I hadn't thought about that...

THERAPIST: Let's take a moment and give you a little time to think about that...

CLIENT: Are you thinking I might be afraid of getting close to someone—like here to you—and want to isolate again?

THERAPIST: What are your thoughts?

CLIENT: Hmm, that's a thought... I do want to be able to have someone to talk to, and often, I just feel so alone. But then I'm sure I'll get worried—how would I ever repay you? I know this is your job but... I'd be leaning on you so much, you'd end up giving me so much... yes, maybe I would worry about coming back because I'd be worried about how I'd be able to give back to you... I'd owe you too much... (tearful)

THERAPIST: I'm glad we are talking about this; I hope you will come back so we can work together on what has been so troubling and caused so much distress for you. Maybe you and I can change the old pattern that has depleted you so much. That pattern helped you as a child—it gave you connection to some caring neighbors, so it makes sense. But—you don't need it anymore, not with me at least. You and I can be connected without depleting yourself and without having to "enslave" yourself to me.

CLIENT: I'd like that—I want to come back, and I'm scared. How will I know if this will be different?

THERAPIST: Yes, that's a reasonable concern. Are you willing to try a few sessions and see how it is for you—if you and I can do this differently and make it work for you?

CLIENT: Yes, how many should we try?

THERAPIST: What seems most reasonable to you? What seems most doable?

CLIENT: Four more would be OK.

THERAPIST: Great, let's schedule for four more sessions and keep checking in to make sure you're not feeling like you "owe" me or have to take care of me—like you've had to with others.

Thus, exploring the client's resistance commonly reveals an *outdated coping strategy*—not a lack of motivation. To begin learning how the therapist can start to change this outdated or maladaptive coping strategy, we briefly introduce a four-step sequence therapists can follow.

First, the therapist is trying to help clients *identify* their outdated coping strategies with others (such as being perfect, withdrawing, intimidating, complying, and so on). Take, for example, the client who always takes care of everyone else, at the expense of her own needs, and feels burned out at work and resentful at home. As part of this sequence, the therapist also helps clients understand why they originally needed to cope or protect themselves in these ways. For example, the client's parent was alcoholic or depressed, and she grew up taking care of her parent rather than being taken care of. Another aspect of identifying this outdated coping strategy is to clarify how clients may be continuing to do this now in their current interactions with others—including the therapist. For example, the client doesn't share difficult feelings because she doesn't want to "bother" the therapist or "burden" him.

Second, the therapist *validates* the protection this interpersonal coping strategy once provided. This strategy (always taking care of others, not meeting her own needs), is now causing problems in the therapeutic relationship—as it is with others. However, once it was a strong and adaptive response to a reality-based problem. For example, "taking care" preserved her attachment to her unresponsive caregiver, and it became the basis of her identity and the source of what little self-esteem she could garner through her ability to meet the needs of others. Affirming the validity and original need for this coping strategy is often an empowering liberation to clients as this new understanding leads them to feel more self-acceptance rather than self-blame (for example, Client: "So, I guess I'm not just stupid and hopeless. I learned to sabotage my successes for very good reasons").

Third, the therapist holds a steady intention to *track the process dimension and ensure that the therapeutic relationship does not repeat this problematic pattern* that has transpired so often with important others. The therapist can provide a corrective emotional experience—rather than an unwanted reenactment—by checking this out with the client and asking:

THERAPIST: At work you take care of everybody and everything, at home you do the same—and understandably feel resentful and exhausted. So, as we begin to clarify this unwanted life-script of yours, it makes me wonder if it ever feels like you are taking care of me in some way, too?

Therapists often sit in disbelief as many clients respond "yes" to process questions like this and go on to explain how aspects of the old coping strategy they are describing with others are indeed being reenacted between them in ways the therapist never dreamed were possibilities. For example:

CLIENT: Well, I guess sometimes I worry that I'm burdening you with my problems. I can see from the picture on your desk that you're a mother, so I know that you go home to a lot of needs yourself...

Fourth, the therapist's ultimate goal is to *help clients transfer this new experience or re-learning with the therapist and apply it with others* in their lives.

For example, this caretaking client could also respond differently and say "no," the pattern isn't repeating with the therapist:

CLIENT: No, you're different than my boss and husband and daughter—I don't have to take care of you. It's nice, I can be responded to.

THERAPIST: Good, I'm glad this is one place where you can let that depleting role go and allow yourself to be responded to. Who else in your life can you break this pattern with—you know, respond to your needs equally?

CLIENT: Well, I could probably do more of this with my sister, and some of my coworkers are pretty responsive.

THERAPIST: Let's try to build on those relationships, and find others where things could be more balanced. But it sounds like home—with your family—is the most pressing place for you right now. I'm wondering if we could work together and try to change your part in this pattern with your husband and daughter?

CLIENT: What do you mean, my part?

THERAPIST: As I listen to you, it often sounds like you do everything for everybody before they have a chance to pitch in—you're so capable and you respond so quickly. What would happen if you asked your daughter to clean up her own room, or your husband to help with the dishes?

CLIENT: Oh, I've done that. You don't understand, it's totally hopeless.

THERAPIST: OK, maybe it is, but let's walk through the sequence together and see how it usually goes. Maybe we can spot some places to try something new—like waiting a moment and giving them time to respond before you jump in and take care of it, or sticking to your request and asking a second time.

By following this four-step sequence, over and over again, therapists provide the client with the real-life experience that, at least here, they no longer need to keep using the outdated coping strategy. More precisely, the therapist can help them learn to *discriminate* when they still need to respond in this old way (for instance, at work with their rigid and demanding supervisor) and when they don't (such as with their supportive friend or cooperative coworker).

There is a great deal of complex but important information here—we'll return to this pivotal topic of outdated interpersonal coping strategies and work further with this four-step sequence of change. To sum up for now, we have seen that clients are far more likely to become stalled in therapy, or to drop out prematurely, if their resistance, ambivalence, or rupture is not addressed in a supportive and collaborative manner (see Bender, 2005; Ogrodniczuk et al., 2005). One of the most important ways to successfully engage clients in a therapeutic alliance, and to identify the key conflicts in their lives that need to be worked on in treatment, is to invite clients to discuss any concerns they may have about any aspect of being in treatment or problems with anything the therapist does. In this regard, Rhodes et al., (1994) asked clients who were satisfied with treatment what their therapists had done to resolve misunderstanding between them. More satisfied clients reported that therapists asked them how they were feeling about what was going on in their relationship, listened nondefensively and were willing to hear what the clients thought they were doing wrong, and apologized if they made a mistake or hurt the client's feelings. In sum, *the guiding principle is that if clients talk about their fear, ambivalence, frustration, or potential misunderstandings, they will be less likely to act on them and drop out prematurely* (Cartwright, 2004). Before going on to illustrate specific ways of implementing this principle, we need to learn more about identifying clients' resistance and how to distinguish it from reality-based constraints that need to be affirmed.

IDENTIFYING RESISTANCE

Especially as they begin their training, most new therapists will have clients who drop out of treatment within the first few sessions. This doesn't feel very good, of course, but there is much we can do to prevent these unwanted premature terminations. Instead, however, therapists often blame the client and conclude that he wasn't really motivated or wasn't ready to look at his problems (which usually isn't the case, unless the client is entering treatment at someone else's behest). Or the therapist may blame herself and think that she unwittingly made some irrevocable mistake in the previous session—which is usually untrue as well. Something else is going on with these premature terminations. Of great significance, researchers have interviewed clients after treatment and found that *clients often report having negative feelings toward the therapist, which they hide from the therapist* (Hill et al., 1992, 1993). All of us need to remember and learn from this. More specifically, clients report that it often *doesn't feel safe* to bring up these negative feelings on their own and, generally, therapists do not ask about potential problems or misunderstandings between them. That is, once clients feel secure, deep sharing is more likely, a finding that has been noted in the research (Romano et al., 2008). As a result, many clients terminate prematurely because they are acting on, rather than talking about, their conflicted feelings about entering therapy, or unspoken misunderstandings or difficulties with the therapist that were not talked through and repaired. This unfortunate but common outcome is not necessary—there is so much therapists can easily do to help clients successfully engage in treatment—and resolve the misunderstandings and difficulties that commonly arise. For example, the therapist can:

* help clients identify when resistance may be occurring,
* approach this issue in a noncritical, welcoming manner that actively encourages clients to express fully any concerns they may have about the therapist or the treatment process, and
* help clients resolve these concerns by listening in an accepting and nondefensive manner, expressing appreciation for clients' willingness to talk about problems, and communicating a willingness to be flexible, take their concerns seriously, and work out mutually agreeable solutions—as described next.

Beginning with the first bullet above, how does the therapist identify or know when resistance is occurring? Therapists never really know what any behavior means for a particular client, but resistance may be operating when clients *consistently* have difficulty participating in treatment. For example, after reality-based constraints have been considered (such as scheduling conflicts or fees), psychologically based resistance (for example, getting help is inconsistent with cultural norms or the familial role of caretaker) is probably occurring when any of the following occur repeatedly or in combination:

- The client misses multiple appointments or repeatedly comes late.
- The client asks to reschedule frequently or often cancels appointments.
- The client has very limited hours available for therapy.
- The client does not want to make a firm commitment to attend the next session.
- The client attends but remains vague or elusive about her problems.
- The client becomes reluctant to talk freely ("I don't feel much like talking about anything today").
- The client avoids certain topics that may have been important to him or brings up highly significant topics only at the end of the session.
- The client has been working productively but suddenly suggests stopping and terminating prematurely.

It is important to stress that the same behavior often means very different things to different clients. However, when clients have trouble attending, or participating fully in, sessions in one of these ways, some form of resistance is often at work (Callahan, 2000; Meissner, 2007). In such cases, the therapist can begin to generate "working hypotheses" about the possible meaning this resistance holds for the client.

FORMULATE WORKING HYPOTHESES

Therapists provide more effective treatment when they can formulate useful case conceptualizations that capture the key problem(s) and clarify for both the therapist and the client what the basic problem really is. These case conceptualizations that clarify what's wrong are used to guide how therapists intervene and what they choose to focus on and work together to change. Therapists also need to write treatment plans for third-party or HMO insurance payments. Looking for more specific and focused treatment plans, some HMOs will require operationalized and measurable treatment goals. In this regard, texts are available to help therapists clarify where they going in treatment and how they plan to get there (Bloom et al., 2003; Fonagy, 2002). The first step in clarifying treatment goals is to formulate tentative working hypotheses about the client's maladaptive relational patterns, faulty expectations and pathogenic beliefs, and core conflicted feelings. One systematic way to do this is to begin formulating answers to three questions during the initial contact. (Further guidelines are provided in Appendixes A and B.)

1. What does the client elicit from others? Even in the initial telephone contact, the client's interpersonal style—and what it tends to *elicit* or "pull" from others—is an important source of information (Benjamin, 1987a; Cashdan, 1988). For example, if the client sounds helpless and confused on the telephone, might he adopt a victim stance and invite others' rescuing behavior? If the client presents in an angry and demanding way, might others often withdraw or tend to become embroiled in power struggles?

If so, perhaps this client is testing whether the therapist can tolerate angry challenges without being run over and still see the underlying need or pain. As therapy progresses and the therapist learns more about each client, many of these initial hypotheses will prove inaccurate and will need to be discarded. As therapists learn to attend to their own reactions and the feelings evoked in them by the client, they will find that many hypotheses do indeed fit the client and can be further developed and refined. Thus, to begin conceptualizing the client and developing initial treatment plans, therapists can follow these four steps:

- Formulate working hypotheses about his interpersonal style and what it tends to elicit in others (for example, his loud and aggressive demeanor tends to elicit a fight response from some or a flight response from others).
- Evaluate these tentative initial hypotheses as the therapist learns more about the client and as they interact further (e.g., as the therapist listens to his narratives, it sounds as if co-workers, family members, and others also seem to react to his dominating style with either their own counter anger and competitiveness or with feelings of intimidation that lead to compliance or withdrawal).
- Revise these hypotheses over time to more accurately reflect each client's personality and problems (e.g., she takes over and tries to intimidate when she is uncertain or feels she has made a mistake).
- Formulate intervention plans (e.g., provide interpersonal feedback about his dominating presentation, and help the client become more aware of the often unwanted impact his abrasive style has on his wife, children, and others).

For example:

THERAPIST: I'm wondering if you are aware that you're speaking to me in a harsh, loud voice right now?

CLIENT: Well, I don't know, I guess maybe.

THERAPIST: Yes, you are quite loud, and the harshness has a threatening quality. How do the people you're close to respond when you talk to them like this?

CLIENT: I don't know what you're talking about.

THERAPIST: What does your wife say and do, or how would you describe the look on your son's face, when you sound this angry to them?

CLIENT: Well, OK, I guess my wife gets a stubborn look on her face—like she's not going to give me anything. Maybe my son looks a little nervous...

THERAPIST: So this interpersonal style isn't working very well for you. I believe you when you said earlier that you loved your family, but I can see how they wouldn't feel safe enough to talk or share much with you. What's going on right now as we talk about this?

CLIENT: *(Slowly)* It's probably true... I feel kinda sad I guess... not sure what to do.

THERAPIST: Yes, I can see your sadness, this does cause real problems in your life. I think the first part of changing this is naming it and talking about it—like we're doing right now. Why don't we focus on this together and see if we can start to understand what that angry voice is about—you know, what it gets you and what it costs you—and if you want, work together to find some ways to change it.

Although the steps bulleted above may feel challenging right now, they can be learned with time and practice. To help, in the chapters ahead we will return to this important concept of "eliciting behavior" in particular.

2. What is the threat? On the basis of information the therapist has started to gather about the client's current and past relationships, the therapist can begin speculating about the different ways that entering treatment could be aversive for the client. Therapists can formulate working hypotheses such as the following about the issues or concerns that are likely to make it difficult for this particular client to enter treatment.

- Does this Christian client believe that her inability to resolve her problems through prayer means that she has failed in her faith?
- Is it guilt-inducing for this highly responsible parent, who grew up taking care of her own anxious and insecure mother, to ask for something for herself?
- Is it awkward or embarrassing for this older Latino male to seek help from a younger African American female therapist?
- Is it shameful evidence of his own weakness or inadequacy if this blue-collar worker has an emotional problem that he cannot solve by himself?
- Is it disloyal for this Asian client to discuss personal problems with someone outside of the family?

Therapists may also keep in mind that resistance and defense are attempts to manage unwanted feelings of shame, guilt, and anxiety. *Although clients may be unaware of it, they often struggle with the worry that, if they continue in treatment or engage more fully with the therapist, the therapist will hurt them as others have in the past.* This schema-driven concern won't come up in everyday interactions, but it will be activated when clients are upset, vulnerable, or in distress (the **"hot cognitions"** activated during crisis). We will see how therapists can use process comments to: (1) help clients clarify and better understand these concerns about their psychological safety with the therapist; (2) learn how to assess whether the therapist, unwittingly, may indeed be responding in a familiar but unwanted way that the client has learned to expect in close relationships (re-enacting); or (3) make overt how they are co-creating a new and different type of relationship than they have come to expect (corrective or resolving).

Let's step back for a moment and consider one of the bigger meanings conveyed in our key question, "What is the threat or danger if you talked with me about that—what could go wrong between us?"

Therapists don't want to "play it safe" with clients and keep things on the surface. Instead, therapists are trying to be someone who is willing to go beyond everyday social conversation. Being friendly and nice is appreciated, especially in the beginning, but it doesn't carry us very far and won't be enough to help most clients make progress with enduring or serious problems and change. Thus we are looking at all times for opportunities to engage clients more significantly. We do this by reflecting the deeper meaning, inviting the stronger feeling, and approaching the central concern implied or alluded to in what the client just said. In other words, therapists need to be willing to risk personally engaging with what is most important in the client's life. Exploring the client's resistance, and asking the here-and-now, immediacy questions suggested in this text, is not for the faint of heart, such as:

- "What's the threat here? What's going to go wrong between us if you disagree with me?"
- "What might I say or do that could make things worse for you if you risked sharing that with me?"

This way of responding will highlight the client's core concerns and put them on the table for discussion. It will also strengthen the working alliance by bringing real engagement, shared meaning, and greater intensity to the therapeutic relationship. And it is the working alliance that is the key to positive treatment outcome (Horvath & Bedi, 2002). Further, as noted by Ackerman & Hilsenroth (2003), there is an association between being active, interested, and engaged with clients and establishing strong alliances. Thus, new therapists are encouraged to take the risk of responding to clients in this stronger, more forthright manner from the start to the end of treatment. If the therapist has held a consistently empathic and non-judgmental attitude toward the client, and has been willing to break the social rules and check out with the client how it is to talk together in this more straightforward way, most clients will welcome such honest communication. In fact, most clients find it reassuring because the therapist is behaviorally demonstrating that they aren't alone with their problems anymore but have found someone who is ready and willing to join them and address what's really wrong. Like all of the ideas suggested here, role-play or practice with a classmate. Thereafter, try it out when you're ready and think it might work, then assess how your client responds and decide for yourself.

3. How will the client express resistance? Finally, therapists can formulate a third set of hypotheses that anticipate how clients are most likely to play out or enact these concerns with the therapist. For example, suppose a therapist has observed that her depressed client feels guilty and believes he is selfish whenever he does something for himself or someone does something for him. The therapist then hypothesizes that this client may withdraw from treatment as soon as he starts to improve in treatment and feel better, or distance himself and physically or emotionally disengage from the therapist when he realizes that she genuinely cares about his well-being and their work together. By anticipating potential ways that clients may express or act on their resistance, the therapist is much better prepared to recognize it in the moment and respond more effectively as it is occurring.

To illustrate further how the therapist generates working hypotheses on the basis of the preceding three questions, let's consider another example. Suppose that a young male client whose presenting problem is substance abuse telephones the therapist to schedule an initial appointment. During this telephone call, the client tells the therapist that his alcohol and marijuana usage is more problematic than it has ever been and that "my mother thinks I need to be in therapy." On the basis of the helplessness and urgency communicated in his telephone call, and his disclosures that his mother wants him in therapy and his drug usage is so problematic now, the therapist begins formulating several tentative hypotheses. First, the therapist hypothesizes (again, the therapist doesn't "know," she is just considering possibilities) that an important part of the client's interpersonal style is to let others know that he is distressed and in crisis, to elicit help from them, but then to avoid taking responsibility for his own behavior and, consequently, follow through poorly or actively resist the input he is requesting.

Second, the therapist hypothesizes that this client may become resistant to treatment when the therapist addresses how he expresses and meets his emotional needs, and when he is given reality-based feedback about how he eschews responsibility for his own decisions (for example, to enter therapy or to continue drinking). Third, the therapist hypothesizes that this client may be impulsive and prone to act out his resistance by bolting out of therapy as soon as these anxiety-arousing issues are addressed (just as, perhaps, he has used alcohol and drugs impulsively to avoid dealing with certain feelings, interpersonal problems, or responsibilities that are anxiety-arousing for him).

The therapist knows that these initial hypotheses may not be accurate and is ready to revise or discard them as more is learned about the client. However, the therapist will be able to respond more effectively if she can try to anticipate the concerns that might cause each client to drop out of treatment and the way each client is likely to express these concerns. Prepared with working hypotheses such as these, student therapists will be able to "get it" more quickly and be able to recognize better in the moment what is happening between them—not two hours or two days later when they are reviewing tapes of their sessions. They are then able to respond more effectively by helping clients explore their own resistance before they act on it and leave treatment prematurely.

Although many of these initial hypotheses will need to be discarded as the therapist learns more about the client, some will be accurate. As the therapist identifies the patterns and repetitive issues that keep arising throughout the client's life, the therapist is better prepared to center treatment on these recurrent themes. The therapist does this by highlighting them tentatively—observing or wondering about them aloud to the client as they occur. These patterns, tentatively suggested so the client is invited to modify or discard them, readily become useful and of interest to most clients when offered collaboratively—creating a shared treatment focus and providing a compass for the ongoing course of therapy (Binder, 2004; Levensen, 2004). This is part of the ongoing process of formulating and refining case conceptualizations (that is, what's really wrong) and treatment plans (that is, what the therapist wants to do to help with this problem).

RESPONDING TO RESISTANCE

The tired cliché is true—life is hard. We need defenses in our lives; they often are adaptive and necessary to cope. We certainly needed certain coping strategies to sustain attachment ties, and to defend or protect ourselves from real threats and dangers, when we first learned them as children. And, in a reality-based way, we often need them now in some situations and relationships as adults. A primary treatment goal for therapists with almost all clients is to help clients learn to *discriminate* when certain defensive coping strategies are indeed needed and when they are outdated and no longer serve us well. Most clients do not do this well— contributing strongly to many symptoms and problems. Throughout this text, we are going to study how the therapist can help clients assess when their coping strategies are still necessary and helpful, and in which situations they are no longer useful or needed. For example, becoming quiet and developing the chameleon-like ability to make oneself almost invisible in a room full of people—in order to go unnoticed and stay out of the line of fire—once was an effective coping strategy. It was necessary and adaptive while growing up with a hot-tempered parent and still is useful at the office for the same client who now is working for a critical, demanding, and volatile supervisor (at least until we can help this client find a different job or supervisor and stop this reenactment of his childhood dilemma). In contrast, however, being compliant and withdrawn in this way is no longer adaptive or needed in his current relationship with the therapist, and will almost certainly cause problems with his girlfriend and in other close personal relationships where he (and most clients) continue to act in this same way, even though it now causes many more difficulties than it solves.

In this section, we learn how therapists can respond effectively to common expressions of resistance. Sample therapist-client dialogues illustrate effective and ineffective responses at three critical points: during a telephone conversation in which the client has difficulty scheduling the initial appointment; at the end of the first session; and during subsequent sessions with a client who has difficulty keeping appointments. As before, student therapists should find their own words to express the principles embodied in these dialogues.

Addressing Resistance during the Initial Telephone Contact

While scheduling the initial appointment, the therapist wishes to obtain a definitive commitment to attend from the client. The client, however, may be ambivalent about entering therapy and, overtly or implicitly, express this during the initial telephone contact. The therapist's aim is to attend to potential signs of resistance and broach this issue with the client if some aspect of resistance seems to be in play.

An Uncertain Commitment.   In the first example, the client expresses only minimal uncertainty about attending the first session.

THERAPIST: It's been good talking with you, and I'll plan on meeting with you on Tuesday at 4 P.M.

CLIENT: OK, I guess I'll see you then.

THERAPIST: You sound a bit uncertain as you say "I guess." Maybe there's nothing to it, but I'm wondering if we should talk together for a minute about how it feels to come in? Do you have any questions or concerns I can answer for you?

CLIENT: Well, maybe I'm a bit cautious about most new things I try, but I do want to come and will be there on Tuesday.

THERAPIST: Good, I'm looking forward to meeting with you. We can talk then about that little bit of cautiousness, if you'd like, and any questions you may have about me or treatment. See you next Tuesday at 4 P.M.

—An effective response to the client's ambiguity ("I guess...") is modeled in this dialogue. The therapist hears potential indecision in the client's vague commitment, and approaches it directly but sensitively by asking about it in an exploratory, non-blaming manner. An ineffective response would be for the therapist to let it pass by on the assumption that it probably doesn't mean much. Perhaps it doesn't, of course, and chances are that most clients will still arrive at the appointment time. However, the chances of the client not showing up for the first appointment are far greater if the therapist does not acknowledge these signs of uncertain commitment and invite a dialogue about this potential resistance. Furthermore, although this ambiguous comment may seem too subtle or small to be significant, it is a statement about the counseling relationship—about you and me. Treatment benefits when therapists track clients' comments about the therapeutic relationship—and offer open-ended bids to explore them further.

In addition, the therapist can mentally note this indecision as a possible sign of resistance—that something about beginning may not feel quite right. This is an instance where the therapist can begin to generate working hypotheses—derived from the first of our three orienting questions, about the possible meaning of this "cautiousness." As the therapist meets this client and listens to the narratives of his life, she can begin to consider tentative working hypotheses about other material related to "uncertainty"—or dismiss it if she doesn't hear this theme in other arenas of the client's life.

A More Ambivalent Client.   Now imagine a similar telephone conversation, but with a more resistant client this time.

CLIENT: OK, I guess I'll see you then.

THERAPIST: You sound a bit uncertain; can we talk a little more about what it means for you to come in and see me?

CLIENT: Well, in my family, we don't really talk to others about family problems.

THERAPIST: I respect your concern about being loyal to your family—I can see that they are important they are to you. But I'm also hoping that we can find a way to balance things. Maybe one of the things we can talk about is how you can ask for help from others and get your own needs met in a way that still respects your family—you know, a way that leaves you with the feeling that you aren't being disloyal.

CLIENT: I'd like that. My family is really important to me. Maybe another part of this is that sometimes it seems likes psychologists just want to blame parents for everything. That doesn't feel right to me.

THERAPIST: Your concern makes sense to me, I'm glad you brought it up. That wouldn't feel right to me, either. I'm hoping that we can better understand what's wrong and causing problems in your life, so you can make the changes that you choose to make. So, I'm looking to understand problems better, not blame. That's a big difference.

CLIENT: I like that. OK, I'll see you on Tuesday.

THERAPIST: Good. I'll look forward to meeting you Tuesday at 4.

‾ An ineffective response would be to accept the client's uncertain commitment ("OK, I guess I'll see you then") and avoid the possibility that an important issue about what it means for this client to have a problem, ask for help, or enter treatment is being played out. In this dialogue, the therapist has tried to "hear" and respond to the client's concern instead. By affirming both sides of the client's ambivalent feelings (balancing personal needs with family loyalties), differentiating herself from other psychologists in the clients worldview—trying to understand not blame, and simply being accepting and respectful of his conflict—the therapist has done much to help this client take the first step and enter treatment. Without following up on his small ambiguous statement, however, this highly workable client likely would not have been able to successfully enter treatment.

Let's look further at why therapists in both dialogues would take this stance and press for a clear commitment to attend. Wouldn't it be more supportive to let clients leave the appointment a little bit tentative, if that is what they need to do? No. Little therapeutic change will occur until clients take responsibility for the decision to enter therapy and work on their problems. Therapists do need to be flexible. As we will see shortly, for example, therapists can give clients permission to make only a very limited commitment—for one session or just a few sessions—to get a better sense of whether treatment, or this particular therapist, is right for them. Without this commitment, however, the therapist has little to work with. In fact, it is better for the client to remain out of therapy than to enter without commitment and have an unsuccessful therapeutic experience. As Yalom (1981) emphasizes, the therapist wants to be concerned about preventing failed hope in certain clients. If clients have one or more unsuccessful therapy experiences, it may discourage them from trying to seek help with their problems again. When a client makes a commitment, the therapy process becomes a joint endeavor and therapist and client become partners or collaborators in the work—which facilitates positive treatment outcome (Tyrone and Winograd, 2002).

Clients Who Try to Make the Therapist Responsible. Let's continue with a different telephone conversation between a therapist and a prospective client. Some clients will try to make the therapist (and others) take responsibility for their decisions.

CLIENT: I'm not sure if I should start therapy or not. What do you think I should do?

It is important to reach out to clients and welcome them into treatment, but the therapist does not want to resolve clients' ambivalence by taking responsibility for their decisions or being a cheerleader and trying to provide the motivation for treatment. With some of these externalizing clients, a helpful response might sound something like this:

THERAPIST: From what you've just been telling me, it really does sound like this has been a difficult time for you, and I would like to try and help. But, since you are uncertain about coming to therapy, I suggest that you decide if you would like to come in for just one session and see how you like it. As we talk about what's going on for you, I think you'll get a good sense of treatment, and me, and will be able to decide for yourself whether you'd like to continue or not. How does that sound?

OR

THERAPIST: I can't offer you any guarantees of course but, yes, I do believe that therapy may be of help. I'd like to try and work with you, but I suggest that we agree to meet for just one time. After we've had the chance to talk together, I think you'll have a good sense of me and how I work, and will be able to decide for yourself what you want to do. At the end of the hour, we can check in on this and talk forthrightly about how it's been to meet and whether you want to continue. If we're not a good match, I can easily give you other referrals.

As we would expect on the basis of client response specificity, such direct invitations may be useful at times. When the therapist discerns that this client may have felt unseen, dismissed, or even rejected while growing up, or otherwise come to believe that others are uninterested and do not really want or care about him, it becomes important for the therapist to express her genuine interest or even enthusiasm in working with the client. However, even when therapists are trying hard not to repeat the problematic relational patterns the client is describing, they still want to respond in ways that ensure clients take responsibility for their own decisions to enter treatment; therapists do not want to cajole, coerce, or win clients into therapy. At the same time, therapists do not want to be aloof or indifferent. Thus the therapist is trying to meet clients where they are—working supportively with their ambivalence or conflicts about entering—yet without taking responsibility for the clients' choices. In many cases, an effective response to the client's question about entering treatment may be as follows:

THERAPIST: It seems as if one part of you wants to be in therapy, but another part of you doesn't. Tell me about both sides of your feelings.

OR

THERAPIST: It sounds like a part of you wants to try this and a part of you doesn't. Why don't we agree to meet for just one or two sessions? Let's see if we can work together and sort through both sides of this, so you can decide for yourself what you'd like to do.

The client in the dialogue below is providing the therapist with potentially important information. As before, the therapist can begin generating working hypotheses about what this behavior might mean. Having generated hypotheses, the therapist can then be alert for evidence of these themes in other areas of the client's life. If, for example, there are narratives about "uncertainty" or avoiding personal responsibility that keep coming up and prove to be enduring relational themes, the therapist can use these working hypotheses, that have now gained further support, to help provide the client with a focus for treatment. For example:

THERAPIST: You know, Marie, I'm hearing the same theme again with your boss that you were talking about last week with your boyfriend—you become quiet and "go along," rather than speak up for yourself when you disagree. What do you think about this possibility?

The therapist does not want to become wed to these early hypotheses, however, and needs to be ready to discard them if the theme does not keep coming up or hold real feeling for the client. As therapists become more experienced, their initial hypotheses will become more fruitful. This developing skill will become an especially important aid to counselors as they begin working in short-term and crisis intervention modalities that require quicker, more accurate assessments during the initial phase of treatment.

As emphasized in Chapter 1, the therapist's overarching intention is to try not to get stuck in an ongoing way reenacting the client's faulty patterns in the therapeutic relationship. However, this can easily occur in a metaphorical or encapsulated way in the initial negotiations between therapist and client. This is another reason why the therapist does not want to take over and tell the client seeking advice what to do. For example, the following response is well intended, but it is not going to be helpful to many clients:

THERAPIST: On the basis of what you've told me, I think therapy will help you and believe you should give it a try.

Assuming such a directive stance reenacts a problematic relational scenario for many clients and confirms a maladaptive schema. Many clients are hampered in life by the faulty belief that they cannot act on what they want, initiate or pursue their own goals, or advocate on their own behalf. Others originally learned in attachment relationships that caregivers needed them to remain dependent to assuage the caregiver's own anxiety and now, as adults, these clients continue to act dependently and allow others to take responsibility and make decisions for them in order to remain connected. Although such clients will also believe that they need to ask the therapist to tell them what to do, paradoxically, the healthier part of them that wants a more independent self will resist the control they have just elicited. Clients with these "compliance dynamics" (sometimes referred to as **"excessive accommodation"**) then have to resist the therapist—and reject the good help the therapist may be offering as well—or they are further complying and losing even more of themselves by being helped and getting better in treatment! As this common compliance scenario continues, clients come to feel badly about themselves—confused as to why they are rejecting the help they have just asked for, and guilty for rejecting the therapist who was trying to help. Therapeutic responses that give the client responsibility for deciding whether to follow through and enter treatment prevent this reenactment from occurring, and will provide some clients with a new and corrective response that challenges the old schema and begins to change a long-standing behavior problem.

EXPLORING RESISTANCE AT THE END OF THE FIRST SESSION

Toward the end of the first session, no matter how well it seems to have gone, the therapist is encouraged to ask clients how the session felt to them and whether they have any concerns about the treatment process or the therapist. Unless the therapist is willing to inquire about potential problems in this straightforward way, such concerns are probably going to remain unspoken. Recall that follow-up research with clients after treatment indicates that many do indeed have some disappointments or difficulties with the therapist or some aspect of treatment, but they usually do not bring these up or address these on their own (Hill et al., 1992). Failing to inquire about, and then sort through whatever concerns the client might be having with the therapist, is an opportunity lost. Many clients feel some reticence about entering treatment. And, at times, almost all clients will feel that the therapist misunderstood something important they have said. Such misunderstandings are an inevitable part of all relationships—and may occur because of the distortions shaped by the client's maladaptive schemas or because the well-intended therapist simply misunderstood or did not grasp something that was meaningful to the client. Similarly, most clients will be uncomfortable at some points in treatment with how the therapist is responding to them (feeling, for example, that the therapist is being too quiet, too directive, and so forth). All of these clients will be less likely to drop out prematurely and will be better able to make progress and change in treatment, if the therapist has:

- explained that these misunderstandings are expectable and given the client overtly spoken permission to talk about whatever concerns may come up for her,
- overtly welcomed hearing any concerns and listened to her nondefensively and fully, and

- taken the client's concerns or complaints seriously and demonstrated a sincere effort to sort through misunderstandings and resolve whatever problems or concerns the client risked expressing.

Addressing Interpersonal Conflict.   To illustrate, five or ten minutes before the end of the first session (to allow clients enough time to talk about their reactions to the therapist and the session), the therapist can check in with the client:

THERAPIST: How was it to be here today?

Most clients will answer, "Fine." Perhaps thinking that the student therapist is feeling insecure and may be seeking some reassurance, clients may go on to say how helpful the session has been. If so, the therapist might respond as follows:

THERAPIST: Good, I'm glad you've found our first session helpful. You've told me a lot about yourself today, and I feel we have gotten off to a good start, too. As part of our work together here in therapy, I think it will help us if we are able to talk about our relationship more directly than others usually do. I'm especially hoping that it will feel safe enough for you to be able to tell me about what does and doesn't feel good about our interaction, and things that we could change or do differently to make our work together here more helpful to you. So, I'm wondering if there was anything about coming today, or anything about our work together, that didn't feel quite right? If so, I would really welcome talking about it—you know, so we could change things and make it better.

CLIENT: Oh no. I was eager to begin, and you have been very nice and understanding.

THERAPIST: Good. In the future, though, if something does come up—and it probably will—I think it will help our work together if we could talk about it. How would it be for you to do that? Would that be OK or maybe kind of hard for you?

CLIENT: Well, I don't know... that might be kind of different for me. We were taught to be polite in my family, you know.

THERAPIST: Absolutely. It is important to be polite, but I'm hoping we can be polite and honest with each other, so we can solve any problems that might come up between us. One of the things I value about therapy is that this a place where we can talk about whatever is most important to us, including our relationship and what's going on between us that does and doesn't feel good. Often that will be things we can't do so easily with others out in the world but, in here with me, I welcome it—that way, we can always be up front... respectfully, and also know where we stand with each other. How does that sound to you?

(Here, clients might respond in a variety of ways)

CLIENT: I like the idea of things being up front... I'm not used to doing that so it might be hard at first;

OR

CLIENT: Hmm, what if I have something to tell you about you that I don't like?

OR

CLIENT: I don't like this. I don't want you telling me what's wrong with me, I get that already from enough people in my life.

OR

CLIENT: Well, I don't know... that might be kind of different for me. We were taught to be polite in my family, you know.

OR

CLIENT: So you're saying we can just get down to brass tacks here and talk about what's really important, and not beat around the bush and worry so much about what you might be thinking—or try so hard to be nice.

In this vignette, the therapist is educating the client about the treatment process and trying to establish an important set of expectations for what is going to occur in treatment. The therapist is telling the client that, in contrast to many others, the therapist wants to hear about the client's wishes and concerns, and address potential problems between them straightforwardly. Because straightforward communication and, especially, talking directly about conflict or personal problems between "you and me," was not allowed in their families of origin, many clients will be surprised, but welcome, this invitation. Here, the therapist is laying the groundwork for a relationship that is more direct and authentic than most clients have enjoyed in the past. Therapists want to establish these important expectations with their actions, and not just their words, early in treatment. As emphasized before, there will be misunderstandings or "ruptures" in the therapeutic alliance—it's a normative part of the treatment process—not a mistake (Safran et al., 2002). Effective therapists repair the inevitable ruptures that occur by making process comments, asking about potential problems, and talking them through.

THERAPIST: You seem kind of distant today, or maybe more preoccupied. Any ideas about what might be going on for you—or between us?"

CLIENT: Well, maybe I am a little distant today, but I don't see what's wrong with that. I don't like feeling judged by you.

THERAPIST: "Judged." I'm sorry that happened, but glad you're telling me about this. I wouldn't like feeling judged either—but I'm not aware of feeling judgmental toward you. Let's sort this through together—tell me more about what I said or did that made you feel judged.

Therapeutic relationships will remain superficial unless therapists are willing to broach these expectable ruptures and take the risk to talk them through. More help with this core issue of rupture and repair follows—it is essential for successful treatment.

A More Assertive Client.   As previously noted, approaching interpersonal conflict in these direct ways is anxiety arousing—it may violate familial and cultural norms. To help, let's continue with another assertive and critical client.

THERAPIST: I see that we're about out of time. How has it been for you to talk with me today?

CLIENT: Well, OK for the most part, but I did feel you were trying to hurry me up a few times.

Clinical training holds many challenges, but for many student therapists none is bigger than learning to *approach* interpersonal conflict—right now, in the moment as it is occurring between the therapist and the client. We do not want to deny, minimize, or avoid it by moving on to another topic, truncate it with a punitive response, or vaguely express hurt and communicate that the client's criticism is in any way wounding, too much for us to handle, or unwanted. Instead, the best way to approach conflict is by encouraging clients to express their concerns more fully and, as hard as it may be to do, strive to remain as *nondefensive* as you can and listen with a receptive, open mind.

THERAPIST: Thanks for being honest—it helps our work together when you bring up problems like that. Let's look at this together and see if we can figure out what happened for each of us. Tell me more about your feeling of being "hurried up" or pushed.

Of course, it is often difficult for new therapists to do this—just as it is for many supervisors in the workplace, teachers in the classroom, and, unfortunately, many highly trained and experienced therapists as well. Such direct, here-and-now conversations about conflict between "you and me" may arouse the therapist's own anxiety, especially when the therapist has strong needs for the client's approval. If therapists are to approach such anxiety-arousing ruptures and restore the therapeutic alliance, however, *therapists must first become aware of their own characteristic responses to interpersonal conflict.* For example, the automatic, knee-jerk reaction for some therapists is to avoid the client's complaint and move on as if nothing significant was said. For other therapists, the initial response tendency is to readily agree with the client and quickly apologize, in order to abate the criticism. Still other therapists will automatically begin to justify themselves and offer lengthy explanations aimed at stopping the criticism. Disturbingly, researchers find that a significant number of highly experienced therapists actually counter with their own hostility and punitiveness, and criticize the client in turn (Binder & Strupp, 1997; Hadley & Strupp, 1979). The first step in learning better ways of responding is to become more aware of how we typically tend to react to criticism, negative evaluations, or unwanted confrontations (see Teyber & McClure, 2010, "Part II: Self-Reflections in Chapter Three of the Student Workbook"). Research shows that rupture resolution is important to retaining clients in treatment (Muran et al., 2009).

One of the best ways to develop this self-awareness and learn more about how we usually respond to interpersonal conflict is to recall how it was dealt with in your own family of origin. Regardless of age, your initial reaction to conflict usually follows closely how your parents dealt with conflict in their marriage and how each parent dealt with conflict with you as a child. Rather than automatically following their initial response propensity, therapists can learn to approach conflicts with more openness. With help from a supportive supervisor, new therapists can learn to do this effectively in a year or two and, before too long, feel comfortable and actually welcome misunderstandings and conflict as a therapeutic opportunity.

After inviting clients to express fully their concerns, the therapist is not in a hurry to explain "what really happened" between them or disagree with the client's perception. Oftentimes the client is right and the therapist indeed was "rushing" the client and, most commonly, the client's perception is partially accurate and partially distorted—there is something to it. Thus the therapist's intention is to be nondefensive, examine her own behavior, and try to see what accuracy there may be to the client's perception. If we look from the client's perspective and see that there may be a little or a lot of sense in her experience, the therapist does not need to be afraid of acknowledging this to the client.

THERAPIST: You know, I think you might be right. I was aware of the time going by, and I wanted to touch on some other issues before we had to stop. I probably was hurrying you there, and I can see why that bothered you. I'm glad you made me aware of that; I'll try not to do it again. You know, I like your honesty—it's a real strength of yours.

We have been keying on how, throughout their first years of training, many student therapists suffer significantly over their fears of making mistakes, hurting clients, failing to help, and so forth. Sadly, these worries keep many from enjoying their clinical training, and others from taking personal risks and trying out new ways of thinking about and responding to clients. Many caring and responsible students simply do not enjoy their years of clinical training because of these worries. It is liberating as new therapists learn that they do not need to be afraid of making mistakes; they need only to remain open to the possibility that they may have erred and be willing to acknowledge mistakes when recognized. In this example, the therapist is responding to the client in a reality-based and egalitarian way. The therapist validates the client's perception, which tells clients that the therapist is willing to have a genuine dialogue and risk entering into a real relationship with them. It also tells clients that the therapist will respond to clients' concerns with respect. Too often, therapists and clients enact a relationship in which the client fulfills the role of the weak or needy one and the therapist is the healthy one who does not make mistakes or have problems. The responses suggested here discourage that illusion. The therapist's willingness to risk having an authentic relationship is a gift to many clients. The client finds that, at least sometimes, problems with others can be talked through and resolved. The relearning from such corrective experiences with the therapist often has a powerful effect on clients and accelerates change with others in their lives. For example, clients routinely come back to the next session with reports of having acted stronger with someone in their everyday lives, as they have just done with the therapist; for example, using their own voice and speaking up for themselves with someone who previously had been intimidating.

In contrast, therapists can handle the negative feedback in a way that keeps the relationship superficial or puts them in a superior position.

THERAPIST: Oh, you thought I was hurrying you up, how was that for you?

OR

THERAPIST: That's interesting. Do you ever feel this way in other relationships, too?

These responses are a misuse of the client-centered reflection or therapeutic "neutrality." With this type of *deflection*, the message to the client is, "It's always your problem. I will not look at my own contribution here. This will not be an honest or real relationship." Such a response limits the relationship to a superficial encounter and sets up a covert power battle between client and therapist. Little therapeutic gain will be realized as long as their interpersonal process continues in this problematic way.

But what if the therapist does not agree with the client's complaint? We don't want to say anything we don't really mean. Thus, the therapist can still accept the validity of the client's perception, yet without agreeing to the comment.

THERAPIST: I'm sorry you saw me as being impatient with you; that wouldn't feel very good to me, either. I wasn't aware of being in a hurry or trying to rush you, but I'll watch out for that in the future. If you ever feel that happens again, stop me right then and we'll look at it together.

In this example, the therapist is telling the client:

* I will take your concerns seriously even if I do not see it the same way.
* We can disagree and have differences between us, and still remain connected and work together respectfully.
* Your feelings about us, and our work together, are important to me.

This type of response doesn't just tell clients that problems between them can be resolved, it behaviorally shows this to them. In turn, this lends hope that the therapist and client may be able to resolve the client's problems with others as well. By addressing conflicts between the therapist and the client directly, the therapist provides the client with an important model of having interpersonal conflicts resolved in a new and more constructive way. Different theoretical orientations use varying terms for this type of experiential relearning: exposure trials, modeling, in vivo relearning, corrective emotional experience, and others. Especially for clients who have learned to avoid interpersonal conflict because it has been threatening or dangerous, such experiential re-learning is empowering indeed. With most, it leads quickly to related behavioral changes with others in their lives

CLIENT: Guess what? I spoke up at my big meeting at work! I didn't even think about it, I just jumped in and said what I thought. It was great—my boss seemed really surprised—and I think he's treated me with a little more respect since.

Finally, it is also possible that the client is systematically misperceiving the therapist's behavior in line with early maladaptive schemas. For example, this client may readily experience the therapist (and most other authority figures) as demanding more than the client would like to produce, and then being dissatisfied with whatever the client does produce. If so, this client may have grown up with a parent who repeatedly demanded that the client do everything on the parent's timetable or ignored the child's own accomplishments. Even if the therapist has gained further evidence to support such a hypothesis, this type of historical or transference interpretation should not be posed before the therapist and client have resolved the dispute in their real-life relationship. Therapists stand to lose a lot if clients see them as sidestepping a reality-based confrontation. However, the therapist can use this information to generate working hypotheses about their cognitive schemas and faulty expectations that can be utilized later.

Clients Who Test the Therapist's Adequacy.   Because it is very challenging for most new therapists to approach interpersonal conflict directly with their clients and work with ruptures in the therapeutic alliance, let's continue with another example. In the following situation, the client challenges the therapist's competence—which cuts deeply into an anxiety-arousing area for most new therapists. As we see, however, the therapist is able to remain nondefensive enough to approach the issue forthrightly and respond effectively by inviting the client to express her concern more fully.

THERAPIST: I'm wondering if there is something about our relationship, or being in therapy, that just isn't feeling good for you right now?

CLIENT: Well, you did say you were a student here. Isn't that what you said when we first spoke on the phone?

THERAPIST: Yes, I'm a second-year graduate student, working on my master's degree.

CLIENT: Well, I don't really feel like a guinea pig or anything, but you really are just practicing on me, aren't you? I don't want to sound unkind or anything, but I'm wondering if you think you've had enough experience to help me?

THERAPIST: Thanks for bringing this up. Let me rephrase and make sure I'm hearing you accurately. Maybe you're saying that you are worried that if you go to all of the trouble of coming here each week, and talk with me about problems that are difficult to bring up, that in the end I just won't know enough to be able to help you?

CLIENT: Well, yeah, that's it. After all, you really are just a beginner, and I must be 15 years older than you. So, what do you think—can you help me?

THERAPIST: I cannot offer you any guarantees, of course, but I will do my best. And as we continue to meet together for two or three sessions, you will get a better sense of me, and what it's like to work with me. I think you will be able to decide for yourself soon whether this is the right match for you—and we can talk together about that honestly. But for now, let's see if there are any questions I can answer for you about my training status, and how our age difference might get in the way.

In this vignette, the therapist was able to approach the client's concern directly. She responded effectively because she remained nondefensive, even though this encounter was anxiety arousing and difficult for her to do. The therapist did not act on her initial impulse, which was to try to reassure the client of her ability to help. Instead of responding to her own personal need to seem competent to the client, the therapist was able to respond to the client's concern—by inviting the client to express her reservations directly and fully. The therapist tolerated her own personal discomfort well enough to be able to discuss the client's concerns and, in so doing, behaviorally demonstrated her competence. This is always more effective than offering verbal reassurances, which would only sound hollow to the client and still leave the burden of proof on the therapist.

Clients' Concerns about Entering Therapy.   In the previous examples, the therapist has inquired about resistance and the client has expressed concerns that personally challenged the therapist. In the following example, the therapist asks about the client's potential resistance (for example, "How is it for you to come in today and begin talking with me about these problems?") and the client expresses concerns about entering treatment—not about the therapist. Student therapists can be reassured that such concerns are far more common than challenges to the therapist and, of course, much easier for new therapists to address.

CLIENT: It's kind of hard for me to ask for help. I guess I'm not used to talking about myself, or maybe even thinking about what I might need or want.

OR

CLIENT: I feel a little awkward talking about problems to a stranger. In our family, we always kept things to ourselves.

The therapist's aim is to let clients know that he has heard their concerns—that he is trying his best to grasp what they are saying and "get" the key message, core meaning, or biggest feeling. The therapist's intention is also to demonstrate that she takes the client's concerns seriously and is willing to be flexible and try to do something about them if possible. Again, the best way to behaviorally demonstrate this empathic responsiveness is to invite clients to discuss their concerns more fully.

THERAPIST: That does sound important. Tell me more about what it means for you to focus on yourself, on your own experience or needs, and talk with me about your thoughts and feelings when you haven't done this much in the past?

CLIENT: I don't even know where to begin… it's like I've always been so concerned about everybody else that I've never really stopped to pay much attention to what was going on for me. Or maybe I just wasn't supposed to do that in my family… you know, be selfish…

OR

THERAPIST: I'm glad you're telling me about this awkward feeling around talking about personal problems with someone outside your family. Let's work together and see what we can do to make this better for you. I'd welcome hearing more about what your parents or others might think about you coming here to see me. And I guess I'm also wondering about how this family value might have been helpful to you, and how you think it might have held you back as well. Those are some of my thoughts. Tell me more about yours.

CLIENT: Well, my parents might think I was being disrespectful… I don't know, it's almost like I'm breaking some rule by coming here and talking to you, even though nobody ever really said anything like that out loud. But I do love my family, and I wouldn't want to do anything they don't want or that might make them mad.

THERAPIST: I can see how important your family is to you, and how much you want to honor them. So why don't we agree that, here in therapy, you decide what we talk about and how much you want to share. You set the pace and we'll go at your speed in here. We don't need to talk about anything that you don't want to. And I'll join you there—with the things that you decide you want to talk about. How does that sound?

CLIENT: Yeah, I like that. That sounds good. If I feel like I'm betraying my family, I don't have to bring it up, but if I decide I need to talk about something anyway, then it's OK with you.

THERAPIST: Yes, that sounds like a really good way for us to work together.

Listening and responding to the client's concerns with such affirming responsiveness is also important when ethnic or cultural differences exist between the therapist and client (McClure & Teyber, 2003). Cultural differences can be a sensitive issue; neither client nor therapist may feel comfortable addressing them. It sometimes seems as if there is an unwritten societal rule against recognizing differences. As before, however, the best way to work with these important differences—which often are central to the client's identity and sense of self—is to acknowledge them in an open-ended way. Therapists want to give clients overtly spoken permission to be able to express any concerns they may have. For example, suppose that the client and the therapist clearly are of different ethnicity, age, socioeconomic status, or religion.

THERAPIST: I wanted to check in with you about what it's like to meet with me. Was there anything about coming today, or talking with me, that could get in the way of our work together?

CLIENT: Well, I don't mean to be disrespectful, but I don't know if you can understand my situation because you're a different race... I mean, you're Latino and I'm African-American.

THERAPIST: I'm glad we're talking about this, because our cultural backgrounds really are different. At times, there probably will be important things about you that I won't understand as well as I want to. When that happens, help me out and tell me when I'm not getting it. We'll talk it through together until we've got it right. Was there something today that I missed or didn't understand that we should revisit?

The therapist responds affirmingly by accepting the client's concern, validating his point of view, and being willing to explore it further. It is important for therapists to be aware of differences between themselves and their clients as they work with clients who differ from themselves in terms of sexual orientation, social class, spiritual beliefs, race/ethnicity, and other factors. For example, some minorities may feel different in a predominantly white school or workplace—especially if few attempts are made to reach out and include them. In some settings, people who are different may be regarded pejoratively as deficient. The therapist's willingness to acknowledge differences and invite clients to express any concerns or misgivings they may have will go a long way toward establishing a therapeutic alliance and keeping clients who may feel different from dropping out prematurely.

In the examples above, the therapist was able to approach the client's concerns. As the therapist and client discuss the issue, some of the client's concerns will abate with a more complete understanding of the treatment process. Other concerns will be resolved as the client's faulty beliefs and expectations about the therapist are corrected. However, some of the concerns that clients express cannot be assuaged so easily (for example, the client says bluntly, "I don't trust people much"). The therapist can offer to be sensitive to these concerns and express a willingness to work with the client on them over the course of therapy. For example,

THERAPIST: I'm sorry that, too often, others have not treated you in trustworthy ways. I hope you find, as you get to know me better, that I am different.

The best ways to respond to clients' concerns about entering treatment is by enlisting clients' help in better understanding their background and experience.

The therapist can then ask clients to share their concerns more fully and begin a mutual dialogue to understand and change the problem.

As we have seen, when obvious ethnic, religious, or other cultural differences exist, this approach of enlisting the client's help is even more apt. Ethnic, class, gender, and religious differences are complex, and it is counterproductive for therapists to labor under the misconception that it is their responsibility to understand everything about a client whose culture of origin differs from theirs. However, therapists do have a responsibility to educate themselves about clients from different social contexts, for example, by reading and consulting with informed colleagues (see Pedersen et al., 2008; Sue & Sue, 2008). In particular, the therapist can invite the client to educate the therapist whenever the client feels the therapist is misinformed or does not understand. By welcoming this clarification instead of feeling threatened by it, the therapist signals acceptance of their very real differences, openness to a genuine dialogue, and offers an invitation for a real relationship. In so doing, the therapist gains credibility with the client and strengthens the working alliance.

The therapist also needs to explore more fully with clients how they believe their cultural or social context influences their current problems and shapes their subjective experience (Comas-Diaz, 2006; Constantine & Sue, 2005).

THERAPIST: Would you help me understand this problem in terms of the expectations from your family or community?

CLIENT: I can't discipline my children the way I want without being disrespectful to my mother. I'm a Latina and my mother lives with us—she thinks I'm still her daughter, you know, and that it's her role to help out by taking care of the children. I love her very much and appreciate her help, but she's spoiling my son and she won't listen when I ask her to stop. She almost gets mad at me when I speak up—like she thinks *she's* their mother and she decides how to raise them—not me!

Harkening back to our core concepts in Chapter 1, even when obvious cultural differences are not evident, therapists will be more effective as they try to enter their clients' subjective worldview more fully (Speight et al., 1991).

To sum up, the therapist's intention is to encourage clients to express any concerns about treatment or the therapist that arise, and then sincerely try to accommodate those concerns. This will go a long way toward resolving them and allowing many more clients to enter treatment and successfully receive help.

RESISTANCE DURING SUBSEQUENT SESSIONS

Understanding Clients' Inability to Change: the Three R's. As treatment progresses, some motivated clients will have phases when they find it hard to come to sessions or simply can't make progress in treatment. Too often, supervisors observe that student therapists disengage—quietly giving up on clients at this point—often doubting whether the client is "motivated" or questioning whether this client "is ready to take action and start making real changes." Instead, more effective therapists welcome these expectable impasses and use them as opportunities to take treatment further. Rather than blame or judge the client, and stop trying as hard as a result, effective therapists use process comments and invite the client to join them in a collaborative, nonblaming exploration to understand the resistance or block:

THERAPIST: It seems to me that we've kind of gotten stuck. You know, like we just aren't moving forward the way we were when we first started. What are your thoughts about this—how do you see things going for us at this point?

As we would expect from client response specificity, the reasons for clients' resistance are complex and varied. However, when therapy has stalled or clients have stopped making progress on a problem, therapists can begin making sense of things by formulating working hypotheses around three issues: *ruptures, re-enactments,* and *resistance.*

First, we have already seen that *ruptures* occur when there has been a misunderstanding or interpersonal conflict between the therapist and client that has disrupted the working alliance. The therapist repairs or restores these expectable ruptures by talking directly with the client about the interpersonal conflict or problem between them. For example,

THERAPIST: You sound irritated with me, James, maybe there's a problem between us? If so, I'd sure like to talk it through—you know, so we can understand what's wrong and work it out.

Again, one of the biggest problems in the counseling field—cutting across clinicians' experience from every brand of treatment—is therapists' reluctance to address the misunderstandings and misperceptions that so commonly occur and routinely disrupt the therapeutic alliance.

Second, *reenactments* also hold back clients from being able to succeed in counseling and make meaningful changes in their lives. As emphasized earlier, therapists are encouraged to consider the possibility that in some actual or metaphorical way, the same type of problem or unwanted interaction that the client is having with others is being played out between the therapist and the client. That is, the treatment block is occurring because the interpersonal process between the therapist and client is reenacting, rather than resolving, some aspect of the client's problem. For example, based on early maladaptive schemas, the client believes the therapist has been critical of her, or privately has become absolutely certain that the therapist has become deeply disappointed in her—just as her spouse and parent have been. As before, the therapist intervenes best by working in the moment and exploring the current interaction between the therapist and client:

THERAPIST: What do you think I might be thinking about you right now, Joan, as you're telling me about this?

Or, suppose another therapist chose to self-disclose a problem of her own—without realizing how pervasively this client had been parentified while growing up. Now, this client feels she is in her old familial role again and has to "take care of" the therapist, too. When such reenactments occur—as they predictably do for all therapists—the therapist can use process comments to help change the therapeutic interaction once it becomes clear that they have become stuck or progress has slowed. In the following example, a student therapist shared some personal family information with a client who was struggling with her parents' recent divorce and was feeling "shame" about this family "failure." In discussing this self-disclosure with her supervisor, the therapist said she'd shared her own experience of divorce to "normalize" the reality of divorce and minimize the stigma the client appeared to be struggling with. Her supervisor cautioned, however, that the client seemed to have been parentified in her family of origin and even now, having left for college, was still called frequently by her mother to get emotional support. The supervisor suggested that she be alert for "caretaking" from this client:

THERAPIST: You asked about my own life and parenting, and I told you about my divorce this past year. I'm not sure, but it sees like you have gotten quieter since. Maybe it's just me, but you seem a little more distant or reserved. Can you help me out here... could my feeling be telling us anything important?

CLIENT: Well, I guess I'm feeling sad for you, and concerned about how your life is going.

THERAPIST: I appreciate your concern for me, but I'm doing fine, and wondering right now if this is putting you in a very familiar but unwanted spot?

CLIENT: Well, this is what I do, you know, I've been taking care of others for a long time.

THERAPIST: Uh-huh, so this has kind of put you in your old, familiar spot of having to take care of me, too. Well, if we did that, it sure would end your opportunity to learn about and change that old role, wouldn't it? I'm so glad we're talking about this, so we can make sure that doesn't happen here. You've been a caregiver for others, at your own expense, long enough in your life and you don't need more of it with your therapist, too. But I can sure see, now, how talking about my own divorce triggered that familiar role for you. Have there been other times when this has come up between us—when you have felt that you needed to take care of me, too, in some way?

By being able to explore and talk together in this way about what *may* be occurring between them, the therapist and client can readily restore their working alliance, stop reenactments from continuing to play out, and resume productive work.

Early in their training, most student therapists will not possess the objectivity necessary to identify these predictable but subtle reenactments and will need assistance and input from their supervisors to help identify when potential reenactments may be occurring. As well, new and experienced therapists alike benefit from personal exploration in their own therapy, so they can become more aware of their own response styles and better anticipate how their own tendencies to respond could interact with this client's interpersonal patterns. When reenactments are occurring between the therapist and the client, clients are set back in treatment because their maladaptive schemas are behaviorally confirmed rather than experientially disproved. When maladaptive schemas are confirmed, clients lose the *interpersonal safety* they enjoyed with the therapist and cannot risk further exploring their problems, so treatment commonly becomes stalled and proceeds on a more superficial or intellectualized level. However, once the ship has been righted and the therapeutic relationship again is enacting a solution to clients' maladaptive relational patterns, they can resume changing as these schemas are being challenged and expanded through the interaction with the therapist, rather than confirmed (Silberschatz, 2005).

Third, *resistance* will be re-evoked throughout treatment. As different issues are explored in treatment, clients will not be fully aware of all of the varying conflicts and feelings that have been activated by a certain topic. Especially for clients whose feelings and perceptions were consistently invalidated as they grew up, they soon become unaware or "confused" about many aspects of their own subjective experience—what they are feeling or seeing. Resistant clients are not lying or deceiving the therapist, and they have not lost their motivation—these are different treatment issues. Resistance occurs when clients are simply unaware of the multiple and often contradictory feelings that have been activated by acknowledging realistically that certain problems really exist, asking for help, exploring difficult topics with more specificity and—surprising to most new therapists—even by succeeding in treatment, making meaningful changes, and getting better. For example:

- Getting further help is relieving but may also arouse detested feelings of humiliation.
- Feeling cared about by the therapist's genuine concern is comforting but can also evoke sadness about the many times this need went unfulfilled.
- Being listened to and understood in a consistent, reliable manner is empowering but also may trigger guilt over being given to rather than taking care of others.
- Becoming more successful is exciting but arouses the fear of being envied, undermined, or retaliated against.

Unless addressed and clarified, these commonly occurring reactions will lead to resistance. They will cause a few clients to leave treatment prematurely, and many more to sabotage the success or undo the changes they have achieved. For others, therapy will stall and becomes repetitive or intellectualized. At the beginning of their clinical training, however, many therapists are not familiar with the ambivalent, push-pull, or paradoxical nature of these conflicts. Let's explore this further by recalling Marsha from Chapter 2.

It was profoundly reassuring for Marsha when her second therapist approached her feelings directly and validated her experience. At the same time, however, this very positive experience also aroused other contradictory feelings that she "hated." Marsha now felt heard and understood but, at times, also felt:

- sad, as years of unacknowledged loneliness welled up;
- angry, at not being heard so many times;
- guilty, for feeling angry at her parents and being disloyal to them; and
- anxious, for breaking familial rules and talking to someone about problems and what was really wrong in her family.

Marsha did not want to experience any of these feelings and, without being very aware of it, it was hard for her to stay with treatment issues related to her family when these feelings were evoked. She did not understand them and felt threatened by them, yet she could not make them go away. Fortunately, her therapist was comfortable with the push-pull nature of conflict and consistently helped her make sense of the different, contradictory aspects of her experience that emerged. His accepting response helped her to integrate these conflicted feelings and gradually resolve them. In these ways, the second therapist was providing

Marsha with **containment.** In attachment terms, this is a "holding environment" that provided her with the psychological safety she needed to address—explore, experience, share—these sensitive issues for the first time. As these changes occurred in the therapeutic relationship, Marsha began telling the therapist that she found herself "feeling more confident" than she had before and that she was getting better grades and making new friends at school.

At different points in treatment, clients will often be unaware of, and threatened by, the conflicted feelings evoked in the change process. As a result, they may feel misunderstood or even blamed when the therapist inquires about potential signs of resistance. Sometimes clients do feel that the therapist's questions imply that they are doing something wrong or are not trying hard enough. Concerned that the therapist is frustrated or disappointed with them, clients may try to justify their good intentions:

CLIENT: Oh no, you don't understand. I really do want to see you and get here on time. It's just that…

As already emphasized, therapists never want clients to feel blamed or judged. Rather, *the therapist is inviting the client to join in a collaborative exploration to understand the threat or danger that some aspect of treatment has aroused.* Once clients understand that the therapist's intention is to explore the threat or danger, and not blame or judge, most readily join in a productive exploration.

Thus, therapists can best approach potential signs of resistance in a spirit of mutual exploration. Although new therapists are often concerned about being too blunt or being misperceived as critical, the problem actually goes the other way. *Because so many therapists are too worried about hurting their clients' feelings or evoking their disapproval, they do not respond as clearly and forthrightly as they could.* Consequently, many new therapists dilute what they say and do with clients to mute their impact. If therapists are worried that their explorations may be seem intrusive or unwanted to the client, they can shift gears and find another way to respond. Better yet, they can check this out by just asking the client directly.

THERAPIST: I like talking about these issues that come up between us, I think it helps our work together. But I wanted to check in and make sure that you're feeling comfortable with this forthrightness, too. How is it for you to talk with me like this—is this going OK for you?

CLIENT: It was different at first, I wasn't sure what you were up to. But now I like it— I feel like I'm finally talking to somebody about what's really going on. My wife likes it, too; she says I'm more straightforward and talk about things more honestly than I used to.

Researchers evaluating the effectiveness of the integrative treatment approach of Motivational Interviewing find that pushing or arguing against resistance is especially counterproductive, while collaboration and empathy facilitated change (White & Miller, 2007). Thus, the therapist's intentions are to inquire about potential signs of client resistance in a respectful manner that invites a collaborative dialogue. As before, a *tentative* approach is needed. Therapists can merely wonder aloud with clients about the possible meanings certain behavior may hold; there is no insistence on exploring a client's resistance or any other topic. Therapists can approach resistance without arousing defensiveness through a series of gradual steps that are progressively more direct; a three-step sequence is often effective.

In the first step, the therapist offers a *permission-giving and educative response* to encourage clients to talk about the positive and negative reactions that come up for them about treatment or toward the therapist. In particular, the therapist explains that problems or misunderstandings with the therapist are inevitable, and the client can help by bringing them up. For many, this is all that is needed. When this invitation is insufficient, the therapist can take a second step and encourage the client to explore the potential threat these feelings hold; this focuses on the *defense*. For example:

THERAPIST: Nadine, again today, you kept jumping around from topic to topic. It was especially evident after you said you had dinner with your sister. I'm wondering what the threat or danger might be for you if you didn't change the topic so frequently but stayed with one issue longer and allowed me to join you there?

CLIENT: I don't know. I guess I just don't want to think about anything for too long...

THERAPIST: You don't want to think about anything for too long—something feels unsafe about that?

CLIENT: Yes... (pauses) I think maybe I'd start to cry or something.

THERAPIST: Something about crying with me isn't safe—it doesn't feel OK. Help me understand that better. If you were sad and started crying, what might I do that would be unwanted or make things worse for you?

CLIENT: Well, I'm not sure why I'm thinking this, but I'm afraid you might put me down or something.

THERAPIST: Put you down?

CLIENT: Yeah, think I'm feeling sorry for myself, acting like a baby...

In the third step, if the client continues to show signs of resistance but cannot talk about it the therapist can draw on previous working hypotheses, tentatively suggest a possible interpretation, and try to get the client to join collaboratively in refining or improving this suggestion. For example:

THERAPIST: It seems like it's been harder for you to get here and really talk to me since we began talking about your wife's illness. And that makes *so* much sense to me! The doctor said she's got cancer, Bob, and it's completely understandable that it's being really hard to face all of that. What do you think about what I'm suggesting right now?

CLIENT: I just can't stand to think that those kids could lose their mother...

THERAPIST: Yes, of course, that possibility is absolutely heartbreaking. Bring me in on this with you, Bob, tell me more about any part you wish.

Let's explore these three steps more closely.

Step One: Permission-Giving and Educative Response

THERAPIST: Can we talk about what it's like for you to come and see me? I noticed that you were 20 minutes late today and had to reschedule your appointment last week. Maybe it doesn't mean anything, but I was wondering if something about our work together might not be feeling right for you. Any thoughts come to mind?

CLIENT: Oh no, I really do want to be here. Things have just been crazy at work. You've been very helpful, really. I had a deadline last week and today it's just that my boss called as I was leaving.

THERAPIST: OK, but if in the future you ever find yourself having any concerns about therapy or the way we are working together, I would really appreciate it if you would speak up and tell me about them—you know, let me know about any problems or misunderstandings that might come up. This can happen between us just as it does in other relationships. How would it be for you to tell me about something I did that you didn't like or something about being in treatment that was uncomfortable for you?

CLIENT: Well, that might be kind of hard for me to do.

THERAPIST: Uh huh... tell me, what do you think might be hard about that for you?

CLIENT: Well, I wouldn't want to hurt your feelings or anything, especially because you're such a nice person. It's hard for me to tell people things that are not nice— I try to be nice to everybody, I guess. It's important for me to be sure that I'm being respectful and nice. Maybe I worry about that too much.

THERAPIST: Yeah, maybe being nice all the time has become too much of a worry for you. I can honestly say that I would welcome hearing any concerns you might have about us or about counseling. I'd think you were working with me to help you. I certainly wouldn't think that you weren't being nice to me. I would welcome that kind of openness between us. But maybe you're telling me that's been different with other people that have been important in your life.

CLIENT: Oh, gosh yes, it's very important what others think. But I like what you're saying—that I can really tell you if something doesn't feel OK...

At this point the therapist and client are off and running together. The therapist can help the client work through whatever issues emerge that could prevent him from expressing any dissatisfaction with treatment or disagreement that might come up toward the therapist. In this example, the client's concern that the therapist may not like or approve of him, or perhaps that the therapist and others will leave him unless he is always "nice," has become overt and on the table for discussion. Not only will this go a very long way toward allowing this client to remain engaged in treatment; the permission-giving and educative response revealed an important problem that needs to be addressed—in this example, the client's ineffective coping style of compulsively pleasing others.

In this first example, the therapist offered a permission-giving and educative response in the hope that the client would then feel free to talk about any ambivalent feelings about treatment or conflicts with the therapist that may emerge. If the client continues to show signs of resistance without being able to talk about it (missing, canceling, rescheduling, coming late, discussing only superficial issues), the therapist moves to the second step by exploring the interpersonal threat.

Step Two: Explore the Danger/Identify the Threat

THERAPIST: It's important for our work together that you and I are able to talk about any difficulties that might come up between us—they're bound to occur in any relationship. I'm wondering if you would be able to tell me if something about our relationship or work together was troubling you?

CLIENT: I suppose.

THERAPIST: Well, OK, but based on what you've told me about some other parts of your life, I'm wondering if that might be kind of hard for you to do. You know, thinking that it might be kind of new or difficult for you to tell me if I did something that you didn't like, or if there was something about being in treatment that wasn't feeling very good for you. Sometimes people believe they shouldn't have frustrated or disappointed feelings, or at least they don't think they should express them.

CLIENT: Well, yes, that's probably true.

THERAPIST: Let's try to figure this out together. It seems that you are having some trouble being in treatment but don't feel comfortable talking with me about that yet. I'm wondering what the threat or danger might be for you if you were forthright and took the risk of telling me about something here that didn't feel right. What's gone wrong with others in the past, or what might go wrong between us?

CLIENT: Well, that's pretty obvious, isn't it! You probably wouldn't want to see me anymore—I'd be too much trouble, too critical or too demanding or too something, and whether you said it or not you'd rather I just went away and didn't bother you anymore.

THERAPIST: You're saying so many important things right there, but maybe the biggest thing you're saying is that I'm going to leave you or walk away from our relationship if you speak up and have your own voice—and don't just go along with what I say or comply with what I want?

CLIENT: Well, yeah, I guess that's kinda how it's been for me.

In this example, the therapist is trying to identify the threat that keeps the client from being able to talk about feelings or concerns that contribute to resistance and are central to the problems she is having with others. Emphasizing process rather than content, *the therapist is not trying to find out why the client is not showing up for therapy, but why the client is having trouble talking about it.* Here, the client reveals her pathogenic belief that if she disagrees with the therapist, or speaks up and expresses what she wants or doesn't like, the therapist (and others) will leave her. This may not seem like a welcome opportunity to the new therapist, but the potential for significant change is present at this moment. The therapist above is beginning to provide a corrective emotional experience by clarifying that he does not wish to end their relationship as soon as they have a disagreement. As this pathogenic belief of "comply or be left" is highlighted and then disconfirmed in their relationship, it will also bring up how this belief had in fact been a reality-based expectation in other important relationships. As we see below, this client goes on to recall how the abandonment or rejection she expects from the therapist is what she behaviorally experienced while growing up in her family and in her first marriage (Client: In my family it was always, "go along or be left"). By first highlighting this outdated coping strategy of compliance in the therapeutic relationship and changing it in their relationship, the therapist can then take the next step in the change process and help the client generalize this to her everyday life. This occurs as the therapist helps this client begin exploring current relationships with others where she is also anxiously pleasing, unable to speak up and say what she thinks, and "goes along" in order to keep others connected to her, and begin to question and challenge this problematic way of relating with them as well. For example:

THERAPIST: Tell me how this poignant problem of either "going along or being alone" might be a problem with others in your life, too, like it's been with me?

CLIENT: I think it makes it hard for me to say "no" to anybody. Right now, as we're talking, I'm thinking about what a bad mom I'm being because I just can't discipline my kids. I guess I don't want them to be unhappy with me either—so, yeah, I guess I'm going along with what they want and just trying to please them, too.

THERAPIST: You're doing great. Now that we're starting to understand this, let's see if we can help you change it. You know, take a stronger stand with them, tolerate their disapproval rather than cave in to that fear of being left, and follow through and enforce the rules you set. How does it sound to start speaking up more assertively in here with me, and to start working together to help you discipline your kids more effectively?

CLIENT: It's like my whole life would change—I'd be a different person if I could do that.

This client continued to work productively in treatment and did not miss or cancel another appointment after this session. Similarly, with most clients, exploring the threat or danger in this way will both resolve their resistance to treatment and uncover key concerns that are central to their presenting symptoms and problems. When repeated attempts to focus on the threat do not work—the therapist and client cannot successfully collaborate together and identify or "name" the threat or danger, therapists can move to the third step. As the next alternative, the therapist can try to interpret the content of the client's resistance. For example, if the therapist thinks the client's resistance is about to pull the client out of treatment, and the client cannot join with the therapist and talk about the difficulty in the ways described above, the therapist can draw on working hypotheses and tentatively suggest possible reasons or interpretations for the resistance.

Step Three: Tentatively Interpret Potential Reasons for the Resistance

THERAPIST: It's been hard for you to get here; you've missed the last two sessions. Maybe you just "forgot", but I'm wondering if there is something about coming here to talk with me that doesn't quite feel safe—or something that might have gone on between us or that I said that didn't feel quite right to you. I know I've asked you this before... but do any possibilities come to mind?

CLIENT: No, not really.

THERAPIST: Well, OK. Let me suggest some possibilities I've wondered about, and you tell me what you think of them. As I think about some of the things you've told me about yourself, I'm wondering if you're concerned that I too am going to "judge" you, like too many others in your life have done?

CLIENT: Yeah, I guess that might be true. But it's not so much that you would judge me, it's more like you wouldn't respect me if you knew more about me.

THERAPIST: Thank you for clarifying that. It's not about judging, it's that I wouldn't "respect" you. I really appreciate that you are willing to risk sharing this with me.

Has that ever happened between us—has there ever been a time when you felt that maybe I wasn't treating you respectfully?

CLIENT: Well, no, it hasn't happened yet, but I haven't told you very much either.

THERAPIST: Good, I'm glad that hasn't happened yet. I really don't ever want you to feel disrespected by me and, if that ever comes up here with me, I want you to stop us right then and tell me so we can sort it out. But I hear you—you're afraid I'm going to lose respect for you if you tell me more about yourself. I see how that would make it difficult for you to come here each week and talk with me—and maybe make it hard to get close to others, too.

CLIENT: Yeah, I think that respect thing has gotten in the way a lot in my life. But I'm just not going to put up with anybody putting me down...

THERAPIST: What would make it safe for you to talk about all you need to say in here?

Why does the therapist take this more interpretive stance (tentative as it is) only as a last resort? The interpretation, accurate or not, is the therapist's issue and puts the ball in the therapist's court. Whenever possible, it is better to try to follow the client's lead, or focus on the therapeutic process, rather than to pull the client along in the therapist's direction. In other words, it is usually more effective to explore why it might make sense that the client is resisting, or make a process comment and address the current interaction, than to make interpretations or tell the client what to do. Why? The goal is to maintain a working partnership and keep the relationship as collaborative as possible. Whereas steps one and two will be highly effective with most clients, step three—tentatively suggesting possible reasons or interpretations for the resistance, and asking the client to modify or improve the interpretations to make them fit better, can be productive at times as well.

SHAME FUELS RESISTANCE

Resistance, defense, ambivalence, and ruptures will occur throughout every phase of treatment—it's just part of the work. The most common source of resistance, however, is *shame*. Only in recent years have therapists begun to recognize the pervasive, trans-diagnostic role of shame in many clients' symptoms and problems. Shame is more apt than any other issue to generate resistance, lead to premature termination, hold clients back from initiating change, and lead to relapses or the undoing of successful changes they have attained. In a highly unfortunate interaction, it is also the feeling that therapists are most likely to avoid with their clients. Simply put, it can be terribly difficult for therapists to bear sitting with their clients as they suffer such an excruciating feeling. Routinely, new and experienced therapists alike feel inadequate to help and painfully uncertain of what to do. Predictably, it evokes their own shame as well—which they wish also to escape as much as their clients do. As therapists read and learn more about this cardinal issue, they will develop the eyes and ears to recognize the *shame motifs* that pervade so many of their clients' narratives. For example, therapists will become better able to recognize the many different faces of shame that clients present: perfectionism; frequently blaming or criticizing others; having temper outbursts; being edgy and explosive; being judgmental or contemptuous of others; being self-critical and having low self-esteem or chronic depression; struggling with eating disorders and other addictions; living anxiously preoccupied about appearances, presentation, or approval; social withdrawal and avoidance; an inability to accept compliments or tolerate constructive criticism of any type; never feeling vulnerable or in need of anyone; being preoccupied about always being seen as strong, powerful, or independent; and many others. With guidance from supportive supervisors, therapists-in-training can begin listening for and recognizing shame-related themes in their clients' concerns about entering therapy, as well as in the issues and concerns that block their progress later in treatment.

SHAME VS. GUILT

Let's start with the basics: shame is different than guilt—and a far more significant problem to deal with. With guilt, clients feel only that they have done something wrong. The guilty client often feels regret and may look for ways to apologize or make reparations. With shame, in sharp contrast, *it's not that I have* done *something bad, it's that I am bad* (Lewis, 1971; Tangney, 1998). The sense of "badness" that is activated with shame generally leads clients to want to hide and avoid others' perceived scrutiny, and there is no thought or hope of resolution or repair as there is with guilt (Wells & Jones, 2000). A more primary and pervasive feeling than guilt, shame is a total or all-encompassing feeling about one's Self. With the toxic or "core shame" discussed here, as opposed to everyday, innocuous experiences of self-consciousness or embarrassment, clients report feeling essentially flawed in who they really are. Throughout this discussion, keep in mind that the key emotion in all forms of shame is *contempt* (Miller, 1984; Tomkins, 1967). To suffer shame is to feel that the true self with all of its defects is exposed, naked, and vulnerable to the damning judgment or criticism of others (Spiegel & Alpert, 2000; Teyber & McClure, 2011). Clients may be without words to communicate their experience when they are suffering a profound shame reaction ("My mind went blank."). Sometimes all they can say is something like "I'm just hideous," "I don't matter," "There's something wrong with me," or simply, "I hate myself." A full-blown shame reaction is agonizing to suffer, and very difficult for compassionate therapists to remain co-present with and share. During these shame-filled moments, when the Self is being annihilated, some clients may express a wish to die.

Oftentimes, shame can be expressed by clients along two different themes or within two distinct domains. First, there is a dimension of *self as bad* that predominately reflects feeling worthless, flawed, unlovable or unwanted, or defective in the core of their being. This client, sometimes referred to as a "hated child," often has been overtly rejected or angrily denigrated by caregivers and struggles with a sense of disgust for, and anger toward, the self. Perhaps the most heartbreaking consequence of child sexual abuse is this *internalized contempt* and resulting sense of self as stigmatized ("ruined," "dirty") (Finkelhor, 1990). The second dimension is one of *self as inadequate* and reflects enduring feelings of being small or weak, helpless or disempowered, incompetent or lacking, or a failure. In particular, this client regularly feels defeated, inadequate, and inferior (for example, "stupid").

Researchers also differentiate "internal and external" shame. Clients who struggle more with external shame are concerned about how one exists in the minds of others—how others are seeing or evaluating me. Those with internal shame hold a devaluing and self-critical attentional focus on themselves, and some clients cope with both internal and external shame (Lewis, 1995). Related, another useful distinction in the rich literature on shame is that some clients are quite unaware of their own propensity to feel shame, whereas other are highly sensitive to it. Some clients, who may be variously described as narcissistic, aggrandizing, or controlling, are unaware of their own shame but often are critical or judgmental and readily evoke shame in others. In contrast, other clients are more aware of their own "low self-esteem." They are intra-punitive, harshly self-critical, and too readily please, take care of, or comply with others (Wells & Jones, 2000). Sadly, these clients also *withdraw* from many promising social relations and good opportunities in life, thinking, "She won't want me," "They won't hire me," and so forth.

SHAME-PRONE SENSE OF SELF

Let's use the term *"shame-prone"* to clarify the difference between the toxic or core shame that is central to so many clients' symptoms and problems, and the everyday embarrassments that are just an awkward part of life for everyone but are not important treatment issues. It is universal to feel embarrassed in awkward situations and common to feel ashamed in certain moments. This is unpleasant, of course, but not important to our work as therapists or relevant to this discussion. In contrast to these situation-specific feelings, clients who have been frequently shamed in their important relationships—that is, *repeatedly held in contempt by their attachment caregivers*, develop a *shame-based sense of self* and are *shame-prone*. This shame-proneness pervades most types of psychopathology and is central to our work with many clients. Shame-prone individuals, for example, will not be able to accept constructive criticism because it threatens to expose their deeper, pervasive, and long-standing feelings of shame. Even good-natured teasing from family or friends can evoke these feelings. Humiliating feelings of being profoundly diminished, rejected or abandoned, or exposed are readily triggered for shame-prone clients. As noted, these intense and sometimes volatile reactions are triggered in many different situations in their daily lives that would not evoke such strong reactions in others. In response to the intense feelings of shame that are continuously evoked, shame-prone clients employ a wide variety of coping strategies to keep their shame concealed from others and, perhaps more importantly, from themselves as well. Routinely, such clients protect themselves from having their underlying sense of *shame-worthiness* revealed by:

- acting arrogantly and self-righteously toward others;
- intimidating, controlling, or inducing shame in others;
- adopting perfectionistic standards around their own work, cleanliness, or religious practices;
- developing eating disorders or other addictions;
- becoming preoccupied with appearances and social approval;
- becoming derisively self-critical and withdrawing socially;
- becoming judgmental and critical of others;
- Withdrawing from others and avoiding many interactions.

SHAME-RAGE CYCLE

Therapists will observe other faulty coping strategies that clients have developed to cope with these dreaded feelings. Perhaps most disturbing to witness is the **shame-rage cycle** that characterizes "road rage" and most domestic violence episodes of battering and physical abuse (Gilligan, 2009; Simon, 2002). This shame-prone individual explodes in rage in the seconds after feeling diminished or demeaned by someone, often in response to a completely innocuous, benign comment, such as his wife asking him to take out the trash (or a child asking "why?"). The temper outburst or rageful response is a desperate but futile attempt to disprove their shame-worthiness ("I'm not weak," "I'm not a piece of garbage;" "You can't boss me around like that!"). This is an artificial or defensive attempt to "*restore*" a sense of personal power and Self—often through aggression or contemptuously diminishing others ("You can't tell me what to do—you can't "dis" me like that. I'll show you who's the trash here, and who's really the boss!").

To emphasize, clients who have grown up suffering *contempt* from caregivers (for example, a parent who often says, in a disgusted tone of voice, "What's wrong with you? Can't you do anything right?" or "You're ruining my life... I wish you were never born"), develop a shame-based sense of self. These clients usually remember the painful developmental experiences that were so crushing of their sense of self but have split-off or defended against the accompanying feelings. For example, the client whose drunken father made him lick the kitchen floor, while frightened siblings looked on helplessly, can usually remember what occurred behaviorally—what was said and done, and can recount this humiliating experience to the therapist. However, *he does so without being able to feel the shameful affect or experience the unwanted emotions that accompany this abuse.* That is, the behavioral events are often accessible and remembered whereas the accompanying feelings are too painful to acknowledge or experience and may be split off or sequestered away. Similarly, this client often cannot recognize or identify the current experience (feeling "bossed around" or told what to do) that has just triggered his temper outburst. In the following dialogue, the therapist is helping this shame-prone client make the connection between the triggering event (feeling put down or "dissed" by someone) and his presenting problem (losing close relationships because he keeps losing his temper).

THERAPIST: As I listen to the sequence here, it sounds like you lost your temper just after your wife asked you to take out the trash. Last week you told me you lost your temper when your coworker decided which part of the job you should do. You said afterwards that in reality it wasn't a big deal but your first response was to "go off" at him. I wonder if that's telling us that you lose your temper when it seems as if others have just put you down or dominated you in some way?

CLIENT: I hadn't thought of that, maybe... You know, that reminds me, I almost got in another fight this week, when this worm flipped me off on the freeway. I wanted to kill him—my buddy, the one who calls me "pit-bull," said my face was so red he could see the veins bursting out of my neck while I was screaming at him.

THERAPIST: That fits with my suggestion that when someone flips you off, or even just tells you what to do in a reasonable way, you feel shamed—and then the rocket goes off. It's like it almost becomes a matter of life and death for you—as if you are saying: I'm being put down, I can't stand it—no matter what, nobody is ever going to treat me like this again—so you explode.

CLIENT: Yeah, that's right. I hadn't thought of the word "shame," but I do go kind of crazy when someone puts me down.

THERAPIST: Yes, I think you do, too, and I'm concerned that you are embarrassing your wife and scaring your son. Maybe you feel shamed sometimes even when others aren't putting you down. It sounds to me like your wife really was just asking you to take out the trash. So, as you think about it now—here with me—do you think she was trying to boss you around or put you down? Let's think about what she might have been thinking—what could have been going on in her mind, as all of this began.

CLIENT: She's a good person... she wasn't trying to put me down. (*long pause*) I've got to change this—it's ruining my life.

THERAPIST: Yes, it is ruining your life, and I'd like for us to keep working together to help you change this. So, let's keep talking about "shame"—that's the feeling that seems to trigger or set off your temper. When I say "shame," what comes to mind?

CLIENT: ... Well, my stepfather used to beat us... he's dead, but I still hate him...

SHAME-ANXIETY

Instead of temper outbursts and rage, other clients who are shame-prone have intense anxiety evoked in many different situations that hold the ever-present threat of revealing their basic unworthiness, inadequacy, or inherently flawed and defective self. As the dreaded feeling of shame is evoked, yet again in this new situation, therapists will commonly witness **shame-anxiety**. For example, something as innocuous as asking someone for a date, applying for a job, needing to ask for help with a flat tire, getting lost and needing to ask for directions, or making a simple mistake at work evokes the anxiety that the client's shameful inadequacy, neediness, demandingness, and so forth will be exposed. Repeatedly, throughout their life, strong anxiety is evoked by the threat of having their inadequate, flawed or bad self revealed. In many shame-prone clients, this occurs so frequently that they live with a pervasive, anxious vigilance that may be expressed in the presenting symptom of anxiety attacks, social phobia, avoidant personality disorder, a generalized anxiety disorder, and others. In other words, the anxiety signals that clients' defenses against their shame are being threatened in this situation, and raising the specter that their shame-worthiness is going to be revealed to others right now. To manage this anxiety and protect against the potential exposure of their shame, clients may cope by withdrawing or avoiding, being pleasing or complying, trying harder to be perfect, never needing anybody or asking for anything, and so forth, which becomes a characteristic or habitual coping response that characterizes their lives.

As we will continue to explore, therapists can help clients recognize and change these problematic coping strategies for defending against their shame (for example, yelling and hitting, avoiding and withdrawing, controlling or dominating others, inhibiting or constricting oneself, complying with others and going along when they want to say no, becoming intoxicated or getting high every day, binging and so forth). Therapists help clients resolve their shame-proneness, and the many varying symptoms that accompany it, by allowing them to experience and share the self-hatred or contempt they feel toward themselves. Change occurs as clients have a corrective emotional experience and find that the therapist is not critical/judgmental but compassionate and the client, in turn, *begins to have some empathy or compassion for themselves and the plight they once were in* (Gilbert, 2009; Gilbert & Procter, 2006). The biggest problem with these interventions, however, is that therapists often have as much difficulty witnessing and responding to the raw experience of shame as clients have in bearing it. Of course, many therapists are shame-prone themselves and have their own unresolved shame dynamics activated by the client's shame. Too often, these therapists try to cope with their own activated shame by reassuring the client, minimizing the problem, offering intellectual explanations or interpretations, changing the topic or avoiding the issue altogether, and providing other well-intended but ineffective responses. Further guidelines for responding to shame and helping clients "contain" these and other painful feelings will be provided in Chapter 5—we need to learn about this difficult topic in manageable doses. For the time being, therapists can begin listening for the presence of shame in their client's narratives and begin considering the possibility that *resistance often serves to protect clients from their shame and from having seemingly unacceptable parts of themselves revealed.* This is a simple concept to say but, because of our own shame dynamics and cultural prohibitions against approaching such taboo issues, takes most new therapists several years to integrate and apply. Having a supportive supervisor who is attuned to shame dynamics is essential to help new therapists begin working with their clients' shame.

> *Susie's boyfriend did not show up for their date. Feeling desperate and empty, she found herself in the kitchen spooning peanut butter out of the jar and into her mouth until the jar was empty. The next day in therapy, telling her therapist about what had happened, she became overwhelmed with shame—feeling like a "disgusting pig." Susie's therapist wanted to look away; her mind raced trying to think how she could make her client feel better and stop this sickening pain. By asking about the binging episode, Susie's therapist felt she would be causing her client incredible self-loathing that only seemed to be accelerating out of control. And, almost immediately, the therapist's shame about her own weight and appearance began to flood her as well. The therapist wanted to look away, get away, and do anything to change the conversation and end this excruciating moment—it was too painful and too close. Fortunately, she was able to recall her supervisor's words of support and take a slow, deep breath instead. Rather than flee from her own shame and try to stop the client from feeling hers, the therapist was able to provide a corrective emotional experience by somehow getting out the words, "I'm honored that you are able to share such a vulnerable part of yourself with me. It is a privilege for me to be with you right now." Susie cried hard and spent the next three sessions sharing her history of experiencing rejection, feeling alone, and her desperate need for more support in her life. Susie and her therapist were able to begin addressing her profound shame related to issues of worthiness as well as how to cope without binging when these feelings were activated.*

Learning to work with clients' shame is not easy, certainly, but therapists will find their clients remain engaged in treatment, and make real progress with their problems, as they become more effective in working with shame. More specifically, clients resolve their shame as therapists acknowledge or "name" it and respond empathically, as Susie's therapist demonstrates so effectively. It's exciting to see the important changes that commonly follow as the client's shame-based sense of self improves ("low self esteem" is often a euphemism for shame). Shame-prone clients, for example, who were too preoccupied with their appearance or weight become more self-accepting; others who didn't care enough about how they looked take pride as they lose weight, begin fixing their hair, or dress better. Anxiety over having their shame-based sense of self revealed, and depression over feeling so bad about one's self, diminish markedly as the client finds that the therapist can see this unacceptable part of them and still respond with compassion rather than the criticism and judgment they expect.

SUCCESS IN TREATMENT CAN BE THREATENING

Moving on, these behavioral changes are not an end point in treatment. Continuing with resistance and the push-pull nature of change, many clients will experience anxiety and conflict over succeeding in treatment and achieving these new behavioral changes. This paradox or self-defeating behavior is perplexing to most new therapists. In other words, resistance predictably returns, even as clients make healthy new changes. For example, some clients will experience binding feelings of guilt or disloyalty over making progress and having success in treatment. Thus therapists are going to observe that some clients retreat from progress in treatment or undo successful changes they have just made. Often, clients cannot sustain positive changes they have just achieved because the healthy new behavior is inconsistent with their cognitive schemas. Following attachment theory and internal working models, acting in this successful new manner (such as stopping binge-eating episodes, leaving a nonsupportive relationship, graduating from college) threatens their attachment ties to caregivers who did not support their independence or success. Becoming stronger or improving in therapy makes some clients feel cut off from parental approval and affection, disloyal to caregivers, or guilty about hurting, leaving, or surpassing the parent. As they make significant behavioral changes, or report feeling better about themselves, therapist can expect some clients to report feeling either: (1) alone, empty, or disconnected following meaningful change or (2) guilty, selfish, or bad. When these reactions occur, clients need to be reassured of the therapist's continued presence and support for this stronger self. If not, they will have to manage their guilt, or restore their insecure and threatened attachment ties, by sabotaging their own success. For example, the therapist challenges the client's maladaptive schema, disconfirms the problematic expectation, and provides a corrective emotional experience by saying:

THERAPIST: You've just had a big success at work—and I'm wondering if that has something to do with why you feel so anxious and alone right now. How does that sound to you?

CLIENT: As you say that, it makes so much sense... You know, I called my brother last night, really excited, thinking he'd be happy. You know, when he was promoted last month I took him out to a fancy dinner. All he said was, "oh... nice" and quickly he had to go. Usually he talks more. It made me sad. I do feel that somehow I should not be succeeding... or succeeding so fast...

THERAPIST: I am glad we are talking about this. I hope your experience with me will be different than with your brother—and maybe from others you have been close to in the past as well, because I am very happy for your new promotion. I'd like to hear all about it and spend as much of the session as you like talking about this success and what it is like for you to be doing so well.

CLIENT: I'd like that; it will be good to have someone who can help me learn to celebrate the good things in my life too.

These commonly occurring sources of resistance and impediments to change will be explored further in the chapter ahead. In particular, we will return to the cardinal affect of shame in Chapter 5.

CLOSING

Listening to the client with presence and respect is the most important intervention the therapist can make in the beginning of treatment. Sustaining this empathic stance toward the client and consistently grasping the most important feeling or key concern in what the client just said is the basic tool the therapist uses to establish a working alliance. The next step in the treatment process is to be attentive for ambivalence and find nonblaming, exploratory ways to inquire about potential signs of resistance in order to maintain the relationship that has just begun. Resistance will be more of an issue with some clients than with others, but it will occur to some extent for every client. As we have seen, if the client can talk about these conflicted feelings, and the therapist responds affirmingly, they will be far less likely to pull the client out of treatment prematurely. Resistance will wax and wane throughout therapy, often reemerging as clients enter more deeply into difficult issues and as they try out new behavior and make successful changes. To sustain progress in treatment, therapists need to continue responding effectively to the many types of resistance that occur. If therapists formulate working hypotheses about potential client resistance, it will help them to anticipate when it is likely to occur and to respond more effectively when it does. In sum, the therapist's goals are:

- to encourage the client's lead and ownership of the treatment process;
- to provide accurate empathy to foster a working alliance;
- to remain alert for potential signs of resistance and address it collaboratively through process comments, and;
- to begin formulating treatment plans by generating working hypotheses.

Resistance, ambivalence, and defense provide a window to observe the fascinating workings of internal conflict. People do not possess a unified self; we are so complex and multifaceted that it often feels as if one part of ourselves is working against another part. By resolving an internal conflict like the ones discussed here, a person becomes a little more integrated or whole. Integrating disparate parts of the self is usually an important part of behavior change and the avenue to greater self-efficacy. Working successfully with resistance shows clients that they possess the internal resources, and have the help they need, to make the changes in their lives they choose.

Suggested Readings

1. A case study illustrating how therapists can respond to a client's shame is provided in Chapter 3 of the Student Workbook: "Shame Dynamics: Case Study of Hank."
2. Readers interested in learning more about how to intervene with clients' shame may read E. Teyber & F. McClure (2011). Shame in families: Transmission across generations, to appear in J. Tagney & R. Dearing (Eds.) *Shame in the Therapy Hour*, American Psychological Association, in press, 2011.
3. Shame plays a bigger role in resistance and blocking change than any other issue, and it is the most sensitive affect for therapists to approach. To get an introduction and overview to the shame literature, readers are encouraged to explore Robert Karen's lucid article, "Shame," *The Atlantic Monthly*, February 1992, pp. 40–70. A highly informative article about shame intervention in couples therapy is "The Systemic Treatment of Shame in Couples" by D. Balcom, R. Lee, and J. Tager, *Journal of Marital and Family Therapy* 21, no. 1 (1995): 55–65. This article illuminates how couples "blame" each other to externalize or defend against their own shame, and how these blaming interaction sequences keep couples from being able to address and resolve any problem—no matter how small.
4. In the interpersonal process approach, we are working with resistance as a relationship issue rather than a fixed personality trait or isolated behavior. Readers are encouraged to explore other texts that similarly reframe the client's "resistance" as an outdated interpersonal coping strategy (see Clark, 1998; Harris, 1995; and especially, Miller & Rodnick, 2002).
5. A process-oriented approach to multicultural issues in counseling can be found in Chapter 1 of the casebook by F. McClure and E. Teyber, *Child and Adolescent Treatment: Cultural and Familial Contexts*, (Pacific Grove, CA: Wadsworth, 2003). Excellent information on ethnographic and multicultural issues in counseling can be found in M. G. Constantine (ed.), *Clinical Practice with People of Color: A Guide to Becoming Culturally Competent* (New York: Teachers College Press, 2007); and in M. G. Constantine and D. W. Sue (eds.), *Strategies for Building Multicultural Competence in Mental Health and Educational Settings* (Hoboken, NJ: Wiley, 2005).

AN INTERNAL FOCUS FOR CHANGE

Year after year as we supervise smart, promising student therapists, we are continually pointing out how they are keeping the conversation on the surface. Almost more than any other issue, we are giving some therapists permission, and challenging others, to go beyond surface, everyday conversations and engage clients more deeply—to repeatedly strive to join the client at the most meaningful point possible. This chapter gives therapists tools to achieve this goal.

Dana was presenting a new client she was seeing to her first-year practicum class. She showed a videotaped segment from her previous session in which her client was complaining about her past two husbands, her mother, her previous therapist, and others. Lost in trying to grasp what all of this meant, Dana half-pleaded with her practicum instructor, "What should I do?" With her usual good sense of humor, her instructor jokingly replied, "OK, you guys are always asking me what to do, and this time I'm actually going to tell you. Next week, Dana, tell your client that she has two choices. Either everyone she has ever known needs to be in therapy—so she should refer every person in her life here to the clinic for treatment. Or, she can just meet with you herself and begin to explore the decisions and choices she makes and her contributions to these problems." The class laughed and got the point.

CONCEPTUAL OVERVIEW

The first stage of therapy is complete when therapist and client have established a working alliance and begun to work together on the client's problems. This collaborative partnership is a necessary prerequisite to the second stage of therapy: the client's journey inward. In order to change, clients need to become less preoccupied with the problematic behavior of others and begin to explore their own role in problems. As real and compelling as these problems with others usually are, clients need to stop focusing exclusively on historical events, past relationships, and the problematic behavior of others in their lives. Instead, clients change when they begin to clarify their own thoughts, feelings, and reactions to the problems they are having. Why? Clients will usually fail in their attempts to change others in their lives, whereas they can often resolve problems by changing themselves—the way

145

they respond and their own participation in problems. In this process, clients examine their habitual response patterns in problematic situations (for example, am I withdrawing and avoiding in this situation—as I learned to do as a child? am I being too controlling and taking over—and acting toward her in the dominating way my father used to act toward me?), evaluate their usefulness in their current lives, and begin trying out new and more adaptive ways of responding. This internal focus on changing their own responses, rather than trying to change the other person, leads clients toward greater self-efficacy. In particular, presenting symptoms of anxiety and depression often diminish as clients are empowered to act more purposefully and live better in the ways they choose.

Thus, the therapist's task is to help clients make the transition from seeing the source and resolution of problems in others to adopting an internal focus for change. This is a twofold process. First, the therapist helps clients begin to look within. That is, without invalidating clients or using language that makes them feel "blamed" in any way, the therapist focuses clients away from complaining about or trying to change others, and toward understanding and changing their own problematic reactions. Psychotherapy process studies demonstrate that—across differing brands of treatment—positive outcomes are facilitated with therapist interactions that encourage clients to engage in self-observation and introspection (Beitman & Soth, 2006).

Second, the therapist's intention is to help clients assume more responsibility for change—more ownership of the treatment process—by becoming active agents in their own therapeutic work. This occurs as clients become more aware of how their cognitive schemas are not always accurate and their usual coping styles are not effective in many current relationships. With greater awareness of their own responses, and how they may be participating in or contributing to their problems, clients increasingly recognize that they do not have to keep responding in old and familiar ways that are ineffective. With this new perspective, clients often join readily with the therapist in exploring new behavioral options and begin working actively to change the problematic ways they are responding to others. That is, as clients become more aware of their contribution to problems and how changing their responses may help change the dynamics that have not been working in certain relationships, they are more able to commit to a change process that is empowering, and this will in turn contribute to behavioral change (see Miller & Rose, 2009, for research on empowering clients to commit to change). Throughout this chapter, we will see that when clients change their part in problematic scenarios, the interaction sequence changes—even if the other person doesn't change and keeps giving the same hurtful or disappointing response. This new option of modifying their role in problematic interactions, coupled with the therapist's support for the client's own self-direction and initiative within the therapeutic relationship, helps to empower clients to change.

Self-efficacy develops out of this collaborative process between therapist and client (Bandura, 2006) The therapist is not fixing or curing the client—the client is sharing ownership of the treatment process. Why is this interpersonal process so important? The therapist's goal is to empower clients rather than just give them answers or tell them how to live their lives. Our goal is to help clients know their own mind—to give them permission to have their own voice and trust their own intuition or gut feelings—and thereby support them to become better able to make their own decisions. When given answers or prescribed solutions, clients do not make them their own—they cannot apply the advice in the next situation or learn how to solve future problems without having someone else tell them what to do. By helping clients look inside at what they are feeling, expecting, and doing, we are teaching them skills that they can carry over to their everyday lives and use on their own when treatment is over. In this way, enduring change results when clients participate actively in treatment and feel ownership of the change process.

When clients "own" insights, and feel shared responsibility for constructing them, they are vastly more successful at translating these new understandings into new behavior. That is, the "aha" experience that can lead to new behavior comes about when the client has been an active participant in achieving insight, not when the therapist has just explained a connection or interpreted what something means. Treatment gains are maintained after therapy has stopped when the client helped to make them happen. However, it may be difficult for some clients to shift their focus away from others and begin looking at their own contribution or role in problems. In order to take this productive journey inward, clients need a relationship with a therapist who is both affirming, on the one hand, and, on the other, willing to take the risk to challenge them to look within. In this chapter, we will see how therapists can do both.

SHIFTING TO AN INTERNAL FOCUS

HOW THERAPISTS CAN HELP CLIENTS FOCUS INWARD

Early in treatment, many clients see the source of their problems in others. Clients often want to spend more time describing others' problematic behavior than discussing their own experience of and response to the problems. For example, many clients begin the first few therapy sessions by announcing that the problem is really with another person:

- My husband won't pay any attention to me.
- My wife is always on my back.
- My children are impossible; they won't do anything I say.
- My boss is a demanding tyrant. I never seem to do enough.
- My mother won't stop criticizing me; I can't do anything right according to her.
- I'm 27 years old, and my father treats me like a child.
- My boyfriend keeps cheating on me.

It is essential for the therapist to affirm these complaints as genuinely troubling concerns and communicate overtly that the therapist wants to help the client improve this situation. The therapist's intention is to address the problem that is most pressing or salient for the client right now—to find and begin with the client's "point of urgency." Even though these concerns and complaints may include overreactions or schema distortions, from the client's point of view, they are the truth. Thus, as emphasized in Chapter 2, the therapist joins the client there, validating and even helping them articulate more clearly how the problematic behavior of others is indeed troublesome.

In some cases, however, the therapist's assessment of the situation is so different that such affirmation would not be genuine for the therapist. If not, therapists can at least affirm the subjective reality of the client's perceptions. For example:

THERAPIST: It sounds as though you feel your boss, like your girlfriend, is overly demanding. So, from your point of view, it seems that they are being unfair again.

If the therapist does not provide this affirmation first, many clients will feel that the therapist doesn't understand the reality of the other person's problematic behavior, doesn't believe them, or doesn't care about what's really troubling them. This unwanted outcome is counterproductive and will often reenact clients' developmental history—they are not listened to or believed again. Thus, to keep this potential invalidation from occurring, the therapist's first aim is to hear and validate the client's concerns. Throughout treatment, there is a consistent intention here: we are repeatedly trying "to meet the client where she is at." Only after making this genuine empathic connection, which often requires the therapist to de-center and see issues from the client's subjective viewpoint, can the therapist take the next step.

The second step is to begin focusing clients back on their own thoughts, feelings, and reactions to the problematic behavior of others, rather than joining the client in focusing exclusively on the external problem. Often, it will be easy for therapists to pair these first two steps together.

THERAPIST: I'm sorry your father does that to you. I can see, as you talk about it, how affected you are by it. What do you say and do when he...?

OR

THERAPIST: I can see how frustrating it can be to have your boss respond that way over and over again, no matter what you do. How do you feel, what's going on inside, when he...?

This principle is especially important when working with individuals who have been disenfranchised, for example, racial minorities, women, gay men, and lesbian women. Often, they are accurately explaining the social inequities that realistically contribute to their problems. Here again, the therapist wants to join the clients in the reality of their social context, as they experience it, before focusing inward on their reactions to, and ways of coping with, these inequities.

Why is therapy more productive when therapists focus clients inward? In many cases, clients who feel anxious, distrustful, angry, or helpless because of the behavior of another person will try to enlist the therapist in criticizing or blaming, complaining about, or trying to change the other person. However, therapy will not progress very far if the therapist merely joins clients in focusing on the other person's behavior, no matter how problematic this behavior may be. Why? As the existential therapists inform us, clients' attempts to change the other person will usually fail; clients are much more likely to resolve the problem by changing their own way of responding and finding their own authentic voice (Jacobsen, 2007; Wheelis, 1974). Yalom (2003) emphasizes that once clients recognize their own role in creating their life predicament, they realize they have the power to change it. Working in a supportive and nonblaming way, therapists are trying to help clients consider the question, How do I contribute to my own distress? Once they grasp this, they are able to recognize that they have the power to change their lives and chart a new course with a different internal dialogue and different behavioral responses.

Thus, the therapist's task is to expand clients' focus beyond the other person and include themselves as well. Note that the therapist should be flexible in adopting an internal focus; the therapist is trying to clarify both the reality of what the other is doing and the client's ineffective response to it. Most clients will benefit greatly from the therapist's help in understanding the behavior or potential intentions of problematic others. For example, if a client has grown up being befuddled by a parent with a borderline or narcissistic personality disorder, it may be deeply validating for the client to read the descriptions of these disorders in the *Diagnostic and Statistical Manual* (DSM-IV-TR). This type of external validation may help this client realize that all of this was not his fault and did not happen solely to him:

CLIENT: My father really was unpredictable and self-centered—it wasn't just me!

In tandem, however, therapists are working to increase clients' awareness of their own thoughts, feelings, and reactions in challenging situations. Recall the externalizing comments cited at the beginning of this section. Each of the following questions can be used to focus those clients back on themselves and begin to consider their role in the problem as well:

- What do you find yourself thinking when your husband is ignoring you?
- How do you feel when your wife is nagging you?
- What do you do when your children disobey you?
- When your boss is being demanding and critical, how do you typically respond?
- What thoughts were you having as your mother was criticizing you?
- How would you like to be able to respond when your father diminishes you like that?
- Tell me about the thoughts and feelings that are evoked in you when you find that your boyfriend is cheating?

Simple inquiries of this type serve two important functions. First, they tell clients that the therapist is listening to their concerns and is taking them seriously. The therapist has not changed the topic and is responding directly to clients' concerns as the clients see them. Second, while inviting clients to say more about their concerns, the therapist is also shifting the clients' focus away from others and encouraging them to look more closely at themselves. For clients, becoming more aware of their own reactions is a powerful intervention that will elicit strong feelings and reveal important links and connections. It is a critical step toward understanding and resolving their problems.

In many cases, clients will welcome the therapist's offer to talk more about themselves. Their active exploration of their own expectations and responses in problematic situations will lead to increased awareness of the narrow range of responses they repeatedly employ, and, with the therapist's help, they can begin to identify and try out a wider range of options that are actually available to them.

This increased self-awareness of their old response patterns, coupled with learning new or more flexible ways of responding, leads to more problem resolution skills and increased feelings of self-efficacy. However, not all clients will respond so positively to the therapist's initial invitation to look more closely at themselves. Some clients may avoid or actively reject an internal focus and continue to talk about the problem out there in others. In that case, the therapist continues to inquire about the personal meaning that this particular situation holds for the client.

CLIENT: My wife is impossible to live with. She complains constantly; nothing ever pleases her.

THERAPIST: It sounds like you've had a difficult week with her. What's been the hardest thing for you?

CLIENT: Do you know how hard it is to live with an angry, demanding wife who keeps trying to tell you what to do all the time?

THERAPIST: You sound very angry. I can see how hard it is for you to have someone after you like that.

CLIENT: She just won't stop—she's shrill and just keeps at it.

THERAPIST: How do you cope when she is doing this? What do you do when she keeps at it?

CLIENT: I get really mad, and I suppose I start to yell back some.

THERAPIST: You get really mad and start to yell back at her?

CLIENT: She just won't stop…

THERAPIST: What does the yelling mean to you? How do you make sense of all her yelling?

CLIENT: That she is just an angry bitch. It's not just her, you know; her whole family is like that.

THERAPIST: Her whole family yells a lot?

CLIENT: Gosh, yes…and always finding fault with others. It's just tiresome. I'm tired of constantly hearing what's wrong with me.

THERAPIST: They are really very critical of you; I can imagine how hard that would be to live with. It seems like criticism, in particular, really gets under your skin. What's it like for you to be criticized so much?

CLIENT: I hate it. I just hate it. They make me feel like I can't do anything right—that I'm doing everything wrong and failing all the time.

THERAPIST: "Doing everything wrong and failing"—ouch! Sounds like her family's constant criticism taps into your own painful feelings about yourself of not measuring up or not being enough. And having those shameful feelings of inadequacy aroused all the time could be infuriating…

CLIENT: Yeah, I hate them for making me feel this way. If I could just make them stop, I think my whole life would be better.

THERAPIST: This is very important for you, and we need to work together on it. I would like to understand better what you do when they criticize you, so that I can help you learn some more assertive, limit-setting responses. But your own feeling of not measuring up is also a part of this problem that we need to work on, so that you can stop charging at the red flag they are waving. If we can change the internal part of you that reacts so strongly—as if you believe what they are saying is true, or that this has to be disproved—it would be much easier for you to handle this than it has been in the past.

CLIENT: What do you mean, "reacts so strongly"?

THERAPIST: Sounds like their criticism leaves you feeling like you can never do anything right. Is that accurate?

CLIENT: Yes, but that's not true. They just expect too much, and I can't do it all. I just never do it well enough for them.

THERAPIST: I believe you—no matter what you do, they will never be pleased. But I don't think that's the whole problem. I can see how their expectations can still upset you but what's more important is the way you end up feeling inside, about yourself; that becomes the bigger problem.

CLIENT: I'm not really sure what you mean.

THERAPIST: There is relatively little we can do about their expectations. But we can work on your feelings about yourself—separate from their feelings about you—and on your responses to them, which could be helpful.

CLIENT: Well, maybe, I see what you mean a little bit, but I'm not sure what to do.

THERAPIST: Tell me more about your feeling of not measuring up.

CLIENT: Well, I guess I've always sort of felt like I'm not really good enough...

In this dialogue, the therapist validated the client's experience while at the same time encouraged the client to look at his own reactions and role in the problem. Although the client kept trying to focus on his wife and her family, the therapist's repeated but patient invitations to have the client look at himself as well soon slowed his defensive, externalizing stance. As a result, the client moved closer to his own problematic feelings and beliefs about himself that contributed to the marital conflict. By gaining a better understanding of his own feelings of inadequacy, the client can become less shame-prone and reactive to others' criticisms. This capability, in turn, will be an important part of learning to respond more assertively and set limits more effectively, rather than merely yelling back or withdrawing—which only exacerbated the conflict with his wife.

Many clients have not been encouraged to focus on themselves before. Some clients may resist this internal focus initially because it asks them to come face-to-face with difficult feelings that they have not been able to understand or resolve alone (in this case, for example, "I feel inadequate."). For many others, focusing on the other fends off the blame they have learned to expect from others (for example, "You never do it right."). However, the paradox is that as long as clients avoid the internal or personal aspects of their relational conflicts, by externalizing their problems onto others, they will feel powerless and frustrated—without any control in the situation.

Many clients who enter treatment are overly invested in changing others as a means of managing their own problems or insecurities, that is, they are seeking an external solution to a problem that can best be changed internally. The reality is that, beyond clearly expressing our preferences and personal limits ("I would like..."; "I will not..."), we cannot readily influence how others think, feel, and act. In contrast, however, we can change problematic interpersonal scenarios that keep playing out by changing our own responses to others. For example, clients routinely fail in their attempts to make their spouse stop drinking, smoking, or overeating. Similarly, some clients try for decades to win the approval or recognition from others that they never received from a parent. Others try unsuccessfully for years to have their grown offspring choose a different mate, religion, or career. As a result of these failed attempts to change others, many clients enter therapy with feelings of helplessness, hopelessness, and depression.

In sum, to gain a greater sense of self-efficacy and feel less anxious or depressed, clients will need to focus on understanding and changing themselves. Feelings of helplessness are inevitable when one tries to change others who do not want to change—whereas empowerment can come from recognizing and changing one's own responses. Thus, *a consistent therapeutic intervention is to pair an empathic or affirming response with an invitation to focus clients back onto themselves* (such as, "I'm sorry that happened. How would you like to be able to respond the next time he does that?"). The following types of questions will help clients explore their own responses.

- What is the main feeling you are left with when...?
- What were the thoughts you were having when...?
- What was the most difficult thing for you when...?
- How would you like to be able to respond when...?

Therapy Deepens and Intensifies When Clients Focus Inward

Focusing clients on their own behavior often reveals how they are contributing to, or participating in, their own problems. As we will see, clients who can focus on themselves and see their own participation in a conflict are usually motivated to change their own part in it. This, in turn, often allows the other person to respond differently as well. To illustrate, let's return to our example of the shame-prone husband who complained about his critical wife and in-laws.

Although the client did begin to explore his own lifelong feelings of inadequacy and the ineffective ways he responded to his wife, he kept complaining about his "obnoxious" wife and "superior" in-laws. As before, the therapist was empathic, validated his experience, and affirmed his anger. At the same time, however, the therapist did not join the client in focusing exclusively on his wife or blaming her as the sole source of his problems. Instead, the therapist continued to focus the client away from his preoccupation with his wife's behavior and toward his own reactions to her as well. For example:

THERAPIST: Rather than just insulting her back, which isn't making things better for you, what could you say instead the next time she criticizes you like that? What would be a realistic way for you to speak up for yourself and say what you don't like?

What would have happened if the therapist had not taken this approach? If the therapist had responded to the client's eliciting pull to blame his wife, the client would have remained an angry but helpless victim. In contrast, if the therapist had emphasized just the client's contribution to his marital conflict, without first validating his experience, it would have repeated the problematic relational pattern for him of always being blamed. Exploring the client's feelings of inadequacy revealed that these shameful feelings stemmed from rarely being supported by his parents and consistently being blamed for whatever went wrong—in his recollections, "everything was my fault." Thus, by validating the client's experience first and, second, focusing him inward, the therapist helped the client adopt a new and more assertive response to the old problem of excessive criticism from his wife. For example, the client rehearsed with the therapist and later tried saying to his wife:

CLIENT: You're saying, in so many words, that I did it wrong, but I disagree. I guess we just see this differently.

AND

CLIENT: I feel put down when you talk like that. Please stop or I'm leaving the room.

By coupling the therapist's validation with a consistent invitation to adopt an internal focus and recognize his own contribution to the problem, the client became more aware of how his responses were contributing to the marital conflict. For example, the client learned that he was quiet and unresponsive to his wife when he arrived home from work. The therapist helped him see that his silence increased his wife's attempts to gain his attention, and, in response, he became even less communicative. This interaction cycle escalated as his wife became more demanding of his attention and he withdrew even further—by taking a nap, for example. By focusing on his own emotional reactions at that moment, the client also realized that his wife's increasing requests for attention made him feel "overwhelmed" by the prospect of trying to meet "all of her needs." He had held the faulty belief that if he did not always respond to her, he was failing as a husband and again being inadequate.

While encouraging the client to become more aware of his own internal and interpersonal responses to his wife, the therapist was also suggesting different ways he could respond that would change his part in their marital conflict. Specifically, the husband began to express more directly to his wife when he would like to talk or interact together and when he would prefer to be alone. For example, he clarified with his wife that he wanted 30 minutes alone to unwind when he returned home from work, after which he would like to sit down together and share their day.

More important, as the husband gained a better understanding of his tendency to feel overwhelmed by others' demands, he became less reactive to his wife's requests. The biggest change occurred as he increasingly realized that he felt shamefully ruled by her because he could neither say no to her requests nor express his own wishes. On the one hand, if his wife wanted something from him, he believed that he was failing if he didn't comply and do it her way. But, on the other hand, he felt ashamed of being "told what to do" by her and responded angrily. And because he couldn't set limits or ask for what he wanted, he resented the "unfairness" in their relationship. The therapist was also able to help the client see that he expressed this resentment to his wife indirectly—in his own critical, shaming, and withholding ways. Over the next few months, this new awareness allowed him to respond more often to his wife's requests without feeling like he was complying with her or being dominated. As he increasingly realized that he too readily felt that he must comply with others, he began to challenge this faulty belief and started trying out more assertive responses, which led to his resentment diminishing. This was not a simple "aha" experience that changed his life, of course, but his situation gradually improved as he continued to set limits with her and speak up for himself, as the therapist helped him to do. When he increasingly recognized his life-long pattern of compliance, and began to change that pattern, in part, by setting limits and speaking up for himself in more direct but modulated ways, he began to feel better about himself and became less reactive to his wife and in-laws—even though they still could be "bossy" and critical at times.

Thus, as clients begin to focus on their own behavior, they often start to feel less "stuck" or powerless. As therapists help them see that more behavior alternatives are available to them, clients begin to feel more hopeful and in charge of their lives. Thus, the therapist's intention is to try and orient clients away from helplessly complaining about or trying to shape and control others' behavior. To emphasize, however, therapists will be more successful if they do this in tandem with affirming clients' legitimate concerns about the problematic behavior of others.

RELUCTANCE TO ADOPT AN INTERNAL FOCUS

Therapists' Reluctance.   Most new therapists feel comfortable in the role of friend and, too often, try to assuage their clients' fears or concerns. This is a benevolent social skill at which they have long been adept, but the role of friend does not apply well to the role of therapist. Thus, in sharp contrast to the role of friend, focusing the client inward often means that the therapist is drawing out and highlighting—rather than reassuring or minimizing—the client's fears and concerns:

THERAPIST: Don't worry, your doctor will figure out what's wrong—it'll be all right.

VERSUS

THERAPIST: Your doctor can't make a diagnosis yet and tell you what's wrong. You just don't know what to expect or prepare for—so it makes sense that you're scared right now. Let's explore the fear and potential diagnoses together.

There are many different reasons that therapists often have difficulty approaching and clarifying clients' concerns. For some, such directness goes against unspoken familial rules or cultural norms. In treatment, however, therapists want to do more than follow the limiting social rules and norms they may have grown up with. Instead, the therapist's sustained intention is to capture the core meaning, key issue, or central feeling in the vignettes the client is relaying. *With every turn of the conversation that seems to hold something meaningful for the client, the therapist is trying to register/reflect/highlight the most important issue or key concern—what's really being said here—rather than keeping things conversational or on the surface.* Doing so helps clients feel heard, seen, understood—that the therapist has the client's thought's, feelings, experiences, and "mind" fully in mind. That is, the therapist is fully present and attending—trying to discern the most salient message or grasp what the client is really saying or trying to convey right here (Allen et al., 2008; Groth, 2008). When therapists can repeatedly capture the most important feeling or key issue in what the client just said, clients leave the session with the feeling "We're really getting down to it" or "I'm not exactly sure what I want to do about this yet, but we're sure talking about what's important." *This is not so much about leading clients toward more threatening material or pressing for difficult feelings; rather, it is more about not backing away from the significance of the material that clients volunteer or already choose to present on their own.* Indeed, to the extent that the therapist has provided the client with a sense of security, entering and exploring issues that were previously more difficult to explore is now safer for the client. Specifically, after the client has experienced the therapist as a "safe haven" and been able to share her distress, focus internally, and become more "reflective," the therapist serves as a "secure base" for exploration. That is, from the safety of this collaborative and affirming relationship, the client can become more "reflective" (that is, expand narrow schemas and consider alternative explanations for their own and others' behavior) and become more flexible in how they respond. The attachment constructs of safe haven, secure base, and reflective capacity, and their impact on psychological well-being have been empirically validated in work with children and parents, as well as in work with adults (Hoffman et al., 2006; Slade, 2008).

Perhaps the biggest impediment to taking an internal focus is the therapist's reluctance to engage the client consistently in this substantive manner. While this approach brings meaning and intensity to the therapeutic interaction, *new therapists often feel unsure of "what to do" as clients focus inward and get closer to their real concerns.* These unwanted performance anxieties or feelings of inadequacy often arise because therapists believe they need to "fix" or "solve" the client's problem, when what is really needed is a partnership that first focuses on clarifying together what the concerns are and then collaboratively sets treatment goals and expectations. This process involves helping the client articulate,

"This is what I need…"

"This is what I want…"

"This is what I would like…"

"This is what I want to change…"

"This is what I would like to stop doing…"

"This is what keeps hurting me…"

Other reasons occur as well. Some therapists may equate being forthright with being intrusive or exposing because they have seen directness utilized only in hurtful ways. Still other therapists are reluctant to hear the emotional message in what their clients are saying or to help clients explore their participation in interpersonal conflicts because they don't want to make others feel bad. These therapists try to make clients feel better by reassuring them about their insecurities and by emphasizing only their strengths and successes. Unfortunately, this doesn't give the client the opportunity to get a more balanced "self-appraisal." That is, stepping back from the accurate reflection of clients' concerns or distress in these ways is usually ineffective.

Although well intended, *it keeps things on the surface and prevents clients from being able to look at what's really wrong.* When this happens, clients lose the opportunity to sort out what's really wrong, explore what they want to do about it, and develop the self-efficacy that comes from being able to address and resolve problems. In this vein, Farber et al. (2009) also write compellingly about the benefits that can accompany client self-disclosure. In particular, they note that clients who have experienced trauma can, through self-disclosure with a supportive and affirming therapist who listens to all aspects of their story, develop greater self-awareness and a more cohesive sense of self.

Still other therapists may avoid an internal focus in order to please clients and maintain their approval. These therapists fear that clients will become angry as the therapist invites them to explore the difficult feelings or choices they have been avoiding. Therapists are also concerned at times that they may not be able to remain present and responsive to the client when strong emotions are expressed. For therapists who have these concerns, it is critical that they address these overtly in supervision. That is, they should bring to supervision those emotions that might impact their work with clients so that they can discuss and role-play with supervisors various ways to manage these. When concerns are very strong and linked to significant personal concerns, addressing these in personal therapy is the most appropriate strategy.

Student therapists frequently need permission to risk not staying on the surface with clients and instead to enter what holds real meaning for the client. That is, to listen to the deeper meaning that the client is trying to convey and risk going there. Fears of the clients' feelings—what they might reveal, or of not knowing what to do or say—often interfere. What is most important to clients is that their therapist *join* them where they are and convey a willingness to grasp what is most difficult and important to them. This willingness to be present in their need or vulnerability means more to our clients than having an "intelligent" or "learned" answer or solution—it provides them with a sense of partnership and feeling that someone wants to understand and meet their need. Attachment theorists have often noted that this type of sensitive attunement "contains" distress. For clients, the safety or security provided by such a response will facilitate an internal focus and will also facilitate their willingness to try new ways of thinking and behaving.

Thus, for many of the reasons noted, therapists often join with clients in looking away from the internal aspect or the client's participation in the problem. Therapists collude with the client in externalizing problems when they repeatedly:

- give advice and tell the client what to do or how to respond to others,
- interpret or explain what something means,
- reassure clients that their problems will go away or are not something to be concerned about, or
- disclose what the therapist has done to cope with a similar person or problem.

As we would expect from client response specificity, each of these responses will be effective at times. However, if they come to characterize the ongoing course of treatment, it will hold clients back from learning about themselves and, ultimately, from being able to resolve their own problems. It is important that throughout treatment therapists repeatedly try to help clients become more aware of their own internal reactions and interpersonal responses in problem situations. Although use of these "self-monitoring techniques" can be challenging for therapist and client alike, they are necessary for real change to occur. It is inevitable, however, that as clients focus inward, unwanted feelings, faulty beliefs, and expectations of hurtful responses from the therapist and others will emerge and need to be dealt with.

Clients' Reluctance.   Why do clients tend to avoid an internal focus? As we have seen, a shift away from their focus on others can make some clients feel that the therapist does not really understand them. Clients can become distressed because it seems as if the therapist is not grasping how difficult the other person really is behaving or is just not sympathetic to their concerns. Other clients fear that if they give up their attempts to change others, they will have to either accept the blame for the problem and be the "bad" one, or remain forever resigned to "defeat." Still other clients, not yet feeling that a working alliance is solidly in place, do not have the security they need to approach the difficult feelings or choices that looking inside entails. Clients also will be unable to enter certain issues if the therapist and client are reenacting some aspect of the client's conflict in their interpersonal process (for example, feeling unheard, misunderstood, invalidated). When this is occurring, the therapist-client process will need to change for the client to progress further:

THERAPIST: It sounds as though you feel misunderstood by your wife, your children, and by others often...

CLIENT: Yes, no one seems to *get* me.

THERAPIST: I want to *get* you, so please tell me more about you—what you wish they would *get* better. What does your wife not know or see in you that you wish she did?

CLIENT: That I like my quiet time...that I can deal with problems better in the morning than I can after a long day at work. At least, I can't handle things without a little time out; it just kills me when she meets me at the door with all the day's problems—she doesn't seem to get that I had a day of problems, too...

After exploring these issues, the therapist can inquire how the client communicates this preference and the focus is now on the client. As the client does this, the therapist assists the client in clarifying and learning about his own communication style. The client can also now be assisted in exploring what things he can control (such as setting limits on when to discuss issues and also his own responses to the discussions) and things he cannot control (such as how his wife responds). Significantly, he can do so having been validated in his initial concern and seeing that he and the therapist are collaboratively engaged in a process that, while at times challenging, is designed to minimize distress in his life, and provide him with a sense of greater control and efficacy.

Cultural factors may also impede an internal focus. For example, Native American, Asian, and other clients from communally centered cultures may initially react negatively to looking within because it may sound self-centered. If clients perceive a difference in values and goals, they may question the therapist's credibility. Therapists can help by educating clients about the treatment process and, more important, by responding in a manner that is congruent with the clients' worldview. Sue and Zane (2009) highlight the constructs of *credibility* and *giving* in working with culturally diverse clients. They note that therapists can achieve credibility by intervening in a way that shows sensitivity and conveys to the client that they are trustworthy and capable of providing assistance. Thus, in a family where roles are very important, for example, it would be inappropriate to set as a treatment goal "confronting" parents. They identify "gifts" as features that convey hope or provide some benefit in the treatment process. This could include helping clarify, or normalize, some aspect of the client's problem, assisting with troubling symptoms (for example, teaching breathing skills to decrease anxiety, setting treatment goals together in a way that gives the client a sense of partnership, assisting with new coping skills, and so forth).

One of the best ways to work with clients who are not ready to look at their own behavior or shared participation in problems is to *explore their reasons for not wanting to look within.*

THERAPIST: When I ask you about your feelings or responses when your wife is critical, you often change the subject. What do you think keeps you from talking about those feelings?

CLIENT: I'm afraid I'd get mad, too mad, you know, like I might explode and not be able to stop.

THERAPIST: Am I understanding you accurately: are you saying that as long as you focus on her nagging, you are able to keep at bay or stay away from your own real feelings? That if you stopped and looked inside yourself, you'd explode?

CLIENT: Yes, I'm afraid I might explode...and afraid I might cry and not be able to stop.

THERAPIST: Angry, and also sad, both very big feelings...

This process then gets at what is being defended against. Externalizing is often a defense against feelings that might, at the surface, include explosive anger but often cover profound sadness or shame. This might be sadness and/or shame about being unseen, not cared about, demeaned, and so forth. However, empowerment comes from owning these feelings and then choosing for oneself how to behave rather than repeat the cycles of behavior that have simply fueled the feelings of hopelessness and helplessness.

In most cases, clients will accept the therapist's invitation to say a little more about themselves. However, if the client remains unable to respond, the therapist can simply give the client permission to disclose or proceed at her own pace. For example:

THERAPIST: It doesn't seem comfortable for you to share very much of yourself yet, but I do think it's important that you choose how much you share.

If the client's externalizing stance does not begin to change, however, the therapist can make a process comment that simply describes their interaction:

THERAPIST: I've noticed that you talk very easily about your husband and your daughter, but you don't say very much about yourself. Are you aware that that happens?

OR

THERAPIST: I think we've been missing each other the last two sessions. I keep asking what you were thinking about or trying to do in a particular situation, and you usually respond by telling me more about the other person. What do you see going on between us?

These process comments will help clients become aware of their externalizing style with the therapist and learn that they probably respond in the same way to other people as well. For example, clients who cannot share themselves in a personal way with others will often be perceived as boring or aloof. In that case, part of the clients' presenting problem, such as the clients' sense of loneliness or lack of meaningful relationships with others, may also be enacted in their relationship with the therapist. While often discouraging for clients to realize that they are recreating their conflict in therapy, this reenactment provides the opportunity to begin resolving the problem by changing the distancing pattern within the therapeutic relationship. The interpersonal feedback provided by the therapist helps clients become aware of how they interact with others and how others experience them. The therapist then offers a new and more effective way of relating in their relationship. Using self-involving statements, for example, the therapist can describe their current interaction and invite a more meaningful relationship:

THERAPIST: I feel that I am being held away from you when you talk about others rather than yourself—I feel like I am missing you, and I don't want that. I would like to learn more about you, or your reservations in talking about yourself. Can we work together on this?

When the therapist uses questions or process comments to focus clients inward, clients often reveal a variety of new and important concerns that they have not talked about before. For example:

* Well, I guess there really isn't very much about me that people would want to know.
* I'm not used to telling people what I'm thinking or feeling.
* Every time I try to get close to someone, I get hurt in the end...
* You wouldn't like me very much if you knew what I was really like.

In this way, the therapist's process comment brings out important new concerns that can now be addressed in treatment. The therapeutic relationship has intensified, and significant new information that is central to the client's presenting problems has been revealed. This process of helping clients focus inward and explore their resistance to looking within provides some of the most important material to be addressed in therapy. These important new concerns are unlikely to be brought out in treatment unless the therapist uncovers them by working in the moment with different types of process comments and focusing clients inward. That is, in the safety of the counseling relationship, clients can "let down" their guard and allow themselves to know and feel things they have tried to keep at bay. This becomes an opportunity for increased personal empowerment—they no longer have to operate based on old coping strategies and old beliefs that were previously needed for survival. Clients can now have an opportunity to choose more deliberately and clearly which beliefs and coping approaches they want to hold on to or discard, and develop new ways of being both with themselves and in their relationships. Thus, one of the most important reasons for adopting an internal focus is to bring out key issues that are contributing to the client's problems and make them accessible for treatment.

DEVELOPING AGENCY: PLACING THE LOCUS OF CHANGE WITH CLIENTS

The first component of the internal focus was to help clients look within and become more aware of their own responses. The second component was to help clients adopt an internal locus for change and help them to begin to act from within. In this way, we now examine how clients can gain a greater sense of self-efficacy in their lives by becoming more active agents in their own change process (Bandura, 1997). First, we will see how the therapist can use the therapeutic relationship to foster the client's own initiative. Second, we will explore interventions that encourage the client's participation in and sense of ownership of the change process. Throughout, the therapist's goal is to foster clients' initiative by responding to their interests and concerns.

Fostering the Client's Initiative

In the course of treatment, a therapist has the opportunity to convey to clients the broader message that they are able to take charge of their lives, make their own decisions, and live their lives more as they choose. Effective therapists of every theoretical orientation nurture clients' own sense of personal dignity and self-efficacy. More important than verbally encouraging clients' initiative, therapists can give clients the experience of acting more effectively during the therapy session. Once clients experience greater agency in their relationship with the therapist, it is relatively easy to help them generalize this greater sense of effectance to other relationships beyond the therapeutic setting (Bandura, 1997; Cervone et al., 2006). Let's see how therapists can help clients become more active, initiating partners in the change process.

   The first way to help clients feel more responsible for and capable of change is to encourage them to shape their own agenda in treatment and talk about the issues they feel are most important or salient right now.

THERAPIST: Where would you like to begin today?

CLIENT: I'm not sure—what do you think would be best?

THERAPIST: I'd like to join you in exploring whatever you think is most important to work on right now. What would that be?

CLIENT: I'm not sure.

THERAPIST: Let's sit a moment, let you take a breath and settle into yourself—and then see what seems most alive for you right now.

CLIENT: *(Pause)* I think I want to talk about my wife. We had the best weekend together we've had in a long time.

THERAPIST: I'm happy for you. What was different this weekend—were you doing anything differently?

CLIENT: I think I talked to her more like you talk to me. I asked about what she thought and what she wanted more, and I think she liked it....

Once clients begin pursuing their own concerns in this way, the therapist can be an active participant who helps clients to explore their own concerns more fully, to understand their problems better, and to generate potential solutions and behavioral alternatives. Why is it essential to encourage clients' own initiative and meet them at their "point of urgency," even in the time-limited modalities that increasingly dominate the therapeutic milieu? In one way or another, most clients have been unable to act on their own interests or pursue their own goals. In past relationships, these clients have not had significant others support their own interests or, simply put, help them do what they want. As a result, these clients will feel encouraged, but often anxious as well, if the therapist supports their own self-direction and cares about what seems to matter most to them. Regardless of treatment length or the therapist's theoretical orientation, therapy becomes a more intense and productive experience when the therapist can successfully engage clients in pursuing their own interests and what matters most to them. When this occurs, clients feel contained and validated and take ownership of the change process, evidenced by investing more fully in the working alliance.

When the therapist can help clients explore and understand the material they choose to focus on, rather than merely direct them to the therapist's own agenda, change usually begins to occur and presenting problems improve. When clients provide the momentum and direction for therapy, they often experience a reparative relationship where, for the first time, they can have their own opinions, act more effectively, and be more in charge of their own lives. Once a client has begun to initiate, the therapist actively participates by providing the types of responses that this particular client can utilize most productively. Commonly, this will include interventions from a wide variety of theoretical orientations—contributing information, explanations, or interpretations; constructing behavioral alternatives; suggesting possibilities about what problematic others may be doing or intending; providing interpersonal feedback about how the therapist and others may be perceiving or reacting to the client; and so forth. Whatever interventions or techniques the therapist uses, however, they will be far more effective when given in response to the clients' own interests or concerns, rather than the agenda of the therapist, employer, or spouse. This process dimension is one of the most important characteristics of the therapeutic relationship, and we will explore it further in the chapters ahead.

Avoiding a Hierarchical Relationship.   We have already noted that clients may resist the therapist's attempts to make them active participants in treatment. Why? Frequently, clients will be conflicted about expressing their own wishes, acting on their own initiative, and achieving or enjoying success. Many clients have not been supported in having their own voice or becoming stronger, and feel guilty or anxious if they speak on their own behalf, act more boldly or demonstrate more self-direction, or achieve success. As a result, they continually elicit advice and direction from the therapist and others. Therapy will not be productive if this helper–helpee mode comes to characterize the therapeutic process. This mode shifts responsibility away from the client and onto the therapist and creates the hierarchical relationship discussed in Chapter 2. Clients will not be able to feel greater self-efficacy and adopt a stronger stance in their lives as long as this dependency is fostered—their *compliance* is being reenacted in the therapeutic relationship rather than being challenged or questioned. And, as long as they remain compliant, their sense of empowerment will be muted and treatment will stagnate if this remains the primary treatment mode.

Too often, therapists unwittingly comply with the clients' subtle or overt requests to tell them what to talk about in therapy and what to do in their lives. While flattering to think that we know what is best and can tell others what to do (it appeals to the narcissism that can emerge in every therapist), it stymies clients' agency. As we would expect on the basis of client response specificity, there will be circumstances in which directives, advice, opinions, and so forth, are helpful for a client. For example, they will often be helpful for clients who grew up with caretakers who were not able or were not interested in helping them solve their problems—perhaps because they were involved with drugs, depressed, or otherwise self-absorbed. However, when directive or prescriptive responses characterize the ongoing process of therapy, clients will usually remain dependent on helpers to manage their lives. In addition, researchers find that many clients tend to be resistant and uncooperative when therapists are too directive, so therapists want to assess carefully how this particular client is responding to the therapist's advice or suggestions (Bischoff & Tracey, 1995; Mahalik, 1994). Because this problematic process occurs subtly (but frequently), we need to examine further this dependency-fostering approach to therapy.

Supporting Clients' Own Autonomy and Initiative.

Effective therapy of any treatment length should foster the client's self-efficacy—we want the client to develop a greater sense of agency through the treatment process (Cervone, 2000; Galassi & Bruch, 1992). Therapists cannot just talk with clients about choice, responsibility, and personal power, however; they want to cocreate a relationship in which clients are behaving in stronger ways with the therapist in the session. When clients can first do this with the therapist, the therapist can readily help them transfer this stronger way of acting to the rest of their lives. Thus, the therapist first gives clients permission to follow their own interests and actively encourages them to introduce the material that seems most relevant to them. As detailed in Chapter 2, the therapist then tries to capture the emotional message or identify the core meaning in their narrative and encourages clients to elaborate it further. In response to clients' increasingly specific exploration, the therapist actively helps clients clarify the repetitive themes that bring the central problems into focus—and work together to understand this theme or issue. They explore how aspects of this theme or problem may be evident in the way they interact and, together, begin to generate more effective ways to respond and change this pattern with others in their lives. When this mutual interchange continues throughout several sessions, most clients become committed to the treatment process and will report that meaningful changes with others are beginning to occur. Facilitating this growth process may be the primary challenge and satisfaction of being a therapist. Let's see how this sequence might actually sound with a client:

CLIENT: I need more direction from you. What should I do here? You're the expert.

THERAPIST: I have a couple of ideas to share with you, but I think it would work best if I heard yours first.

CLIENT: *(impatiently)* If I knew what to do, I'd just do it and wouldn't ask you or be coming here to see you!

THERAPIST: All right, let's swap ideas. You tell me what you think is going on; then I'll tell you what I see occurring; and let's see what we can put together.

CLIENT: Like I just said, I don't know—I don't really have any ideas.

THERAPIST: Let's wait for just a minute and see if anything comes to you. If not, I'll be happy to go first and start us off.

CLIENT: *(20-second pause)* Well, maybe I'm afraid of being alone or something like that.

THERAPIST: That's interesting and fits with what I'm thinking about. Tell me more about what it means to be "alone"; it sounds important.

CLIENT: I think I've always been worried about that.

THERAPIST: From other things you've said, it makes me wonder if your parents cut off from you emotionally whenever you disappointed them by doing what you wanted rather than what they expected. I wonder if that created the feeling of being all alone, even though others were physically present—which would have been very confusing to a child.

CLIENT: Yeah, maybe...and the worst thing about it was that it seemed like it was all my fault. They were going away because I let them down. What do you think?

THERAPIST: I think you have very good ideas, but something often seems to hold you back from expressing them in the strong way you have just been doing with me. I have noticed that happening in here with me sometimes, and I hear it in your relationships with others as well. I wonder if there is some connection between holding yourself back in this way and being afraid of being left. What comes to mind as I wonder aloud about this?

This collaborative interaction, in which the therapist and client each build on what the other has just produced, is an independence-fostering approach to treatment. Clients have the experience of sharing responsibility for the course of treatment, as the therapist actively encourages them to define and address their own concerns. However, as therapists support clients in achieving this more active stance, they are simultaneously contributing their own ideas and suggestions in a way that creates a working partnership. *The therapist in the dialogue above is highly engaging and active—but not very directive.* She is following the client's lead, but she is not being nondirective, either. The interpersonal process we are striving to achieve is this collaborative effort—which has not been developed very well in the counseling literature.

Taking this point further, a common misconception is that longer-term, dynamic, or relationship-based therapies are dependency fostering, whereas short-term, problem-solving, or strictly behavioral approaches are not. Actually, whether the therapy fosters dependence or independence is determined by the therapeutic process and not by the length of treatment or the theoretical orientation of the clinician. In short- or longer-term therapy, the client's dependency is inappropriately fostered when the therapist repeatedly directs the course of therapy, gives advice, and prescribes solutions for the client. Let's examine this collaborative alliance further—it offers an effective middle ground between the less productive polarities of directive and nondirective control.

Shared Control in the Therapist–Client Relationship. As we have begun to see, treatment will be most successful when the therapist and client share control over the agenda and direction of therapy. In most cases, it is overly controlling for the therapist to play the predominant role in structuring and directing the course of treatment, and it is ineffective to nondirectively abandon clients to their own confusion and conflicts. In a more productive relationship, therapists will encourage the client to take the lead but will actively contribute their own understanding and guidance about the material that the client has produced. It will be therapeutic for many clients to experience a relationship in which both participants share responsibility rather than a relationship in which controls are held by one party or the two parties compete. This collaboration in the therapeutic alliance is the key to greater self-efficacy and provides clients with the support needed to act in new and stronger ways with others in their lives. As discussed previously, the therapist is encouraging the client to take the lead and initiate whatever she would find most useful to discuss.

THERAPIST: I would like to begin each session by having you bring up what you want to talk about. I would like to join you in working on what seems most important to you. I will be sharing my own ideas, reactions, and suggestions as well, but I think this is the best way to begin a good partnership.

Some therapists will be frustrated by this approach and direct clients toward taking action and finding solutions, especially if they are feeling pressed to fix the problem quickly because of the limited number of sessions available. Although a highly prescriptive stance will be helpful at times, such as when clients are in crisis, it imposes several important limitations. First, as we have seen, a directive approach prevents the therapist and client from discovering key issues that were not evident in the client's presenting problem. Second, the goal of therapy is not merely to fix clients' presenting problems but to do so in a way that leaves clients with a greater sense of their own personal resources and ability to live their own lives meaningfully. Working collaboratively, clients gain an increasing sense of self-efficacy by participating in treatment progress, which directive approaches do not provide. That is, clients share ownership of the success and change that is achieved,—which is quite different than feeling that the smarter or stronger therapist cured them.

Responding to these arguments, the solution-oriented or directive therapist might counter, "What if you follow the client's lead and it takes you nowhere?" Certainly, a nondirective approach requires an unrealistic amount of time from the therapist and often results in a disorganized therapy that is lacking in focus. Moreover, a purely nondirective approach is likely to reenact problematic relational patterns for clients who grew up in a permissive home, or with passive/disengaged caretakers, just as a purely directive approach may reenact such patterns for clients who grew up with controlling parents in an authoritarian home. As we have seen, however, there is an effective middle ground of shared therapist–client control, although it has not been delineated so well in the counseling literature, with exceptions of Miller & Rollnick's Motivational Interviewing (2002). For example, the therapist below does not wait nondirectively for the client to bring up more relevant material or directively lead the client to a new topic. Instead, the therapist makes a process comment about their current interaction and invites the client to join in reshaping their interaction:

THERAPIST: This doesn't seem to be taking us anywhere right now. Is there a better way to use our time—or another way to talk about this—that could get us closer to what's really wrong or matters most to you?

Therapists need treatment plans and intervention strategies with short- and longer-term goals. However, these are most effective in producing enduring or sustainable change when they develop out of a collaborative interaction, where the client shares responsibility for successes and disappointments and feels ownership of the treatment process. To implement this interpersonal process, the therapist needs to be able to tolerate ambiguity and refrain from subtly controlling the interaction. Instead, the therapist is trying to create opportunities for clients to voice their own concerns and to act on their own initiative. The therapist's task, then, is to be able to decenter; enter into the client's subjective worldview; and empathically highlight the core meaning, central feeling, or relational pattern that this issue seems to hold for the client. This approach requires an active therapist who is neither directive nor nondirective but has the flexibility to tolerate ambiguity and share control (that is, not know what the client is going to say or do next but be open to whatever he presents). In this approach, therapists respond actively by:

- communicating their understanding and affirming the client's experience;
- joining the client in exploring further the key meaning in what the client just said;
- using self-involving comments or their own experience of the client to provide interpersonal feedback about how the client affects the therapist—and may be contributing to problems with others;
- affirming new behavior that steps out of old patterns and demonstrates change with the therapist or with others; and
- providing a focus for treatment by highlighting the repetitive interaction sequences, central feelings, and faulty beliefs that recur throughout the client's narratives.

Furthermore, the therapist is using process comments to address treatment impediments—such as resistance or an externalizing focus—and to prevent unwanted reenactments between the therapist and client (for example, the client is silent about her faulty belief that the therapist is impatient, bored, or burdened by her—as others often have been). Therapists do this by working in the here-and-now and asking about what may be going on between them in their relationship or current interaction. For example:

THERAPIST: What's it like to be talking with me about this?

CLIENT: Well, you're probably feeling impatient with me—like everybody else does—because I'm still stuck on this.

In this way, the therapist intervenes by taking the client's problems out of abstract discussions about others "then and there." Instead, the therapist is looking for opportunities to create immediacy by linking the client's problems with others to what is occurring between them in the way they are interacting together right now. For example:

THERAPIST: No, I'm not feeling "impatient" with you at all. Is there anything I'm doing or that is happening between us that makes you feel that way? If so, let's sort that through together. I can see this is a complicated problem for you, and I appreciate how hard you are trying to do something about it.

CLIENT: Really, you're not frustrated with me? You know, that makes me almost want to cry—it feels like I've been a disappointment to just about everybody in my life....

Working in the "here and now" with clients is a big new step for most therapists. It brings therapists and clients out of the mode of everyday social interaction, and it quickly takes us to the heart of what's wrong. Significant feelings come up and key issues that the therapist didn't know even mattered to the client now take center stage. And, just as important, the relationship becomes more intense and begins to matter more to both the therapist and the client.

As we will see in Chapter 6, an "authoritative" parenting style that encompasses both parental warmth and firm limits reflects a more effective middle ground between the better known and more widely adopted authoritarian and permissive parenting styles. In parallel, therapists will find that shared control offers most clients a more productive alternative than either directive or nondirective approaches. New therapists need role models of how therapists and clients can work together in this actively engaged and collaborative way, so let's examine interventions that foster a strong working alliance and illustrate this middle ground of shared control.

INTERVENTIONS THAT PLACE CLIENTS AT THE FULCRUM OF CHANGE

As we have seen, the therapist's goal in the interpersonal process approach is to encourage the client's lead while still participating actively in shaping the course of treatment. The key is for the therapist to intervene without taking the impetus away from the client. The following examples illustrate ineffective and then effective ways to do this.

Ineffective Interventions.   Suppose that the client is filling the therapy session with seemingly irrelevant storytelling. The therapist cannot find a common theme to any of the client's narratives or understand the personal meaning that these vignettes hold for the client. It seems as if nothing significant is occurring. At this point, it is easy for the therapist to stop the client and direct her toward a specific topic that the therapist thinks would be more fruitful. This refocusing can be effective at times, but therapists will succeed more often if they revitalize the therapeutic interaction without shifting the impetus away from the client and onto the therapist. In the following dialogue, the onus for therapy comes to rest with the therapist, and the client loses an internal focus for change:

THERAPIST: I'm wondering where this is taking us. Have I lost the focus here?

CLIENT: I'm not sure where I'm going with this, either. What do you think I should talk about?

THERAPIST: You've had trouble asserting yourself in the past, and I think we need to look more closely at that. Last week, you said you wanted to ask your boss for three weeks of vacation instead of two. How are you going to handle that confrontation?

CLIENT: I'm not sure. What do you think I should say?

THERAPIST: To begin with, you need to arrange a face-to-face meeting with him. It is important that only the two of you are present, so that you can have his full attention and there is less threat for either of you to lose face. Then use the "I" statements we have practiced to directly state what you want.

CLIENT: Sounds good, but what would you say to him first?

Assessing the client's responses in this dialogue, it is clear that this intervention is not being productive, even though the therapist has moved the client to a more salient topic and provided useful information about effective negotiations. The fulcrum of therapeutic movement has tipped from the client to the therapist, and a hierarchical teacher–student process has been established. Most clients will not be able to utilize the therapist's useful information until they pick up the momentum and begin to actively participate again. Furthermore, because the client remains in a passive role as the therapist continues to inform, the client does not gain the increasing sense of self-efficacy that comes from participating in success or progress in treatment. A more productive intervention might be to ask the client whether he feels another topic might be more relevant, and to wait until the client becomes actively involved again before offering additional information or further shaping the direction of treatment. However, *if the client does not become actively engaged again, therapists do not want to just sit and wait nondirectively.* Instead, they can actively intervene by simply observing this with a process comment. For example,

THERAPIST: I'm having a little difficulty following you right now. Wherever you go, it often seems to trail off and get lost for me. What do you see happening here between us?

Effective Interventions.   Let's look at how the therapist can refocus the client toward more productive material—but in a way that keeps the momentum for treatment with the client. Again, process comments that make the current interaction between therapist and client an overt topic for discussion are often effective in these circumstances.

THERAPIST: I don't have the feeling that what you're talking about is really very important to you. Am I missing the point here, or is this really important? Help me out.

CLIENT: Yeah, you're right, I'm not sure where I'm going with this, either. What would you like me to talk about?

THERAPIST: I think that we should try to identify what would be most important to you and talk about that. What might that be right now?

CLIENT: I'm not really sure.

THERAPIST: Let's just sit together quietly for a moment, then, and see what comes to you.

CLIENT: *(pauses)* Do you think it would be a sin if I was bisexual, or if I had sexual feelings toward men sometimes?

New therapists often feel uncomfortable with silences and may find themselves filling them. If therapists can refrain from doing so, however, these are opportune moments for clients to establish their own agenda or introduce new material that is far more meaningful—as in the dialogue above. This type of process comment invites the client to approach more substantial material but, in line with our goal, without taking the impetus away from the client. The therapist has directly intervened by sharing her observation and asking about the client's perceptions of their current interaction, but she has still left the client an active participant in revitalizing the discussion. Routinely, clients produce far more significant material when they are invited to lead in this way rather than to follow. As always, try it out with your clients, assess their reactions, and see what works best for you.

Discovering More about the Problem.   Important new information, central to understanding and changing the client's problems, is often identified when the therapist intervenes without taking the locus for change away from the client. Imagine the following situation: A 19-year-old client in a college counseling center had been telling his therapist about his attempts in early adolescence to observe his stepmother undressing. The client had been detailing his voyeuristic efforts at great lengths, but the therapist did not feel that this was genuinely of much concern to the client. Unless the therapist could find the relevant meaning for the client, he wanted to move on to other, more salient issues in the client's current life. In the following dialogue, the therapist uses a process comment to refocus the client in a way that places the locus for change more fully with the client and, in the process, reveals a new and more salient treatment issue.

THERAPIST: Does this feel like an important topic that you want to discuss with me? As I listen to you, I don't get the feeling that you are really very interested in what you are telling me. Is that the case, or am I not understanding the meaning this holds for you?

CLIENT: I thought therapists were interested in this oedipal stuff. I figured you would want to hear about it.

THERAPIST: I'm struck by the fact that you are telling me what you think I want to hear, rather than working on what is most important to you. Maybe that's something we should talk about. I wonder if you find yourself doing this in other relationships as well—trying to sense other people's needs or responding to unspoken demands at the expense of expressing your own interests or concerns?

Here again, the therapist has used a process comment to revitalize treatment without taking the impetus away from the client. In this vignette, the therapist has also used their current interaction to discover a more important aspect of the client's problems—compliance issues. Because this new issue arose from their joint interaction, and not from the therapist's own agenda, most clients will be highly motivated to explore it. Thus, the therapist has effectively focused the client internally, identified a key concern to be explored, and helped the client become a more active participant in shaping the course of treatment.

As we see here, adopting an internal focus reveals new aspects of their problems and clarifies related issues that are shaping and contributing to their problems. As the client begins to work with these salient new issues, and with the themes and patterns that have been identified in the way the therapist and client interact together, important progress occurs on the client's presenting problems. How does this occur? The therapist makes links between understanding and changing what is going on between them in the therapeutic relationship, and making the same types of interpersonal or behavioral changes with others in their lives. If changes occur with the therapist but do not generalize to the client's everyday life, treatment has failed, of course.

To illustrate this transfer, let's continue where we left off with the therapist and client in the previous dialogue. As we will see, the therapist is making connections between what is going on between them and other relationships in the client's life (such as his girlfriend) where the same faulty patterns are occurring.

CLIENT: *(sarcastically)* Well, duh! Of course—isn't that what everyone has to do with teachers and parents and therapists like you? Figure out what you want and go along with it?

THERAPIST: Well, it's what some people have learned they have to do with some teachers and some parents and some therapists. But it's not what you have to do with everyone, and it's not what you have to do with me. I like people who have their own mind.

CLIENT: *(sarcastically again)* Do you now?

THERAPIST: Yes, actually, I do. So, tell me, who in your life can you be honest with and talk about what you really care about—and who do you have to comply with and talk their talk?

CLIENT: Well, I guess I can be the way I want to be with my guitar teacher—he's cool.

THERAPIST: I'm glad you have him—you need good friendships in your life. Who's the most important person in your life that you have to go along with—like you thought you had to do with me?

CLIENT: Probably my mom.

THERAPIST: What happens if you to talk to her about what's real or what matters for you?

CLIENT: She's mad—like I embarrass her by being who I am.

THERAPIST: Ouch! I'm sorry for both of you. Have you ever tried talking to her about this? You just said it so clearly.

CLIENT: *(with disdain)* Of course not—do you believe in Santa Claus and the tooth fairy, too? She's not going to change!

THERAPIST: As you talk with me like that, especially in that contemptuous tone of voice, I feel insulted—as if you think I'm an idiot or a fool. Was that what you meant to convey?

CLIENT: *(pauses)* Uhh...

THERAPIST: I wonder if you are aware of talking to people in that sarcastic or contemptuous way?

CLIENT: I guess I sort of do and sort of don't—but as you say it, I guess it could be a problem.

THERAPIST: Yeah, I don't think others would like that much. Some might respond in kind, and some might step away a bit—or even permanently end their relationship with you.

CLIENT: Yeah, all that's happened...My mom and I do this to each other sometimes too.

THERAPIST: I can see how things may not go very well for either of you sometimes. I don't know if your mother can change with you or not, but I think it's worth testing the waters to see. Maybe she could respond better if you began talking to her differently—in a respectful and straightforward way about what's really going on between you, rather than in the sarcastic way you were just talking to me. That improves her chances of being able to respond better. What do you think?

CLIENT: Yeah, maybe, but how can things get better for me if she still thinks I have to do everything her way and be the way she needs me to be all the time?

THERAPIST: I have several ideas about what you could do differently with her— whether she changes or not—that could leave you feeling better about yourself.

CLIENT: How would things be better for me if it turns out to be hopeless with my mom? I think she'll just say what she always does—everything's my fault; I'm always doing it wrong.

THERAPIST: She very well might, sadly, but if you tried talking with her about this in a more respectful way, you'd feel differently about your part in the interaction. Maybe you'd even stop believing that all of the problem is always your fault. So, other relationships in your life might go a little better, too, even if it doesn't change with your mom.

CLIENT: Like with my girlfriend?

THERAPIST: Yes, exactly. Tell me how this goes with your girlfriend....

In summary, the therapist in this dialogue is providing interpersonal feedback about the client's insulting manner and challenging the client to take more responsibility for causing and solving his problem. Carkhuff (1987) calls this **"personalizing,"** and encourages therapists to respond in ways that help clients realize that they usually have some shared responsibility for creating and maintaining their problems. Therapists who avoid making this challenge for greater personal responsibility generally lose credibility with the client, and the working alliance is diminished. The therapeutic dialogue may be supportive and understanding, but it doesn't go very far—it isn't enough to lead to change for a client such as this. Thus, therapists are encouraged to respond in this personalizing way and say, for example, "You're upset because they took advantage of you, and because you didn't stand up for yourself when it happened."

ENLIST CLIENTS IN SOLVING THEIR OWN PROBLEMS

As noted earlier, clients benefit most from looking within, sharing their inner world, and engaging in understanding and resolving their own problems; that is, taking ownership of the change process where they can learn from their mistakes and gain from their successes. A common misconception about therapy is that the therapist assumes responsibility for figuring out what is wrong and prescribes what clients should do. New therapists often hold unrealistic expectations that they must be experts who possess insightful solutions to their clients' problems; they then suffer under these performance demands with painful feelings of inadequacy. Here again, this misconception places the impetus for change in the therapist's lap and gives clients the role of passive recipients, waiting to be cured. Rather than adopting this hierarchical model, a more effective approach is to extend the collaborative alliance into the problem-solving phase as well. Therapists will be more effective when they elicit clients' active involvement in resolving their own problems. For example, Therapist:

• What have you tried in the past that has worked for you?
• Looking back, what have others done to make things better—or worse—for you, when you were in this situation before?
• What could you and I do together to try and help with this?
• How do you understand the way things have gone in your life to get you in this position?

In contrast, therapists too often establish expectations that set the stage for the therapist to tell clients how to lead their lives—by explaining what it all means, giving advice, and prescribing solutions. Instead, when clients participate in a collaborative partnership, it is empowering and gives them the opportunity to become more capable of managing their own independent lives. Rather than just providing answers, therapists are most effective when they help clients learn how to think about and explore their own problems, and generate their own solutions and alternatives. Real change occurs when therapy has not only resolved clients' presenting problems but has fostered their self-efficacy in this way. To illustrate this, let's look at two different ways of responding to a client's dream. Although the interpretation of the dream remains the same, the therapeutic approaches differ and enact very different interpersonal processes with the client.

RECAPITULATING CLIENTS' CONFLICTS

Anna, a 24-year-old client, lived at home with her embittered and chronically embattled parents. For several years, Anna had been struggling with the developmental transition of emancipation from her family of origin. She could not establish her own adult life. Anna had few friendships, dated little, and had no serious educational or career involvements. Anna's "martyr-ish" mother and alcoholic father wrangled constantly, and Anna felt that it was her responsibility to stay home and help her mother cope with her failed marriage. Anna entered therapy complaining of depression.

After several sessions, Anna recounted to her therapist a dream that was of great importance to her. As the dream began, Anna was riding a beautiful horse across an open savannah. She felt as one with this graceful animal as they glided effortlessly across the broad grasslands. They sped toward a distant mountain, past sunlit rivers, birds in flight, and herds of grazing elk. Anna felt strong and free as she urged the tireless animal onward.

The distant mountains held the promise of new life amid green meadows and tall trees. Anna felt their promise quicken inside her as she urged the horse onward. But as the mountains drew near, Anna and the horse began to slow. The horse's legs became her own and grew heavier with each step. Anna desperately tried to will them on, but their footing became unsure and they began to stumble. At that moment, she was surrounded on all sides by menacing riders, ready to overtake and capture her. As she awoke, Anna choked back a scream.

"What do you think my dream means?" Anna asked.

Her bright, concerned young therapist offered a lengthy and insightful explanation. The interpretation focused on Anna's guilt over leaving home. The therapist suggested that, if Anna went on with her life and pursued her own interests, she would feel powerful and alive—just as she had in the dream. But before Anna could reach her goal and experience the satisfactions of having her own adult life, she would have to free herself from the binding ties of responsibility that she felt for her mother and her parents' marriage. Her loyalty to her mother threatened to entrap her in guilt and prevent her from being able to live her own adult life.

ANNA: *(enthusiastically)* Yes, you're right, I do feel like I'm doing something wrong whenever I leave my mother and do what I want. Is that why I dreamed that?

THERAPIST: We've been talking about whether you are going to move into an apartment next month. I think the dream reflects your guilt over taking this big step on your own. You know, like we've been talking—that by becoming more independent—in a normal or appropriate way, you're doing something wrong and selfishly hurting your mother.

ANNA: I want to move out, but I can't leave my mother with my father. She says she will divorce him if I move out, and it'll be my fault. What should I do?

THERAPIST: I can't make that decision for you. It's important for you to be responsible for your own decisions.

ANNA: But I don't know what to do, and you always know what's best. You're so much smarter than me. I thought about that dream a lot, and I didn't know that's what it meant.

Throughout the rest of the session, Anna continued to plead for advice and expressed her discouragement about being able to resolve her own problems. The well-intended therapist did not want her to become dependent on therapy and kept refusing to tell her what to do—giving her a lengthy explanation about autonomy, independence, and the need for Anna to find her own solutions. At the end of the hour, Anna felt agitated and depressed, and the therapist was still trying to explain the need for Anna to make her own decisions.

Although the therapist was astute in linking Anna's dream to her separation guilt and the current manifestation of her emancipation problems, this was an unproductive session. How did their therapeutic process go awry?

In this session, Anna experienced a relational pattern with her therapist similar to the pattern she was struggling with at home. In her family, Anna was indeed trained to be dependent and believe that she didn't have the wherewithal to live successfully on her own. Further, she was made to feel guilty whenever she did act independently or on her own behalf. In the beginning of their session when the therapist interpreted the dream so accurately, the therapist acted as the knowing parent who gave all the necessary answers to the needy child. This interaction behaviorally communicated the message that Anna will be dependent on the therapist's superior understanding. On another level, however, this message was contradicted by the therapist's verbal message about independence. This mixed message from the well-intentioned therapist (being told what the dream meant and then being told to act independently) immobilized Anna. The therapist's direct interpretation could have worked well for some clients, but (recalling the concept of client response specificity) such an approach was problematic for Anna.

Anna reacted so strongly because this interaction cut right to the quick of her problem. On the one hand, she needed and greatly wanted permission from the therapist to become more independent but, on the other, expected that the therapist—deep down—really needed her to remain dependent—just as her mother did. When the problematic relational pattern of having to remain dependent on the authority was reenacted with her therapist, Anna's faulty belief that she was incapable of making her own decisions and having her own life was confirmed. Anna remained depressed until the therapist successfully reestablished a more collaborative interpersonal process with her. With the help of his supervisor, the therapist did this in the next session by soliciting Anna's ideas about something they were discussing. The therapist expressed his genuine pleasure in watching her be so insightful and, following this, shared some related ideas of his own. This time, Anna readily picked up on his ideas and used them to further her own thinking—it was a partnership.

PROVIDING A CORRECTIVE EMOTIONAL EXPERIENCE

Recalling "rupture and repair," Anna and her therapist soon corrected their reenactment and restored their working alliance, and treatment successfully progressed. (When therapists can acknowledge a misattunement on their part, this allows the client to enter the relationship more fully and authentically. The client is given the opportunity to clarify her experience and her need. In this way the client becomes further empowered to elucidate what is going on for her and what she wants to have change—she is now more actively engaged as an agent of the treatment process [Benjamin, 2009].) With this in mind, let's examine other, more effective ways of responding to Anna's question, "What do you think my dream means?" Interventions that invite a collaborative partnership—engage clients in exploring issues with the therapist—will usually be more effective than explanations or guidance provided by the therapist, no matter how accurate. Therapists can choose to respond in many different ways and can engage clients in exploring their own dreams (and any other material they present) as follows:

- It sounds like a very important dream to me, too. Let's work on it together. Where should we begin?
- What was the primary feeling you were left with from the dream? Can you connect that feeling to anything going on in your life right now?
- What was the most important image in the dream for you? What does that image suggest to you?
- Let's exchange ideas. I'll give you a possibility I'm thinking about, and you share one with me. It'll be interesting to see what we can come up with together. Who should go first?

Each of these varying responses encourages the client to participate actively with the therapist. Ultimately, the therapist may want to give the same interpretation of the dream (but with "skillful tentativeness" and collaboratively invite the client to modify or refine the interpretation to make it fit better). Even though the therapist provides an explanation, however, the process will be significantly different and far more enlivening for clients if it evolves out of their joint efforts and is integrated into their continuing collaboration. This approach gives clients a relationship in which they are not one down, or dependent on the therapist, but are encouraged to exercise their own abilities. When this occurs, clients feel more ownership of the therapeutic process, their motivation to work in treatment increases, and their self-efficacy is enhanced. Facilitating such self-efficacy provides a corrective emotional experience for clients like Anna, who get the permission they need to grow out of their dependence and act capably in their relationship with the therapist—something they may not have experienced in the past (Silberschatz, 2005; Weiss, 1993). To illustrate this more concretely, let's return to the same dialogue and see how quickly the therapist can make it more collaborative and more productive:

ANNA: *(enthusiastically)* Yes, you're right, I do feel like I'm doing something wrong whenever I leave my mother or do what I want. Is that why I dreamt that?

THERAPIST: We've been talking about whether you are going to move into an apartment next month. I'm wondering if the dream has something to do with you feeling guilty about taking this big step on your own. What do you think?

ANNA: I want to move out, but I can't leave my mother with my father. She says she will divorce him if I move out, and it'll be my fault. What should I do?

THERAPIST: Tell me what you think as I slowly reflect back what you just said: It's your fault if your mother leaves your father.

ANNA: *(pauses)* It sounds crazy...it's making me crazy.

THERAPIST: That's right—it makes you feel crazy, and it's not true. Children are not responsible for their parents' marriages, and it's not fair when they are pulled into marital conflict or rescuing a parent like this—it keeps you from having your own life.

ANNA: I feel better when you say that...but what should I do?

THERAPIST: What would you like to be able to do about this?

ANNA: I think I want to speak up, but I just don't want to hurt her feelings...

THERAPIST: Yes, I get the maddening conflict you've been talking about: you want to be able to have your own life, but you don't want to hurt your mother because you love her. So, tell me about both sides of "speaking up." What would you want to say, and how would it be for you if your mother felt hurt by that?

ANNA: I'd like to say that I don't want her to complain to me about my dad anymore. I don't want to hear what's wrong with him. I don't want to have to take sides. But she'd definitely feel hurt and betrayed if I said anything like that.

THERAPIST: It would be great to set boundaries with her like that, but the guilt over hurting her would be very hard for you. You've struggled with this dilemma for a long time now...saying and doing what you need "hurts" her.

ANNA: Yeah, it does, but I would like to find a nice way to speak up—you know, respectfully.

THERAPIST: Sure, let's find respectful ways to do this, but also keep realistic expectations about how she's likely to respond—and how that's going to affect you.

ANNA: Right, I know what you're thinking. I can probably stand up better for myself, but my mother's probably not going to give me much support for moving out or doing what I want with my life. She's going to feel hurt and betrayed no matter what I say.

THERAPIST: I'd love to be wrong but, sadly, that's what I expect too.

ANNA: Yeah, it is sad, really sad. But let's get on with this...I like it when we role-play things.

THERAPIST: OK. Why don't we practice this by role-playing how the conversation might go, and anticipate the trouble spots that are likely to come up. I could role-play you, and you could be your mother and say the things she will probably say when you talk to her about this, and we can both watch out for the guilt, and the sadness, as we go along...

TRACKING CLIENTS' ANXIETY

Another way to help clients turn inward is to develop the skill of being able to *track the client's anxiety* (Sullivan, 1970). Although, the subjective experience of anxiety is uncomfortable and may even be painful, it is still the therapist's ally. Anxiety serves as a signpost that points to the heart of the problem—a signal that the real threat or danger is at hand. That is, tracking what makes this client anxious will help therapists identify the client's key concerns. Therapists do not fully understand clients' problems—or what needs to occur in counseling to resolve them—until they understand what makes each client anxious. Why do certain situations or interactions make a particular client anxious, and how has the client learned to cope with this threat or insecurity? Therapists are encouraged to begin formulating working hypotheses to these questions in order to clarify treatment plans. As we will see, focusing clients inward on their anxieties—exploring together what activates or triggers them—will lead the therapist and the client to the client's central concerns.

Therapists can use a four-step sequence to track clients' anxiety and better understand, at core, what's really wrong for them. In this sequence, the therapist:

1. identifies manifest and covert signs of client anxiety;
2. approaches signs of client anxiety;
3. notes the topic presently under discussion and considers the interpersonal process currently transpiring that may have precipitated the client's anxiety; and
4. focuses the client inward to explore the meaning of the client's anxiety.

Let's look at these four steps more closely.

IDENTIFY SIGNS OF CLIENTS' ANXIETY

Clients will become anxious in the session when the issues they are discussing cut close to home and touch on their key concerns. Clients develop certain interpersonal coping strategies, which they employ over and over again in different situations, to ward off this anxiety that has been triggered. The therapist is trying to identify what makes this particular client anxious (for example, needing to ask for help and expecting the other to resent or feel burdened by their need; not knowing how to fix or do something and expecting the other to be impatient or disdainful; and so forth). When therapists can do this, they are better prepared to recognize the unifying themes or repetitive patterns that have triggered the client's anxiety. Thus, throughout treatment, therapists are continually asking themselves, "What makes this client anxious? When does the client feel threatened or insecure? Where does safety lie, and where does danger lie?"

To identify the patterns and themes that are causing problems in clients' lives, the therapist tracks their anxiety and remains alert for signs indicating that something they are experiencing is making them anxious right now. Clients may express their anxiety in a thousand different ways—nervous laughter, nail biting, hand gesticulation, agitated movement, stuttering, hair pulling, speech blockage, and so on. These "anxiety equivalents" will be expressed in endlessly varied ways across clients, and therapists will have to learn how each particular client tends to express their anxiety. (This insight will be especially important when working with clients from cultures in which certain behaviors—such as gaze aversion—may be normative rather than a signal of anxiety.) When they become anxious, most clients utilize the same interpersonal coping strategies in a repetitive or characteristic way. Routinely, these strategies include pleasing others, becoming critical or controlling, acting helpless or confused, withdrawing from people or avoiding situations, and so forth. We will examine these interpersonal coping strategies closely in Chapter 7. For present purposes, it is enough to note that the first step in tracking the client's anxiety is to observe and mentally note when clients become anxious (for instance, when they have to assume a leadership role, when they succeed, when they have to disagree with or challenge others, and so forth).

APPROACH CLIENTS' ANXIETY DIRECTLY

In Chapter 3, we suggested that therapists ask about and work with the feelings and concerns that clients may have around asking for help, having a problem, or entering treatment (resistance). Further along in treatment, therapists similarly are encouraged to approach clients' anxiety about an issue that they are discussing or about what may be occurring with the therapist at that moment. Therapists can focus clients on their immediate experience in an open-ended way (for example, by asking, "What are you feeling, right now, as you talk about this with me?"). Therapists can also label the anxiety more directly. For example:

THERAPIST: Something seems to be making you uncomfortable right now. Any idea about what that may be?

CLIENT: No, I'm not sure.

THERAPIST: What's the first thing that comes to mind when you ask yourself, "I'm afraid that _____."

CLIENT: You're mad. I think I'm afraid that you're mad or upset with me somehow...

By focusing clients inward on their anxiety as they are experiencing it, and helping them explore it further and discern or "name" it more precisely, the therapist is both following the clients' lead and leading clients closer to the source of their problems. For example, Greenberg (2002) writes compellingly about the importance of attending to emotions, with the therapist and client together using these to facilitate change. Although, some clients may be ambivalent about the therapist's request to look within and explore their anxiety further, some clients may welcome the opportunity to share their anxiety with the therapist—and the sadness, shame, or pain that usually follows close behind. On the other hand, clients may also recognize that approaching their anxiety will bring them closer to faulty beliefs about themselves ("I'm too demanding or needy") and unwanted expectations of others ("They really don't want me") that, yet again, evoke the same unwanted feelings.

Thus, though at times a client may experience the therapist's focus on his anxiety as validating and reassuring, at other times it may intensify anxiety and even arouse resistance. As discussed in Chapter 3, therapists need to respond to the client's resistance when it emerges by exploring what about this anxiety-arousing topic is threatening or does not feel safe to approach. This approach will be far more effective than ignoring the client's reluctance as if nothing significant is occurring, or pushing through it and trying to persuade the reluctant client to talk further about the difficult issue. This process, of respect and honor of the client's process and autonomy, is collaborative, fosters motivation and empowerment in the client, and contributes to positive outcome (Guame et al., 2008; Miller & Rollnick, 2002). Before the therapist and client can address the problem itself, they can explore the resistance and together learn what about this topic or issue doesn't feel safe for the client to discuss further:

THERAPIST: OK, don't talk anymore about this with me but, instead, help me understand what's the threat or danger for you if you did talk to me about that. What's going to go wrong between us—or how are you going to end up getting hurt again—if you did?

Perhaps the client fears that the therapist will not like or respect the client anymore if he talks further about this; he will start crying and won't be able to stop, and so forth. As already emphasized, the therapist's intention is to respond in a way that honors the client's resistance. Recall from Chapter 3 that the therapist does this by (1) trying to understand the original, aversive experiences that led the client to behave in this particular way; (2) helping the client appreciate how this "resistant" or "defensive" response was once a necessary and adaptive coping strategy; and (3) disconfirming the schema by providing the client with a different and more satisfying response than the client has come to expect, as in the following example:

THERAPIST: No, I don't find you "needy" or "demanding" for asking me if I would call you and touch base for a minute before you make this big presentation on Tuesday. It would be easy for me to take a few minutes to do that, and I'm honored that you are willing to risk sharing this with me.

OBSERVE WHAT PRECIPITATES CLIENTS' ANXIETY

As we have seen, the therapist first observes when the client is anxious, and then helps the client focus inward to explore and try to understand what she is experiencing. At the same time, the therapist has a third task: to try to recognize (1) what issue was just being discussed (content) or (2) what type of interaction between the therapist and the client (process) may have precipitated the client's anxiety. If the therapist can identify what it was that made the client anxious, the therapist will be better able to identify the faulty belief, interpersonal pattern, or difficult feeling that may have generated the anxiety. If therapists can discern the precipitating or triggering event, it will go a long way toward helping them understand what's really wrong or troubling for this client.

The client will be able to discuss many different issues comfortably with the therapist. However, when the client becomes anxious, the therapist's aim is to try to identify the central issue or concern that has just precipitated the anxiety. What was the client just talking about? Was the topic about sex, loss, intimacy, money, vulnerability, success, divorce, inadequacy? The answer will highlight the key concerns that are most central to the client's problems. As we know, therapists can help the client explore this more effectively if they have already formulated working hypotheses about the possible issues that may be generating the anxiety. Keeping process notes after each session, as suggested in Appendix A, will help therapists formulate their working hypotheses more effectively.

In addition, the client's anxiety is often a signal that the interpersonal process between the therapist and client is reenacting a developmental problem or cyclical pattern for the client. For example, the therapist might observe client anxiety in the following circumstances:

- As the therapist was expressing uncertainty about what was going on in the session at that point, the client became anxious, perhaps because her alcoholic father would start demeaning her whenever he felt uncertain, confused, or inadequate.
- The client had just made an important insight and the therapist acknowledged his achievement. The client then became anxious, perhaps because he believed that he would always have to be so insightful and perform so well, as an aggrandizing and demanding parent had always expected of him.
- The therapist had just been supportive, and perhaps the client felt anxious because "strings" had always been attached to what she had been given in the past. Following her schemas, the client may have become concerned that the therapist would make her pay for his support in some unwanted way, just as a caregiver used to do.

The therapist can generate working hypotheses such as these about how the current therapist–client interaction may have triggered a developmental problem or unwittingly reenacted a maladaptive relational pattern for the client. These hypotheses can then be used to help therapists and clients explore what evoked their anxiety, understand what this threat or danger really means, and find more effective ways to respond in their lives.

FOCUSING CLIENTS INWARD TO EXPLORE THEIR ANXIETY

As we have seen, the therapist is alert for signs of client anxiety and approaches it whenever it seems to be occurring. Simultaneously, the therapist tries to identify what precipitated the client's anxiety—especially what has just transpired between them—and begins to generate working hypotheses about the relational patterns, faulty beliefs, or difficult feelings that may have triggered the anxiety. Finally, the therapist also focuses the client inward—to explore more specifically what the threat or danger seems to be for the client right now. If the therapist can help clarify the thoughts and feelings that are associated with the anxiety, the client's vague discomfort will usually become more specific. That is, once the client's basic concern is highlighted, the client can address the anxiety-arousing problem more directly. (Client: "I guess I'm afraid they won't like it if I speak up, disagree, or say 'no'— but maybe that's a consequence I can live with.") The following dialogue illustrates this process:

THERAPIST: It seems like something's going on for you right now. I wonder what's happening?

CLIENT: You know, this doesn't sound very nice to say, but I don't think my mother really wants me to change very much. I don't think she's a mean person or anything, but I do think I'm getting better in therapy, and she's not completely happy about that.

THERAPIST: How so?

CLIENT: Well, I haven't been depressed for a while now, and I've actually been feeling pretty good the last month or so. Maybe it's just coincidence, but it seems that, as I've gotten better, my mother has withdrawn and been harder to talk to. And I think this has gone on between us before.

THERAPIST: You know, I think you're right. You have been doing very good work in here and getting better, and that may be kind of hard for your mother sometimes.

CLIENT: (fidgets, begins picking at her nail, and then starts talking about another topic)

THERAPIST: Maybe something just made you feel a bit uncomfortable?

CLIENT: I don't know. How much time is left?

THERAPIST: I'm wondering if something we're talking about, or something that might be going on between us, could be making you uncomfortable right now?

CLIENT: *(pauses)* Well, I don't really know why I'm saying this, but maybe I'm afraid that you're going to go away or something?

THERAPIST: That sure makes sense to me. We were just talking about how your mother seems to act hurt and withdraws when you are feeling stronger, and I just told you that I thought you were doing very good work in here.

CLIENT: Well, I guess so, but that just makes things worse. You're not going to stay with me either if I get better—isn't that the point of this whole therapy thing anyway?

THERAPIST: That sounded important, but I didn't understand it as well as I wanted to. Can you say that again or say it differently?

CLIENT: If I get better, then we stop—just like with my mother. Then I won't have my mother, or you, or anybody. Don't you see? It's hopeless.

THERAPIST: You're saying this so clearly. I didn't understand at first, but you're right— the issue we are struggling with right now is like what's gone wrong before. To feel better and act stronger has meant that you have had to be alone. In the past, you have had to be sad and depressed and needy in order to be close or connected to others. I think you've been struggling with this dilemma most of your life, and, right now, it feels like it has to be the same old story again for you and me.

CLIENT: Yeah, it does.

THERAPIST: But I wonder if you and I can do this differently for once. Can we work together and try to find a way to make it come out better this time?

CLIENT: It's not working very well so far.

THERAPIST: Well, one difference I'm thinking about is that we're talking about it—naming it and sharing it—and that hasn't happened before.

CLIENT: What difference does that make?

THERAPIST: Maybe it's different because I can see what's happening for you and get how discouraging it's been for you to be undermined in this way so many times. And right now, I'm still feeling connected to you—still "for" you—as you are changing this and acting stronger in here with me.

CLIENT: Well, that's probably true, but these don't really seem like huge differences to me.

THERAPIST: OK. I see these differences as more significant than you do right now, but let's keep talking about this.

CLIENT: Why? What's the point?

THERAPIST: If you can stop holding yourself back in here with me, and find that I, and maybe some other people in your life too, don't go away like your mother does but stay connected to you when you succeed, it might help you act stronger with others in your life. Like your boyfriend, and your professor, and others we've been talking about.

CLIENT: Do you really think that could happen?

THERAPIST: Yes, I do. How do you think your boyfriend will respond to you if you let yourself be as smart with him as you are in here with me? Would he be threatened, like your mother has been and some others in your life will be, or would he enjoy that nice part of you, as I do and some others would?

In this way, tracking the client's anxiety can help the therapist and the client focus inward and clarify the client's key concerns. In particular, it will often reveal how the same problem that the client is having with others has been activated in the therapeutic relationship. By clarifying how aspects of the client's problems with others seem to be reoccurring between the therapist and client, the therapist has the opportunity to intervene in two important ways. First, therapists can talk about and change their interaction with the client to ensure that the therapeutic relationship does not repeat the problematic patterns that have occurred with others. Second, as we have just seen, they can link the client's positive new behavior with the therapist to others in the client's life. We will return to these basic, orienting constructs in the chapters ahead and explore further how clients can begin to try out, with the therapist, new ways of relating, and generalize this to others in their lives.

CLOSING

The basic theme in the interpersonal process approach is that therapists from every theoretical orientation can better help their clients change by living out or enacting solutions to their problems in their real-life relationship with the therapist. Helping clients adopt an internal focus for change is one important way in which therapists provide clients with a different type of relationship than the problematic patterns they have experienced and come to expect with others. By codetermining what issues and goals will be pursued in treatment, and by actively participating in finding solutions to their problems, clients are far more likely to change, and gain an increasing sense of self-efficacy in the process. Perhaps Bowlby (1988) has captured this interpersonal process most succinctly by saying that the basic therapeutic stance toward the client is not "I know; I'll tell you" but "You know; you tell me."

As clients begin to reflect on their own reactions and become less preoccupied with the problematic behavior of others, not just their anxiety but their entire affective world opens up and becomes more accessible. This affective unfolding is a pivotal step in the change process. However, the strong feelings aroused by pursuing an internal focus can be distressing to clients and uncomfortable for therapists as well. Thus, the next chapter explores how therapists can help clients come to terms with the significant feelings activated by looking within.

SUGGESTED READINGS

1. As we are seeing, therapists in training are trying to develop new interpersonal and process-oriented skills that are not easy to acquire. To help develop these important but often-challenging skills, a self-assessment measure is provided in Chapter 4 of the Student Workbook. Completing this self-assessment will help readers evaluate their current level of skill along several important therapeutic dimensions.

2. Several microskills texts are available that teach new therapists how to focus inward, use immediacy interventions, and provide interpersonal feedback (see Egan (2010, Chapters 11 and 12) and Hill (2009, Chapters 13–16). Better yet, watch a skilled therapist on DVD using immediacy interactions with a real client and discussing these interactions after the compelling interview, in the APA Psychotherapy Series I, with Dr. Jeremy Safran.

3. Years ago, Albert Bandura argued convincingly that perceived self-efficacy is an underlying mechanism of change across all treatment approaches; see his classic article "Self Efficacy: Toward a Unifying Theory of Behavioral Change" (1977) or his recent book, *Self-Efficacy: The Exercise of Control* (2006). He emphasizes that little enduring and generalized change results from verbal suggestion or persuasion. Instead, clients change when they have authentic experiences of "enactive mastery"—that is, when they have the experience of performing adaptive new coping responses with the therapist.

4. Alan Wheelis (1974) discusses the internal process of change in his book *How People Change*. In particular, see the chapter "Freedom and Necessity," which uses the language of responsibility to clarify how clients must focus on themselves rather than others in order to change. Therapists can learn much from other existential therapists as well. In particular, see Irvin Yalom's highly engaging book *The Gift of Therapy* (2003), which helps therapists work in the here-and-now and facilitate clients' taking more responsibility for their own lives.

5. As therapists work with diverse clients from varying cultural contexts, it is helpful to become more aware of normative verbal and nonverbal behavior. In particular, readers may wish to examine Chapters 1 and 2 of *Counseling the Culturally Different* (Sue & Sue, 2008), which address language differences among cultural groups. For a discussion of how to assess anxiety in Asian clients by focusing on nonverbal behavior, see the case study "Rape Trauma Syndrome" in *Casebook in Child and Adolescent Treatment: Cultural and Familial Contexts* (McClure & Teyber, 2003).

HELPING CLIENTS WITH THEIR FEELINGS

CHAPTER **5**

Jennie, a first-year practicum student, was off to a good start with her client, Sue. At their third session, however, Sue entered in true crisis: "I have breast cancer!" she cried. "My doctor says I have breast cancer!" With undisguised anguish, she went right to her biggest fear: "My girls may not have a mother to help them grow up!" Taken aback sharply by the intensity of this raw emotion, Jennie didn't know what to do at first. Urgently thinking of what she could say, Jennie jumped in and began doing what she had always done best: getting the facts, making plans, and solving problems: "Have you got a second opinion...?" More questions and answers followed: "Have they talked to you about your treatment options yet? They can do so much more now than they used to; chemo has become so much more effective...."

In supervision the next day, her supervisor tried to help Jennie see that she had moved right in to a problem-solving and reassuring mode without responding very directly to Sue's profound fear: "Such big feelings, Jennie, such vulnerability. It sounds like you wanted to help by helping Sue figure out what to do. Instead, what do you think would have happened if you had just tried to stay with her feelings longer?" Jennie was rather quiet, and vaguely suggested that she wasn't sure how to do that. Wanting to help, her supervisor tried to be more specific: "I'm thinking of several ways you might be able to do that in the future—like expressing more directly your own compassion or concern for her. Or perhaps helping her clarify or name her feelings that seem so overwhelming, or maybe just affirming how frightening all of this is right now? I think the sequence is important here. If you respond to her feelings first with this type of understanding or empathy, I think that then she could better use your suggestions about what to do."

As their discussion unfolded, however, Jennie disagreed and expressed that "the client wasn't ready to go that deep yet—we haven't been working together long enough to push her into feelings like that." Jennie looked puzzled as the supervisor replied, "But you weren't 'pushing' her to 'go deeper' than she wanted. Sue initiated this—she brought these feelings to you."

It was too new and too much for Jennie to absorb now, but, by the end of her practicum year, Jennie had become better at letting her clients feel what they were feeling.

CONCEPTUAL OVERVIEW

Painful feelings lie at the heart of enduring problems, and therapists help clients change when they respond effectively to them. When the therapist focuses clients inward on their own experience, clients will begin to feel more intensely, and express more directly, the difficult feelings that accompany their problems. This affective unfolding is a pivotal point in treatment. It highlights or reveals the conflicted feelings that are central to the clients' problems—clarifying what's really wrong; and it makes these important feelings, that have not been responded to well in the past, accessible for the therapist to respond to and work with. On the one hand, clients will welcome the promise of having therapists make contact with them on this deeper, more personal or feeling level. But on the other hand, clients still fear and may want to avoid feelings that have been unacceptable in other relationships— or that just seem too painful, shameful, or hopelessly unsolvable. Thus the therapist's response to the challenging feelings that clients bring will have a pivotal impact on the outcome of treatment, and this chapter will help therapists learn much more about helpful ways to respond.

Therapists facilitate change by providing a more helpful response to the client's feelings than he has come to expect from others (that is, the therapist validates these, takes them seriously, and does not minimize them or offer superficial reassurances). Furthermore, when therapists avoid or do not respond to the client's feelings, the therapeutic relationship loses its vitality and meaning—treatment is reduced to an intellectual pursuit. Thus, the purpose of this chapter is to help therapists respond effectively to the feelings evoked when clients look within.

Before starting, however, let's begin this chapter with a note of encouragement and an eye toward realistic expectations for new therapists. The material presented in this chapter is more personally evocative and challenging for many therapists than the material presented in other chapters. The issues discussed here often evoke personal concerns for therapists in training, and these principles are difficult to readily apply with clients. Understanding these processes intellectually is a realistic goal for first-year therapists; being able to utilize these interventions effectively in ambiguous therapeutic interactions is not. Be patient: In a year or two, you will be able to integrate these suggestions and make many of these interventions your own. Like so many things, learning this comes with practice.

RESPONDING TO CLIENTS' FEELINGS

WORK WITH CLIENTS' AMBIVALENCE ABOUT ADDRESSING THEIR FEELINGS

Many of the problems that clients present can be resolved by trying out new coping strategies or behavioral alternatives, questioning faulty beliefs, or reframing problems in a new perspective. These brief interventions are sufficient to help with many of the difficulties clients present, and they are an important aspect of change for most problems. However, for the more enduring and pervasive problems that clients often present, the conflicted emotions that accompany their difficulties need to be addressed as well. Psychotherapy process studies have demonstrated that positive outcomes are associated with therapist interventions that facilitate clients' emotional experiencing and encourage verbalization of affect (Diener et al., 2007).

At times, clients may be reluctant or ambivalent about exploring more fully the difficult feelings that emerge in treatment. On the one hand, clients hope that the therapist will be able to help with their emotional reactions; on the other, they may want to avoid the difficult feelings that they have not been able to resolve on their own. This reluctance emerges in the beginning for many different reasons. To the client, it may seem pointless:

- To share the same painful feelings again when—based on what has happened before—they have scant hope of receiving help;
- To suffer shame by revealing secrets, failure experiences, or feelings of inadequacy;
- To risk unwanted but expected judgment from the therapist for having feelings that have always seemed unacceptable;
- To struggle with their own fear of "losing control."

As a result of these and other concerns, clients sometimes resist the therapist's attempts to touch the emotional core of their experience. At the same time, however, they often long for this empathic understanding as well. The hope is that:

- Someone may, at last, truly listen to them with deep understanding or compassion rather than criticism or blame.
- Their experiences might, in fact, be validated.
- They no longer have to be alone with the shame, anxiety, anger, guilt, or other form of distress that they have felt so alone with.
- They can begin the process of connecting meaningfully with another person and have a relationship characterized by trust.

Once the therapist recognizes the client's relational schemas and understands how significant others have characteristically responded to the client's vulnerability in the past, the fear or shame that motivates her ambivalence will make sense. As outlined in Appendix B, therapists can **generate working hypotheses** and try to anticipate clients' potential response to conflicted feelings and attempt to work with those feelings rather than avoid them. If not, treatment will be just an intellectual exercise that will not be sufficiently potent to propel meaningful and enduring change. (Following Goldfried, extrapolations from cognitive science and cognitive neuroscience all point to the need for affective arousal in order to reorganize meaning structures (Goldfried, 2006; Samoilov & Goldfried, 2000).) In sum, the therapist can appreciate the client's reluctance to approach certain feelings yet also keep in mind the client's simultaneous wish to be responded to in this deeply personal way.

Although difficult emotions are central to most clients' problems, the therapist's affirming response to these feelings also provides the avenue to resolution and change. The corrective emotional experience that the therapist provides to clients—the experience of sharing important feelings with another who remains connected and validating—loosens the hold of faulty beliefs and expectations. As rigid schemas become more flexible and expand, clients are empowered to make the changes they desire. At this critical juncture when clients experience a reparative or corrective response to such significant feelings, the treatment process springs forward. At this point we will see that clients typically become more receptive to a wide range of interventions from varying theoretical orientations.

CLIENTS AVOID FEELINGS BECAUSE OF UNWANTED INTERPERSONAL CONSEQUENCES

In Chapter 3, we saw that both therapists and clients may be reluctant to address the client's resistance. Similarly, therapists and clients alike may steer away from the client's difficult feelings, as Jennie did at the beginning of this chapter. As we have just seen, however, treatment doesn't progress very far—it soon stalls and loses meaning—when strong feelings are not dealt with. As with resistance to treatment, the therapist does not simply want to push through these defenses in order to reach clients' sadness, anger, or other significant feelings. We don't want to press clients to feel or disclose anything they don't want to share. So what do we do instead—just wait and try to be patient? No, as before, *therapists can join collaboratively with the client in trying to understand the threat or danger if they shared a certain feeling*, rather than press for it. Let's see how therapists can take this more effective approach.

Although the common notion is that clients avoid sharing or experiencing feelings because "it hurts too much," we suggest something different: we suggest that the defenses are against *expected* but unwanted *interpersonal consequences*. For example,

CLIENT: "You wouldn't respect me if you really knew how afraid I am to leave him, and how much I put up with just so I don't have to be alone. It would be humiliating for you to see what a doormat I really am."

This contrasts to the more traditional notion of *intrapsychic dynamics:*

CLIENT: "It hurts too much; I can't stand it."

We propose that *clients often minimize or avoid difficult feelings because they expect to receive familiar but unwanted responses* from the therapist (e.g., judgment, criticism, withdrawal, advice) and others in their lives that evoke fear, shame, or guilt. Therapists create a more affirming interpersonal process, with greater safety for clients, if they enlist clients' collaboration in trying to understand the very valid or realistic reasons why it has been threatening to experience or share certain feelings. For example, rather than pressing the client to disclose a difficult feeling, the therapist may use responses such as these to explore the relational threat instead:

- OK, let's not talk about how you feel about this. Let's talk instead about what might go wrong for you, or between us, if we did.
- If you did cry or let me see your vulnerability, what might go on between us? I'm wondering if there is something that has happened with others that you don't want to happen again here with me?
- What is the danger for you if you let me see your sadness? Help me understand how you have been hurt in other relationships—what have others said or done that wasn't good for you when you shared your sadness with them?
- Growing up, how did your parents respond to you when you were feeling _____? What did each of them typically say and do when you were feeling that way? Can you recall the look on their faces, or talk about

what they might have been thinking or feeling toward you as you were feeling _____?

In these ways, the therapist does not push for the content—the specific feeling—but works with the client to clarify the threat or danger—the reality-based reasons for why it has not been safe to experience or share the feeling. As in Chapter 3, the therapist is honoring the client's resistance by collaborating with the client in trying to understand why this reluctance or defense was once a necessary and adaptive coping mechanism. The therapist's basic orienting assumption is that the client's resistance to certain feelings makes sense in terms of what has transpired in other relationships, although it is no longer necessary or adaptive in many current relationships. *If the therapist and client can clarify how the client's defense was necessary and adaptive at another point in time, and in some current relationships as well, the therapist can make it clear that she (and later, that some others in the client's life) will not respond in these familiar but unwanted ways.* With this new understanding, it will soon be safe enough for the client's threatening or unacceptable feelings to emerge.

To reiterate, there are often interpersonal reasons why clients want to avoid difficult feelings. Though clients may tell therapists that they don't want to explore their underlying feelings because they are "too painful," "overwhelming," or it "doesn't do any good anyway," *clients usually avoid difficult feelings because of their expectations that certain unwanted interpersonal consequences will result,* expectations of being ridiculed or judged, being invalidated, having their feelings dismissed as unimportant or unrealistic, being ignored or not taken seriously, seeing that others are hurt or burdened by their feelings, having others overreact or dramatize their feelings, and so forth. Such interpersonal threats are usually far more important than their own subjective discomfort. Although it is very difficult for most beginning therapists to grasp this, *clients also hold these same expectations toward the therapist—even though the therapist has never responded in any of these problematic ways.* The key issue is for therapists to be empathic to clients' challenging feelings and expectations, and help clients clarify the very good reasons they learned that certain feelings were not safe to experience or share. Therapists can also help the client "differentiate then from now" and make overt how they would like to respond differently now in their current relationship than others have in the past. Although it may seem to the therapist that these distinctions are obvious to the client, they are not and the client will need to have these differences punctuated. For example:

THERAPIST: I'm wondering if it has ever seemed like I was responding to you in that kind of judgmental way—you know, like your husband does sometimes, and your father did, at times?

OR

THERAPIST: If you decide to share that feeling with me, what do you think I might I be thinking about you or feeling toward you?

OR

THERAPIST: How do you think I am going to respond if you risked sharing that feeling and letting me see that part of you?

OR

THERAPIST: If you chose to share that with me, how would you hope I would respond? What response from me would you be most afraid of receiving?

Of special interest, one the interpersonal reasons clients defend against certain feelings is to protect their caregivers. That is, many clients avoid or minimize their own painful feelings in order to protect their caregivers from seeing the hurtful impact they are having on them. However, this loyalty to the caregiver is carried out at clients' own expense. As adults, these clients are still *complying* with familial rules and colluding with their caregivers in denying the impact of hurtful parental actions. By so doing, these clients are preserving insecure attachment ties ("Everything's fine—no problem") at the cost of disavowing their own fear or deprivation. When clients lose the validity of their own experience in this way, they also lose their own voice. This protects the (internalized) caregivers from seeing what they did that hurt the client (such as embarrassing or ridiculing him, repeatedly ignoring him or favoring a sibling, needing the child to be "perfect" or fulfill the family role of "hero," being afraid of getting hit or witnessing domestic violence, and so forth). If clients participate in this denial and dismiss the problems that actually were occurring, homeostatic family rules are maintained. The client remains attached to internalized parents who often remain idealized or "good" while, in this splitting defense, the client remains "bad" and feels that something is wrong with him. However, if painful feelings are permitted expression in therapy, and *the therapist can affirm the clients' experience without blaming or rejecting the parent*, chronic symptoms such as lifelong dysthymia or generalized anxiety often change. For example:

THERAPIST: Yes, it makes sense to me that you feel so anxious and worried "all the time." It must have been frightening to hear your parents fight like that—threatening and demeaning each other. Your mom might get really hurt—your dad might go away, and you couldn't stop them or fix this big problem. And you and your sister listening to them at night through your bedroom door like that...it's such a heartbreaking spot for a little girl to be in. So, no, I don't think your making a "big deal out of nothing."

To sum up, the therapist's intention is not just to help clients "get their feelings out" in a global way that does not help them connect their feelings to the specific person with whom they are sad, mad, or hurt, and just what it was about that interaction that evoked their feelings. Just helping clients "vent" in this general way does not help most clients make progress on their problems. Too often instead, it serves only to camouflage or divert clients away from the real feelings or concerns they are struggling with. Thus one more useful approach can be for the therapist to help clients identify the *problematic interaction sequences* that repeatedly occurred and gave rise to their difficulties with experiencing or sharing certain feelings (for example: the client feels upset about something; tells her mother who exaggerates the issue and then subtly shifts the focus to her own distress; and the client—feeling confused and alone, finds herself walking toward the refrigerator to eat ice cream). By clarifying what actually did occur over and over in other important relationships in this way, but is not going to occur in the therapeutic relationship, the therapist provides the interpersonal safety that clients need to enter and experience their feelings more fully. Further, with the safety that comes from this empathic understanding, therapists can also help clients identify the problematic relationships in their current lives where these same unwanted scenarios are recurring, and discern other relationships where better responses are occurring or could be developed.

APPROACH THE CLIENT'S MOST SALIENT FEELING

With each turn of the conversation, therapists are challenged to select how they want to respond right now to what the client has just said. For example, the therapist might respond by seeking more information, clarifying what the client has just said, making connections between this issue and other material the client has presented, or linking it to what is occurring between the therapist and client at that moment. As a general principle, however, *the most productive intervention is to approach the most salient feeling in the material that the client is talking about right now*. In other words, the therapist's initial focus or priority is to acknowledge the affective component—the feeling or emotion—in the client's response.

To illustrate, imagine that the therapist and client have just sat down to begin their first session:

THERAPIST: Tell me, Mike, what's the difficulty that brings you to therapy?

CLIENT: I'm having a lot of problems with my 15-year-old son. We disagree about everything and can't seem to talk to each other anymore. He doesn't do what I ask him to, and I don't like his values or his attitude. I guess I'm pretty angry with him. His mother and I are divorced and have been sharing custody, but I'm thinking it may be time for him to go live with her. Do you think it's OK for a teenage boy to live with his mother?

THERAPIST: I don't think either of us understands what's going on between you two well enough to make any decisions yet, but you said you were "pretty angry" at him. Tell me more about your anger.

The client quickly has presented many different issues that the therapist could have chosen to pursue. Following the general guideline, however, the therapist approached the primary affect or feeling that the client presented. The therapist could have responded differently and inquired further about the issues that the father and son disagreed about, worked on values clarification or communication skills, obtained more background information about the father-son relationship or the divorce, provided research findings about the effects of mother custody versus father custody on boys, and so forth. These and many other responses will often work well. However, *responding first to the central or most salient feeling that the client expresses will usually produce the most meaningful information and intensify the interaction*.

In this dialogue, the therapist responded when the client spoke directly about his anger. Therapists also need to respond to the covert or unverbalized feelings that clients experience sometimes. For instance, the therapist in our example might have chosen to respond instead to a salient but more covert affect. In the example presented, the therapist could do this by tentatively wondering aloud, "As I listen, it sounds like you're feeling really discouraged about being a dad right now—maybe so discouraged that you just feel like giving up sometimes?"

When clients begin talking about what matters most to them, important feelings usually come up. Such nonverbal affective signs as tearing, sighing, grimacing, or blushing are signals to the therapist that the client has entered a significant topic that holds real meaning. Responding to these nonverbal cues, the therapist invites the client to share the feeling or deeper meaning that this situation holds for the client. For example, imagine that the client has been discussing her marital problems:

CLIENT: I don't know if I should stay married or not. I haven't been happy with him for a long time, but I can see how hard he's trying to make our relationship work. And our four-year-old son would be devastated if we broke up. I don't know what to do, and I have to make the right decision.

THERAPIST: As you speak about this, your face tightens. I'm wondering what you might be feeling right now as you tell me about this?

When the therapist responds to the client's feelings in this open-ended way that invites exploration, it often serves to clarify the client's key concern. Often, clients will respond by expressing more specific concerns. Continuing with the previous dialogue, the client may respond:

- I'm so sad about hurting the people I love.
- I'm afraid that everyone will think I'm selfish for leaving, and they'll blame me for the divorce. They'll think I haven't been a good mother.
- I don't want to be alone. I don't want to be married to him anymore, but I'm afraid of leaving and trying to make it on my own.
- I'm furious that I'm the one who has to make this decision. I'm responsible for every decision we make, and I always pay for it in the end. It's not fair!

Throughout each session, clients have emotional reactions to the issues and concerns that trouble them most. Sometimes the client's affect will be presented forthrightly; at other times, it will be subtle or more covert. Cultural, class, and gender factors will further shape how each client displays affect (Sue & Sue, 2008). For example, some clients may overtly state their reactions, whereas those from other cultural or familial backgrounds may simply divert their eyes or become noticeably quieter when certain feelings are evoked. Or, when clients from some cultures feel angry and want to disagree, they feel anxious, become quiet, and shut down because it would be disrespectful to express their disagreement. The therapist's abiding intention, however, is *to keep giving invitations or open-ended bids to explore the feelings that seem most important to the client right now*. In other words, we are trying to reflect or highlight the most important feeling—the central feeling that seems to hold the most meaning at this moment. As the therapist joins the client in sharing and understanding her feelings more clearly—as in the examples just listed—it will:

- facilitate further disclosure from the client,
- highlight her key concerns, and
- provide a clearer direction or focus for treatment.

To illustrate, Ella is a 20-year-old client who is talking about her on-and-off again boyfriend and two-year-old son's father, Edward.

CLIENT: I'm just like my mother…

THERAPIST: How so, Ella? You have tears in your eyes as you say that.

CLIENT: I choose men who hurt me, dis me, and I can't seem to leave them.

THERAPIST: You see this pattern yet something keeps you going back…

CLIENT: Yup, that's me, I just don't know what's wrong with me.

THERAPIST: Let's just be together with this and see if we can figure it out together. Maybe if we explore how the sequence works we might understand what need or needs are being met.

CLIENT: OK.

THERAPIST: Tell me about the last time you and Edward had a problem. What happened? How did you feel after it happened?

CLIENT: It was my son's birthday and he didn't even call to say "happy birthday" to my son, our son. So I called him the next day to ask why he didn't call.

THERAPIST: How was it for you on your son's birthday when you realized that he wasn't going to call?

CLIENT: Sad, alone, no support.

THERAPIST: Alone, very alone.

CLIENT: *(tears)* Yes, I just feel so alone, I do everything alone, I am alone.

THERAPIST: You have been alone a lot. I am so glad that you are letting me in on your experience right now so you don't have to be alone with your sadness.

CLIENT: I think I keep taking him back when he "sweet-talks" me because I'm so afraid of being alone. Yet I know he is not right for me. He is so much like my dad. I don't want to end up like my mother, all doped up to deal with the abuse from him.

THERAPIST: You are getting help now and that takes a lot of courage. These feelings of being alone and of being afraid to be alone make sense. You have had to take care of yourself for so much of your life and have had to do so much on your own. (This led Ella to talk about her history of multiple foster care placements during the times when her mother was incapacitated by her drug use and that she sees how her fears of being alone are impacting her willingness even to tolerate Edward's behavior. In this way, Ella was able to address more clearly her needs and how these impact her relationships.)

To emphasize, the therapist wants to find *collaborative* ways to invite and explore the client's affect. We don't want to "push" for feelings, of course, or become invested in the client having sad, angry, or any other feelings during the session. For most therapists in training, however, the problem usually goes the other way. That is, the client introduces a feeling and the therapist does not "hear" it or, even if the therapist does register it, the therapist doesn't approach or respond to it. When the therapist does not respond to important feelings that the client presents, it keeps their interaction—and the therapeutic relationship—superficial. Oftentimes, it also confirms the client's expectation that his feelings are unimportant or unwanted to others (for example, "Well, wait a minute now, I'm not so sure it's really THAT bad"). Instead, we want to invite whatever feelings the client seems to be experiencing by trying to reflect, clarify, and affirm them.

All of this may sound easy enough to do, but it is not. Researchers find that even experienced therapists often do not approach sensitive feelings and concerns very effectively (Hill & O'Brien, 1999). Recalling the discussion of "rupture and repair" in Chapter 2, we have already seen that clients routinely hide or withhold their negative feelings and reactions toward the therapist (Regan & Hill, 1992), and that many clients leave one or more important things unsaid during sessions (Hill et al., 1993). Clients reported leaving important things unsaid because (1) their own emotions felt "overwhelming" to them, (2) they wanted to avoid dealing with the therapist about the disclosure, and (3) they feared that their therapist "would not understand." Related, in another study, about one-half of the clients sampled had major life secrets that they had not told their therapists, even in long-term therapy (Kelly, 1998). They withheld these secrets for two reasons. First, returning to our old friend, they reported feeling embarrassed and ashamed. Second, these clients also believed that either they themselves would be "overwhelmed" by their emotions, or that their therapists "would not understand" or could not handle the disclosure!

Further in therapy, Ella tells the therapist:

CLIENT: I have been afraid to tell to tell you about Edward and me having unprotected sex because I was afraid you'd think, "How stupid is she." But I really need help talking to him. I know, after all the ways he's treated me, how could I take him back, and even have sex with him? And you know, I didn't even ask him if he is having sex with other women. I'm afraid if I ask him if he is sleeping with other women, he might get mad and, you know, what if he decides to leave me?

THERAPIST: Being alone is really frightening for you...I can see that. You are also worried about safe sex—that's a very reasonable concern, Ella. It sounds as though you've also been afraid of being judged by me if you talk about how you have managed your relationship with Edward.

CLIENT: Yeah. I've let him walk all over me. And I never used to worry about safe sex before but now—since I've been talking with you, I want more for myself, and I do worry about my son too. But, I really don't want Edward to leave me; I don't want to be alone. I'm ashamed to admit that I'm risking getting HIV because I'm afraid to be alone. But when I think about it, that could leave my child with a sick mother—that would make me just like my mother, ending up sick and unable to be a mom and maybe abandoning my child because of sickness.

THERAPIST: You've just said a lot here Ella. You are afraid of being alone. You want more from a relationship than you are getting. And you want to be responsible to yourself, your health, and to your child. These are all important, Ella. Your needs and your feelings do matter. I'm glad we are talking about these things together.

CLIENT: I'm glad you make it OK for me to talk about it. I don't want to be alone but I also don't want to fail my child. Can you help me find the words to talk to Edward? I want to talk to him about having sex only with me, even though I know he could become mad about it. My son does need me, and you're right, my fear of being alone is keeping me from making good choices for me and for my son. But I don't want to end up like my mother—with a disease, and my son could end up in foster homes if I can't take care of him.

THERAPIST: How is it for you to be talking to me about this right now, Ella?

CLIENT: It's really helpful.

THERAPIST: What am I doing that's helpful to you?

CLIENT: It helps that you haven't made me feel like I'm a horrible mother. I feel like I can tell you things I can't talk about to most people. They would think I'm just a loser. What should I say to Edward?

THERAPIST: Lets think about what you want him to hear and practice that together...

The troubling research findings referred to earlier that clients fear therapists' responses to their revelations tell us a great deal about approaching clients' feelings. Recalling Jennie and her client with a breast cancer crisis at the beginning of this chapter, we note that *therapists often shy away from strong feelings and keep things on the surface by not acknowledging or capturing the most significant feeling in what the client just said.* Actively intending to "hear" and approach the feeling that seems to be the most sensitive, meaningful, or intense is a new way of responding for most therapists in training. For example:

THERAPIST: You're saying so much here. Out of all of this, what's the most important feeling you're left with right now?

It often feels awkward to approach feelings so directly because it breaks the social norms that most of us grew up with—we are not used to talking to friends and family like this in everyday interactions. However, with help from supervisors and practice with colleagues, new therapists can learn how to approach their clients' feelings more directly and respond effectively to the sensitive material they choose to share.

EXPAND AND ELABORATE THE CLIENT'S AFFECT

Let's look at **open-ended questions** to explore further how therapists can best respond to clients' feelings. At times, it can be helpful to suggest or tentatively name a client's feeling (for example, by saying, "It sounds like you were just furious" or "I'm wondering if you're feeling disappointed right now?"). However, as a general guideline, a more effective response is to give clients a more open-ended invitation to explore the feeling further:

- I'm wondering what you might be feeling right now?
- Can you tell me a little more about that feeling?
- How do you feel now as you are telling me about what happened with them?
- Help me understand some of the feelings you're experiencing as you tell me this.

These open-ended invitations encourage clients to explore whatever it is that they feel is central in their affective experience, rather than what the therapist thinks is most important (Fitzpatrick et al., 2001). In particular, these open-ended responses also bring more *immediacy* to the therapeutic relationship. That is, the client's feelings are dealt with in the here-and-now with the therapist, which usually intensifies them.

In contrast, we don't want to ask too many "closed questions." They lead us to get all the facts and details about a problem and away from the feelings and what this event really means to the client (Barkham & Shapiro, 1986). For example, suppose the client says, "I got my Chemistry mid-term back." Much more is gained if the therapist responds with the open-ended bid "Tell me your thoughts about it" than with the closed question, "What did you get on it?" This takes practice, of course, because most of us have not been socially trained to respond in this open-ended, exploratory way that "unpacks" or expands the personal meaning this holds.

An open-ended invitation also is more effective than trying to label the client's feeling in an either-or fashion—for example, by saying, "Were you feeling _____ or _____ in that situation?" Such a choice may restrict the client rather than invite a free range of responses. Clients often experience something that the therapist has not anticipated (no matter how much experience we have, clients always seem to surprise us), and the client may have several different feelings about the same situation. Thus, the open-ended queries are more effective in drawing out exactly what the client is feeling, and does so with an interpersonal process that is more collaborative.

Additionally, open-ended responses usually are more effective than asking clients why they are experiencing a particular feeling. For example:

THERAPIST: Why were you so upset and crying?

CLIENT: I don't know.

Although *why* questions can be effective at times, as a rule of thumb, therapists are advised to minimize them. Clients often don't really understand why they feel, think, or do something. Asking them may only make them feel "on the spot" or defensive—as if they are failing because they don't have the answer. The therapist will usually elicit more information and collaboration from clients by offering a more open-ended invitation—"Help me understand your anger better" or "Tell me more about your anger"—than by asking, "Why were you angry about that?" By inviting clients to express more fully whatever feeling they are experiencing, therapists are giving the message that they are interested in clients' own subjective experience and are comfortable with whatever clients say. As we have seen, clients frequently withhold affect-laden material from their therapists. Most clients have not had permission to share their emotions so genuinely in other relationships, which is why these open-ended invitations to explore whatever the client might be feeling is one of the most important responses therapists can offer. As they discover that they can be authentic with the therapist, and that it is safer than it has been in other relationships to feel whatever they are feeling, clients become willing to risk disclosing further, trying out new behavior, and investing further in the therapeutic relationship.

Giving clients an open-ended invitation to explore their feelings will also help them clarify what each particular feeling means to them. Too often, therapists assume that a particular affective word, such as angry or sad, means about the same thing to the client as it does to them. Therapists should not assume that they understand a particular affective word without clarifying what it means to this particular client. Although this may be especially apt when there are cultural differences between therapist and client, it applies with every client. In this process of mutual exploration, just as we did with empathy as a collaborative exchange in Chapter 2, therapists share their understanding of what the client meant and then encourage clients to correct or further clarify the personal meaning that a feeling holds for them. This clarification and mutual exploration is especially important if clients repeatedly use the same "compacted" affective word to describe themselves, such as boring, overwhelmed, and so forth. When this occurs, often several different and important meanings are contained or compacted in that single affective word. For example,

THERAPIST: Here again, you're describing yourself as feeling "erased" and "dismissed." It's sounding to me like you feel totally insignificant when he does that—you are so completely unimportant that nothing you say or do matters. Am I saying that right— or can you help me capture it better?

Having the client elaborate her experience in this way—*clarify the subjective or personal meaning it holds more specifically*—will also safeguard therapists against becoming overidentified with the client and misperceiving the client's situation as being the same as their own, as so commonly occurs.

Not listened to or invalidated in many relationships, many clients do not know very well what they feel. In many families, children grew up learning that it was not acceptable to feel a certain feeling—such as being sad, angry, or even happy. As a result, many clients are uncomfortable with, or unaware of, what they are actually feeling. As a result, many clients need to clarify the broad, "undifferentiated" feeling states that they often experience (Client: "I don't know, guess I'm just mad all the time"). Too often, new therapists try to approach or invite their clients' feelings by repeatedly asking, "How did that make you feel?"—which frequently leads to a dead end. Instead, therapists can help clients learn more about themselves, their own inner life, and the emotional reactions that are linked to their problems by asking questions such as these:

- Can you bring that feeling to life for me...help me understand a little better what it's like for you when you're feeling that way?
- Do you have an image that captures that feeling or goes along with it?
- Is there a particular place in your body where you experience that feeling?
- Is this a familiar or old feeling? When is the first time you can remember having it? Where were you? Who were you with? How did the other person respond to you?
- How old do you feel when you experience that emotion? Can you attach an age to it, such as seven years old or 13 years old?

Some of these exploratory or uncovering questions will work well with a particular client and, of course, not at all with another. Following client response specificity, therapists want to assess the client's reactions to each way of responding and find what works best for each client. In general, however, these responses are all ways of approaching clients' emotions, entering their subjective worldview, and clarifying the subjective meaning this feeling holds for them. These types of responses also tell clients that the therapist is concerned about their feelings—who they are and what is most important to them. Furthermore, they behaviorally inform clients that the therapist is different from many other people they have known. The therapist is comfortable sharing whatever emotions the client has and, in addition, wants to know the client's authentic experience more fully—as good or as bad as it actually is for them—without minimizing or exaggerating it as others have so often done.

When the therapist works with clients to clarify their emotions, it creates an opportunity for the therapist to see and respond to the clients' most personal experience. This also provides clients with the opportunity to *know* themselves, especially since significant others have repeatedly invalidated, ignored, or disdained certain feelings that the client has had, such as being sad, mad, or afraid. This occurs, for example, when caregivers say:

CAREGIVER: Oh, come on sweetheart, you're not really mad about that.

OR

CAREGIVER: One more look like that, and I'll give you something to be sad about.

As old expectations of judgment, invalidation, and so forth are disconfirmed by the therapist's empathy and validation, the client receives an important, real life experience of change—a corrective emotional experience.

Clients Make Progress When They Experience Rather Than Simply Talk about Feelings.

As stated above, our treatment goal is not to just help clients "get their feelings out." Of itself, "venting" does not usually help much or lead to behavior change (Lohr et al., 2007). On the other hand, however, therapists want to do much more than just cognitively label, explain, or interpret a feeling—they want to help clients experience it more fully in the moment, and then respond in a corrective way that provides validation and containment. Little change will occur for most clients until they are able to stop talking about their emotions in an intellectualized manner and actually experience or feel their feelings in the presence of the therapist (Carkhuff, 1999; Gendlin, 1998).

Gendlin (1996), an early research collaborator with Rogers, has linked treatment outcomes with the importance of clients tuning in on their own emotions and their level of "focusing" on their own experiencing process (see Hendricks [2002] for an extensive research review). With consistent validation and accurate empathy, the therapeutic alliance continues to develop, and clients feel more *safety*. In turn, this **interpersonal safety** allows clients to be able to risk "feeling what they feel" and "knowing what they know"—something that often has been too threatening or anxiety arousing in the past. The exploratory responses that we have been discussing, especially the interventions that create immediacy, will intensify clients' experience and allow them to enter their own experience more fully—as long as the therapist can "contain" the affect and it doesn't disrupt the therapist. That is, clients need help from their therapist to access the emotional and mental states that they may not yet have words or constructs for (which may be disowned or even dissociated) and then assist them in making sense of these by highlighting how the past has shaped their experience and expectations. In this process, the therapist provides the client with a relationship that is characterized by, and cultivates, being **reflective**—that is, the therapist resonates with, is attuned to, and experiences the client's experience as fully as possible in the here and now, and in doing so provides a restorative experience for the client (Fonagy et al., 2002).

Therapists may also draw out the client's affect by addressing any **incongruence** that they perceive between the client's narrative and the accompanying affect. That is, in relaying a story that the therapist finds poignant or disturbing, the client may display no feeling at all or an incongruent affect, such as laughter. Therapists are subtly but powerfully "pulled" by social norms to match the client's incongruent affect. New therapists, in particular, often feel they are violating unspoken social rules if they do not respond in kind and laugh—even though the story seemed sad to them. However, clients are usually relieved when the therapist risks breaking the social rules and makes a process comment by acknowledging this discrepancy between what was said and the feeling that accompanied it. By working with the process dimension and finding a diplomatic way to make this discrepancy overt, the therapist makes an important intervention: she begins to invite the client's true feelings or congruent experience. For example:

THERAPIST: I'm a little confused. What you are saying seems so sad to me, yet you are almost smiling as you tell me this. What are you thinking as I share this with you?

OR

THERAPIST: You're telling me how angry you are about this, Lee, but your voice is so pleasant as you speak—so soft and quiet. What you're saying doesn't seem to quite match the way you are saying it. What do you think—is there a discrepancy here that might be telling us something?

Our intervention goal here is to provide *interpersonal feedback* that helps clients register the feelings that accompany their actual experience—to become more congruent. So many symptoms and problems stem from clients' inability to feel what they feel: to be sad about what was lost or missed, afraid of what really was once frightening, or angry about what actually was hurtful. No small intervention, this is about identity, personal power, and the ability to be who you really are. *Simply allowing clients to be able to have the feelings that are commensurate with what actually happened to them is a lifelong gift to many.* Recalling the concept of invalidation, it allows clients to reclaim their own experience. It means nothing less than empowering clients to know their own mind so they can live their lives without feeling confused, uncertain, or "mixed up" anymore. Working within different theoretical traditions, many forerunners in the field have written eloquently about this far-reaching outcome: "congruence" for Carl Rogers, "voice" for Carol Gilligan, "authenticity" for Irvin Yalom and other existentialists, "identity" for Erik Erikson, "coherence" for Main et al. and other attachment researchers.

The very best opportunity to provide a corrective emotional experience and help clients change occurs *at the moment when they are experiencing the full emotional impact of their problem.* Unfortunately, however, this is also the moment when therapists are most likely to feel anxiously uncertain about what to do. If therapists can communicate their accurate understanding and genuine compassion to clients in the moment, as they are experiencing difficult emotions, it will help clients resolve the affective component of their problems. As a result of these real-life experiences of change with the therapist, clients move forward and make behavioral changes with others as their schemas expand. In particular, this in vivo relearning changes their expectations of how some others in their lives can respond to them, and it changes their selective biases in filtering and distorting how others are actually responding to them. However, as the following case example illustrates, clients cannot make such progress when they are psychologically alone with their feelings or can't let their true feelings emerge.

During one session, Jean had been telling her therapist about her hopelessness. She described her life as a merry-go-round of endless ups and downs that always returned to the same discouraging spot. This past week she had dropped out of school, as she had done "so many times before," and had given in to pressure from her self-centered, jealous, and possessive boyfriend to let him move back into her apartment again. Jean described her hopelessness about ever being able to do or have anything for herself.

The therapist could feel Jean's despair and was moved by it. The therapist tried to work with Jean's feelings by acknowledging her despondency and validating her experience.

THERAPIST: You sound sad and hopelessness right now—I hear you say how painful it is to feel so defeated. You haven't been able to follow through and do what you wanted to do for yourself again, and that feels so discouraging. You are again feeling that you have to go along and meet someone else's needs rather than say no and do what you want for yourself.

As in their previous sessions, however, Jean was not able to let the therapist be emotionally connected to her while she was experiencing this. She talked about these feelings, but, both agreed, she held the therapist at arm's distance. Because the emotional connection to the therapist was missing, it was hard for her to use the therapist as a secure base. Even though she was talking to the therapist, she was still alone with her feelings and hopelessly alone in her life.

Developmentally, Jean had grown up with a self-centered mother whom Jean had often found herself "parenting." Jean could never take her own needs to her mother—Jean's mother had needed Jean to stay close to her and stifled her ability to launch and enjoy an independent and successful life. This had made it difficult for Jean to share her struggles or concerns with her mother (or anyone else) because her mother regularly used these to highlight why Jean should live at home, attend the local college, and in other ways foiled Jean's trajectory toward an independent life. Jean's inability to stay in school, or to set boundaries with her boyfriend, were thus part of her struggle with making choices for herself because her mother often called her "selfish" when she began to make choices that were for her own advancement. However, when she acted on others' behalf and squelched her own needs—as her mother dictated—she experienced depression and feelings of hopelessness, such as she was now experiencing. Thus, *separation guilt* kept her from advocating on her own behalf; yet, *compliance* kept her disempowered and depressed. Clearly, Joan was stuck.

The setbacks of the previous week had intensified Jean's distress. The therapist recognized that this crisis provided an opportunity to be more connected to Jean in her distress—as she was feeling her feelings—for the first time:

THERAPIST: Jean, you've dropped out of school and let Josh move in. Right now it seems your needs have evaporated. It makes sense that you feel discouraged and hopeless.

JEAN: *(long pause, nods)* Yes, it seems hopeless. I'm just hopeless. I don't matter.

THERAPIST: You feel unimportant...that you don't matter...too discouraged to even try.

JEAN: *(nods)* I always have to do it their way.

THERAPIST: This is familiar...

JEAN: Yeah, my mother, my boyfriend, even at work—I'm just pathetic...

THERAPIST: Yes, you've felt this way a lot before. I can see how painful this is for you right now.

JEAN: *(looks away, puts hands over eyes)*

THERAPIST: I'm feeling with you in your sadness right now...

JEAN: Don't you think I'm just pathetic for letting this jerk move back in with me again?

THERAPIST: No, I don't see you as pathetic or weak. But I do see your sadness right now and how much all of this has hurt you. I think you are showing great courage in sharing this with me; and I see how hard you are trying to sort out what you need and want—I respect that.

JEAN: I just don't know what to do next, I feel so hopeless...

THERAPIST: Yes, it strikes me that "going along" leaves you feeling disempowered—it evokes feelings of hopelessness.

JEAN: I hate all of this. I wish I felt different...was different.

THERAPIST: What would you like to be different? What would you like to change if you could?

JEAN: I wish I could tell Josh to just leave but I don't know how to stand up for myself...and I worry about hurting his feelings.

THERAPIST: It's been hard to let yourself matter...

JEAN: *(hands over face)* Yes, yes.

THERAPIST: You do matter, your needs are important.

JEAN: It is OK to stand up for myself, isn't it?

THERAPIST: Yes, Jean, it is. I know it has been hard to set limits and it has been hard to let your needs be more important than other people's needs. Let's spend more time together exploring the feelings that come up when you let yourself matter, when you let yourself be a priority in your own life. It's absolutely OK and actually terrific that you are standing up for yourself.

JEAN: *(looks at therapist and slowly begins to cry)* Well, maybe I'm not completely hopeless...I want more for myself. I do want to make more of my own choices. I've been worried about being called selfish. But I don't want to live the rest of my life taking care of everyone else and feeling so "used" and depressed...

For the first time, Jean was able to let the therapist be emotionally connected with her while she experienced her despondency. The therapist acknowledged the risk that Jean had taken, and they went on to discuss how it felt for her to share her sadness, discouragement, and disdain for herself so openly. Jean said that she was surprised to hear that the therapist didn't judge her and still respected her, and that made her feel "lighter." At their next session, Jean said that her depression had been a little better and mentioned, almost in passing, that she had told her boyfriend she wasn't ready to live together yet.

IDENTIFY THE PREDOMINANT AFFECT

Clients often feel confused by their emotions; their feelings seem to occur for no reason and often are stronger than, or different from, what the situation realistically seems to call for. Clients' inability to understand their own emotional reactions undermines their self-confidence and engenders doubt. Helping clients make sense of their emotional reactions is an important pathway to their greater self-efficacy (Bandura, 1997). One way therapists can do this is by working with the client's **predominant affect**. As we will see, the client's experience often revolves around a central feeling state, such as being shame-prone or guilt-prone, and therapists help when they can identify and highlight this predominant feeling.

Clients often enter therapy in response to one of two situations. First, a current life crisis echoes an emotional problem that originally occurred years before—the current stressor taps into an *old wound*. Second, clients also enter therapy in response to feeling overwhelmed by *too many stressors* in a short period. These multiple stressors have overwhelmed the clients' usual coping strategies, and symptoms such as anxiety, depression, sleep disturbances, and increasing conflict in relationships develop.

AN OLD WOUND

Many clients enter therapy because of an "old wound." For example, a client who has suffered a painful loss through divorce or death will find it more difficult to cope with the stress of that loss if it simultaneously evokes unresolved or painful feelings from previous losses in the client's life ("compacted grief"). The current loss may be associated with a predominant feeling—perhaps regret, loneliness, or sadness. *The feeling becomes understandable, and far more manageable, when it can be linked to similar feelings that have been aroused by other losses in the client's life.*

For example, Nora entered therapy at the student-counseling center for "depression." She had been having intense crying spells since her aunt died two months ago—even though she hadn't known this relative very well or been especially close to her. Things began to make sense for Nora and when her therapist began to ask about other times in her life when she similarly felt like "crying all the time." Nora surprised herself by replying to the therapist, "when my Mother died... when I was eleven."

In this way, the therapist is trying to identify the core or primary affect that the current crisis has evoked and explore if it might be linked it to the original wound. Connecting this primary affect to the way clients felt in the past enables them to make sense of their seemingly irrational feelings. When this occurs clients change by feeling more accepting or forgiving of themselves. They are better able to appreciate the bigger problem they are actually dealing with—a story playing in two times—and gain a greater sense of self-efficacy. Although this is important in all treatment modalities, it is one of the most significant interventions therapists can make in crisis intervention or short-term therapy.

MULTIPLE STRESSORS

When his company downsized, Jim was laid off from work. Unable to find a new job during the economic downturn, he felt like a failure—but his problems were just beginning. As his self-esteem decreased and his financial worries increased, his wife described him as "negative" and said he was "impossible" to put up with. His marriage had never been the best, but now, living all day together at home, day after day, with this increasingly short-tempered, pessimistic husband became too much for her. Six months after his being laid off, she had had enough of his complaining and irritability and moved out. Because his wife had always created their social life, and without coworkers to visit with at his office anymore, Jim became even more isolated and lonely. Eating and drinking too much to cope with his anxiety and depression, Jim's blood pressure rose markedly. At his doctor's prompting, Jim reluctantly entered therapy, describing himself as a "loser."

Most individuals have adaptive coping mechanisms that allow them to manage a single stressful event, unless it taps into the type of preexisting vulnerability or old wound described above with Nora. However, if a second or third stressful event follows from the first, as it did for Jim and many others, it becomes much harder for individuals to cope. The client's usual coping strategies fail and the cumulative stressors precipitate symptoms. In this way, most significant and enduring problems are not caused by simple, isolated events. They usually develop over time out of a sequence of events or interaction cycles—as they did for Jim.

When subjected to multiple stressors, clients may describe themselves as "overwhelmed," "burned out," or even "broken." They tend to repeat a **compacted phrase** that encapsulates their primary emotional reaction to the stressful events:

- I want to go away and just be alone.
- It's too much for me; I can't stand it.
- I don't care anymore; there's no point in trying.
- I've had it, I'm "done," I can't face any more.

As noted in Chapter 2, therapists are looking for **recurrent affective themes** and trying to discern the most central or unifying feeling that is encapsulated in these compacted sentences. Underlying each compacted sentence are one or two core feelings that keep coming up and capture the essential impact these stressors have had on the client. Generally, the various stressors are arrayed around one or two overriding feelings, like spokes around the hub of a wheel. For example, in response to losing too many people or things, the client may have an overriding feeling of being alone or powerless, or the client may be afraid of experiencing further losses or changes or taking on new commitments. When therapists are able to articulate and capture the underlying feeling that links the impact of several different crisis events, clients often feel seen and deeply understood. For example:

THERAPIST: As I listen to what happened over the weekend, it seems that in each case—with the soccer coach, the telephone call from your boss, and then the argument with your husband—you were not heard. It seems like you became so upset because you were never really being listened to—you weren't being taken seriously.

When the therapist has been accurately empathic in this way, and able to identify and articulate the integrating theme, a measure of hope and security may be engendered. In response to this attunement—which many clients have not found elsewhere—their motivation to explore and risk in treatment increases and they invest further in the therapeutic relationship. To succeed in short-term treatment, therapists need to develop the skill to make this type of engaging empathetic connection, relatively consistently, early in treatment in order to readily establish credibility and initiate a working alliance. Of course, this ability only comes in time with supportive training and the benefit of observing role models; it is not a realistic goal or expectation for beginning therapists.

A CHARACTEROLOGICAL AFFECT

As clients relay their narratives, the therapist's intention is to try to discern and respond to the most important or meaningful feeling that clients are experiencing. Therapists can also listen for a recurrent feeling that keeps coming up in the client's life—a **characterological affect**.

As treatment proceeds, the therapist can often identify a predominant feeling that captures the client's current conflict and also characterizes her life more broadly—the client has long experienced this familiar feeling as nothing less than part of the fabric of her life. That is, if therapists are accurately empathic, they can often identify a central feeling that comes up repeatedly—a familiar feeling that pervades the client's life or expectedly returns when she is distressed or in crisis. Clients often have one or two of these core feelings that have been continuous throughout much of their lives. For example, when the therapist can identify these core feelings that come up again and again in different situations, clients often respond emphatically:

- Yes, that's how it's always been for me.
- That's what it's like to be me—that's my life.
- That's who I really am.
- That's it! That's exactly how it is.
- You've got it!
- Bingo!
- That's my experience, precisely.

Clients may describe these core feelings as their "fate," because it feels as if they have always been there and seems like they always will be. One of the most significant interventions therapists can make is to identify these characterological feelings and accurately reflect them to the client. For example, after listening to the client for a while, *and grasping the integrating affective theme*, the therapist may respond:

- As I listen, it seems as if you've always felt burdened by all the demands you feel you have to meet.

- I'm getting a sense that you've always been afraid—that you live with the fear that people are going to find you out and point out your true inadequacies.
- Are you saying that you've always felt resentful; that no matter how much you do, it's never enough?
- It sounds as if you've always felt wary...perhaps that others are trying to put you down or take advantage of you.
- I'm wondering if you're worried that if people really knew the true you, they couldn't possibly love or choose you?

Therapists are holding a sustained intention to consistently hear and respond to each of the varying emotions that clients present as they relay their narratives. However, therapists will be most effective when they can identify one or two repetitive feelings that have recurred throughout a client's life and that the client considers to be central to his or her sense of self. This often will be one of the most important interventions that therapists can make to establish their credibility with clients and strengthen the working alliance—it demonstrates that the therapist can help.

Twenty-year-old Jose came to counseling after being mugged. He struggled with revealing his characterological affect to others—his lifelong feelings of "inadequacy":

THERAPIST: What do you think I'm thinking as I see you cry?

CLIENT: That I'm weak.

THERAPIST: I'm thinking how strong it is that you can allow yourself to have such deep feelings.

CLIENT: *(silent for a moment)* Most people see the "in control" part of me...

THERAPIST: Yes, I get that...thanks for showing me the *feeling* part of you.

CLIENT: *(crying more)* No one knows this part of me...In public, I only show the social part of me, the competent part of me...I feel so alone with my real feelings...

THERAPIST: Yes, in public they see you, but only the part that is fully social, yet in private you often feel lonely and have feelings that you are alone with...that you wish you could share with at least one other person...

CLIENT: *That's it!* I'd like to be able to be more myself with others. I do get along good publicly, yet I go home and know I'm all alone. I want to be more open with others...I need to do it differently. I don't want to only show the strong part of me.

THERAPIST: I think it is great that you feel so deeply and want more balance, more relationships where you can be open and share more of yourself. We can work on that together if you like. Actually you have already started that process—right here with me. You are already sharing more of yourself to a new person, and I'm respecting how courageous you are right now.

CLIENT: I told my parents how scared I've been feeling after being mugged, but they... I don't know...I was hoping for something different from them.

THERAPIST: How so?

CLIENT: Well, my dad did his usual thing and started talking about being tough—and how tough he is. And my mom did her usual thing, and went on and on about how she was scared too. And then they started talking about themselves and all the things they always worry about. I wished they had listened to me and my fears, you know, let me speak more.

THERAPIST: That makes sense, you needed to be heard, your fears acknowledged, rather than hear about them and their problems?

CLIENT: Yes! How did you know? It *always* becomes about them…

THERAPIST: I didn't know, I was inquiring, wondering out loud with you.

CLIENT: That's exactly what I wanted—them to have just listened to me, to my needs, given me the feeling that I could come to them, have a need, and not always seem so much in control; that they would just be with me without going into their own stuff so much.

THERAPIST: You are putting it so well. You know, here with me, you can say exactly what you need. It makes sense to me that you were afraid. Being mugged would make most people really afraid. Wanting to be heard and responded to by your parents makes a lot of sense to me.

CLIENT: I'm glad you understand. *(Tearing up more)* Thank you for not making me feel like a wuss for feeling this way. I did need to be heard and not have them go into their stuff or have it become all about them. I feel like I always have to put on being tough and hide when I feel upset or scared or I'm not doing OK…

This interaction was especially significant because Jose came from a Mexican-American family. While gender roles are important for men from most cultures, they are especially pronounced for men of Mexican descent. For men from this culture, there is a greater struggle with expressions of vulnerability because there is a need to preserve a sense of respect and dignity—a fear that showing vulnerability may be misunderstood. That is, men of Mexican descent are often expected to assume a dominant, "strong," provider role (sometimes included in the "machismo" definition) in the family. Crying in front of others is generally seen as shameful, and sharing or talking about problems can also be seen as "weak." Thus, while there is variability in men from this culture, many feel the need to be "on guard" and typically present as "strong." This means presenting as stoic ("in control" for Jose) and not showing vulnerability to others—even in response to violent assault (Archiniega & Anderson, 2008; Fragoso & Kashubeck, 2000; Sue & Sue, 2008).

RECOGNIZE THE CONSTELLATION OF FEELINGS THAT CLIENTS FREQUENTLY PRESENT

It is deeply rewarding but sometimes challenging to work with the painful feelings that clients bring. Some clients want to downplay or avoid their emotions altogether; others may be overly preoccupied with them and talk obsessively about the same feelings or worries without any progress. When clients are not progressing in treatment, one reason may be that therapists have only responded to the clients' single presenting affect. In this section, we will see what to do about this by learning about the constellation or sequence of feelings (the "firing order") that clients often present.

If the therapist acknowledges the client's current affect and invites the client to explore it further, *a sequence of interrelated feelings will often occur together in a predictable way.* For example, suppose the client feels anger, which occurs in response to feeling hurt or sad about something that happened. However, the hurt or sad feeling is unacceptable to the client and is dismissed or disavowed because it seems "weak" to this client and evokes shame. To ward off the unwanted feeling of shame, the client artificially or defensively "restores" his self-esteem and sense of personal power or worth by returning to the "stronger" or safer feeling of anger again (which usually was an acceptable feeling or even scripted role for him in his family). As a result, others oftentimes perceive him as being an "angry person." This constellation of emotional reactions is central to the client's problems, and each successive affect in the sequence needs to be addressed. Although different sequences occur for each client, the therapist can often identify a triad of interrelated feelings that cycle repeatedly when clients are distressed. Making a similar point, Lazarus (1989) writes informatively about tracking a client's "firing order." The process of change involves coming to terms with each of the feelings in this affective triad. In this section, we examine two **affective constellations** that commonly occur. We will see that therapists help clients change by responding to the entire sequence of feelings in their affective constellation, rather than just responding to the first or initial feeling in the sequence—the client's primary presenting affect.

ANGER-SADNESS-SHAME

The first affective constellation we will examine consists of anger, sadness, and shame. The predominant or characteristic feeling state for some clients is anger. These clients often experience and readily express a range of angry feelings including irritation, impatience, criticism, and cynicism. However, the client's anger is usually *reactive*; that is, it is a secondary feeling that occurs in response to an original experience of sadness or pain.

The therapist needs to acknowledge the client's anger, collaboratively work to clarify more specifically who and what the client is really angry about, and help the client find appropriate ways to express it. For example, becoming more appropriately assertive, setting limits, or telling someone what it is that you don't like rather than losing your temper, yelling, or demeaning them. However, this approach is not enough. If the therapist stops at this point, the client may still remain stuck in an angry feeling state and more substantial change will not result. Instead, *the therapist's intervention goal is to help the client identify and stay with the primary emotional response of sadness or hurt that led to the reactive feeling of anger.* This primary feeling is more threatening or unacceptable for the client than the anger that it prompts.

To reach the original affect, the therapist might wait for clients to ventilate their anger spontaneously and, immediately afterward, invite them to attend to what they are feeling now. Following the expression of the reactive feeling, there is often an open window to the primary feeling of hurt or sadness. For example,

THERAPIST: OK, you've just been sharing how angry this made you, and that sure makes sense to me. But let's stop here for a moment and check in on what else you might be feeling right now, on the other side of your anger.

Often, the original feeling (for example, vulnerability, embarrassment, helplessness) will often surface next. By joining the client in this feeling and working together to accurately name it, the therapist can often dislodge the client from the familiar (but safer and nonproductive) reactive feeling of anger.

Alternatively, if this approach does not work, the therapist can inquire directly about the original experience (such as her husband's criticizing her in front of friends, and so forth) that led to the client's reactive or secondary feeling of anger.

THERAPIST: Something must have hurt you very much for you to remain so angry. Tell me how you felt when that first happened.

If the client again begins to express anger or indignation about what occurred ("It really made me mad when he put me down like that..."), the therapist focuses the client back on the original experience:

THERAPIST: Yes, I can see why it made you so mad, but I'm also wondering how it felt right then, when he was putting you down in front of everybody?

CLIENT: Well, I guess maybe it hurt my feelings a little.

THERAPIST: Yes, it hurt your feelings. Of course it did. I'm so glad you can share that hurtful feeling with me. You've just made a big change, you know: you've never been able to risk sharing the hurt part of this before. How does it feel to take that big step with me?

CLIENT: (*emphatically*) I just hate feeling so weak that he can hurt me like that....

Thus, in response to this query, the client's original feeling of "hurt" was expressed for the first time, rather than continuing on—almost obsessively, only about how angry she is. The therapist responds affirmingly as the client begins to experience and share the internal aspect of her problem—that is, the feeling of hurt or sadness that is so unwanted or unacceptable for her. Recalling our internal focus for change, *it is this internal aspect of her problem—her sense of herself as "weak" or shame-worthy because he has been able to hurt her—that has kept leading her back to her safer, reactive feeling of being angry all the time.* This is a complex sequence for new therapists to follow in the beginning; let's keep going and try to make it clearer.

As we have just seen, when this client experiences anger, hurt or sad feelings about what happened to her often follow. Although her anger is easily experienced and readily expressed, the client has learned that it is not safe to experience or share the original hurt or vulnerability. To avoid the original painful feeling, the client defensively returns to the reactive feeling of anger in a repetitive manner. Again, this defensive coping pattern often leads others to regard this client as an "angry person"—and contributes to the problems she keeps having with others at work and at home.

The goal of the therapist's intervention is to help such clients experience the sadness, hurt, or vulnerability that precipitates their reactive/repetitive anger. However, as soon as the original hurt is activated, the client feels ashamed and will defend against this unwanted feeling by reflexively returning to the safer expression of anger that, in her belief, is "stronger." Thus, the therapist needs to explore the client's resistance to the original feeling of being hurt. Often, clients avoid the original painful affect because it arouses a third, aversive feeling of anxiety, guilt, or—most frequently—shame, as we saw here.

For example, as the therapist inquires about clients' reluctance to feel the original hurt, they will often reveal faulty beliefs and say something like this:

- If I let the pain be there, it's admitting that he really can hurt me.
- If I acknowledge that it hurts, it means he won.
- If I feel sad, it shows that I really am just a baby feeling sorry for myself.
- If I admit that part of this was my fault, then EVERYTHING becomes my fault!

Thus, if the therapist draws out the original pain that underlies the reactive anger, a third feeling of shame will often be aroused. To defend against the original sadness, and the shame associated with being sad, the client has learned to automatically return to the "stronger" feeling of anger that is safer or more acceptable for this client to reveal.

Although shame is a common reaction, other clients will experience anxiety or guilt in response to feeling the original hurt. For example, some clients will say that if they let themselves feel sad, hurt, or vulnerable, then "no one will be there," "others will go away," or "I will be alone." These clients feel painful separation anxieties upon experiencing their hurt or pain and avoid this anxiety by reflexively returning to their anger. Still other clients will say that, if they let their sadness be real, they would be admitting that they really do have a need, which would make them feel selfish or demanding. For these clients, guilt is the third element of their affective constellation.

Thus, this triad of feelings exists for many clients: *Frequent anger defends against unexpressed sadness, which, in turn, is associated with shame, guilt, or anxiety.* Enduring change results when therapists can help clients better come to terms with each feeling in the triad. More specifically, clients resolve such conflicted emotions when they can:

1. allow themselves to fully experience or feel each feeling;
2. share or be able to disclose the feeling with the therapist so they are no longer alone with it or hiding it; and
3. with the therapist's support, be able to stay with the difficult feeling for a moment or two (that is, tolerate, hold, or contain it) rather than have to move away from it, as they have in the past.

This is what it means to integrate or come to terms with unresolved feelings.

Thus, clients integrate their conflicted emotions when they can progress through these three steps for each feeling in their affective constellation; they are internally resolving their conflict. With this resolution, clients no longer need the outdated coping strategies or defenses that they have employed in the past. Furthermore, at this point clients are prepared to adopt new, more adaptive responses with others in their lives that they were previously unable to incorporate. For example, rather than always get mad and alienate others, they can opt to respond differently sometimes. In this example, becoming more direct and assertive without elevating:

CLIENT: (*to husband*) I'm feeling criticized right now and I don't like it—I'm asking you to find a better way to talk about this.

HUSBAND: (*contemptuously*) Umm, I'm not really so concerned about what you think. Why don't you just settle down and be quiet.

CLIENT: I'm feeling disrespected right now and it's not good for me to continue this conversation. I'd like to talk about how to manage our money with more civility. Let me know when you are willing to hear my ideas without telling me that everything I say is stupid.

Clearly, this affective integration is a pivotal experience in the process of change.

Shame.   Before going on to examine another common affective constellation, let's pause to highlight further the special role of **shame**. Although much has been written about anxiety, depression, and guilt, shame has long been eschewed in clinical training. Correspondingly, clients could be anxious, guilty, or depressed with their therapists, but not ashamed. Thankfully, shame is no longer the taboo topic it has been; indeed, this most unwanted and painful feeling state is now regarded by many theorists as the "master emotion" (Nathanson, 1987; Scheff, 2000). The cardinal role of shame in most resistance and defense, and in so many of the symptoms and problems that clients present, was first illuminated by the groundbreaking work of Helen Block Lewis (1971) and later developed by many others (Karen, 1992; Kaufman, 1989; Tracey et al., 2007).

As we began exploring in Chapter 3, new realms of understanding will open up to therapists as they allow themselves to begin "hearing" the family of shame-based emotions that pervade many of their clients' narratives. To feel shame is to feel fundamentally bad as a person—inadequate or worthless in the essence of your being (Kaufman, 1989). Clients directly reveal their shame-based sense of self by repeatedly using words such as *bad, stupid, worthless, self-conscious, sheepish, embarrassed, low self-esteem, humiliated*, and so forth. Less direct indications of shame may include the repetitive use of words such as *should, ought to, must*, and *perfect*.

On an interpersonal level, some clients defend against their own shame, and defensively attempt to "restore" their injured sense of self, by inducing shame in others. That is, these shame-prone clients routinely try to "disprove" their feelings of shame through rage and violent temper outbursts ("You can't tell me what to do. I'll show you!"); showing contempt for others or giving strong judgments ("You're stupid" or "What's wrong with you!"); provoking others with bravado ("Just try me"); and automatic blame or counter-criticism of the other—especially in chronic marital conflict ("No, it has absolutely nothing to do with me, it is *all* your fault"). In contrast, other clients also defensively try to restore themselves and deny their feelings of shame by rigidly striving for perfection, withdrawing from others or going away inside (e.g., "I just wish I was invisible"), and other avoidant and shame-anxiety cycles (Balcom et al., 1995; Tangney & Dearing, 2002).

When therapists are able to break cultural taboos and begin approaching shame reactions sensitively but directly, significant progress follows readily. However, most therapists have been socialized strongly to avoid or to rescue clients from these shame-based affects (that is, ignore, make light of, or change the topic). As a result of these social norms, new therapists will need the support and guidance of their supervisors, and additional reading, to begin responding effectively (Bromberg, 2006, writes elegantly about the importance of approaching and addressing shame—so it is not given legitimacy—in the safety of a supportive therapeutic relationship). In a nutshell, however, therapists help clients resolve their shame when they demonstrate compassion for clients' feelings of inadequacy or incompetency, harsh or contemptuous judgments toward themselves, and feelings of worthlessness. Often, these shame-based emotions are referred to euphemistically as "low self-esteem," masking the far-reaching emotional wound that is actually at hand. Therapists help clients resolve such shame-based reactions in two steps: first, by providing an unambiguously kind and accepting response to clients as they are experiencing these feelings of shame, and then following up and processing this with clients by clarifying that the therapist was not feeling judgmental or critical toward them in any way (as they expect and typically misperceive); second, the therapist is working (over time) to help clients *internalize* the therapist's compassion for their shame and begin to feel the same empathy for themselves and their own suffering that the therapist felt toward them. Said differently, the client breaks the identification with the attachment figure who was rejecting or contemptuous and, instead of despising herself in the same way the caregiver once did, comes to adopt the same compassionate or empathic stance toward herself and her own humanness that the therapist provided. This sequence resolves the internalized contempt at the core of shame dynamics. In closing, it is very difficult for most of us to bear witness to our client's shame, especially as we begin our training. With supportive supervision and role models, however, therapists can soon learn how to offer clients this great gift.

SADNESS-ANGER-GUILT

Whereas some clients lead with their anger and defend against their hurt, others characteristically lead with their sadness and avoid their anger. This type of client may present with a general ("undifferentiated") feeling state of sadness, helplessness, vulnerability, or depression. These clients do not experience or express anger, they avoid interpersonal conflict, and they tend to respond to others' needs at the expense of their own. In addition to this caretaking role, some may present as acting like a martyr or a helpless victim. To illustrate this affective constellation, we return to the case of Jean.

Earlier in this chapter, we examined a dialogue in which, for the first time, Jean was able to let the therapist respond to her sadness and be emotionally connected to her while she was experiencing it. Subsequently, the therapist explored what had made it difficult for Jean to share her vulnerable feelings with others. Jean explained that historically her mother would listen to her. However, her mother would then tell her she "shouldn't be that way," especially if Jean wanted to act on her own behalf. She'd tell Jean to be "considerate" of others and put others' needs first—that doing so was the "Christian" thing to do. (Jean subsequently clarified in treatment that she partly agreed with her mother and truly desired to be a good Christian, but her mother's dictum was out of balance and led her to deny other important parts of herself—it didn't support her wish to honor both God and herself.) Thus, in these ways, Jean had come to keep her feelings to herself, because she felt sharing them was no use anyway—she'd just be criticized, told she was being selfish, and, in the end, her needs would not be met, since the needs of others would always be deemed more worthy than hers. Following these early schemas, Jean had continued in adulthood to reenact the affective themes and relational patterns of her childhood in the problematic love relationships that she kept reconstructing.

Recall that when Jean returned to therapy the following week, she had made a significant change: she had told her boyfriend she wasn't ready to live with him yet. Two weeks later, Jean's boyfriend borrowed her car for the day and arrived late to pick her up from work, and this time she expressed her indignation. Further, in an uncharacteristically strong move, she told him that she deserved to be treated with more respect and consideration and, soon afterward, ended the relationship. Jean had remained locked in her sad and helpless victim role, in part, because the direct expression of anger and the assertive expression of her own needs were unacceptable—selfish and hurtful to others. As often happens, Jean began to claim her own personal power when her feelings were expressed and she received affirming responses to them from the therapist. However, we need to look more carefully at the sequence of feelings (sadness-anger-guilt) that unfolded for Jean.

Although Jean began the next therapy session by sharing her good news about ending the problematic relationship, she had difficulty sustaining this stronger stance. As the session progressed, Jean retreated from her indignation at how badly her boyfriend had treated her, began to feel concerned about how she may have "hurt" him, and wondered whether she had been "selfish" to end the relationship because he "needed me so much." However, the therapist's working hypotheses had prepared her for the possibility that Jean would feel guilty upon experiencing her anger, setting limits, or responding to her own needs. With this awareness, the therapist was able to quickly recognize Jean's guilt and engage her in questioning its validity. Jean had defensively avoided her anger because it aroused guilt, which served to reflexively return her to her characteristic affect of sadness and interpersonal presentation of helplessness. As the therapist helped Jean identify and consider all three of these interrelated feelings, she was able to relinquish her victim stance for increasing periods.

With the therapist's support, Jean could acknowledge the painful reality of the deprivation and invalidation that she had originally experienced in her family. *As she stopped denying the reality of these painful experiences, she stopped reenacting them in her current relationships*; she no longer recreated these familiar relational patterns that kept evoking the same disheartening emotions. By responding to each feeling in clients' affective constellations, therapists help clients make sense of themselves and their lives with a new and more realistic self-narrative, and change how they respond to others.

Again, this is a lot of material for most new therapists to absorb—let's pause and try to tie things together a bit. Stepping back for a moment, let's see how the previous topic, the predominant characterological affect, relates to the client's affective constellation. The predominant **characterological affect** will typically contain the leading edge of the affective constellation or triad. For example, clients who present affectively a predominant sense of helplessness can easily access and express sadness, which is first in their triad. Often, *the first emotion in the triad is the one that was allowed—or even prescribed as a part of the child's familial role—in the client's family of origin.* Other feelings became subsumed under, or covered over by, this emotion. Oftentimes, the second affect in the triad was considered too threatening to a caregiver—and therefore was not allowed to be experienced or expressed. Over time, the client also came to see these legitimate feelings as unacceptable, and the client's own authentic voice was muted. The third affect in the sequence, shame or guilt in the two examples here, is often avoided because it is too disruptive or aversive for the client. Hence, the client repetitively returns to the first or leading affect that is safer and more familiar.

The therapist's intention is to validate the legitimacy of each feeling in the affective triad and remain empathic or emotionally connected to clients as they are experiencing them. This gives clients the experience of expressing themselves authentically without receiving the unwanted responses they have come to expect (on the one hand, for example, the therapist withdrawing, rejecting, controlling, or criticizing the client; or on the other hand, the therapist feeling burdened, resentful, uninterested, or overwhelmed by the client—as the client learned as a child to expect from her parent). Meaningful changes are set in motion for many clients as they find that their previously unacceptable feelings can be shared without the familiar but unwanted consequences occurring. In particular, many begin to feel that they can center in themselves and "live inside their own skin," rather than being so anxiously preoccupied with appearances and trying to manage how others see and respond to them. This is especially problematic for clients who are shame-prone because shame involves such insidious self-consciousness, negative self-evaluation, and the fundamental feeling that something is wrong with one's *self*, which is being negatively viewed by others (see Parker & Thomas, 2009; Tangney & Dearing, 2002). Thus, outgrowing these limiting familial rules about affective experience and expression—which often are embedded within cultural expectations—brings clients into a more coherent personal identity. Feeling safer and stronger, they are able to risk more with the therapist and empowered to initiate changes in their own responses to others in their lives.

PROVIDE A HOLDING ENVIRONMENT FOR THE CLIENT'S DISTRESS

In attachment terms, the best way for therapists to help clients resolve challenging feelings is to provide a **holding environment** that contains their distress. However, when stressful life situations activate clients' most troubling feelings, their schemas will shape faulty expectations and distort their perceptions of the therapist. That is, based on the early maladaptive schemas that dominate in these affect-laden moments, clients anticipate that they are going to receive the same problematic response from the therapist that they have received from others in the past. It is very difficult for most beginning therapists to grasp that clients believe this because the therapist knows that she has never responded to the client in these hurtful or unwanted ways. However, the client fully anticipates this and *routinely misperceives the therapist's benevolent responses*—slotting it inaccurately to fit the expected but problematic old response. It might be helpful, therefore, before addressing affective containment, to review briefly the developmental attachment experiences that set the very profound templates for clients' emotional regulation and establishment of a coherent sense of self.

CONTAINMENT: DEVELOPMENTAL PERSPECTIVE

Children are born with a set of sensory and social capacities that develop based on the how they are responded to by their primary caretakers. Children who receive sensitively "attuned" responses—responses that are mindful or reflective of their needs—and take into account their ever-changing states of arousal in an organized and consistent way, are provided with a sense of security and safety. When parents and children function in a *synchronous* or mutually attuned emotional "dance," the children come to feel that they are seen and understood, learn that they matter to someone, and that their feelings are important and will be responded to. Within a well-functioning interactive dyad like this the parent sees or registers the child's experience and responds to the child's need before the child becomes overly distressed. This child successfully learns to deal with stress and distress, and this nurturing and attuned relationship also teaches the child that *moments* of mis-attunement between them are not fatal but can be overcome. That is, that recovery from ruptures or mis-cues is possible in their relationship and, as a result, that emotions are manageable. Indeed, short-term "dysregulations" (where a caretaker and child "miss" each other) can be used as opportunities for repair (for example, Mom: "You were trying to tell me that you were upset and I thought you were just being difficult and mad; I was in a hurry and wasn't listening very well and I'm sorry that I got so impatient with you. Let's get back on track"). In contrast, long-term or chronic states of mis-attunement are highly problematic for most children—and result in deficits in affect modulation and self-regulation that characterize most of the symptoms our clients present. For the child to feel safe and contained, her caretaker needs to *sustain* the consistent intention to provide "holding"—*characterized by responsiveness to the child's distress and the willingness to engage in repair when mis-attunement occurs.* Parents who provide this attuned responsiveness, where children can bring any need or feeling, are considered a **safe haven** for the child—which provides the secure attachment that helps the child modulate and manage his affects—buffering him from symptom development.

Closely related, an attachment relationship with a caretaker who is consistently attuned also provides the child with a **secure base** from which to venture out and explore the world with increasing confidence and autonomy. The child often uses his caretaker's emotional expressions as a "signal" or reference of what is safe and what is dangerous in the environment. When the child feels secure and is "cued" that it is safe or permitted to venture out, he is more likely to explore, develop mastery, show flexibility in how he deals with others and the world, and develop a more coherent identity or self-narrative. That is, he is able to individuate or become more autonomous, and have a more balanced approach to affective experiences, such as being able to accept imperfections in himself and others. In a similar way, *full and successful engagement in treatment is associated with clients' experience of their therapists as both safe havens (support for their distress and emotional needs) and secure bases (support for their differentiation and independence)* (Parish & Eagle, 2003).

In contrast, clients who have not had secure attachment histories often come to therapy having had developmental experiences that were painfully isolating, anxiously unpredictable, aggrandizing or disempowering, and so forth—because their caregivers did not consistently respond to their fear or distress with the affirmation and understanding that children needed to learn how to manage or self-regulate their own emotional arousal (Brisch, 2002; Cassidy & Shaver, 2008). To be more behaviorally specific, **containment** or "holding" meant that the child's concerns were heard, taken seriously, and understood by a caregiver who remained emotionally present and responsive. The child's distress registered with the caregiver and was taken seriously, without being diminished or exaggerated. The child who felt secure had learned that the caregiver would remain available to help manage the problem, if only by expressing interest and concern. That is, *the securely attached child is secure in the expectation of getting help or being responded to when she is distressed, vulnerable, or in need.* It doesn't mean that the child is indulged or spoiled in any way or that the caretaker can always solve the problem—it does mean that the child is not alone with her distress or need. Before children (and clients) can develop the capacity to manage disruptive feelings on their own and function more independently, someone must first provide this holding context for them. However, student therapists are quickly going to recognize that, with most of their clients, this essential developmental experience of **affect regulation** was not provided. For example, suppose the child is sad or hurt. In some families, the child's sadness:

- arouses the parent's own sadness, and the parent withdraws, which leaves the child emotionally alone;
- makes the parent feel anxious or guilty about not knowing how to respond, and the parent denies the child's sad feelings and tries to cheer the child up, which leaves the child confused and alienated from his own authentic experience; or
- makes the parent feel inadequate as a mother or father, and the parent responds punitively or derisively toward the child's sadness, which leaves the child ashamed of her vulnerability.

In such commonly occurring scenarios, the child's sadness cannot be heard and responded to; as a result, these feelings cannot run their natural course and come to their own natural resolution. Young children cannot contain a strong, painful affect by themselves—such as fear—without support from an emotionally available caregiver. As a result, the child will have to find ways to deny or avoid this painful feeling. Years later, when this child is an adult client and risks sharing the unacceptable or threatening feeling with the therapist, *the therapist's intention is to provide the holding environment that was missed developmentally.* Working within this attachment framework, the therapist's role is to provide a safe harbor for a while—the secure haven and secure base that the client missed developmentally. Providing clients with this attachment security is so important because it is crucial to the client's affect regulation—that is, the client's ability to manage anger, sadness, loneliness, anxiety, and other disruptive feelings without acting out or developing symptoms such as depression, over-eating, alcohol or substance abuse, and so forth (Shaver & Cassidy, 2008). We have already seen two instructive examples of psychological holding in Chapter 2: Marsha's second therapist heard and responded to her experience of "emptiness," and the therapist of the depressed young mother sensitively articulated the mother's own "unanswered cries." Both therapists successfully provided a holding environment—which allowed their clients to make important changes in the next few sessions.

CONTAINMENT: INTERVENTION GUIDELINES

With the concepts of a holding environment and affect regulation now in mind, the therapist's next concern is how to put this in practice and respond most effectively. What can we say and do to help our clients in these special moments when they are feeling deeply their pain and distress? Perhaps nothing evokes the new therapist's insecurities more than this moment:

*   I'm so uncertain about what to do…how can I make him feel better…
*   What if I make a mistake or say something wrong—and make things even worse for her…
*   How can I help with these feelings—I want to help her so much but I really don't know what to do…

As we have just seen, attachment-informed therapists respond effectively to clients' distressing feelings by providing a **holding environment** of empathic understanding and attuned responsiveness to "contain" the client's affect. One way they provide this *is by maintaining a steady presence in the face of the client's distress*. The therapist's verbal and nonverbal communications are:

*   I can register or hear these difficult feelings and register how important they are to you;
*   I can tolerate the feelings and stay steady with you without becoming overwhelmed or disrupted myself;
*   Together, we can get through this; you don't need to be alone with it; and,
*   I understand that this has felt undoing for you. I'm OK hearing it, feeling with you, and choose to stay present and supportive as we explore and clarify further what you need and want.

A corrective emotional experience occurs when the therapist responds in a new and safer way that resolves rather than metaphorically reenacts old relational patterns—*especially around affective experiences*. As emphasized above, however, *unless the therapist clarifies or makes overt what the therapist is thinking and feeling as clients experience or share these conflicted feelings, clients routinely misperceive the therapist's response to fit their old schemas*. For example, by working in the here-and-now with what is going on between them, the therapist can clarify or correct these expectable misunderstandings or transference distortions:

THERAPIST: Right now, as you are feeling so upset about this, what do you think is going on inside for me?

CLIENT: I don't know what you mean—what are you asking for?

THERAPIST: As you share this with me, what might I be feeling or thinking about you?

CLIENT: Well, since you're a therapist, you have to act nice and supportive so you'd probably never say anything. But you're probably thinking I'm overreacting again. You know, acting like a child, and I should just stop feeling sorry for myself and grow up. Like my husband says.

THERAPIST: Oh, no, I'm not thinking that at all. Actually, I was feeling how sad it is that this has hurt you so much for so long; and that you have had to be alone with it.

Without such exploration and clarification, clients routinely hold the inaccurate belief that the therapist is privately disappointed, frustrated, burdened, and so forth, just as significant others have been so many times before.

In sum, as an emotionally responsive and affirming therapist continues to provide these corrective emotional experiences, clients' feelings are contained, and distressed clients will discover that they can find or restore their own emotional equilibrium. Real comfort comes from being seen and psychologically "held" in this way, which leaves clients feeling secure and empowered. They do not become dependent and their feelings do not continue to escalate out of control, as some therapists and clients were taught to fear in their families-of-origin. In sharp contrast, clients actually are becoming independent and more coherent as they learn how to "self-regulate" their emotional reactions through the experience of feeling containment with the therapist (Greenberg, 2002). We will continue to explore how, in the moments following such containment experiences, clients often make progress in treatment. For now, however, let's look further at how therapists can provide containment for their clients.

Too often, therapists minimize, reassure, explain or simply move away from clients' painful feelings. In so doing, they commonly reenact the clients' developmental problem and again leave clients disconnected from others while they are experiencing their emotions. The most common reason for this is because *therapists assume too much responsibility both for causing the client's pain and for alleviating it.* Erroneously, new therapists may think, "I got the client into this, so now it's up to me to get him out of it." Both parts of this belief are false. By responding in the ways suggested here, therapists *reveal* feelings that are already present in the client; they do not *cause* the pain (for example, **therapist**: "I made her feel terrible. I made her cry so hard when I asked about that"). Moreover, it is patronizing for therapists to think they can "fix" clients and "get them back together again." In actuality, the therapist never has the power to manipulate another's feelings in this way, and attempts to do so commonly reenact the caretakers' controlling or dependency fostering stance with the client. When these ineffective responses occur, there is no holding environment. Therapists' inability to hear and join clients in their core conflicted emotions, as reflected in these ineffective responses, replays the client's developmental predicament. For example, by trying to talk her out of her feelings, Marsha's first therapist reenacted her experiences in her family. Marsha's first therapist went even further to avoid Marsha's feelings by, in effect, sending her away: When Marsha got near her sad, empty feelings, the therapist suggested that she join a group and talked to her about seeing a psychiatrist for antidepressant medications. Such referrals are certainly necessary at times, yet in this case they were used more to distance the therapist from his own discomfort with her feelings. Unfortunately, this occurs commonly and, as we will explore closely in Chapter 6, is more likely for therapists who themselves have an Avoidant or "Dismissive" attachment style.

If these types of responses are ineffective, what should the therapist do instead? The therapist's aim is to welcome the client's feelings and approach them directly. For example, the following responses would be more effective ways to respond to Marsha's strong feelings.

THERAPIST: I'm glad you're taking the risk of telling me about these sad and empty feelings. Let's keep talking, I want to understand them better.

OR

THERAPIST: I know that talking this way and being together with such painful feelings breaks the norms you grew up with, but I appreciate being able to share all of this with you.

Therapists also want to find genuine ways to express their care and concern. Therapists can do this in words, of course, but what we communicate nonverbally— in our tone of voice and our manner—is equally important. In particular, therapists want to be *emotionally present* with clients while they are experiencing their feelings. There are many ways to do this, of course, and much of this communication is nonverbal or involves only a few words. Below, we'll provide some guidance for successfully providing a holding environment. However, therapists will want to tailor their responses to fit each particular client and to make them congruent with their own personal styles.

Initially, the therapist's primary goal is to stay emotionally connected or present with clients as they are experiencing painful feelings. As discussed in Chapter 2, the therapist may also want to reflect what clients are feeling and to affirm the reality of their experience. This can be done as simply as by acknowledging, "You're very sad right now." With some clients, it may be helpful to further validate their experience by saying, "Of course you are feeling mad right now. It hurt you very much when he did that." An important part of developing their own professional identities, therapists in training are encouraged to explore and try out their own ways of expressing their compassion or concern. Role models are very important, yet instead of trying to be like your favorite supervisor or therapist, new therapists are encouraged to ultimately find their own style or natural way of being with clients in these special moments:

THERAPIST: I can see how much you miss your grandfather and how sad you are about his passing—he was so important to you. I feel honored that you choose to share such a special part of yourself with me—thank you.

To provide a secure holding environment, therapists also want to demonstrate behaviorally that they can tolerate the full intensity of clients' feelings. In other words, they must communicate by their manner—and overtly in words with some clients—that they are in no way hurt, burdened, or undone by clients' feelings (as most clients expect).

THERAPIST: I'm privileged that you choose to share this with me. I'm with you right now, and we're going to stay together in this until we've worked our way through it.

As clients re-gather their equilibrium from a strong emotion, the therapist may also want to clarify what caregivers and others have done when they felt this way. The therapist may want to highlight how their current relationship is different from the client's relationships with others (for example, the therapist remains emotionally present and affirming rather than becoming impatient, withdrawing, or dismissive as others have tended to do) and emphasize that the therapist is accepting of this aspect of the client that has been revealed. Thus, *it is essential to debrief clients by asking them how it has been to share these feelings or personal disclosures with the therapist.* As emphasized above, too often therapists miss this important opportunity. Genuinely touched by the client, they fail to appreciate that it is at these intense, sensitive moments that the client is most likely to misperceive the therapist's caring response in terms of earlier schemas. To correct these expectable misunderstandings, it is necessary to ask clients what they think the therapist was thinking, feeling, or going through while they were feeling sad, angry, and so forth. Early in their training, it can be awkward for new therapists to broach this. However, long-standing transference distortions, which also cause problems and disrupt relationships with others, can be identified and resolved at these moments. For example, the therapist might clarify the client's distortion by means of a self-involving comment:

THERAPIST: No, I wasn't thinking that you "looked silly" or "sounded weird" in any way. In fact, I was touched by how alone you felt in that awkward, embarrassing situation. Your feelings sure made sense to me.

When offered sincerely, simple human responses of validation and respect mean much to clients and help to restore their dignity and composure. Unfortunately, new therapists often believe that they have to do much more than this to respond adequately to the client's pain. These concerns may reflect the therapist's own unrealistic expectations that there is an absolute right or wrong way to respond. In reality, the client does not need the therapist to provide eloquent, crafted responses in a calmly self-assured manner. Such perfectionism is as unnecessary as it is unrealistic. It often serves only to make the therapist more preoccupied with evaluating her own performance and less focused on just being present with the client and grasping what he is really saying. As Winnicott (1965) says, the child needs only "good enough mothering." Therapists provide a "good enough" holding environment when they find their own genuine ways to enact these guidelines. Clients will be appreciative of sincere efforts and find them helpful, even if they are not always expressed articulately.

AFFECTIVE CONTAINMENT FACILITATES CHANGE
FROM THE INSIDE

When clients feel psychologically held, they experience safety. In this safe relationship, they can bring their needs, clarify their beliefs about themselves and the world, and begin to forge with selected others the kind of supportive relationship they are having with their therapist. Indeed, as a result of clarifying who they are, their likes and dislikes, and how they want to be in the world, they are now empowered to become their own agents—to develop their own "voice." The acceptance and safety they experience from the therapeutic relationship fosters a sense of coherence and autonomy—where they are now able to embrace increasing aspects of themselves (including their imperfections and insecurities, as well as their strengths). In contrast to how they presented at the initial sessions, they no longer have to be detached, frozen, or cut off from parts of themselves, or flooded, chaotic, or hyperactivated. Being psychologically held provides the safety to sort through the feelings, thoughts, and experiences that were previously too threatening to examine and modulate. Now, however, they become more manageable and can become part of a coherent life narrative.

Many clients will be profoundly relieved to discover that neither they nor others need be hurt, burdened, or overwhelmed by their feelings. When the therapist can encompass clients' most threatening emotions, *and it is made clear to the client that the unwanted consequences they expected did not ensue this time*, clients forge a more authentic connection to the therapist—and to themselves. They discover that it is safe this time to be seen and known as they truly are. This reparative experience also allows clients to begin responding in new and more adaptive ways with others in their lives—pursuing their own interests and goals, and choosing more actively how they want to live their lives and shape their relationships.

In this supportive context, clients' feelings can run their course and come to a natural close, usually in just a few moments. Clients then recover, by themselves, and readily reestablish their own equilibrium. The issue here is containment and self-regulation. Clients (and therapists) who believe that feelings are dangerously unending are describing a child who could, in fact, not stop crying when distressed because there was no safe haven (no one responding to their needs)—which is a humiliating, out-of-control predicament. When the therapist provides the corrective emotional experience of a holding environment, however, the client finds affect containment, can modulate her affect, and begins to make far-reaching changes in treatment as her capacity for affect self-regulation begins to develop.

As clients continue to receive these types of corrective emotional experiences, they disconfirm pathogenic beliefs about themselves. Clients learn that their feelings—and they themselves—are not needy, irritating, contemptible, demanding, overwhelming, and so forth. As a result, they no longer feel compelled to withdraw and be isolated, feel ashamed, strive to be perfect, or fall back on other interpersonal coping strategies that, sadly, just bring more symptoms and problems. Thus, the feelings that previously had to be sequestered away can now be experienced, understood, and integrated. Furthermore, when the therapist can hold clients' threatening feelings and allow them to tolerate experiencing and sharing it, they, in turn, will become more able to contain these feelings on their own, which greatly changes how they can respond to others. This is change from the inside out. This interpersonal process often propels a far-reaching trajectory of enduring behavioral change, and longstanding symptoms often abate at this point. Ironically, when clients risk exposing their pain, vulnerability, or shame, and the therapist responds with compassion, they feel more powerful. They are more flexible, more accepting of their vulnerabilities and imperfections, and approach relationships in a more "reflective" way (Fonagy et al., 2002). Being able to facilitate such self-acceptance and self-efficacy is one of the most satisfying aspects of being a therapist. To illustrate this approach, let's look at a brief case study.

Sherry, a first-year practicum student, was struggling to assimilate all the new ways of thinking about therapy that she had been learning. It was an exciting but disorienting year as her entire worldview was shifting. Part of the difficulty was that her practicum instructor and her individual supervisor were proposing very different ways to approach her clients. Having grown up with wrangling, divorced parents, she was often dismayed about her consistently failed attempts to try to please "both sides." Now, this distressing familial scenario was being activated again by this situation, which made the already complex challenge of clinical training even more difficult. During sessions with her clients, Sherry was almost immobilized at times as she tried to sift through their competing ideas, decide what made sense to her, and come up with something that she thought would be useful to say to her clients. Two-thirds of the way through her first practicum, she felt more self-conscious and inhibited about responding to people than she had been when she started the program!

Despite her insecurity, Sherry actually was learning a lot and had been responding quite helpfully to her client, Maria. Near the end of one of their weekly sessions, Maria began to disclose a painful memory. With tears in her eyes, Maria recounted the sickening violation of a date rape 15 years earlier, when she first went to college. Listening to Maria's story, Sherry felt sad and angry. However, as she began to think of all the things she should and should not say, and the different things her instructor and supervisor would probably want her to say, she felt frozen and could not find the words to say anything at all. Although it lasted less than a minute, the silence grew painfully awkward as Maria finished speaking and waited for Sherry to respond. Unable to speak, Sherry just sat there. Deeply moved by Maria's painful experience, but immobilized by thinking of all the different ways she "should" respond, she just couldn't find any words at all.

After an excruciating pause, Maria tried to take care of Sherry and started talking again, but this did not last long. After a few minutes, Maria chided, "Don't you understand? He hurt me a lot when he did that!" Sherry's inability to respond had ruptured their alliance. Although Sherry felt she had made a "terrible mistake," she was able to restore their breach by validating Maria's anger and talking through what just happened between them.

SHERRY: Yes. He did hurt you very much—it was frightening and infuriating. I'm very sorry that happened. And I understand why you are angry at me for not responding. I was touched by what you said, but I just couldn't find the right words for a minute.

MARIA: You couldn't?

SHERRY: No, I was trying to so hard, because I felt for you. But I just couldn't find the words I wanted to say, and the harder I searched for them, the worse it got. I'm sorry this happened; it must have been hurtful for you.

MARIA: It's OK. I don't know what to say either sometimes…(*pauses*) I don't know why, but this makes me think of my daughter when she was little. If I would take the time to stop and pick her up and really listen to her when she needed me, we would get through her problem pretty quick. My husband called it "collecting her." But if I was in a hurry and tried to ignore her or speed her up, she would escalate and then the whole episode could go on for a pretty long time.

SHERRY: You're a very good mom. I think you're remembering that right now because that's what you needed from me—to be attended to, cared about, collected. You needed me to give you what you gave your daughter.

MARIA: Yeah, I think that's just exactly what I needed. (*tearing*) You know, I never dreamed of telling my mother what happened to me—you know, about the date rape. She needed me to be "happy" all the time; she didn't want to hear about problems…not even big ones.

Obviously, Sherry recovered from her "mistake" very effectively, and began providing the holding environment that the client herself had articulated so well with her own young daughter. As in this example, being *nondefensive* and validating the client's dissatisfaction with us often leads the client to act stronger in the next few minutes and make progress in treatment. For example, clients may bring up related issues in other relationships—as Maria did; achieve meaningful insights that lead to behavior change; risk bringing up significant new material—such as other problems they have been having with the therapist or others; and so forth. Moreover, by recovering and affirming Maria's feelings at the second attempt, Sherry gave Maria a corrective emotional experience that helped her progress in treatment.

As previously noted, in secure relationships that function well, moments of mis-attunement provide opportunities for repair. Ruptures are inevitable in all relationships—the key issue is to convey the intention to be attuned and the wish for repair, and then to follow through and work collaboratively with the client to find ways to repair. As Sherry is teaching us, we're often too afraid of making mistakes—while the reality is that misunderstandings will occur but can be addressed and resolved. If the therapist is willing to take the risk of making problems overt and talking them through with the client, the expectable ruptures that will occur in every therapeutic relationship can be restored (Hovarth, 2000; Safran & Muran, 2000). This is so important in the therapeutic relationship because, as children, most of our clients were not able to restore the ruptures that they regularly experienced with their caregivers. When conflicts, mistakes, or misunderstandings occurred, the child often was left feeling helplessly alone and unable to reconnect. In secure parent-child relationships, in contrast, parents and children have developed rituals or learned ways to restore and reconnect with each other when problems disrupted their relationship. This is why it is so profound for many clients to have this new experience with the therapist—they never dreamed that ruptures could be repaired. As the therapeutic relationship becomes more secure for clients, because they find that ruptures can be repaired, they become stronger and better able to make changes they were unable to achieve before.

To sum up, when difficult feelings emerge, therapists have the best opportunity to help clients change. *The most important corrective emotional experiences usually occur around the therapist's new, more satisfying response to the client's emotions. On the other hand, however, familiar but unwanted patterns are most likely to be reenacted with the therapist in these affect-laden moments as well.* Why? Vulnerability activates our schemas and leads clients to expect the familiar but unwanted response from the therapist they have received in the past. Per Beck's "hot cognitions," it is when strong feelings have been triggered that clients are most apt to distort or misperceive the therapist's response and slot it to fit old expectations. Compounding this situation, the therapist's own personal issues or countertransference reactions are also most likely to be evoked in these sensitive and intense interactions. Thus, all therapists need to pay attention to their own countertransference issues, so we turn now to personal factors in the therapist's own life that can make it difficult to respond to clients' feelings.

THE THERAPIST'S OWN COUNTERTRANSFERENCE IS THE PRIMARY OBSTACLE TO HELPING CLIENTS WITH THEIR FEELINGS

In much the same way that it is sometimes easier for clients to externalize and focus on others rather than themselves, it can also be easier for therapists to focus on the client than to look within at their own issues and dynamics. If, however, therapists are not willing to work with their own emotional reactions to the material clients present, they will tend to avoid dealing with clients' feelings and engage clients only on an intellectual level. The purpose of this section is to help therapists responsibly manage their own reactions to the evocative material that clients sometimes present by highlighting potential countertransference propensities that therapists bring to therapeutic relationships.

Neither a working alliance nor a secure holding environment will occur unless therapists are accurately empathic and can communicate their respect and compassion for the emotional conflicts clients are struggling with. Realistically, however, it is difficult for therapists to respond to all of their clients' feelings because all therapists bring their own developmental histories and current life stressors to the counseling session. Let's explore how these personal factors can keep therapists from responding effectively to clients' feelings at times.

THE THERAPIST'S OWN NEED TO BE LIKED CAN INTERFERE

Most trainees who are drawn to clinical practice are generous people, genuinely concerned about others, and ready to give of themselves. Often nurturing by nature, they can readily empathize with others and easily enter their experience. At the same time, however, many therapists also have strong needs to be liked. As a result, many may have difficulty responding nondefensively to certain clients who present in treatment with the same hostile/challenging interpersonal style that these clients have been enacting with others in their lives. That is, some clients will be characteristically critical, confrontational, controlling, competitive, angry and so forth with others—and begin to replay these same provocative/alienating interpersonal patterns with the therapist, as well. In fact, researchers report that clients who present as "negative" often act in the same insulting, demanding, or distrustful ways toward the therapist that they have with others, and that *most therapists respond poorly to this hostility with their own countertransference reactions* (Strupp & Hadley, 1979). More specifically, Binder and Strupp (1997) report that therapists of every theoretical persuasion routinely respond to clients' negativity personally, with their own emotional reactions of anger, withdrawal, and, most frequently, a pejorative attitude toward the client! Unfortunately, most clinical training programs do too little to help new therapists learn how to respond nondefensively to this negativity that trainees are going to encounter, and which is especially problematic for therapists who need their clients to like them. Under these circumstances, what is generally most useful is to adopt a nondefensive stance that conveys an interest in understanding the client's experience, as well as openness to working with the client on whatever may be associated with his anger, negativity, or insulting presentation. As noted earlier, clients' overt affective presentation may be a defense against other, more threatening affects (such as sadness or shame). Adopting such a nondefensive and curious/explorative stance will go a long way toward establishing a collaborative alliance rather than yet another reenactment, providing psychological safety for the client that facilitates his ability to enter the deeper issues underlying the overt and off-putting presentation.

While many therapists are overly concerned about being liked, some have strong needs to nurture others who are dependent on them. These therapists enter clinical practice, in part, to fulfill these unrecognized needs, and they sometimes will have difficulty supporting the client's healthy individuation (for example, welcoming clients' disagreements with the therapist, being genuinely interested in their differing points of view, celebrating their successes, taking pleasure in their increasing independence, and supporting their termination and transition out of treatment). Most commonly, perhaps, are challenges presented to therapists who grew up fulfilling the role of "caretaker" in their families of origin. Some, for example, are drawn to clinical practice in part by their anxious childhood need and family role to rescue their vulnerable or distressed caregiver, and later others, from a life of alcohol or substance abuse, depression and an unhappy marriage, dependency and a generalized anxiety disorder, and so on. This frequently found countertransference issue of **parentification** can make it harder for therapists to address clients' vulnerable feelings of sadness or hurt. Those feelings, in particular, can evoke for therapists the familiar old predicament of needing to protect or take care of their caregiver. Because, in reality, this is always an impossible task for children to succeed with, it threatens them once again with the painful prospect of disappointing their caregiver (and now their client). They feel inadequate because they have failed somehow, and their anxiety is evoked in much the same way it was when their attachment ties to their vulnerable parent were threatened.

For these and a thousand other countertransference issues, all therapists need to assess their own motivations for becoming clinicians. We need to reflect on the roles we played in our families of origin and how our own personal needs, for good reasons and bad, are being met in this work. The issue is not whether therapists have their own needs and countertransference tendencies, but how they deal with them. Every therapist has certain countertransference reactions—it's just human—and self-examination on our own or with a therapist are responsible ways to manage them.

At times, therapists' attempts to approach the client's affect are not received well—especially when clients with these "negative" interpersonal styles respond defensively with anger, distrust, argumentativeness, and so forth. Too often, therapists absorb the client's rebuff, go along with the defensive side of the client's feelings, and stop approaching the client's feelings. This unwanted response is most likely from therapists whose needs to nurture or to be liked have been frustrated by this client's interpersonal style. To be helpful to this client therapists want to use this opportunity by finding ways to process the client's negativity that has made the therapist feel like retreating or disengaging—just as it is discouraging and distancing others in their everyday lives as well. Thus the therapist's aim is to be flexible and look for other ways to approach the client's feelings and begin changing the maladaptive relational pattern that usually ensues (Safran & Muran, 1995, 2000). That is, *rather than responding automatically and reacting like others usually do when they act this way (that is, responding in kind with hostility, defensiveness, or more commonly internal withdrawal and emotional disengagement), the therapist tries instead to be mindful of the thoughts and feelings the client is eliciting in her and others.* With this reflectiveness, therapists can try to find other ways of responding that do not repeat the familiar, problematic pattern in their interaction, such as observing or wondering aloud with a process comment:

THERAPIST: "Bob, I'm wondering if you're aware of how loudly you are speaking right now, and how angry you sound? How do other people in your life usually respond to you when you talk to them like this?"

To respond in this more flexible and nondefensive way, therapists also can adopt the strategies for working with the client's resistance that were presented in Chapter 3. That is, therapists should not press clients to express a feeling they are reluctant to share but, instead, explore what the threat or danger may be for them if they do so.

* What does it say or mean about you if you feel _____?
* What could happen between us, or what could go wrong for you, if you let yourself feel _____?
* How have important people in your life responded to you in the past when you felt _____?

Therapists do not want to win the client's approval by trying to be a "nice" person who does not approach difficult feelings or talk about awkward interactions. On the other hand, therapists do not want to demand that clients enter into and share, or stay with, a certain affect. Instead, they repeatedly offer invitations or "bids" and allow clients to enter and leave difficult emotions as they wish. Letting clients have it their own way avoids the issues of coercion and compliance that so many clients had in their families of origin, and that people who have been marginalized often have experienced in the dominant culture. Therapists provide an effective middle ground when they work with the client to understand the resistance by exploring together the threat or danger that once existed. *With understanding comes safety*, and clients soon will be able to explore feelings that needed to be held away in the past.

To sum up, all therapists are going to have difficulty responding to certain feelings at times, and therapists are especially likely to have difficulty working with clients who are critical, angry, provocative, or have other negative feelings toward them. Learning to respond nondefensively to the client's negativity (anger, irritability, competitiveness, criticism, control, and so forth) is challenging for most trainees—and especially difficult for therapists who need their clients to always like or approve of them. New therapists need practice with their instructors, such as role playing these difficult interactions and learning from seeing how their supervisors respond, to keep from **personalizing** this negativity and reacting defensively with their own countertransference. Therapists can help many clients by:

* listening for "negativity" in clients' interpersonal patterns with the therapist and with others;
* making process comments to intervene and begin changing this problematic pattern in their current interaction; and
* exploring collaboratively how this negativity that is occurring with the therapist is likely to play out and cause problems in clients' relationships with others as well.

For example:

THERAPIST: As you say that to me again, James, I'm feeling kind of criticized. Maybe it's just me, but it seems like that happens pretty often between us. You've been talking about problems being close with your wife and getting along with your son. I'm wondering if they might be feeling criticized or put down by you more often

than you are aware of—you know, like I sometimes do? But maybe I'm wrong…
what do you think?

THERAPISTS OFTEN ASSUME TOO MUCH RESPONSIBILITY
FOR THEIR CLIENTS' FEELINGS

When therapists take on responsibility for the client's feelings and place on themselves the unrealistic expectation that they have to fix or solve the problem, the
usual outcome is for therapists to feel inadequate. In addition, therapists may be
uncomfortable with their own sadness, anger, shame, or other feelings that have
been evoked by the client—that is, countertransference. In such situations, therapists are likely to respond to the client's feelings in ineffective ways:

- Interpreting what the feelings mean and intellectually distancing themselves
- Becoming directive and telling the client what to do
- Reassuring the client that everything is OK or will work out all right
- Becoming anxious and changing the topic
- Falling silent or emotionally withdrawing
- Self-disclosing and moving into their own feelings
- Diminishing the client by trying to rescue him/her
- Over-identifying with the client and becoming controlling—pressing the client
  to make some decision or take a particular action in order to truncate the
  therapist's own unwanted feelings

If therapists do not shoulder this inappropriate responsibility for causing or for
fixing the client's feelings, what should they do instead? The therapist's response to
the client's feelings is threefold: to identify, to join, and to affirm. First, the therapist
tries to help clients *identify and express more fully* whatever the significant feeling
they may be having. Second, the therapist's intention is to be *empathic or emotionally responsive* to clients, so that clients can share their feelings with a concerned
other rather than having to experience them alone or have them dismissed, as they
often have in the past. Third, the therapist wants to *validate* clients' feelings, by
helping clients understand or "contextualize" their feelings so they can make sense
of why they are experiencing this feeling now in this particular situation (White,
2007). If the therapist accepts this threefold challenge, but does not assume responsibility for causing or changing clients' feelings, the therapist will be free to respond
effectively to whatever feelings clients present.

Even therapists who are well aware of the pitfalls of assuming inappropriate
responsibility for clients' feelings are likely to struggle with this issue when the client's affect becomes intense. To illustrate, a graduate student therapist was treating
a young boy who was in the midst of a potentially life-threatening battle with leukemia. The therapist would sit with the boy in the hospital and, among other topics
the child would bring up, they would sometimes talk about the possibility that he
could die. In response to the child's questions about death, the therapist would try
to discern the central or key concern for him: "What would be the scariest thing
about dying? What are you most afraid of?" As the therapist approached the feelings the child was expressing, the boy responded, "I don't want to die! I'm afraid to

die!" Seeing the fear in this young child's eyes and feeling totally inadequate in this moment, the therapist questioned what she was doing and even felt "cruel" to be evoking such anguish.

But the child went on to say, "I don't want to be alone. I'll be all alone if I die!" Inviting the child to explore his feeling more fully, rather than rushing in with her personal beliefs and reassuring explanations about death and the afterlife—which she wanted to do—revealed his key concern that he would be "all alone." Hearing the boy's intense separation anxiety, the therapist now understood the subjective meaning this fear held for him and was able to talk with him in more helpful and specific ways. For example, with this clarification, the therapist was able to intervene by helping the child's mother "hear" and understand his basic fear and respond by sharing it with him more directly. With the therapist acting as "coach," his mother was able to reassure him that, if he were to die someday, their love for each other would last forever and that she would talk to him and hold him closely in her heart every day. By talking with the mother and son together in this and other ways about his fear of being left alone, the therapist helped them to become closer during his illness and hospital stay. The felt sense of security that came from this shared understanding tangibly alleviated the child's (and the mother's) distress.

When working with such profound emotions, therapists almost inevitably feel to some extent that they are responsible for causing the client's suffering. As a result, they tend to avoid or explain away these intense feelings. To prevent this, therapists need to consult with a supervisor or colleague to help them manage their own exaggerated feelings of responsibility at times. Otherwise, these countertransference reactions will stop therapists from responding effectively and providing clients with the holding environment they need. In the example of the boy with leukemia, the therapist could not have stayed with the intensity of the boy's anguish, or the mother's sadness and fear, without regularly scheduled consultations with a supervisor to help her contain her own profound emotional reactions. In fact, one of the very important ways supervisors contribute to their supervisees' development is to address how the supervisees' emotional responses to their clients impact the treatment process (Falender & Shafranske, 2004).

THERAPISTS OFTEN BRING RULES ABOUT EMOTIONAL EXPRESSION FROM THEIR OWN FAMILY OF ORIGIN TO THE THERAPEUTIC RELATIONSHIP

Therapists may not be able to respond effectively to certain feelings that the client presents because of family rules about emotional expression that therapists learned in their own families of origin. Most families have unspoken rules that govern emotional expression, and the more dysfunctional the family, the more inflexible and limiting will be the affective milieu. A **rule-bound homeostatic system** prescribes which feelings can be expressed and which are unacceptable. This system often also prescribes when, how, and to whom certain feelings can be expressed. The permitted intensity or degree of emotional expression is also controlled by homeostatic family mechanisms. In some families, for example, if too much anger or conflict, differing viewpoint or independent voice, or affection or closeness is expressed between certain family members, a corrective homeostatic mechanism may be activated to bring the level of emotional expression back within acceptable limits. The following example illustrates how homeostatic mechanisms operate to keep family emotional expression within prescribed limits.

Suppose that in the Smith family a moderate degree of anger can exist between the older sister and the younger brother. A lesser degree of conflict can be expressed between the father and both children. However, anger or even modest disagreement is rarely expressed between any family member and the mother, and is never expressed between the father and the mother. What is the homeostatic mechanism that serves to keep this system operating within these boundaries? Whenever a child begins to express angry feelings or have direct conflict with the mother, she stops the child's protest by taking the role of martyr and looking sad and hurt. If the child's anger continues, the father intervenes and says, "Don't talk to your mother that way," or else the mother employs the father as a regulatory influence by saying, "I'm going to tell your father about this when he comes home."

In contrast, if anger or differences of opinion begin to escalate between the mother and father, the children serve as homeostatic regulators and respond in predictable, patterned ways to terminate the parental conflict (Guerin et al., 1996). For example, the oldest child, eight-year-old Mary, is **triangulated** to serve the role of **go-between** in the parental relationship. She tries to make peace between her parents by carrying messages back and forth between them. This mechanism usually succeeds in terminating parental conflict, except in periods of high family stress. If the parental conflict continues, the role of the younger brother may be to provide a diversion to escalating parental conflict by acting out in some predictable way—such as breaking something or having an accident. When the son makes himself a problem, which mother and father must jointly address, their rule-breaking level of conflict is terminated, and the family's emotional equilibrium homeostatically returns to acceptable levels.

Imagine that, fifteen years later, Mary, the oldest child in the Smith family who used to serve as the mediator to assuage parental conflict, is now a graduate student in clinical training. The family rules that Mary has learned in her family of origin are going to influence her ability to respond to her clients' emotions—just as the family rules that all therapists have learned in their families will shape how they respond to their clients. The importance of understanding family roles is underscored in working with families from diverse cultural backgrounds (McGoldrick & Hardy, 2008).

For example, Ann, a depressed and irritable client whom Mary was seeing at her internship site, became frustrated with Mary because her dysthymia was continuing and not improving. Just as she accused her husband and others of always letting her down, Ann angrily criticized Mary: "I keep coming here every week, but you're not doing anything to make me better. Sometimes I don't even know where we're going or what we're trying to accomplish here!"

Mary was taken aback by Ann's angry charge, even though it was short-lived. Mary was especially surprised because she thought they had been working together more effectively in recent weeks. Throughout the rest of the session, Mary backed away from Ann's anger and frustration by offering reassurances and trying to convince Ann that she would get better if she would continue to come to therapy and

work with the different issues and sometimes difficult feelings they had been sharing. Mary felt "terrible" that Ann had been so critical of her and anxiously sought
help from her supervisor the moment the session was over.

Mary's supervisor was responsive to her distress over Ann's angry disapproval
and was able to help Mary sort through what went on for her during the session. In
reviewing a videotape recording of the session, the supervisor helped Mary recognize how she had tried to assuage Ann's criticism with reassurances, rather than
acknowledging Ann's anger and inviting her to bring out and discuss her dissatisfaction more fully. Supportive but also challenging, the supervisor was encouraging
Mary to join Ann in figuring out what wasn't working and collaboratively explore
what they could change to make treatment more helpful. She suggested, for example, that Mary say:

> Your depression isn't getting better and it doesn't feel like I'm helping you. I respect
> your honesty and appreciate how you're bringing this up so directly. Let's talk more
> about what hasn't felt useful here and see if we can come up with some different
> ways of working together that might be better. Tell me more about what has been
> frustrating or disappointing for you.

With her supervisor's help, Mary soon realized that she had been unable to tolerate Ann's anger. She had indeed tried to assuage it, just as she had always done
with her parents and others. Mary's role in her family was to take care of the conflict, rather than to allow others to talk about their dissatisfaction and sort through
what they could do to resolve the problem. In addition to clarifying Mary's old response pattern, and role-playing together some different responses she could try out
in the future, the supervisor helped Mary generate working hypotheses about what
this critical response may have meant for Ann. The supervisor observed that Ann
had expressed her anger just after sharing her hurt, and hypothesized that the anger
may be a part of her affective constellation. In other words, Ann needed the experience of being able to express her anger at someone who would take her concerns
seriously but also who would not be run over or intimidated by her—as she had
learned to expect. Thus, Ann may have been "testing" to see whether she could be
angry or critical of Mary and still remain in a collaborative or egalitarian relationship with her—which Ann had not been able to do in other important relationships.

Without this helpful consultation, Mary and Ann may have become stuck at
this point in their relationship and treatment might not have progressed. Ann could
not move forward and make progress with her depression, and its interpersonal
fallout in her marriage, as long as Mary could not tolerate the anger component
in Ann's affective constellation. In their next session, however, Mary was able to
recover and restore their working alliance by inviting Ann to discuss her frustration.
Although Mary still felt an anxious compulsion to reassure Ann and move on, she
was able to contain her discomfort this time and discuss Ann's concerns more fully.
Through this frank discussion, Mary learned about things that weren't working for
Ann in their relationship, and together they came up with changes that each
thought could help them work more effectively. In particular, Ann wanted Mary
to be more active in sessions, "say more" about what she thought was going on,
and offer some suggestions about what Ann could do differently at home or with
others each week—especially with her husband.

Most clients will not be able to resolve long-standing problems unless the therapist can respond to each component in their affective constellation. *Treatment often stalls and becomes repetitive at the point where the therapist cannot respond to an important feeling the client is experiencing.* With this in mind, all therapists are encouraged to think about how emotions were dealt with in their own families and reflect on how they are likely to bring their own familial history to their clinical work. For many therapists, their own therapy will be the best way to recognize and change limiting family rules and roles that still govern how they act with clients.

THERAPISTS OFTEN HAVE DIFFICULTY SEPARATING THEIR OWN PERSONAL ISSUES FROM CLIENTS' PERSONAL ISSUES

Situational stressors in therapists' own lives may also prevent them from responding effectively to clients' feelings. For example, when personal problems in therapists' own lives are similar to those their clients are experiencing (such as marital conflict, caring for aging family members, parenting problems, or health concerns), therapists may have difficulty approaching their clients' affect or may not be able to allow clients to experience or explore fully whatever feelings they are struggling with.

Therapy is such a deeply human and personal interchange. Although the therapist's own personal qualities and emotional responsiveness are the greatest asset in helping clients change, they will also be an obstacle at times. When the therapist repeatedly fails to be accurately empathic and doesn't get the key concern, avoids rather than approaches the client's feelings, or, most significantly, *becomes personally invested in the choices clients make* (for example, whether or not to leave a relationship, get married, quit the job, file a complaint, do or don't have a medical procedure, and so forth), the therapist's own countertransference is operating.

At times, countertransference reactions will occur for every therapist. The issue is not whether they will occur but how they are dealt with (Bergin, 1997). When countertransference reactions are in play, therapists want to consult with a colleague or supervisor to better understand and manage their own personal reactions. When therapists find that they are repeatedly having difficulty with the same type of affect (such as tears or sadness), or when supervision does not help or free them up to respond effectively with more neutrality, therapists are encouraged to seek treatment for themselves. *Because change is predicated on the relationship therapists create with clients, therapists are encouraged to make a lifelong commitment to working on themselves and their own personal development.* Without this nondefensive openness to the possibility that their own countertransference issues may be activated, and ongoing willingness to work on their own personal reactions to clients, the help therapists can provide will be limited (Robbins & Jolkovski, 1987). Furthermore, therapists who are unwilling to acknowledge or work on persistent countertransference reactions are those most likely to have a negative therapeutic impact on their clients (for instance, "It's the client's problem—it doesn't have anything to do with me"). It is a privilege to be able to help people change, and therapists honor that privilege by acknowledging their own limitations and personal involvement in the therapeutic process.

CLOSING

As we saw in Chapter 4, clients' feelings often do not emerge unless the therapist focuses clients inward. An externalizing focus will lead therapists into a superficial problem-solving and advice-giving mode that precludes clients' affective exploration. As clients talk about the problem out there with others, the therapist's complementary social role is to give advice, reassure, or self-disclose what the therapist has done to solve similar problems. Although these responses can be helpful at times, limited change occurs if they characterize the ongoing course of treatment.

But if the therapist does not tell clients what to do, or wait nondirectively for clients to figure it out on their own, what does the therapist really have to offer? An essential component of facilitating change is to help clients resolve difficult emotions by integrating or coming to terms with feelings that have been too painful, shameful, or unacceptable to tolerate in the past. Many clients have experienced tragedies in their lives, and painful feelings will be entwined with the symptoms and problems they present. The therapist helps clients resolve their problems by providing a holding environment for feelings that others have not been able to understand, accept, or encompass. *Perhaps the most important relearning occurs when clients risk sharing certain feelings and find that, this time, they do not receive the same problematic response they have come to expect.* Thus, a corrective emotional experience occurs as old expectations and unwanted relational patterns are disconfirmed. Yet, therapists also need to be aware that the biggest obstacle to providing clients with a relationship that can produce change will be their own discomfort with certain feelings. Thus, therapists are invited to make an on-going commitment to working with their own countertransference propensities and how they influence their work with clients.

What's ahead? In the next three chapters, we will focus on case conceptualization and treatment planning and see how therapists can clarify the client's core problem and better understand what's really wrong and needs to change. In particular, we will learn more about how clients' problems originally developed and how they are being enacted now with the therapist and others. For now, this chapter has presented a great deal of complex and personally evocative information that often takes several years to integrate and make your own. Realistically, this is challenging work—be patient with yourself and find help when you need it.

SUGGESTED READINGS

1. New therapists need help identifying their countertransference propensities. In Chapter 5 of the Student Workbook, a self-assessment scale is provided to help therapists become more aware of their own response tendencies (see scale for "Sensitivity to Countertransference Propensities" in part III).
2. For training in basic communication skills, to help therapists approach or inquire further about clients' feelings, see Chapters 5 and 6 of Egan's (2010) *The Skilled Helper*. These two chapters also help therapists listen without judging and become more aware of their own cultural filters and biases.
3. Readers are also encouraged to examine Leslie Greenberg's (2002) *Emotion-Focused Therapy*. This informative book teaches therapists how to

help clients express emotion in sensible ways that are appropriate to context. Also, *Working with Emotions in Psychotherapy* (Greenberg & Paivio, 2003) helps therapists learn more about working with specific emotions, such as distinguishing between adaptive sadness and maladaptive depression. Greenberg's far reaching contributions in this area assist therapists in understanding and responding more effectively to their clients' feelings.

4. One way for therapists to increase their effectiveness with clients is to explore their own motivations for becoming therapists and explore how their own personalities are expressed in their clinical work. It is just as important to focus on the personhood of the therapist as it is to focus on the client and interventions. One good source to help therapists explore these issues is J. Guy's (1987) *The Personal Life of the Psychotherapist*.

5. An excellent resource for understanding affect regulation from an attachment/relational and neuroscience perspective is Allan Schore's (2003), *Affect Dysregulation and Disorders of the Self*. A useful website for the attachment literature and workshops is the *johnbowlby.com* site. In particular, the "circle of security" work on this website is an excellent source for those interested in understanding the "safe haven" and "secure base" constructs. This website also provides information about training opportunities on the Adult Attachment Inventory and other attachment-related work.

CONCEPTUALIZING CLIENTS AND DEVELOPING A TREATMENT FOCUS

6 | FAMILIAL AND DEVELOPMENTAL FACTORS

CONCEPTUAL OVERVIEW

The previous chapters in Part One and Part Two have focused on the interaction between the therapist and the client and provided guidelines for working with the process dimension. The following three chapters explore more fully how therapists conceptualize their clients' problems and develop a treatment focus. That is, we are trying to become clearer in our thinking about what is wrong for this client and what the treatment focus should be to provide the most help possible and make a difference in this client's life. Therapists are more effective when they can formulate a case conceptualization that leads to specific treatment plans and goals. To do this, therapists are trying to understand:

- how the clients' problems originally developed (Chapter 6);
- how these developmental problems are being played out now in problematic interactions with others and leading to the symptoms and problems that are bringing clients to treatment (Chapter 7); and
- how these problems are brought into the therapeutic relationship and played out in the way the therapist and client are interacting together (Chapter 8).

The more specifically therapists can clarify their thinking and understand their clients in these three ways, the easier it will be to formulate the **treatment focus** therapists need to help clients change. However, most new therapists have had little training or experience conceptualizing clients. Early in the therapist's training, it is difficult to clarify a focus to guide treatment, or to write the treatment plans and case reports that third-party insurers and treatment agencies often require. To help with this, guidelines to meet these professional demands will be provided in the next three chapters.

As indicated, one useful way to help therapists conceptualize their clients and formulate treatment plans is to recognize the familial and developmental antecedents of clients' problems and why some merely situational or modest stressors may become amplified and engender significant symptoms for clients. Let's begin with the burgeoning literature that is applying attachment theory to clinical practice in new and helpful ways.

ATTACHMENT STYLE AND CLINICAL PRESENTATION IN ADULT TREATMENT

In previous chapters, we have already begun discussing basic constructs in attachment theory, such as attachment styles in children, safe haven/secure base, and attuned responsiveness. Currently, an exploding clinical and empirical literature is linking Bowlby's basic attachment theory constructs to clinical practice with adults (see Obegi & Berant, 2009). We begin with the research on four categories or types of adult attachment and how these clients will present in treatment.

FOUR CATEGORIES OF ATTACHMENT

Following Mikulincer and Shaver (2007), Daly & Mallinckrodt (2009), Brennan et al. (1998), and others, adult attachment researchers measure **adult attachment styles** along two orthogonal dimensions of *Anxiety* and *Avoidance*. Anxiety is the "privileged" or primary emotion that attachment-informed therapists are attending to in their clients (Slade, 2004). This anxiety over the security of attachment relationships refers to worry or concern about the availability and willingness of our significant others to meet our attachment needs. That is, clients can be rated high or low on a continuum of *Anxiety* over the availability or responsiveness of their "go to" person or significant other. The other measured dimension is *Avoidance*, which refers to discomfort with intimacy and avoidance of both one's own, and others', feelings and needs, vulnerability or distress, and emotional engagement. The intersection of these two dimensions portrays four adult attachment prototypes: Secure, Preoccupied, Dismissive, and Fearful (Bartholomew & Horowitz, 1991). Adult clients with Secure attachment styles are low in anxiety about, and avoidance of, attachment needs and relationships; Dismissive clients are low in anxiety and high in avoidance; Preoccupied adults are high in anxiety and low in avoidance; and Fearful individuals are high on both anxiety and avoidance. This two-dimensional schema is presented in Figure 6.1 and the four clinical subtypes are introduced below.

SECURE ATTACHMENT STYLE IN ADULT CLIENTS (QUADRANT I: LOW AVOIDANCE/LOW ANXIETY)

Secure clients (sometimes called Autonomous) are valuing of relationships and provide consistent memories and judgments of their childhoods in undefended ways. They acknowledge the impact of childhood experiences realistically and, when these experiences were not positive, they have worked through these and "earned" security. In this way, they don't engage in idealization of parents or denial of real problems in the past—as we will see is prevalent in most insecure adults. They have an internalized sense of themselves as being worthy of care and efficacious, and are more

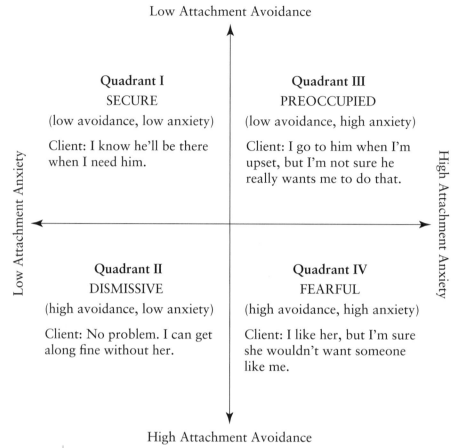

Low Attachment Avoidance

Quadrant I
SECURE
(low avoidance, low anxiety)

Client: I know he'll be there when I need him.

Quadrant III
PREOCCUPIED
(low avoidance, high anxiety)

Client: I go to him when I'm upset, but I'm not sure he really wants me to do that.

Low Attachment Anxiety

High Attachment Anxiety

Quadrant II
DISMISSIVE
(high avoidance, low anxiety)

Client: No problem. I can get along fine without her.

Quadrant IV
FEARFUL
(high avoidance, high anxiety)

Client: I like her, but I'm sure she wouldn't want someone like me.

High Attachment Avoidance

FIGURE 6.1   ATTACHMENT STYLES IN ADULTHOOD

flexible and capable of dealing with stress. Thus, when Secure adult clients come to therapy, they are less conflicted about having a problem and seeking help, are more able to work collaboratively with the therapist and readily establish a working alliance, enter into their exploration of their conflicts with feeling and depth, and are more capable than the other three subtypes to risk exploring ruptures and negative reactions toward the therapist (Rubino et al., 2000). Thus, in Figure 6.1, the Secure client in Quadrant I is low on avoidance and low on anxiety. That is, when distressed, these clients do not avoid relationships or their emotional needs. Rather they appropriately seek out help or understanding from significant others in healthy, adult ways, and they do this without anxiety or excessive worry about being controlled, abandoned, or other threats that the other subtypes will exhibit. When working with Secure clients, you will hear prototypic sentences like these that capture their felt sense of trust, safety, and security:

- I know that he will be there when I need him.
- I know that she won't hurt my feelings when I get close to her.

- He is able to comfort me when I'm distressed.
- I enjoy it when she gets emotionally close to me because I feel close to her.

Leaving Security, the three adult attachment categories that follow reflect the different subtypes of insecure attachment. One of the key concepts to help therapists understand the symptoms, clinical presentation, and defenses that these clients present is *a hyperactivating strategy vs. a deactivating strategy to cope when distressed*. Depending on their attachment histories and subsequent relational experiences, adults who have developed negative internal working models of self and others will implement either of these two strategies (Daly & Mallinckrodt, 2009). A **hyperactivating strategy** is more likely to be used by a person who has a negative model of themselves as being fundamentally unlovable and unworthy of others' love and responsiveness. This usually results in an *exaggeration* of needs so that this individual constantly seeks closeness with others in order to get as much responsiveness as possible. Conversely, a **deactivating strategy** is usually implemented by an individual who has a negative model of others—they are seen as being potentially harmful and unreliable attachment figures. This results in the perception that seeking closeness with others is futile at best and can be emotionally injurious, so that minimizing or denying having a need is the safest strategy. Thus, as we are going to see, Dismissive clients (the adult version of the Avoidant child) damp down and minimize their distress with a deactivating strategy. In contrast, the Preoccupied client (the adult version of the Ambivalent/Resistant child) uses the hyperactivating strategy to heighten affective expression and elevate distress in order to elicit attention and reassurance. These two defensive coping strategies are important to know because they are going to shape many clients' interactions with the therapist and others (Mikulincer & Shaver, 2008).

Dismissive Attachment Style in Adult Clients (Quadrant II: High Avoidance/Low Anxiety)

These adult clients were often classified as Avoidant in childhood and present with behaviors that parallel those observed in children with avoidant attachment styles. **Dismissive adults** tend to have difficulty remembering specific aspects of their childhoods (Client: "My parents…my family…I don't know—I don't remember much about growing up."). Usually, they also present their childhoods in overly ideal or positive terms (Client: "I had a great family—everything was just great"). Even though it may readily become apparent that their caregivers may have been overtly rejecting or even abusive, these painfully problematic experiences and their impact are minimized or "dismissed." In contrast, a few Dismissive clients may overtly devalue their caretakers and attachment experiences (for example, "They're just shitheads and I don't want anything to do with them"). Like Avoidant children, Dismissive adults fundamentally do not trust that emotional or social support will be available when they need it. A common presentation is one of competence, independence, self-sufficiency, and strength; indeed, those who experienced overt rejection and abuse may go so far as to say that they were unaffected by this, or that "it made me stronger." These experiences contribute to detachment from relationships and emotional closeness. Again, the central theme for these counterdependent clients is that showing vulnerability and need (or even experiencing distress privately—even in response to highly stressful circumstances) is unacceptable.

If you are seeing a Dismissive client in therapy, you may initially feel bored with him or her in sessions. These clients avoid close relationships and reliance on others and with the therapist as well. When speaking about themselves, they will most likely be intellectualizing, disclose little, and share few feelings. They are reluctant to come to therapy in the first place, because they learned long ago that it was not a good idea to reach out to others for emotional support. If other individuals get them to come to therapy (for example, a lonely spouse who threatens to leave if s/he doesn't go to therapy and try to become more responsive), Dismissive clients will focus on others' problems rather than their own. These clients will not want to use the therapist to work on themselves—that is, they will not "look within" with much depth and do not want to consider their own contribution to problems. Instead, therapists will feel a strong pull to join them in blaming or criticizing others, which is especially problematic if the therapist shares a Dismissive attachment style and too readily joins in the externalization that is being elicited. Dismissive clients also inflate their self-esteem and competence in response to threat and minimize or deny real dangers (Mikulincer et al., 2003). Finally, pervading many of their coping strategies and responses to others is their *deactivating* style of distancing from emotional needs or vulnerabilities and minimizing distress both in themselves and in others.

Thus, in Figure 6.1, the Dismissive client in Quadrant 2 is high on *Avoidance* (that is, the client avoids her own emotional needs and avoids seeking help from others) and low on *Anxiety* (she does not present with overt anxiety or concern when faced with real problems). If you are working with a Dismissive client, you are going to hear prototypic sentences like these:

- "I don't care if she doesn't love or want me."
- "I don't tell him when I'm upset because I take care of my feelings myself."
- "I'm used to doing things on my own, so I don't ask her for help."
- "I can get along just fine without him."
- "No problem. Everything's fine."

In response to these prototypic Dismissive comments, therapists may find it useful to try making modest empathic bids, such as:

- "It seems as though you've had to do a lot on your own."
- "I'm wondering if, just sometimes, it might have been a little hard for you to be so alone with your problems?"
- "Maybe others haven't always been able to be so helpful to you, or responded quite in the ways that might have worked best for you?"

PREOCCUPIED ATTACHMENT STYLE IN ADULT CLIENTS (QUADRANT III: LOW AVOIDANCE/HIGH ANXIETY)

Adult clients who are **Preoccupied** tend to be "all caught up in" or entangled in angry, idealizing, or worrisome preoccupations about others in current and past relationships. They are often *preoccupied* with whether their significant others will be reliably available and will not leave them or otherwise disappoint or let them down. Their anxious preoccupation with issues related to the dependability or availability of their significant others (for example, "Why won't he text me back? I haven't heard from him since lunch!" or "I just can't stop thinking that she might be cheating on me") is so all-encompassing that *they don't seem to have room in their minds for their own mind*—their own thoughts, goals, and interests (Wallin, 2007). That is, worries and concerns about what is going on with the other person and what s/he may be thinking or feeling about them dominates their subjective experience, oftentimes leaving them feeling "overwhelmed." Additionally, Preoccupied individuals lack self-confidence and doubt their own abilities. They perceive themselves as weak and believe neither in their own capacities to cope nor in others' ability to help in a trustworthy and reliable way (Mikulincer et al., 2003).

The cardinal issue here is the fear of losing relationships, and these abandonment threats readily disrupt Preoccupied clients and consume their lives with drama-laden crises. Thus, they will often enter treatment in response to a recent breakup, the inability to end a relationship that is highly problematic, or the inability to "get over" a relationship that ended long ago. More than most, Preoccupied clients are often self-absorbed and vacillate between a self-centered obliviousness toward the other at some times and, at other times, an intrusive and demanding overinvolvement with the other. Many Preoccupied clients grew up enmeshed (and often parentified) with an unpredictable parent who was too often caught up in his/her own emotional upheavals to be able to be a safe haven and provide containment and affect regulation for the child.

If you are seeing a Preoccupied client, he or she will often be intense and engaging in the initial session with you. There will often be a high level of communication and emotional expression as part of a strong press to engage and "feel close" with the therapist. These clients readily feel overwhelmed and they can feel overwhelming to the therapist, just as they tend to overwhelm others in their lives. Although they will often engage in considerable self-disclosure, it may be almost too indiscriminant or too unfiltered for someone they have just met. During treatment, they will often want to rely heavily on the therapist, just as they want to rely excessively on others. Similarly, these clients will often press the therapist for reassurances rather than joining with the therapist and going to work on their problems in earnest. Because of their insecurity about the therapist's responsiveness as well, they will focus on the therapist (anxiously extending to keep the therapist interested in and engaged with them) at the expense of their own exploratory work. In sum, these Preoccupied clients with a hyperactivating coping strategy will have overly intense reactions to perceived threats, pervasive fears of abandonment, and even though they are often communicating a high need for reassurance, can *resist the help they are seeking* and cannot easily be comforted or contained. Clearly, this client can readily evoke feelings of frustration or incompetence in new therapists who have high needs for their clients to find them helpful.

Thus, in Figure 6.1, the Preoccupied client in Quadrant III is low on *Avoidance* and high on *Anxiety* (that is, he does not avoid his own emotional needs or avoid seeking help from others, although he is overtly anxious and worried when he approaches others with his problems). If you are working with a Preoccupied client, you are going to hear prototypic sentences like these:

- "I can't get along without him, even though being with him isn't working."
- "I'm often wondering whether she really cares about me or not."
- "I often feel really dependent on him for emotional support."
- "I find it hard to forgive her when she lets me down."
- "I turn to him when I'm upset, but it doesn't really help me feel much better."

In response to prototypic statements like these that Preoccupied clients will often make, therapists may find helpful comments such as the differentiating, boundary-setting reflection below, which goes right to the core problem of not having room for a mind of one's own:

THERAPIST: Jane, it's almost like all of the people in your life sort of fill up or take over your thoughts so much that there isn't room for your own mind. You know, like you're so caught up in how other people are seeing you, what they might be thinking about you, or if your boyfriend is going to return your call soon enough, that there isn't room in your head for your own thoughts. Why don't we get those others out of your mind for a while and set them aside while you're in here with me and let's focus instead on what it is that *you* are thinking and feeling and wanting. How does that sound to you?

CLIENT: (*slow deep breath and relaxing sigh*) Ahhh, I really like that. It makes me feel like I can let down and just breathe for a minute.

FEARFUL ATTACHMENT STYLE IN ADULT CLIENTS (QUADRANT IV: HIGH AVOIDANCE/HIGH ANXIETY)

Preoccupied and Dismissive clients may present with serious problems, but their de-activating and hyperactivating coping strategies are "good enough" defenses that allow many of them to function with only modest symptomatology. With the **Fearful client**, in contrast, we don't see the stable defenses or patterned coping strategies that provide a way for less troubled clients to adapt. The hallmark of this category is that there is no patterned defense structure or coping strategy: therapists are going to see a wide range of symptoms and contradictory, back-and-forth behavior. Sadly, we are in the realm of maltreatment and abuse, where two themes predominate. First, many Fearful clients suffered significant parental hostility and overt rejection (for example, "I wish you were never born...we just hate you"). Second, some clients with a Fearful attachment style have suffered physical and/or sexual abuse but have not "resolved" or come to terms with their maltreatment or traumatic loss (this is the adult version of the Unresolved/Disorganized child). Understandably, these clients often struggle with serious psychopathology. This may include acting out symptoms such as substance abuse and addiction or self-injurious behavior, or experiencing altered states such as dissociation or depersonalization. These symptoms often occur under conditions of attachment stress and when cues (such as the smell of alcohol on someone's breath or the threat of invasive medical procedures) evoke frightening memories or feelings about their unresolved trauma or unresolved loss. Fearful clients are challenging in treatment because they often evoke strong countertransference reactions in therapists, and therapists may need to manage acting-out behavior to keep these clients safe. These clients certainly can be helped in treatment but clearly are not appropriate for new therapists.

In treatment, clients with Fearful attachment styles avoid close involvement with an intense fear of rejection. They are low in self-disclosure, intimacy, and reliance on the therapist and others. These individuals cannot easily use the therapist as a secure base for working though and exploring new ways of relating. They tend to move slower in treatment and, because they cannot easily trust, also struggle with finding a safe haven in the therapist. Shame-prone, they possess a sense of personal unworthiness and social insecurity that readily leads them to mis-hear the therapist's benevolent responses through this distorting lens of expecting to be rejected or exploited. This client can look similar to the Dismisive client at times but at other times, confusingly, may look Preoccupied as s/he become immobilized by intense anxiety at the prospect of being rejected or exploited again upon approaching an attachment figure. Under stress, some Preoccupied clients with sexual abuse histories could develop more serious dissociative symptoms, and some Dismissive clients could seem depersonalized in their extreme disconnection from their own experience as they talk about a traumatic experience such as physical abuse. Although trauma, abuse, neglect, and unresolved loss may be found within any of the four attachment categories (including Secure), it is central in the Fearful attachment style.

Thus, in Figure 6.1, the Fearful client in Quadrant IV is high on *Avoidance* and high on *Anxiety* (that is, these clients avoid their own emotional needs and they present with overt anxiety when they approach attachment figures). If you are working with a Fearful client, you are going to hear prototypic sentences that communicate their profound shame—the legacy of their rejection and abuse histories, and the maddening double binds they have been struggling to decipher:

- "There's something wrong with me."
- "I'm losing it—I don't want to live without her, but I'm not sure I can live with her either."
- "I don't matter—I just hate myself."
- "They wouldn't want someone like me."

In treatment, therapists are going to observe Fearful clients enact their core conflict—a maddening double bind that leaves them stuck in a seemingly unresolvable approach-avoidance conflict. The Fearful client has often grown up with the childhood dilemma of sometimes being genuinely helped by the same person who, at other times, is frightening (or disturbingly frightened) and rejects or betrays them. Fearful clients want to approach others at times but are quickly made anxious (or terrified) by the fear of further rejection or abuse, and this back-and-forth will play out with the therapist as it does with others in their lives. Thus, they cannot get close and they cannot get away and so live their lives in a constant state of stress and physiological arousal, leading them to suffer from all of the medical and psychological problems engendered by this "fear without resolution." To capture their difficult lives most simply, *the rigidly controlled Dismissive client holds fear of connection, the impulsive Preoccupied client holds fear of differentiation, and the Fearful client fears both.*

CONCLUDING THOUGHTS ON ADULT ATTACHMENT STYLES

In closing, attachment styles teach therapists much about their clients but, like all classification systems, have real limits. Rather than thinking of your clients as pure types or rigidly attempting to fit them neatly into one of the categories described above, it is best to appreciate the variability that stems from many factors: clients will have these attachment styles to different degrees; there are two to three different subtypes within each of the four primary attachment categories presented; many clients will share characteristics of more than one attachment style; and some children and adults simply do not fit any category. Thus, some clients will present Dismissive or Preoccupied features in different relationships and at different points in their lives. This presentation can also be influenced by cultural features such as ethnicity and religious background (Fiori et al., 2009). Further complicating matters, and confusing themselves and others, we have seen that some Fearful clients will alternate between Preoccupied and Dismissive styles. When distressed, however, many individuals are going to present with one predominant style of attachment in most important relationships.

Having a secure attachment style is a profound buffer or resiliency factor, but it does not mean that Securely attached adults will not have symptoms or problems that result from challenging life circumstances. However, having secure attachments in childhood does mean that you are protected from threat and buffered from real dangers in the world in a way that insecure children are not. Children simply grow up safer when they have a responsive caregiver to go to when distressed. In contrast, children with Ambivalent/Resistant, Avoidant, and Disorganized attachment configurations are less protected and more vulnerable to peer pressures, invitations from predators, and miss danger signs in the environment, that can lead them to walk too readily into threatening situations and relationships that hurt them. (Excellent reviews of the attachment literature can be found in Cassidy and Shaver, 2008; and Obegi and Berant, 2009.)

PARENTING STYLES

In addition to the attachment experiences described above, another dimension of parent–child interaction that significantly influences children is parenting styles (see Baumrind's research: 1971, 1989, 1991). Baumrind and others (for example, Bender et al., 2007, Berk, 2005, Slicker and Thornberry, 2002) have conducted extensive research programs on the impact of different parenting styles on children's self-esteem, development of autonomy and initiative, social and communication skills, and so forth.

To better understand the different styles of child rearing that parents use, and how many adult clients' problems originally developed, consider the two dimensions Control and Affection, as diagrammed in Figure 6.2. As reflected on the horizontal axis, parents can vary along a continuum from firm discipline (high control/structure) to permissive or lax discipline (low control/structure). The vertical axis represents how parents can vary on a continuum from much warmth, emotional responsiveness, and communication (high affection/support) to little approval, acceptance, or interest

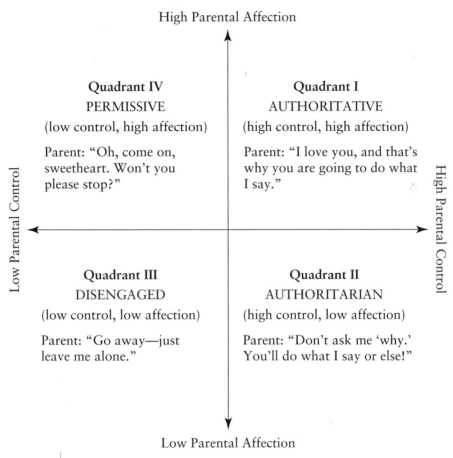

High Parental Affection

Quadrant IV
PERMISSIVE
(low control, high affection)

Parent: "Oh, come on, sweetheart. Won't you please stop?"

Quadrant I
AUTHORITATIVE
(high control, high affection)

Parent: "I love you, and that's why you are going to do what I say."

Low Parental Control

High Parental Control

Quadrant III
DISENGAGED
(low control, low affection)

Parent: "Go away—just leave me alone."

Quadrant II
AUTHORITARIAN
(high control, low affection)

Parent: "Don't ask me 'why.' You'll do what I say or else!"

Low Parental Affection

FIGURE 6.2 | FOUR STYLES OF PARENTING

(low affection/support). Using the high and low points along these two dimensions, parents typically discipline their children using one of four approaches. The first three that we will discuss are problematic and the fourth is effective: *Authoritarian (strict/cold), Permissive (indulgent), Disengaged (neglectful), or Authoritative (loving and firm).*

AUTHORITARIAN PARENTING

One of the most common but ineffective methods of discipline is the **authoritarian** approach, which is punitive, controlling, and unreasoning (see **Quadrant II: high control, low affection**). These parents are low in warmth and responsiveness but high in control and demands. They are strict disciplinarians—they give children clear expectations about what behavior is acceptable and unacceptable. Parental rules and expectations are clear and the consequences for violating them are consistently enforced. Authoritarian parents are demanding—they have high expectations for their children to behave in a responsible and mature manner. However, they rarely engage in mutual, respectful communication that involves explanation for their rules—their children are simply expected to obey and conform to parental dictates. They are also expected to be competent and perform up to their abilities and to be responsible and contributing family members. However, without the emotional support and affection so important to the development of a sense of belonging and security, especially during moments of distress, children of authoritarian parents learn to hide any signs of vulnerability from their parents and sadly, eventually from themselves as well. They develop an interpersonal strategy that in effect communicates to others, "I don't need you or anyone else." While many are able to attain success in school and later in their chosen careers, they tend to keep people at a distance and their own emotions under tight control. Experiencing vulnerability of any kind may activate fear or a Disapproval/Rejection schema, something these clients have often struggled with throughout their lives. They often internalize anger and have difficulty coping with frustration and may respond with anger outbursts. Since they do not have close, affectionate relationships with their strict and rather "cold" parents, unless they have compensatory adult mentors, they will lack healthy role models with whom to identify. Given the extent to which their emotional needs were not of importance in their families of origin, they often become detached from relationships or appear pseudo-independent and some may end up with Avoidant attachment styles in childhood or Dismissive styles as adults.

In addition to a lack of emotional support, authoritarian parents often instill a fear of rejection in their children by their methods of discipline. Authoritarian parents do not give children reasons or explanations for the rules they set. The child is expected to obey without questioning or trying to understand why the parent has set these limits. Children cannot ask why a rule is set; they must simply obey. Children who grow up in an authoritarian household regularly hear their parents make statements such as "Don't ever ask me *why* you can't go out. I am your father, and you'll do what I say or else!" Furthermore, there is no room for compromise or verbal give-and-take between the parent and child. Children are not encouraged to suggest alternatives or to explain their side of the story. Authoritarian parents discipline their children strictly and expect much from them. They are committed to the parenting role, however, and often provide an "instrumental" form of love in the sense that they responsibly feed and clothe their children, help them with their homework, and even play games with them. Most important, however, they do not give their children much warmth, approval, or affection. Clients who grew up in authoritarian households often describe their parents accurately as moral or "good people" but usually experience them as cold, distant, and even intimidating.

There is no idle threat or bluff here—the authoritarian parent means business, and their children know this and behave. The threat of parental power and the fear of rejection keep children in line, especially while they are young. This strict, nononsense approach is far better for children than having no discipline at all, but it has major drawbacks. *These children are obedient and achieving, yet they are also anxious and insecure; they comply out of fear* (Wenar & Kerig, 2006). The insecurity that children feel with their parent often generalizes to teachers, coaches, principals, and other adults in their lives. This insecurity also carries over into adulthood and infuses many clients' relationships with anxiety. Although most children remain intimidated by their authoritarian parents, some become aggressive and defiant as they grow older and more verbal (especially when marital conflict escalates or parental divorce occurs). As adult clients, however, they suffer low self-esteem and lack interpersonal confidence, even though they can be achieving and successful in academic and work environments.

Another drawback to the authoritarian approach is that it limits the growth of children's intellectual abilities. When children are not given reasons to help them understand why parents have set certain rules and are not encouraged to suggest alternatives or compromises, they do not learn to exercise language and reasoning skills. Children of authoritarian parents score lower on verbal tests of intelligence than children who are given an opportunity to interact with parents over rules and directives.

In contrast, within healthy families, children learn to obey without sacrificing their own initiative and positive self-regard. Healthy children become self-controlled and self-reliant without losing their sense of being prized by their parents. However, for children of authoritarian parents, the trade-off between acting on their own wishes and maintaining parental approval is too severe. Because authoritarian parents provide too little nurturance and affection, *too much of their children's initiative and positive self-regard is lost in order to try to win parental approval.* They are anxiously worried about doing everything "right" and never making a mistake, in order to ward off parental criticism and painful disapproval. By the time the children are of school age, these relational patterns with the authoritarian parent have been internalized as a cognitive schema: What was originally an interpersonal conflict becomes an internal conflict that shapes subsequent relationships with other adults, and with themselves, in problematic ways. Specifically, these well-behaved, insecure children become harsh, critical, and demanding toward themselves, just as their parents have been with them. Many will seek therapy as adults, presenting with symptoms of guilt, depression, unassertiveness, anxiety, and low self-esteem. These personal and emotional problems will be present even though these clients typically are responsible, hardworking, and successful adults. Because many clients and therapists come from authoritarian backgrounds, we will explore further aspects of this topic later in this chapter.

PERMISSIVE PARENTING

Although authoritarian parents recognize that children need to know the rules and that the consequences for violating them will be enforced, their discipline is rigid and their parenting lacks empathy. In contrast, other parents may err on the side of permissiveness. Although **permissive** parents (also called laissez-faire or indulgent parents) provide more warmth than authoritarian parents, they are not able to take a firm stance, consistently follow through, and place appropriate controls on children's behavior (see **Quadrant IV: low control, high affection**). These parents are often warm, nurturing, and communicative; however, they are also indulgent and set few rules or constraints. Children in these families make their own decisions and receive too little guidance from their parents. This inability to take charge and discipline effectively, which causes so many significant problems for children and later adults, may occur for many reasons. Some permissive parents may falsely believe that if they are firm they are acting harshly and, in their own minds, being "just like" their own (authoritarian) parents, who were too harsh or intimidating (for example, Adult Male Client: "My father was a tyrant and I knew when I was 10 years old that I wasn't going to be anything like him"). Or, sadly, some insecure parents fear their children will not love or be close to them if they say no and mean it. Still other parents, perhaps disempowered by their own parents in childhood, don't really believe they have the right to be the one in charge or to instill their own values and standards and teach their children how they want them to behave or act in life. For these and other reasons, *the balance of power has tipped in permissive families, and children wield too much control in the parent–child relationship*. These parents often feel disempowered vis-à-vis their children and may be heard negotiating, bargaining, or even pleading with their children to behave ("Oh, come on, sweetheart, won't you please stop?"). Parent and child both begin to suffer when control shifts from the parent to the child in this way. Problems begin to develop as they become "bossy" toward other children and angry with playmates when they can't dominate, and argumentative and disrespectful toward adults when they don't get their way. And, of course, these same relational patterns will be tested in the therapeutic relationship, as these clients will try to evade the same limits and rules that apply to others and to manipulate and control the therapist—as they have learned to do with their caregivers.

Children of permissive parents do not know what behavior is expected of them or what will happen if they violate parental norms. Furthermore, permissive parents do not consistently enforce the few rules they set. As a result, their children learn that they do not have to obey because their parents will not *consistently* enforce the rules. Thus, permissive parents are often loving and communicative, but their children are not disciplined and are not expected to behave in a mature, responsible manner. Without parental expectations that they perform to the best of their abilities, children do not develop the skills or internalize the discipline necessary to succeed on their own. Dependent and demanding, they are often described as "spoiled" children.

Offspring of permissive parents will also have adjustment problems, but their symptoms and interpersonal problems are different than those of children from authoritarian families. These offspring may enter treatment with anxiety, depression, and other "internalizing" symptoms but for different reasons than children from authoritarian families. Children and adolescents with permissive parents are right in knowing that they are not safe and cannot be protected by a parent who cannot say no to them, cannot tolerate their disapproval, or gives them too much power and control over the parent–child relationship. *Underneath the demanding/angry presentation is an anxious child.*

Additionally, children of permissive parents are especially prone to develop acting-out or externalizing problems. Because these children did not have an adult in charge to assist them in developing emotional regulation (temper tantrums were a successful strategy for controlling the parent and getting what they want), and they could escape the consequences of their poor decisions or rule-breaking behavior, they are often lacking in self-regulation. Thus, behavior problems may develop involving school authorities for truancy, the police for reckless driving, or drugs and alcohol. Children do not respect parents (or, later, therapists) whom they can manipulate or parents who cannot say no and discipline effectively because they need the children's approval. In addition to being demanding, selfish, and angry, these children are likely to fail with peers and friendships. Researchers describe these children as dependent, immature, demanding, impulsive, and unhappy.

Later, as adults who enter treatment, they tend to be self-centered, demanding, and dependent in their relationships and less capable of making commitments and following through responsibly on obligations. Oftentimes, these clients will be mandated to treatment by judges and courts and will be seen in alcohol and substance abuse treatment programs. Therapists working in Employee Assistance Programs will also work with this client. Typically, they will be referred for treatment by a supervisor at work who is dissatisfied with their inability to be a good team player and get along with others in their work group. When seeking therapy as adults, they tend to avoid taking responsibility for themselves and try to blame others for their problems. They also have learned that they can break the rules and escape the consequences of their own behavior by manipulating others. Of course, these same interpersonal themes that are causing problems with others will quickly appear in the interpersonal process they begin to play out with the therapist.

Finally, in one subtype of permissive families, parents are acting more as friends than as parents to their children. In this subtype, a client may describe her family of origin as a group of siblings living under the same roof, signaling the absence of clear intergenerational boundaries. When parental guidance or help is required, such parents may appear inept, even incompetent. To remain attached to such parents, children in these families often take on an adult role. During emergencies, for example, these parents may "fall apart," requiring the child to come to their assistance. Children in these families become responsible for the emotional (and sometimes physical) well-being of one or more parents. In families with more serious problems, a child may be given the role of "rescuer" in a family where one parent is physically abusing the other parent. Children may receive love and attention in the form of gratitude; however, *the love is conditioned on the child's ability to aid the parent*. It is a love that ultimately smothers children and prevents them from developing a secure attachment, a viable identity, and from being able to "differentiate" from the parent and successfully emancipate from the family and establish their own independent adult life.

Such clients may recall a parent for whom they were excessively important and still feel guilt for not adequately meeting this parent's needs. To illustrate, Will grew up watching his alcoholic father get drunk, pick a fight with his mother, and sometimes end up "pushing her around." Unable to sleep without nightmares, Will was an anxious child—worried about his mother's safety. But Will was also a big kid, and, as puberty came on, he also became a strong kid. Now he no longer listened anxiously through his bedroom door to hear if his parents might be wrangling. Instead, he learned the best way to protect his mom was to just walk out and provoke a fight with his dad before he could get started with her—which his mother appreciated and for which she seemed to thank Will by aggrandizing and indulging him. Continuing this role as the strong rescuer, Will eventually become a police officer. Regularly promoted, Will would risk his own safety and step into the line of fire to help someone, but too often without good judgment or realistic assessment of his own or others' vulnerability in the situation—as if caution wasn't necessary because he couldn't be stopped or hurt. Over the years, physical injuries accumulated from this expansive, cavalier approach, of course, but the emotional toll was even greater. As an urban police officer, he regularly saw children die in accidents and youth shot on the streets. Will couldn't forgive himself when he saw this kind of tragedy; he felt guilty, as if somehow he should have done more or been better able to prevent it. By age 40, Will was "burned out" and, almost unable to work. He came to treatment on orders from his captain for "drinking too much" and increasingly responding to policing incidents with too much aggression.

DISENGAGED PARENTING

The third parenting style is **Disengaged** (also termed Neglectful). An especially problematic parenting style, these caregivers are disengaged from their children (**Quadrant III: low control, low affection**). Looking at Figure 6.2, one sees that these uninvolved parents are low on both axes—in other words, they are doing little for their children. Whether passively unresponsive or overtly rejecting, this neglectful parent says in word and deed, "Go away—just leave me alone." These parents have little investment in their role as caretaker and engage in inconsistent and erratic discipline. Within different subtypes of Disengaged families, children may be rejected, suffer neglect, be physically or emotionally abandoned, and some will end up in foster placements. For some, their very existence seems to irritate or burden their parents. In order to adapt to these painful situations, children of disengaged or neglectful parents often learn how to hide and not make waves, often limiting their ability to form a personal identity.

How does this disengaged/neglectful parenting style come about? Often caught up in their own drug or alcohol problems, disengaged parents may be too self-absorbed to attend to their child's needs. Similarly, another type of disengaged parent may be chronically depressed and unresponsive to their child, who grows up unnoticed, and in later years describes herself in treatment as having been "invisible." Still other children, who may have grown up with a caregiver who had a personality disorder such as a narcissistic, borderline, or paranoid personality disorder, may be overtly rejected (for example, "I wish you were never born!"), pushed away by an angry parent, or blamed for all of their parent's problems. Heartbreaking to observe, these and other reasons cause these children to be maltreated and grow up without care. Not surprisingly, significant cognitive and emotional delays are evident in children from these families. Unsupervised or unwanted, the disaffected child with neglectful parents becomes at risk for many different problems, including antisocial behavior and the peer influences of drugs, delinquency, and early sexual contact. These clients are often in the child welfare system, frequently being seen in group treatment homes and juvenile probation programs. Because they have not had a caregiver who monitored them and tried to keep the child's "mind" in mind and be aware of their needs, activities, and decisions, they have more accidents, and they are unprotected from exploitative individuals in the environment and too often suffer sexual molestation and physical abuse.

These at-risk youth bring strongly held schemas for rejection and distrust to the therapeutic relationship, and they challenge the therapist to establish the working alliance they need but cannot trust. Their greatest need—and perhaps their worst fear—is for the therapist to "see" them, because often they have learned how to be invisible. In addition to communicating, "I don't need anyone" (a characteristic they share in common with some of those who grew up in more severely authoritarian families), these clients may also communicate, "I don't really exist" and "I'm not worthy of your time and effort." As the therapist consistently passes the client's "tests" (that is, doesn't respond with the rejection, disdain, or disinterest that they expect and elicit), the therapist will become better at establishing a working alliance based on trust. As this occurs, clients will feel more secure and typically will move in the direction of building a personal identity, with the therapist providing the kind of "benevolent guidance" that Erikson thought was so essential in helping a person achieve a viable sense of identity.

AUTHORITATIVE PARENTING

Finally, the **authoritative** child-rearing style is most effective and produces well-adjusted children (see **Quadrant I: high control, high affection**). The authoritative approach combines strict limits and reliably enforced rules with much parental warmth and overtly expressed affection. These warm and nurturing parents are sensitive to their children's needs. They establish relationships that involve clear communication, clear guidelines, and provide explanations and reasons for parental rules and decisions made. Parents have high expectations for responsible and mature behavior, and where possible, give the children the opportunity to be involved in decisions and have choices. This process helps children become confident, secure, and learn to become independent decision makers themselves. In addition, because these parents set reasonable limits and are actively engaged with their children in warm and mutually respectful ways, the children feel loved and valued and are more likely to take on the values held by their parents.

Thus, while the authoritative parents believe in strict discipline, unlike the authoritarian parents, they combine this with physical affection and spoken approval. For example, authoritative parents expect good behavior yet also tell their children stories, roll with them on the floor, hold them in their lap, praise them when they do well, and look in their eyes and say, "I love you." Children more readily cooperate with requests from an affectionate parent than from one who is threatening or distant.

Although authoritative parents are firm about discipline, they also invite children's participation in the process. They encourage children to offer alternatives or compromise solutions. They also tell their children what they would like them to do and explain why certain behavior is encouraged or discouraged. In contrast, the authoritarian parent provides clearly defined and enforced limits but no room for compromises, alternatives, or explanations. Even very young children may be more willing to cooperate if they understand the reasons for the rules. Adult authority seems less arbitrary or unfair when children can participate in the discipline process.

Authoritative parents consistently enforce the rules that they set. Permissive parents may offer reasons and explanations to lessen their children's disapproval, but ultimately they do not take a firm stance and convincingly enforce limits. Children are astute judges of how serious their parents are about enforcing rules. If parents enforce rules inconsistently, children will continually test and try to break them. Thus, whether or not the child has been able to understand or agree, authoritative parents follow through and enforce the rules that have been set.

Clearly, the authoritative parent exercises a wide range of parenting skills. We have seen that the authoritarian parent sets limits but does so harshly, the permissive parent cannot follow through or does not believe in establishing controls, and the disengaged parent has given up on attempts to manage children. Although many parents falsely believe that they have to be either strict or loving (that is, authoritarian or permissive), *authoritative parents are more effective because they have the flexibility to be both at the same time.* Despite the best of intentions, however, balancing these two domains is challenging for most parents. Indeed, researchers find that only about 10–12 percent of parents provide authoritative parenting. Most parents do not possess the wide experiential range necessary to be both firm and loving—just as many therapists find it difficult to be supportive or empathic with their clients and, at other times, forthright or challenging. Although it is not easy for caregivers to provide this authoritative child rearing, it produces the most well-adjusted children. Following them over time in longitudinal studies, researchers find that these children tend to have a greater sense of security and self-esteem and to be more independent, self-controlled, and successful with friends.

As therapists listen to their clients' narratives and learn more about their developmental histories, they will hear relational patterns and themes derived from authoritarian, permissive, and disengaged parenting styles. What about children of authoritative parents—is this a "perfect" family without any problems? Parenting and development are always challenging, even in the best of circumstances, and conflict-free families and individuals are a myth. On average, however, children of authoritative parents are better adjusted and, like Secure clients, they will be less likely to be seen in treatment for enduring problems. More likely, they will seek help in crisis situations (such as a child's major illness) or when negotiating developmental hurdles, such as seeking premarital counseling. And because they have learned that some others can respond responsibly to their emotional needs, they are able to seek help and enter treatment when necessary.

In thinking about these four different styles of child rearing, we also need to recognize the enormous complexity of raising children and all of the different factors that contribute to family interaction. For example, birth order, gender, and temperament greatly influence how parents respond to children. Family functioning is fundamentally shaped by cultural values and beliefs, as all child-rearing practices are embedded in a social context (Domenech-Rodriquez et al., 2009; Huntsinger & Jose, 2009). Because of these and other influences, the four parenting styles we have discussed become far more complicated in everyday life. To illustrate, suppose a 10-year-old child grows up with an authoritarian father and a permissive mother. Not the best circumstances, perhaps, yet between his two parents, this child is benefiting from both consistent discipline and affection in his life and probably adjusts satisfactorily in this developmental situation. But let's play out just one of the thousands of different developmental possibilities that could ensue and that a subsequent therapist will need to understand.

What if the strict but cold father and the warm but permissive mother wrangle over their differing parenting styles and, because of this and other disagreements, eventually divorce? Pressured to "take sides" in the ongoing parental battle, the son chooses to live with his mother and feels disloyal to her when he visits his father. Fed up with infrequent and superficial visits with his son, the father remarries in two years, starts a new family, and, in effect, drops out of the son's life. Now the child is living with his permissive mother, where, among other problems, he no longer has any effective limits. Angry at his father for leaving him and without discipline in his life, this boy begins to act out at home and at school. In particular, he becomes defiant and disrespectful toward his mother. Soon, she no longer feels like expressing the warmth and affection she once did to this son who has become demanding, bossy, and self-centered. An escalating negative interaction cycle ensues as the mother becomes increasingly frustrated with this angry boy that she can't control and who is "ruining my life." By age 13, the exasperated mother feels "defeated" by this "impossible" son and, transitioning toward the Disengaged parenting style, essentially gives up on this relationship. Three years later, a therapist enters the picture as the adolescent boy is court-mandated to treatment for speeding and reckless driving while intoxicated.

LOVE WITHDRAWAL, AS FOUND IN AUTHORITARIAN AND DISENGAGED PARENTING STYLES

Many caregivers, especially authoritarian and disengaged/neglectful parents, often employ discipline techniques based on **love withdrawal**. Instead of communicating that they disapprove of the child's behavior, parents respond with anger or rejection and communicate disapproval of the child's basic self. While disciplining, these parents withdraw their warmth and *emotional connection* to the child, engendering in the process the anxiety of an attachment disruption (even though the parent and child are not physically separated). This parental communication is often nonverbal; the withdrawal of love from the child is expressed as much by tone of voice and inflection, gesture, and facial expression as in words. In the child development literature, researchers call this "love withdrawal" disciplinary techniques. Central to his theory, Carl Rogers (1959, 1980) referred to this far-reaching developmental issue as **conditions of worth**—what children had to do to maintain their parent's approval.

Parents respond in this way to punish the child and communicate their anger—which too often is accompanied by the deeply more wounding affect of contempt or disgust by more severely authoritarian or disengaged parents. They may say, for example, "Get out of here! I don't even want to have to look at you. What's wrong with you anyway?" Better-functioning parents may respond in these hurtful ways occasionally, when they are tired or upset, and may later apologize or clarify that they have overreacted ("Daddy got too mad and shouldn't have said that. I'm sorry").

In contrast, for some parents, withdrawal of love and emotional disengagement occurs routinely, and the child's emotional ties to the parents are disrupted regularly. For example, the parent may say overtly, "I can't stand to be around you. Get away from me." In contrast to these overtly rejecting messages, caregivers may emotionally withdraw in more covert ways. For example, the martyrish parent may say nothing, sigh painfully, and turn her back and walk away from the child, silently shaking her head in disappointment or disgust. Of necessity, the child develops symptoms and defenses to cope with the painful separation anxieties and the shame-based sense of self that these attachment disruptions engender. For these children, self-schemas develop that leave them shame-prone throughout their lives as they come to believe "I'm a bad boy," "There's something wrong with me," or "Mom doesn't want me."

Whereas authoritarian and disengaged parents tend to use love withdrawal disciplinary techniques frequently and create insecure attachments and separation anxieties in their children, permissive parents, in contrast, often become emotionally entangled or overinvolved with their children, at times smothering them with unwanted love and attention. The interpersonal strategy of authoritarian and disengaged parents is to distance the child from them, especially when they, themselves, are feeling anxious or distressed. The permissive parent's strategy, conversely, is to draw the child closer to the parent, but often with the purpose of making the parent (not the child) feel better. Instead of instilling a sense of shame in their children, guilt is often the legacy of children raised in permissive homes.

A Continuum of Love Withdrawal.   Love withdrawal techniques and "conditions of worth" occur on a continuum of severity. In some families, disruption of ties from love withdrawal may not be severe. And, as noted earlier, they may occur infrequently—only when caregivers are stressed or fatigued. If love withdrawal does not occur routinely, and if other opportunities for emotional connection are available, ties can soon be restored. Although the anxiety may still be painful, the child can often fashion a reconnection through **compliance** and taking full responsibility for causing the disruption. For example, the child of a disengaged parent may adopt the interpersonal coping strategy of being quiet and "going away inside" so she needs nothing from her parent. Or, the child of an authoritarian parent may learn that by compulsively achieving or attempting to be perfect, this coping strategy will help these children preserve as much as they can of their insecure attachment. As new therapists gain more clinical experience working with diverse clients, they will begin to recognize *the varied coping strategies their clients have adopted to use in their efforts to maintain ties to important others.* Such interpersonal coping strategies (or "rules of attachment") may sound like one of these:

> It's my fault they don't like me. If I could just play baseball better; be nicer to everyone all the time; do more for my dad; always get A's in every class; be thinner and look prettier; help my mom stop being unhappy all the time; be invisible and not ask for anything, then I'll be OK, and then they'll be happy with me.

In families that are more dysfunctional, however, the attachment disruption will occur more often and more severely. In daily interactions around discipline and control, *children in these families will be exposed regularly to experiences of interpersonal loss and emotional isolation, even though the parent and child remain in physical proximity.* As we move further along this continuum, ridicule and rejection occur more overtly. In disengaged families, gross neglect and even actual abandonment may ensue. In highly authoritarian families, these behaviors may erupt into physical abuse. In these painful situations, the child experiences his parent's anger and contempt as assaults on his basic sense of self, leaving the child ashamed of who he is and psychologically alone. It is important for new therapists to know that they will see many clients with these developmental experiences, and that *these clients will hold the pathogenic belief that they are justifiably to blame* for their parent's neglect, rejection, physical domination, disgust, and so forth. In other words, these clients assumed responsibility and blame themselves for what their parents did and will tell their therapists:

- "If I hadn't fought with my brother so much, my dad would have stayed with my mom. It's my fault she cries all the time now."
- "If I had done all of my chores, like Mom asked me, she wouldn't have hit me so hard. I was bad."
- "If I hadn't dressed like that, he wouldn't have touched me down there. I feel dirty."

Routinely, when parents act irresponsibly in these ways, they overtly blame the child for their own inappropriate behavior. Compounding this, unfortunately, the attachment researchers suggest that this sense of shame and feeling blame-worthy are adaptive if these children are to experience some sense of psychological control in their lives. That is, clients hold on to such false beliefs tenaciously because the unwanted alternative would be to view the parents more realistically as rejecting and punitive, which would leave the child feeling even more powerless and unattached. In sum, when the child experiences this severely authoritarian or dismissive style of parenting, it is the genesis of a shame-based sense of self.

Effects of Severe Love Withdrawal.   Severe love withdrawal as a means of discipline occurs in many ways and in different types of families. For example, variations of this approach can be seen in parents who act like martyrs. These parents often communicate their hurt and disappointment nonverbally, by turning away and withdrawing emotionally or by providing a sigh or long-suffering look. Other parents demand perfection from the child, exert excessive control over the child in order to obtain it, and withdraw their warmth, approval, and emotional presence when the child does not fulfill their perfectionistic demands. In these moments of parental love withdrawal, however they occur, the child in effect loses the parent. The child's emotional connectedness to the parent is temporarily broken, and the child's attachment ties are situationally disrupted. As described earlier, *this engenders separation anxieties and shame until the child can find a way to comply and restore the tie.* Of course, all children's ties to their parents will be threatened at times. However, significant problems occur when disruption of such ties is so frequent as to characterize the relationship. Enduring problems also ensue when these rupturing interactions are *disavowed* by the parent—acting as if nothing significant happened—rather than acknowledged and resolved (for example, a parent saying, "I'm in a bad mood today and got too upset just now; that was my fault, not yours"). Thus, therapists will find that many clients who grew up with these problems also lived with an **unspoken family rule** that hurtful interactions like these could not be talked about—named or made overt—as if the child could not say or even know what just transpired. Just as therapists and clients need to do in the therapeutic relationship, *parents and children need to be able to sort through and restore ruptures in their relationship.* Unfortunately, so many clients were unable to do this in their families.

A constellation of significant emotional reactions occurs when parents cut off their emotional connection to children in anger or disgust. As we are seeing, even though the caregiver is physically present, the child is psychologically alone and suffers painful separation anxieties. This withdrawal also stifles the child's sense of self-efficacy and gives rise to feelings of helplessness and hopelessness because the child in this predicament cannot really win or earn the parent's love. This also leads to dysthymia and other forms of depression, and a shame-based sense of self as ineffectual or feeling unworthy of being loved. Thus, this child is made to feel bad and alone, desperately wants to restore the relationship and renew emotional ties, yet is ultimately helpless to do so until the parent chooses to reengage. Moreover, the child's inefficacy is further exacerbated by being made to feel responsible for the parent's withdrawal and deserving of it. In this immobilizing double bind, children believe that their behavior (spilling a glass of milk, asking a question, crying, feeling angry, needing help, and so forth) has caused the parent to go away, which is the response the child fears most because it threatens already insecure ties.

The child is angry at being abandoned and wants to **protest**, of course, but this reaction would only elicit further domination or intimidation from the authoritarian parent. Through power assertion, the authoritarian parent usually does not allow the child to disagree, let alone find appropriate means of expressing anger. The child may be told, for example, "I'm your father. You are never angry at me. Do you understand that? Look at me and say, 'Yes, sir.'" Thus, the child cannot protest behaviorally, or even experience anger internally, because such reactions will further threaten already tenuous ties to the parent. As a result, this child often becomes intrapunitive by turning the anger inward—and then frequently struggles with a low frustration tolerance and difficulty controlling anger outbursts. It also is expressed through self-deprecation, dysthymia, and having a shame-prone self. This tendency toward self-blame is exacerbated as children come to identify with the parent and *adopt the same critical or contemptuous attitude toward themselves that the parent originally held toward them.*

Just as this type of parent loses touch with the child's feelings or experience in these angry moments, the child, in turn, loses the clarity and authenticity of her own internal experience; such children also lose touch with important aspects of themselves. That is, when authoritarian parents are so rigidly demanding of conformity and obedience, and disengaged parents are so invalidating or non-empathic to their children's feelings and needs, children soon lose touch with their own internal experience and become "incongruent" in Rogers' term (1980). As they present in treatment, they may not know what they like and dislike, and may even be uncertain about what does or does not feel good to them. Their own subjective experience can be so completely overridden that they have no basis for later developing their own belief systems and clarifying their own spiritual, political, or sexual values, or formulating occupational interests in late adolescence and young adulthood. In other words, they haven't been allowed to develop an identity or, more basically, a self and "have their own mind" or "find their own voice." New therapists often will find themselves working with clients with these identity problems and family dynamics. With these clients, the therapist's initial goals are:

1. to validate their subjective experience, which has been so pervasively invalidated;
2. to encourage their initiative, which has been undermined—for example, by following their lead in treatment and supporting their own iniative with others; and

3. to provide a treatment focus by helping them clarify their own interests and act on their own goals whenever possible.

A Continuum Toward Trauma and Abuse.   Moving further out on this continuum of love withdrawal, some authoritarian and disengaged parents become so completely removed from themselves and the child that they become dissociated. For a few moments, usually while in the midst of their own shame–rage cycle or under the influence of alcohol or drugs, these parents lose all vestiges of self-awareness or connection to their own experience. More importantly, they lose any awareness of their child's experience and do not feel or register the frightening or humiliating impact that they are having on their child at that moment. It is terrifying for children when threatening caregivers become disconnected from them in this way; they may describe their parents in these moments as being "possessed," "not there," or "somebody else." Most clinical trainees will not be familiar with these *dissociative ego states* and, early in their training, will be understandably upset by the shame and terror these states engender in some of their clients. However, by learning about these highly dysfunctional families (also known as the Category D, Disorganized) attachment style (one subtype of the Fearful, Quadrant IV, attachment style in figure 6.1), therapists also learn important principles for working with clients from less troubling backgrounds.

Caregivers abuse or mistreat children for many different reasons. Some abusive parents were indulged by their permissive parents when they were children and experienced no consistent limits or consequences for their behavior. Whenever they did something wrong or got in trouble, their caregivers externalized all responsibility for the problem and blamed the teacher, coach, or another child for the problem. Others mistreat their own children in the same ways they were once mistreated. *This multigenerational reenactment is most likely when the parent–child relationship was highly ambivalent; that is, the parent who hurt them also was authentically caring or affirming at other times* (Rocklin & Levitt, 1987). By reenacting the abuse and simply engendering in their child what they once suffered, rather than remembering, feeling, and talking about what they once experienced, some individuals defend against the painful feelings evoked by their own mistreatment. Typically, *one of their own children comes to represent themselves or, more specifically, the unwanted or disowned aspects of themselves* (the emotional needs, vulnerability, or anger) that originally evoked the abuse (Bowen, 1978). Which child is chosen will depend on age, gender, birth order, temperament, and other characteristics. However, most victims of abuse do not reenact what happened to them, although many live with the debilitating fear of being like their parent in any way, even though they are actually very different.

This disturbing reenactment has received different labels within different theoretical systems: projective identification (object relations theory); identification with the aggressor (psychoanalytic theory); the family projection process (family systems theory); and turning passive into active (control/mastery theory). However labeled, it is a defense—an attempt to cope with trauma and manage or gain control of unwanted, painful feelings that once were intolerable. This intergenerational transmission of pathology stops if abusive parents can not only remember the events but also tolerate experiencing the fear and shame they cut off from feeling in their own past (Eron & Huesmann, 1990). This psychological task is not easy to accomplish, however.

Thus, both object relations theorists and family systems theorists inform us that the more unresolved or unintegrated we are, the more likely we are to instill or project unacceptable aspects of ourselves onto others—especially our children. As we discussed in Chapter 4, managing our own personal problems through others is a way of seeking an external solution to an internal problem. Thus, in this type of projective maneuver, the abusive parents' reenactment with their own child is a psychological defense—an attempt to gain some control over unacceptable feelings of being helpless, afraid, bad, and so forth that is associated with their own past abuse. By externally reenacting some version of their own mistreatment/trauma over and over again with their children, some caregivers who are abusive do not have to remember or experience internally their own disavowed feelings. Said differently, by evoking the same terrified or otherwise unacceptable feelings in their child that they were once made to feel, some abusive parents do not have to experience the fear and shame-laden defeat of abuse as their own; their fear and shame, instead, is expressed vicariously through the child. For children in extremely authoritarian and rigidly "hierarchical" families, such abuse is a persistent threat that organizes their daily experience and, ultimately, their psychological adaptation to life. In disengaged families, there may be less physical abuse and more neglect and abandonment, but the legacy of shame instilled in children raised in such families is still very real and debilitating.

Finally, as therapists work with clients who grew up with these tragic developmental experiences, they will find that in many cases these clients can remember and talk about the physically, sexually, or emotionally abusive events that happened to them, yet the fear and shame that accompanied their experience are not available to them. That is, *therapists will observe that these clients remember what happened and relay the story, but without the commensurate emotion.* As student therapists move along in their clinical training and experience, they are going to find that the developmental problems discussed here are far more common in everyday life and clinical practice than they anticipated when they began this work. However, new therapists should also know that they can learn to help clients greatly by expressing compassion for the mistreatment they suffered, validating their experiences without demonizing the caregiver, and helping them find and integrate the feelings that more realistically fit their experiences and what actually happened to them.

Interpersonal Strategies to Cope with Insecurity.   As we are seeing, there is already too little affection in authoritarian, disengaged, and Dismissive families, and maddeningly inconsistent and unpredictable responses from Fearful, Preoccupied, and permissive parents. These thin and wavering threads of connection are repeatedly disrupted when angry, distressed, or demanding parents emotionally disconnect from children. When young children are unable to maintain parental affection and affective ties, they will not grow up to feel love-worthy or secure. These children have missed the essential experience of constancy in their attachment bonds. Let's explore two important clinical implications of this developmental deficit.

First, these children missed the experience of someone actively reaching out and choosing them. In many cases, they do not feel loved or, especially in disengaged families, wanted. Children from authoritarian families often feel variations on the theme, "If only I did better..." In these circumstances, the child attempts to win or earn the parent's attention or approval by adopting interpersonal coping strategies, such as striving to achieve, being perfect, taking care of the parent, and so forth. Children from disengaged families, on the other hand, are more likely to think, "I don't really matter," and cope by withdrawing or becoming invisible and never needing anything. In either case, *the life-shaping problem for the child is that the parent's responsiveness is dependent on the child's efforts to elicit the caretaker's attention.* If the parent's responsiveness to the child is dependent on the child's efforts in one of these ways, the child is responsible for eliciting the caretaker's attention and affection and does not develop a secure attachment. This routinely occurs for children in families where children must find ways to cope with the anxiety generated by their insecure ties.

Let's explore this subtle but important concept more closely. To establish more secure bonds, *children attempt to control or manipulate their parents' feeling for them.* They do this by learning to employ certain interpersonal coping strategies—for example, by being compliant and pleasing, by compulsively striving for achievement, by perfectionistically trying to be good or not getting in the way of the parent, and so forth. As we will see in Chapter 7, this attempt to win attention and affection often becomes a *pervasive coping style* that clients continue to use with others throughout their lives. These coping strategies come at a personal price and take a toll, however, and become central to the symptoms and problems that clients subsequently present in treatment. Furthermore, because of the schemas that have developed from these childhood experiences, these clients also are going to believe that they similarly have to elicit, or be responsible for, the therapist's interest in them—as they have with others in their lives.

Second, whereas secure children do not fear that they can do something wrong to disrupt the parent–child tie, insecurely attached children often are told, erroneously of course, that their parents' anger or frightening temper, rejection or contempt, or physical or emotional withdrawal is occurring because the child is being bad and causing the parents' unwanted response. In this way, the child often is led to believe that she deserved and is responsible for disrupting the parent's love or commitment to her and, as an adult client, will still hold this pathogenic belief. For example, children in highly authoritarian and disengaged families, or with parents who have Dismissive or Fearful attachment styles, learn that their anger, their tears, or even their questions can provoke their parent to cut off emotionally from them and disrupt the parent–child tie with threats such as these:

- Stop those tears right now or I'll give you something to really cry about!
- Wipe that angry look off your face. You'll do what I say, and like it, or else!
- Never ask me "why." Just do what I say, when I say it.
- Now you've really done it! This time I've had it with you—for good!
- Stop bothering me with all those stupid questions. Leave me alone!

Even as adults, the sons and daughters of such parents who used severe love withdrawal techniques may continue to be subjected to the same threats. If they do not do what parenting figures demand, they will (seemingly) destroy the relationship:

- If you marry him, we won't come to the wedding or visit you anymore.

- If you get a divorce, we are going to disown you and take you out of the will.
- If you do that, no one in this family will ever speak to you again.

The theme here is that parents who use severe love withdrawal techniques to control their children are not just setting limits on unacceptable behavior, as authoritative parents do; they are instead *threatening to cut off fundamental relational ties*. As attachment-seeking children try to cope with the intense anxiety this arouses, compliance becomes a generalized trait, pervasive personality constriction and inhibition occur, and obsessive/compulsive symptoms and other control issues, such as eating disorders, often develop.

Further, highly authoritarian and disengaged parents, and caregivers with Dismissive and Fearful attachment styles, break off emotional contact with children, intimidate or frighten their children, and—in words and, especially, through tone—often engender shame by communicating contempt toward them. Children who suffer such developmental experiences—*especially internalized shame from parental contempt*—typically struggle with anxiety and depression throughout their adult lives. They often report feeling guilt, loneliness, and low self-esteem when they enter treatment, but without understanding why, and may describe their parents in idealized or problem-free terms. As we are seeing, these clients develop interpersonal strategies to cope, such as always being good, taking care of the parent, being quiet and "going away inside," and so forth. In the next chapter, we are going to clarify how *the therapist provides a treatment focus by highlighting how these faulty interpersonal coping strategies are still in play and contributing to symptoms and problems*. Although these coping strategies once were necessary and adaptive, the therapist's approach is to help clients explore how they are no longer necessary or effective in many current relationships, and are disrupting relationships with others:

THERAPIST: As you describe this argument with your wife, Bob, it sounds like she is complaining that "you always have to be right." I've heard this theme in problems you've described with others—think there's something to it?

In addition, the therapist is focusing on how these problematic coping strategies may be being reenacted in the therapeutic relationship along the process dimension:

THERAPIST: You know, Bob, it feels almost like we're having an argument right now— like one of us has to be right and the other wrong. What do you see going on between us here? Any ideas?

These interpersonal defenses—always having to be right or on top—originally helped protect these clients from experiencing the painful feelings that resulted from hurtful interactions, like those described earlier, that occurred repeatedly for them. If therapists begin to ask about these interpersonal coping strategies when they see them occurring with others or with the therapist, this accompanying pain will soon emerge. When therapists respond with affirmation and compassion, however, clients have the safety and support they need to be able to reexperience these feelings. They will benefit greatly from the therapist's validating response and begin integrating these feelings that previously had been sequestered away. Only then can these clients stop protecting their caregivers at the expense of their own symptoms, denying their own feelings, and maintaining family myths of happiness and togetherness that only serve to confuse and disempower them.

RESPONDING TO CLIENTS WITH A DIVERSITY OF PARENTING AND ATTACHMENT STYLES

In their caseloads, new therapists will work with many clients who grew up struggling with the problematic experiences discussed here. For therapists who themselves enjoyed better developmental experiences with Authoritative or Secure caregivers, it may be hard to appreciate the emotional severity of highly authoritarian parenting, the anguish of rejection in disengaged families, the aloneness in severe love withdrawal, or the identity confusion and lack of a coherent sense of self from growing up with Fearful or Preoccupied caregivers. These therapists may wonder how the ostensibly normal and, in many other ways, decent parents of these clients can be so rejecting or obliviously self-centered on occasion, and can have caused such profound insecurities or even self-hatred in their clients. Other graduate student therapists may have difficulty with these client dynamics because they evoke therapists' painful feelings about their own developmental experiences and how they were raised. Thus, this material can be challenging to work with because it evokes such strong countertransference reactions in many therapists. To help therapists understand and respond more effectively, let's illustrate further what a range of secure and insecure attachment configurations looks like.

While growing up, Molly benefited from secure attachment ties without fear of love withdrawal. For example, she recalled an incident from when she was seven years old when her mother was angry about something she had done. Her mother made strong eye contact with her, reached out and touched her on the shoulder, and said in a calm but firm voice, "I love you, but I don't like it when you act this way. Go to your room and stay there until you decide you are ready to come back out and get along." Testing the limits, Molly protested, "But you can't send me to my room. You have to want to be with me because you're my mom!" Her mother went on to clarify that it was Molly's behavior toward her sister that she did not like, and Molly recalled that she began to understand that concept.

Her mother was angry, and she got her point across that she did not like what Molly was doing, but Molly also felt secure in her mother's love. Her behavior wasn't acceptable, but she was—Molly was not coping with "conditions of worth." Looking back, Molly thought that her sense of secure attachment in this conflict was maintained primarily by the nonverbal messages that accompanied her mother's restrictions and explanations. Because of these secure relational ties, Molly was able to internalize her mother's loving feelings for her and to develop "object constancy," as discussed in Chapter 1. As an adult, Molly now is able to establish friendships in which she similarly feels respected and affirmed, and was able to establish a marriage in which aspects of this same loving affect are present.

Often, children will be secure in their emotional ties with one parent but struggle with a lack of constancy with the other parent. For example, Ellen recalled that she felt secure with her mother—even when her mother was mad or disapproving of something she had done. Her father was highly inconsistent, however. At times, he was affectionate and responsive, but at other times he could lose his temper and be overtly rejecting. Ellen recalled happy memories of her father patiently painting ladybugs on her roller skates, as well as painful memories of her father blowing up at her and yelling angrily, "Get out of here. I can't stand being around you!"

Ellen's secure base with her mother allowed her to cope with this intensely ambivalent relationship with her father. With the consistent or reliable emotional support provided by her mother, she could learn to anticipate her father's moods and stay away from him when necessary. Her feelings about herself were not based on the unstable fluctuations of her father's moods. Now, as an adult, Ellen is an especially perceptive and aware person; in this way, personality strengths often develop from such conflict-driven demands for coping. Ellen's brother, however, was not so fortunate. Ellen's mother seemed to enjoy raising girls more than boys, and her brother did not receive the same secure base with his mother that Ellen enjoyed. As a result, he was left to bob about unconnected at the mercy of his father's stormy emotional seas. Ellen describes her brother as feeling "very bad about himself" and being depressed a lot. Now in his late 20s, he has been unable to make commitments to relationships or a career and cannot find a life for himself. Ellen says she worries about him a lot.

How can therapists explore and assess parenting styles and attachment histories with their clients? With many clients, therapists may wish to explore by asking directly about them:

• How did your parents usually respond to you when they were upset with you or disciplining you? What did each parent tend to say and do, and how did that usually leave you feeling?

• How did you expect your parents to respond to you when you were distressed or needed help? Would you seek help from your parents and talk to them when you had a problem?

In response to such queries, many clients will relay narratives of troubling parent–child interactions that include themes of anxious love withdrawal, painful isolation, and shame. Again, therapists can anticipate that oftentimes their clients will be able to remember and talk about these very difficult interactions, yet the painful feelings accompanying them are blocked off or unavailable. Therapists will also see that many learned not to expect help and, after a while, *didn't even think of seeking assistance from their caregivers as an option or possibility*. To return to the heart of the attachment story: they were not secure in the expectation that their caregiver would be their ally when they were distressed and try to help them with their problem—even if that meant only to hear their concern and be with them in it.

Additional Clinical Guidelines

Parenting is probably the most challenging task in life. The influential family therapist Salvadore Minuchin says, forgivingly, that parenting has always been more or less impossible. Almost all parents are trying to do the best they can for their children. Even many of the highly ineffective parents we have been discussing, who do indeed engender significant, lifelong problems for their children, are not cruel or ill-intended people in most cases. In child rearing, as in other aspects of personality, people are uneven in their development. Most of these parents do other things well for their children, live by certain moral standards, believe they are trying to do what is best for their children much of the time, and usually are treating their children as well as or better than they themselves were treated. Although children certainly are hurt by such developmental experiences, almost universally they still love their parents and seek their approval.

Therapists will do well to appreciate adult clients' willingness to give their caregivers "another chance" and their lifelong efforts to improve or repair these flawed but all-important relationships. Therapists want to help clients *realistically assess* whether or not caregivers have changed over the years and to what extent they are capable of responding better now than before. That is, some caregivers have gotten better and can respond more constructively now than when clients were young, whereas many others continue to respond in the same problematic ways. The therapist's goal is to foster realistic expectations in their clients and help the current relationships between parents and adult offspring become as good as they can be. It is important that therapists encourage clients to express in treatment the full range of their positive and negative feelings toward their caregivers. Therapists who "foreclose" on this process by quickly echoing and remaining stuck on just the negative aspects of the caregivers, or just the positive characteristics, will rob their clients of the important experience of both integrating the beneficial and resolving the problematic aspects of their caregivers. Over the long term, clients who cannot resolve their ambivalence and integrate the rewarding and the problematic aspects of their relationships with their caregivers will also have difficulty accepting the good and the imperfect parts of themselves—and of their own children.

In sum, the therapist's role is not to foster splitting defenses by bashing parents who have been hurtful and making them bad, by encouraging clients to reject them or break off contact, or by subtly inducing clients to replace the parents with idealizations of the therapist. Nor, on the other hand, is the aim to deny or in any way to minimize the real impact of the hurtful interactions that have occurred. Instead, the therapist:

1. helps clients come to terms more realistically with the good news and the bad news in their family of origin,
2. helps clients change their own responses to problematic others in current relationships, and
3. facilitates clients' current attempts to establish new relationships that do not repeat the problematic relational patterns that have come before.

Therapists provide a corrective emotional experience when they can sustain a working alliance and remain emotionally available to their clients. Doing so provides many of the clients discussed in this chapter with the secure base that they missed developmentally. Such consistent emotional availability, week in and week out over the course of treatment, usually has more effect on client change than do more dramatic but isolated incidents of compelling insight, important self-disclosure, or other significant therapeutic interventions. As noted, however, countertransference issues derived from the therapist's own family-of-origin experiences often make it challenging for therapists to provide this consistent, corrective presence.

The clients whom we have been discussing share the experience that caregivers withdrew from them in one way or another or were not emotionally available to them at important times of need. This developmental deficit—and the maladaptive relational patterns that result from it—can readily yet subtly be reenacted in the therapeutic relationship. For example, some therapists may have trouble being emotionally available to clients' pain and vulnerability and they may not be able to remain present with certain affects, such as the client's raw shame or intense sadness. The client's individuation may make other therapists uncomfortable because success or independence was not supported in their own development or because the client's success evokes the lack of fulfillment in their own work or marriage. As their clients improve in treatment or progress toward termination, these therapists may also feel sad or anxious about losing the caretaking role that they were scripted to assume in their family of origin. Most important, perhaps, many therapists disengage or "give up" on clients when they don't change or somehow disappoint or frustrate the therapist (for example, Preoccupied clients who simultaneously *resist* the help they urgently seek)—routinely saying that "the client wasn't ready to change yet." Therapists may feel they are failing when the client isn't changing, and too often therapists respond ineffectively to this countertransference reaction about their own competence by initially becoming critical and eventually withdrawing from the client—and oftentimes reenacting the client's developmental pattern in this unwanted sequence.

When any of these countertransference reactions occur, as they commonly do, clients are again left alone and unconnected in their experience. Clients' pathogenic beliefs are confirmed rather than resolved by this problematic interpersonal process: feeling sad or mad, or being "stuck" or successful, causes others to be hurt or angry, to disengage, and so forth. When such reenactments occur, without being addressed and rectified, they generate further insecurity for clients and impede change. Rather than being resolved, old schemas and problematic relational expectations are confirmed by what is being played out in the therapeutic relationship.

In addition to their own countertransference tendencies, certain features in the client also make it difficult for therapists to remain consistently available. For example, many clients will report that they like and trust the therapist but they may also believe that, "if the therapist *really* knew me," the therapist would not respect or care about them. In line with their faulty schemas and problematic coping strategies, these clients believe they have deceived the therapist or manipulated the therapist's positive feeling for them. This is why it is necessary for the therapist to help clients adopt an internal focus and to draw out the full range of clients' feelings toward the therapist—including the clients' perceptions of the therapist's reactions to them. For example:

THERAPIST: What do you think I am feeling toward you as you tell me this?

Unfortunately, clients often hold the pathogenic belief that their despised, weak, or otherwise unacceptable emotions constitute their real or true self. This far-reaching problem is resolved when the therapist sees the vulnerable, dependent, shameful, or other parts of themselves, which clients believe are the damning proof or confirming evidence of their basic unacceptability, and still remains affirming and emotionally connected. *Clients' encounter with compassion and understanding, rather than the emotional withdrawal, rejection, and so forth that they have learned to expect, is one of the most powerful relearning experiences that therapists can provide*. Thinking about conditions of worth, Carl Rogers (1951) emphasized long ago that clients do not change until they accept themselves and that they begin to accept themselves when they feel such acceptance from the therapist. Thus, by providing a safe holding environment that contains these conflicted feelings, as clients are experiencing them, the therapist helps clients to resolve them.

Finally, *some clients are adept at eliciting the same problematic responses from the therapist that they have received in the past*. As we have already emphasized, a basic tenet of the interpersonal process approach is that most clients' problems will be temporarily reenacted with the therapist along the process dimension and need to be identified and reworked. It is easy to be empathic, warm, and genuine with clients who are usually cooperative, friendly, and respectful. However, it is far more challenging for therapists to remain committed to a working alliance with difficult clients, whose interpersonal coping strategies—for example, being aggressive or intimidating, critical or controlling, passive or dependent, and so forth—may eventually alienate, intimidate, or frustrate the therapist. With these "negative" clients, it is especially important for therapists to keep process notes and formulate working hypotheses about the clients' maladaptive relational patterns; specifically, *to hypothesize what the clients have tended to elicit from others and what they are likely to evoke in the therapist*. When therapists prepare such conceptual formulations, they increase their chances of sustaining empathy and then providing a corrective experience, rather than responding automatically to the client and merely reenacting the problematic scenario with which the client is familiar (see Appendixes A and B). The following case study illustrates how the therapist was able to resist reenacting the client's maladaptive relational scenario and, instead, provide a corrective relational experience by remaining emotionally connected to him during a challenging disagreement.

John, an 11-year-old client, was being treated in a strict, physically punishing manner by his authoritarian stepfather. His mother reluctantly accepted her husband's harsh corporal punishment and tried to ignore his derision of the boy; she often walked away and left the room when the stepfather was angrily deriding him. Regarding the therapist as "weak" like his mother, John defied his female therapist, responded contemptuously toward her, and repeatedly tried to push her away. Near the end of an especially frustrating session, in which he had repeatedly pushed every limit, John was unwilling to speak. Trying to find some way to salvage something of a working alliance with him, the therapist agreed that they would not have to talk together if he did not want to and that he could just throw a ball as he wished. Trying to join him in his chosen activity, she asked whether she could silently play catch with him. He agreed but, of course, began throwing the ball too hard. This process continued throughout the session: John tested, and the therapist set limits by saying, "You have to throw the ball below my shoulders," while still working hard to try and find some way to connect with him.

Near the end of the hour, the therapist invoked their standing rule and asked him to help her pick up the room. John refused; the therapist insisted. Exasperated, she took his arm lightly and directed him toward the mess in the room, whereupon he recoiled. At this moment, the same conflicted emotions that John struggled with at home with his stepfather were evoked toward the therapist and, in his mind, were about to be reenacted in the session. The therapist had been sorely pushed by John for weeks; but even though she was frustrated with him, she was still trying to find another way to relate. She did not withdraw in resignation as his mother did, although she felt like it. She also did not want to physically dominate him as John felt she was beginning to do, and thereby reenact a version of what his stepfather did. Instead, she met his eyes and slowly said,

> "John, I like being able to play and talk with you. Lately, you are throwing toys more, talking less, and being sullen—like things are not going well. I want to understand and help, and I'm not understanding what you need most right now. Let's work this out together."

Even though she felt challenged by John's behavior, the therapist could still communicate that she felt for John and was remaining emotionally connected to him. This was a turning point in treatment: John was not able to reenact with the therapist the same painful scenario of domination by his stepfather, withdrawal by his mother, and the angry but helpless isolation that he had experienced so many times at home. John grudgingly helped the therapist pick up the room; at their next session, however, he was more respectful of the therapist and began to talk about how angry he was with his stepfather and discouraged that his mother "didn't do anything." John had responded to the therapist's offer to be a "partner" in addressing whatever was troubling him.

As noted, clients will present with an extraordinary diversity of developmental and familial experiences. Some clients' problems will be unrelated to the issues discussed here. However, faulty child-rearing practices, insecure attachments, and love withdrawal techniques will be heard pervasively throughout the narratives clients report in therapy, and they will inform therapists about the experiences that originally shaped the symptoms and problems that clients are presenting in treatment.

Next, we will see that parents are much more able to provide secure attachments and authoritative parenting when they themselves have a strong marital and parental coalition. Let's examine several family systems concepts and see how therapists can use them to better understand their clients.

STRUCTURAL FAMILY RELATIONS

Structural family relations are the relatively enduring patterns of alliances, coalitions, and loyalties that exist in the family. These relationships are the basis for family organization and define how the family operates as a social system (Minuchin, 1974; Minuchin & Nichols, 1998). Such family patterns are strongly influenced by cultural values and beliefs that in turn shape norms and define acceptable behavior and family functioning. Thus, structural family relations, and the cultural context within which they are embedded, shape family communication patterns and the roles that family members adopt. Among the different structural family relationships that exist, the parental coalition is the basic axis of family life and shapes much of how the family functions (Cowan & Cowan, 2005; Teyber, 2001). In particular, the nature of the parental coalition will be pivotal in determining how well children adjust (Beavers & Hampson, 1993). Parental cohesion and harmony are strongly associated with the warmth, support, and quality of parenting they convey to their children (Kitzmann, 2000; Krishnakumar et al., 2000; Sturge-Apple et al., 2006).

SHIFTING LOYALTIES AND ESTABLISHING A MARITAL COALITION

In nuclear families that function well, the marital relationship is the primary two-person relationship in the family. Each partner has a **primary loyalty commitment** to the other, and the marital coalition is stable and cannot be divided by grandparents, children, friends, or employers. Parents will still have differences or problems between them, of course, and have substantive personal relationships with others, yet the marital relationship remains a stable and secure alliance. Researchers find that its cohesiveness is a defining feature of healthy families (Beavers, 1982; Minuchin, 1984). In contrast, two-parent families in which the primary emotional bond or involvement is not between mother and father often have more conflict and are more likely to produce offspring with enduring problems. In these more symptomatic families, the primary alliance is between a grandparent and a parent, between a parent and a child, or within some other dyad. Cultures vary widely in how strongly they emphasize family loyalties, and extended family members may play a more important role in child care responsibilities in some cultures (for example, in Native American cultures). Within varying cultural contexts, however, family researchers find that a cohesive parental coalition is a mainstay in families that function well.

How does a primary parental coalition develop? Recalling Erikson's (1968) stages of development, we note that both parents have been able to **individuate** within their own families of origin. This means they were able to achieve an authentic sense of their own personal identity in late adolescence and early adulthood—a sense of self that is based on their own genuine feelings, interests, and values. In this developmental transition of "leaving home," establishing new commitments in love relationships, and forming their own marital coalition, they need to be able to establish a new balance of **family loyalties**. This complex juggling act involves maintaining ties and responsibilities toward the previous generation, while shifting their primary loyalties or emotional commitment to the new peer relationship with a partner or spouse. Successfully negotiating this developmental hurdle or "crisis" means that young adults have been able to transfer their primary loyalty commitment from their parents in the previous generation to a spouse. When negotiated successfully, this developmental transition is accompanied by increased self-efficacy and self-esteem, permitting these young adults to further define their own values and beliefs. This **developmental transition** in the family life cycle does not mean in any way cutting off relationships with parents or losing their sense of responsibility to their own caregivers or love and commitment to their family of origin. Instead, it signals increased flexibility in their relational capacities and involves transferring their primary relational commitment to a peer relationship characterized by mutual empathy and respect.

If young adult married offspring have not been able to shift their primary loyalty commitment from their family of origin to their spouse, a primary marital coalition will not develop. The primary loyalty of one or both spouses will remain with their parent, grandparent, or other caregiver. When this occurs, the primary dyadic relationship in the family is a **cross-generational alliance** between a parent and the adult offspring, rather than a primary marital coalition between the husband and the wife. With the birth of children, this cross-generational alliance (for example, maternal grandmother–mother) usually repeats in the next generation and results in a primary parent–child coalition as well (for example, mother–oldest daughter). Before learning how these cross-generational alliances become so problematic for offspring and are expressed in treatment, we need to better understand the important developmental transitions that occur while forming a primary parental coalition. To illustrate this, let's examine the wedding ceremony as a cultural rite of passage.

The wedding ceremony is a familiar ritual; its central purpose is to publicly mark a transition in loyalty bonds from the parents and family of origin to the new spouse. How does this occur? In the traditional Christian ceremony, once family, friends, and clergy are all assembled, the father walks the bride down the center aisle to the front of the church. In front of all, the father symbolically gives the bride away by placing her hand in the hand of the waiting groom. The father then leaves the couple by stepping back from the front of the church and sitting down beside his own wife. The bride and groom, who have now been demarcated physically from the parental generation, turn away and step forward to be married. In this ritual, the bride and groom shift their primary loyalty commitments and publicly define themselves to family and friends as an enduring marital couple.

This shifting of primary loyalties from the previous generation to the new spouse is a challenging developmental task. According to Erikson (1968), this transition is a **normative developmental crisis** that arouses difficult feelings of loss and requires a complex redefinition of individual identity, as well as familial roles, membership, and boundaries—even in healthy, well-functioning families. This developmental challenge becomes far more conflict laden, and often leads to symptoms, when the young adult offspring are trying to emancipate from families that have not had a primary marital coalition. The question of who comes first, parent or spouse, may be painfully enacted in conflicts that surround planning for the wedding day. Offspring who grew up embroiled in cross-generational alliances often find that their wedding day, and the preparations for it, are a wrenching battle between conducting the ceremony as the couple being married would wish and doing it the parents' way. Usually, the underlying conflict is over the shift of loyalties from the previous generation to the spouse, but it is played out in arguments over who will be invited to the wedding, where it will be held, how it will be conducted, who is financially responsible, and so on. In contrast, young adults who enjoy their wedding day usually have been allowed to expand and shift their primary loyalty from parents to each other. These couples are not being pulled apart by competing loyalty ties and the accompanying feelings of *guilt*.

How does this model of a primary parental coalition apply to single-parent families? Are they inherently problematic because there is not a primary marital coalition? Of course not. However, this structural model can still help us understand why some single-parent families function very effectively whereas others experience more problems.

In single-parent families that function well, as in nuclear, blended, and same-gender families that function well, clearly defined **intergenerational boundaries** separate adult business from child business. By necessity, children in single-parent families usually need to take on more household duties, and often more emotional sharing occurs between parent and child than in two-parent families. Even so, in healthy single-parent families, *roles and responsibilities are clearly differentiated between adults and children.* This means that the single parent still performs "executive ego functions" for the family, such as providing an organized household with predictable daily routines for children, making decisions and plans for the family, and setting limits and enforcing rules. Furthermore, *although parents have legitimate needs for companionship and support, these emotional needs are met primarily in same-generational peer relationships rather than through the children.*

In contrast, single-parent families (and intact, nuclear families as well) will have more problems when adult and child roles are not clearly distinguished, when intergenerational boundaries are blurred, and when too many adult needs are met through the children. With this general framework in mind, we now examine why primary marital coalitions are more effective than cross-generational alliances and see how the parental coalition helps shape the symptoms and problems that clients present in treatment.

THE PARENTAL COALITION PROVIDES STABILITY AND SHAPES CHILD ADJUSTMENT

Why is the nature of the parental coalition so crucial for family functioning and child adjustment? To answer this question, we look at four common adjustment problems in children from families where the primary coalition is not between mother and father.

1. Problems with limits and authority. When mother and father have a stable alliance, children cannot come between their parents and play one parent against the other. For example, if a primary coalition exists between the father and an 8-year-old son, the mother may have trouble setting and enforcing limits with her son. If the mother tries to put her son to bed at 9 p.m. or to have him pick up his room, he can often defy these attempts at discipline by appealing to his father. The son learns that he does not have to do what his mother says because his father is likely to take his side. When children can play one parent against the other in this way, they learn that rules do not apply to them. Having learned that they do not have to conform to adult rules in the home, they often will disobey teachers and try to manipulate other authority figures in their lives as well. These problems with limits and authority will carry over into adulthood and are especially likely to lead to problems in the work sphere, where adults need to be able to conform to rules and cooperate with others. A tragic illustration of this principle is found in Woodward's (1984) biography of the late comedian John Belushi.

When Belushi was an early adolescent, his authoritarian father would work long hours in the restaurant business. Belushi was supposed to stay home and do chores for his mother. By making up humorous but hostile imitations of his father and performing these routines for his mother, he could get her to laugh at his father and manipulate her out of enforcing household duties. With his mother and siblings in stitches, Belushi would escape his chores and head out the door to meet with his friends. Among other problematic messages here, Belushi learned that if he could successfully play one parent against the other, limits and rules did not apply to him. Despite his talents, his life was a study in compulsively pushing limits and being out of control with food, money, sex, and especially drugs. After years of chronic drug use, he died from an overdose of cocaine and heroin.

Therapists who work in alcohol and substance abuse programs are likely to see large numbers of clients with this background. Clients who learn that they can *manipulate* others, and escape rules and limits by dividing the marital coalition, are more likely to develop impulsive acting-out symptoms. When counselors see them in treatment, these clients often will push limits, avoid taking personal responsibility, and be demanding of the therapist. For example, they will often arrive late for appointments, expect to be able to stay beyond the scheduled time, be delinquent in or try to avoid payment of fees, be insistent on calling the therapist at home, and so forth. These challenging clients receive a corrective relational experience when therapists can tolerate the clients' disapproval, set firm limits, and not be manipulated out of their usual treatment parameters. At the same time, however, therapists want to do this in a nonpunitive way that also communicates their continuing commitment to the client and compassion for the insecurity and loneliness that has been engendered by these developmental experiences. Like Authoritative parents, when therapists have the **interpersonal flexibility** to combine clear limits with empathy and compassion, they establish secure boundaries for these clients for the first time. This new experience of containment creates safety for these clients and allows them to begin focusing inward rather than so insistently testing the therapist, manipulating others, and acting out in self-destructive ways.

2. Exaggerated self-importance.

2. Exaggerated self-importance.   Children in families with a primary marital coalition are more likely to gain a realistic sense of their own personal power and control than are children raised in cross-generational alliances. When children form the primary emotional bond with a parent, they become too important to the parent and can exert too much influence over the parent's well-being. As a result, *these children garner an exaggerated sense of their own importance and a grandiose sense of their own ability to influence others.* Children cannot gain a realistic sense of their own limits and capabilities when they are encouraged in the illusion that they can prop up a parent's sagging self-esteem, maintain their parent's emotional equilibrium, or make important decisions for the parent.

It is highly reinforcing for children to feel so powerful vis-à-vis their parent, and they will be reluctant to give up their special role. Following their schemas, these clients will presume this special status with the therapist and are often effective in reenacting this relational pattern. In particular, such clients may offer the therapist a subtle but all-too-easily accepted invitation to establish a "mutual admiration society." Growing up, these sophisticated clients learned well how to make others feel better and are often adept at flattering the therapist and establishing an elite sense of mutual, shared superiority. For example:

CLIENT: Wow, no one has ever understood me like you do—this is amazing!

In this cozy relationship, the client has "antennae" out to grasp nuances in the therapist's mood, is skilled at making the therapist feel special, and, in turn, expects to be treated as special by the therapist. However, once this unspoken deal is struck, the therapist feels constrained because challenging or disagreeing with the client, or focusing the client inward on his participation in conflicts with others, feels like a betrayal of this special relationship. Of course, this type of reenactment will keep the therapist and client from addressing the client's real problems.

Turning over the "other side of this nickel," we find that, commensurate with their grandiosity, such clients also hide a deep sense of inadequacy. This inadequacy is a pervasive source of anxiety that arises because, as children, *they were never actually capable of successfully meeting their adult caregiver's emotional needs*—it was just a family myth in which they had been enscripted. These clients often suffer from strong performance anxieties and feel inadequate to meet the exaggerated demands they now place on themselves. As before, the therapist's task is to remain empathic toward both sides of their dilemma. On the one hand, these clients struggle both with feelings of inadequacy and anxiety over all of the things they must try to control but ultimately cannot (for example, their father's drinking or temper; their mother's unhappiness or loneliness). On the other, if they relinquish the demands they are struggling to meet, then clients surrender their family role and identity as someone who is special. In so doing, they dissolve the illusion of safety or secure attachment that this coping strategy has provided. Furthermore, to become healthier and relinquish these attempts to control or rescue caregivers also threatens to undo or betray their caregivers, who have, convincingly but covertly, communicated for decades that they cannot get along without them (for example, Parent: "I saw my doctor this week and he says he just doesn't like the way my heart is sounding...."). Thus, therapists can anticipate the paradox that making progress in treatment and doing better in their lives will evoke two intense conflicts for these clients. First, these clients will feel *guilt over leaving their attachment figures* to manage their own lives—as if they are selfishly abandoning their caregivers who cannot cope without them. Second, *anxiety is evoked over disrupting their lifelong coping strategy* of attempting to fashion more secure ties by meeting the emotional needs of their caregiver—and others as well.

What intervention guidelines does this type of case conceptualization suggest? The therapist does not want to overreact to such clients' compliments, criticisms, or distress. The therapist needs to be emotionally responsive, of course. However, recalling client response specificity, these clients will lose ground in treatment if the therapist's own personal equilibrium can be too readily influenced by their defensive style of hyperactivating. In the past, these clients were able to exert too much control over their parents' well-being, and—staying with our core concepts—the therapist does not want to reenact this unwanted pattern in his interpersonal process with the client. In other words, these clients feel safer and are reassured when the therapist is responsive but without becoming overreactive—even as they try to elicit overreactions from the therapist.

3. Emancipation conflicts. Many children caught in cross-generational alliances have problems with individuation and emancipation. If the mother and father have a primary marital coalition or if a single parent has supportive peer relationships, children are free to grow up and successfully launch their own independent lives. When young adults grow up with clear **intergenerational boundaries**, they can "leave home" successfully while still preserving a lifetime of close ties and shared involvement with their families. They do not have to emotionally disengage or physically break off contact with the family in order to have their own independent life or, to use a different strategy, engage in more risky or extreme behavior such as drunk driving or teen pregnancy to separate and "get away." This is possible because the caregivers do not need the child to remain dependent or centered on them in order to fulfill their own lives or manage their own problems.

In contrast, if the primary coalition is between parent and child, these offspring may feel **separation guilt** over leaving their parent alone or forsaking their parent to the unfulfilling relationship with the other spouse (Weiss, 1993). As a result, they are especially likely to feel dissatisfied with their achievements, to be unable to establish fulfilling relationships with others, or to be chronically depressed. Guilt over emancipation frequently underlies academic failure in college, as well as many other symptoms and problems that late-adolescent and young adult clients present in college counseling centers (Teyber, 1983; 1981). Thus, cross-generational alliances impose binding loyalty ties that make young adults feel guilty about leaving home, successfully pursuing their own career interests, and establishing satisfying love relationships. Although emancipation issues are expressed differently in varying cultures, symptomatic guilt and depression often ensue when offspring are not permitted culturally sanctioned avenues of individuation and couple formation. This issue is especially challenging for many first and second generation immigrants whose new family and cultural values may differ in significant ways from those of their adopted country (Varghese & Jenkins, 2009).

Many clients, like Anna in Chapter 4, struggle with binding separation guilt and survivor guilt. These clients, who often grew up with Preoccupied caregivers, may feel guilty about being happy, succeeding in life, or even getting better in therapy. It is important with all clients, but with these in particular, that *therapists enjoy clients' happiness and express pleasure in their success*. Some trainees are unsure of how to respond or what their role is when clients begin the session by saying happily, "I feel really good today; I don't have any problems to talk about right now." When that happens, clients with separation guilt are apt to assume, on the basis of their relational templates, that their happiness, success, or independence has somehow "hurt" the therapist. Other clients, with narcissistic or competitive caregivers, may be deeply concerned that the therapist feels envious of the happiness or success they have just communicated. The therapist disconfirms such clients' guilt or faulty expectations—and expands the healthier ways in which they can be connected to others—by welcoming the client's good feeling and enthusiastically responding.

THERAPIST: "That's great! What's the best thing going on for you this week?"

Without receiving such unambiguous affirmation of the successful aspects of their lives, however, these clients do not find a corrective emotional experience. Treatment will bog down and become repetitive as clients with these developmental experiences begin to retreat from their success and, in line with their old expectations, keep sharing an inability to activate or initiate; feelings of discouragement, uncertainty, or inefficacy; a lack of follow-through, or ability to complete undertakings; and other problems without any progress or resolution.

As we will explore in Chapter 10, conflicts over independence and success will be activated for many clients during the termination phase of treatment. In particular, clients who struggle with separation guilt will feel especially bad or "worried" about no longer needing the therapist and heading off successfully on their own. *Therapists are looking for multiple ways to give these clients permission not to have to need them, to leave when they are ready to go, and to be able to enjoy their own successful lives.* Guilt over growing up and becoming stronger, succeeding in work and love, or even surpassing the parent or the therapist with more happiness or success is a common treatment issue, although few clients will be able to recognize it as such on their own. Fortunately, it is relatively easy to help many clients dispel these binding guilt-related beliefs by communicating in words and behavior that the therapist enjoys their success, takes pleasure in their competence, and is in no way hurt or threatened by their stronger stance.

Finally, cultural factors play an especially important role in balancing these complex issues of individuation and family loyalty. Family connectedness and enduring family ties may be of special concern to clients from strong relational cultures—for example, Asian or Latino. Entering the subjective worldview of these clients, therapists can learn to appreciate the subtle balance between being loyal to family and culture yet still possessing their own authentic self. For example, one young Mexican-American woman, strongly identified with her family and her culture, spoke poignantly of the pressure she felt from her family to be "home more" and "participate in more family events." She also felt increasing pressure to help take care of her aging parents—these pressures coming just as she had been accepted into a prestigious doctoral program. She struggled with how to respectfully convey to her parents that she loved them, always wanted to remain close and connected, yet also wanted to pursue her passion of becoming a clinical psychologist. Even though they were proud of what she had accomplished as the first child in the family to go to college and get a degree, they struggled with the idea of her leaving home to further pursue her studies. On her part, she struggled with not "buying" the accusation from some family members (her older brother and several aunts and uncles) that her "priority" should be staying home and taking care of her parents. However, with help from her therapist and from some supportive family members, she was able to see the value of being able to develop and embrace a meaningful personal identity and follow her dream while still sustaining a respectful, loving, and honorable stance toward her parents and family. One aunt, in particular, helped "bridge" this with her parents by explaining how proud she was making all the family. This young woman was realistic in anticipating that balancing her needs and theirs was something that would need to be revisited and negotiated repeatedly over the years, but she felt security in being more connected to her own needs and wishes and was able to pursue her own goals.

To become familiar with the culturally sanctioned avenues for individuation that exist, therapists will find it helpful to consult with others who are knowledgeable about the client's cultural context. Similarly, it is very important to ask clients how they self-identify, especially since variability exists within groups and in ways that individual members relate to their families and identify with their cultures (Cardemil & Battle, 2003). With such consultation, therapists are then able to help clients find ways to individuate that are acceptable within their own cultural context. Although these individuation and emancipation issues will be expressed differently in each familial and cultural context, therapists who are open to these differences can help clients find acceptable avenues and role models within their culture for pursuing their own goals and making new commitments.

4. Parentification of children. When the primary coalition is between a parent and a child rather than between the mother and father, the result is frequently the **parentification** of children. A role reversal occurs: Rather than the parent responding to the child's needs, the child takes on the role of meeting the parent's emotional needs. That is, when a parent's emotional needs are not met by his or her spouse or other (same-generation) adults, the parent inappropriately turns to one or more of the children to meet his or her own adult needs for affection and intimacy, for approval and reassurance, or for stability and control. Therapists should be alerted to the possibility of such parentification when they hear adults describe their children as their "best friend," "lifeline," or "confidant" (Teyber, 2001).

It is problematic for children when they become responsible for, or take care of, the emotional needs of their parent. The problem with this role reversal is that such children must give to their parents rather than receive, and their own age-appropriate dependency needs go unmet. These parentified children grow up to feel overly responsible for others, insecure about depending on others, and guilty about having their own needs met. As adults, parentified offspring often describe themselves as feeling empty or "having a hole inside" as a result of having given rather than received throughout their childhoods. Parentified children initially enjoy having such a special role vis-à-vis their parent. As adults, however, they often come to resent having been deprived of their own childhood. They also become angry and anxious in current adult relationships because (1) they do not trust that others will be there for them in times of distress, and (2) they are concerned about losing relationships when they cannot meet the needs of the other.

Parentification is pervasive in the background of clients and therapists alike (DiCaccavo, 2006). Many clients seeking outpatient treatment have been parentified, especially by Preoccupied caregivers. Because their basic relational orientation is to take care of others, parentified offspring select careers such as nursing and therapy that extend this caretaking role. Although kind and capable, they often feel guilty about saying no, setting limits, and meeting their own needs. Because they do not draw boundaries well, they tend to become overidentified with others' problems (for example, student therapist describing her new client to her supervisor: "Her divorce was *exactly* like mine!") and are prone to experience burnout. Because they grew up having to take care of their parent, it is now threatening to relinquish this exaggerated sense of responsibility and need for control in their current lives. For example, it may be hard for them to let others share in meeting obligations, even though simultaneously they are resenting having to handle everything themselves. These control issues may also be evidenced in symptoms such as airplane phobias, for example, where these individuals must temporarily relinquish control to the pilot. These clients also have problems in close personal relationships because it is too anxiety-arousing to relinquish the control necessary to be emotionally intimate with someone.

Note, however, that in some families—for economic, single-parent, or other reasons—one child may temporarily assume a parental role with younger siblings. This is not the same as parentification because it remains clear that the parent is in charge when the parent is home. This child's role—to assist in the functioning of the home in the parent's absence—is temporary or situational. Furthermore, the child is caring for siblings, not for the emotional needs of the parent. This responsible and helpful child, who still remains a child and is nurtured and given to without the role reversal that occurs for parentified children, often grows up to be an especially resilient and capable person with significant personal strengths.

Parentified clients will be highly sensitive and responsive to the therapist—as they once were to their caregiver. It can feel great to work with these clients in the short run, because they astutely discern what they need to be or do so that the new therapist will feel competent or secure (for example, the client talks extensively about the positive changes occurring in her life, and even exaggerates them, so that her student therapist's supervisor, watching on videotape, will be approving of the therapist!). However, if the therapist does not collude in reenacting this relational pattern, clients can begin to explore the far-reaching consequences of being parentified and having missed their own childhoods. Simultaneously, it will be relieving for these clients— yet anxiety arousing at times—when the therapist begins this exploration by making process comments that highlight when this relational pattern may be occurring in the therapist–client relationship as well:

THERAPIST: I really appreciate the genuine concern you just expressed for me right there—I know it's caring and genuine. But thinking about the caretaking patterns we have been talking about, and how much that role has been costing you in your life, it leaves me wondering what happens to *your* needs here if you get oriented to what I may be needing or wanting? What do you think—does my concern make any sense?

To further clarify parentification, let's consider a case study. Carol was a highly regarded psychiatric nurse. An utterly dependable and take-charge person, she could seemingly handle every situation that arose. In a hospital emergency setting where she had to deal with seriously disordered patients in crisis, her rapid and accurate assessments, good judgment, and compassion for highly disorganized patients had earned her the respect of the entire staff. Although considered a superstar at work, Carol sought therapy for her recurrent depression and "burnout."

Carol was a bright and engaging client who was adept at getting the therapist to lead, talk about herself, and become involved in collegial discussions about interesting clinical issues. However, the therapist had formulated working hypotheses about this potential reenactment and was usually effective in recognizing Carol's strong "pulls" to repeat this pattern with her. Repeatedly, she looked for opportunities to focus Carol inward by saying, for example:

THERAPIST: Help me understand what was going on for you when...

or

THERAPIST: What were you afraid was going to happen if you said no and... ?

In response to these repeated invitations to consider what she was thinking, feeling, or wanting, instead of being preoccupied with what the therapist might be wanting,

Carol began expressing that it was new for her to pay attention to herself and her inner experience in this way. And, although she enjoyed this very much, she also found that it made her feel "uncomfortable." The therapist did not press for this internalizing focus when Carol did not want it, but instead asked Carol on occasion how it was for her to look within at her own experience. It soon became clear to both of them that guilt over being "selfish" and shame over being "the center of attention" were evoked by this internal focus and paying attention to herself. Thanks to the therapist's affirming responses to these reactions (for example, "Oh no, I don't think you're being selfish here at all..."), both soon agreed that important new material was emerging.

Four weeks into treatment, Carol disclosed something that she had never told anyone: Her alcoholic stepfather had sexually assaulted her in her early adolescence. Carol successfully fought him off, although she was scratched, had her blouse torn open, and suffered a bloody nose when she literally pushed him off her. Carol recounted her ordeal in detail, but without any emotion, and the therapist responded effectively in caring and validating ways. The therapist, herself a mother of two daughters, soon asked what was for her the burning question.

THERAPIST: You have kept this painful secret for almost 20 years. I'm glad that you can share it with me now, but it's heartbreaking that you've had to be alone with this for so long. What kept you from telling your mother and asking her for help?

CAROL: I didn't want to put more on her—her life was kind of overwhelming already. She couldn't have done much anyway, and I didn't want her to worry.

Carol's poignant response illustrates the plight of the seriously parentified child. As in most aspects of her relationship with her mother, Carol's own profound needs for protection and comfort were set aside in order to meet her parent's need.

To sum up, there are many ways in which children function better in families that have a primary marital coalition with clear boundaries between parent and children than in families with cross-generational alliances. Therapists will see the problematic consequences of these structural family relationships and blurred intergenerational boundaries operating with many of their clients. This is one reason why effective therapeutic relationships always honor treatment parameters and provide a corrective experience by maintaining clear boundaries between the therapist and the client. One of the best ways for new therapists to utilize supervision is to have their supervisors actively help them monitor this process dimension of clear therapeutic boundaries with their clients.

BALANCING THE CONTINUUM OF SEPARATENESS—RELATEDNESS

Another basic dimension of family life that helps therapists understand their clients' problems is the **separateness–relatedness dialectic.** An awkward term but a far-reaching concept drawn from the family systems literature, it was a forerunner of the consonant constructs of safe haven/secure base in the attachment literature. The separateness–relatedness dialectic refers, in part, to *the family's ability to respond to the child's simultaneous need for both relatedness (that is, secure attachments and family cohesiveness) and separateness (identity, autonomy, and personal* *authenticity).* At times, the separateness side of this dialectic has been misunderstood. In this context, separateness does not imply cutoff, selfish, or uninvolved individuals who are insensitive to the needs or feelings of others. Rather, separateness (also referred to as differentiation or individuation in other theoretical contexts) connotes being connected to our genuine Self—our own voice, genuine feelings, needs, and wishes. It is also the ability to make choices based on the accurate self-awareness of our own inner experience within our own relational and cultural contexts. Thus, the autonomy side of this dialectic connotes the ability to recognize and act on our own interests and goals—responding purposively, and choosing, instead of merely complying and going along with what others expect us to do.

As children develop, each family faces the complex task of having to provide both secure attachments (a safe haven with a sense of belonging and shared family loyalty), on the one hand, and support for **differentiation** (autonomy and personal identity), on the other hand. In other words, children need to feel close, protected, and that they belong (relatedness or safe haven) while still remaining a separate person "with her own mind" and her own ideas (separateness or secure base). Both of these behavioral systems—of attachment and of differentiation—are present at every stage of the life span and are interrelated. In this regard, attachment theorists believe that having a secure base as a child (or an internal, mental representation of a secure base later as an adult) is necessary to have the capacity to venture out confidently to explore and function independently in the environment Weinfeld et al., 2008).

Although there will be varying expressions and timetables within different cultural contexts, the attachment system holds more importance during childhood than adolescence. Toddlers, for example, look for a parent or other important caregiver for reassurance when they find themselves in new situations or around strangers. If the parent is in close proximity during these times, children gain the confidence needed to explore the environment and begin to take steps away from the parent, trusting the parent will be available if something goes wrong. When a parent or other attachment figure is available to provide physical and/or emotional support, the attachment system provides a secure base from which infants and toddlers are free to engage in other activities, such as exploration and play. While the attachment system remains primary in young children, the exploration or differentiation system becomes primary as children mature and become adolescents (Cassidy & Shaver, 2008; Colin, 1996).

As children's cognitive skills develop, they are able to form mental representations of attachment figures that allow them to interact independently in new environments, such as school and the neighborhood, without requiring a parent's physical presence in order to feel secure or attached. If a parent has provided a good enough secure base during early childhood, the school-age child is better able to meet the developmental crisis of "industry versus inferiority" (Erikson, 1968) by mastering a number of skills and engaging in cooperative relationships with peers. This, in turn, builds a foundation for forming a strong sense of personal identity during adolescence. Although older children and adolescents will continue to seek a safe haven during moments of distress, they are now more secure and able to move into larger circles of the neighborhood as the "world" takes on increasing importance. Teens thus take the initial steps toward emancipation and will eventually leave home, yet they continue to need some reassurance that their parents will support these steps toward greater autonomy and remain emotionally available to them when needed (that is, provide a secure base). *Maintaining a proper balance between attachment and differentiation, one that honors the changing developmental needs of children and adolescents, requires great flexibility on the part of parents.* Therefore, healthy families are characterized by **experiential ranging** between these two poles of intimacy and individuation (Farley, 1979). Effective caregivers have the flexibility to range back and forth along this separateness–relatedness continuum and meet children's changing needs both for closeness and for autonomy throughout their formative years (Thompson, 2008). Easy to say, yet, even for committed and skillful parents, so challenging to do.

In contrast, less effective families do not have the interpersonal range to function well at both ends of the continuum. These less effective parents cannot meet the changing needs that children present at different developmental stages. As we will explore below, **enmeshed** or overinvolved families cannot provide a secure base and support older children's growing independence and development toward greater differentiation, whereas **disengaged** or emotionally cutoff families cannot provide a safe haven for young children and meet their basic attachment needs within a cohesive family context.

This conceptualization provides a useful model for therapists, who similarly need a broad interpersonal range to respond flexibly to their clients' needs. This wide interpersonal range is a critical component of client response specificity. Depending on their developmental history, some clients will need help with issues related to separateness (becoming more assertive and independent; actively exploring different values and *choosing* for themselves their own career paths, lifestyles, spiritual convictions, and political views). In contrast, other clients will present with symptoms and problems in relatedness (issues related to safety, trust, or commitment in close personal relationships). Furthermore, these needs will change during the course of treatment. For example, as clients receive empathic understanding and validation from the therapist for the hurt or vulnerable aspects of themselves, they often begin to improve in treatment. As this leads to improved functioning and they develop an increased sense of self-efficacy or personal power, however, clients often need a different kind of response from the therapist—unambiguous support for this new, stronger self that was not encouraged in other relationships. Keeping in mind this flexibility or wide interpersonal range for both the therapist and the client, let's see how the separateness–relatedness continuum relates to the parental coalition. The following example shows how families without a primary marital coalition tend to be more restricted or less flexible in their experiential range.

Suppose that both members of a young married couple have not made much progress with Erikson's developmental tasks of Identity in late adolescence and Intimacy in early adulthood (Solomon, 1973). As a result, they have not been able to separate psychologically from their families of origin sufficiently—they cannot shift their primary loyalty ties from their parents to their partners and establish a successful marital coalition with each other. Such newlyweds are likely to have a high degree of marital conflict that they are both unable to resolve. In many cases, the couple will attempt to deal with their ongoing relationship problems by having children. As Bowen (1978), Beavers et al. (1993), and other multigenerational family therapists illuminate, these "less differentiated" parents—who have not been successful in completing previous developmental tasks—are more apt to experience the children as extensions of themselves. Without being aware of this overidentification, they tend to project their own unresolved developmental conflicts and marital problems onto one or more of their children, in an attempt to distance and externalize those issues. Additionally, these parents may select a certain child who is similar in some way to them—perhaps an oldest or youngest child who matches their own sibling/birth order, or a boy or a girl like themselves—to carry or express certain unmet needs, unacceptable feelings, or unwanted aspects of themselves. Through this **family projection process**, described by Murray Bowen (1978), Virginia Satir (1967), Satir & Bitter (2000), Ivan Boszormenyi-Nagy et al. (1991), and other multigenerational family therapists, children are scripted into **family roles** (for example, the "good" or the "bad" child, the hero, rescuer, caretaker, or invisible/lost child), restrictive **family rules** (for example, no one in this family is ever sad, the oldest female child cannot leave home and emancipate), and **family myths** (for example, Mom is very happy, Dad doesn't have a drinking problem). Such common family roles, rules, and myths are established, in part, to help parents defend against their own unresolved problems that began in their childhoods and now are being amplified by the stressful life cycle transitions to marriage and child rearing.

To illustrate, if the mother's need for understanding and support cannot be met by the father, she may turn to a particular child to have her emotional needs met. This solution may work well enough in the short run, yet it establishes the basis for long-term family problems. In many cases, the mother could not emancipate psychologically from her own parents, and she is made anxious and/or guilty by normative or healthy steps toward increasing individuation. As long as the child remains dependent on her, her need for closeness is met and her anxiety or guilt over individuation is managed. On the other hand, the father in this prototypical scenario is uncomfortable with closeness; it has held real threats in the past and now makes him anxious. The primary mother–child coalition allows him to gain greater distance from the anxiety-arousing demands of closeness from both the spouse and the child. In this way, the child's role is to hold the marriage together and, at the same time, to ensure distance between the couple.

This situation eventually poses a dilemma for the child's continuing development, however. The child's innate push for growth and increasing autonomy will destabilize the marital relationship, and this will arouse each parent's own internal conflicts. Thus, these parents need to undermine the child's increasing independence or competence in order to maintain the marital status quo. At the same time, the parents may express anger and frustration to the child about the child's refusal to grow up and act responsibly. This family's limited experiential range is most likely to come to a head during adolescence, when offspring face the developmental hurdle of "emancipation," or beginning to launch from the family of origin. The mixed messages that the child has received about growing up and remaining dependent will often erupt in psychological symptoms and acting-out behavior at this time. Furthermore, these underlying issues have been overlearned as maladaptive relational patterns and often continue to be expressed in symptoms and problems with others throughout adulthood.

The separateness–relatedness dialectic is a useful construct to help therapists conceptualize basic familial and developmental processes. It also helps with some of the most common problems that new therapists face in their clinical training. As in this parenting model, therapists need to be *flexible* enough to respond to clients at both ends of the separateness–relatedness dialectic. Regarding the separateness side, for example, most beginning counselors find it far easier to be understanding and empathic than to challenge or disagree with their clients. Understandably, most therapists are reluctant to "confront" clients—it often connotes the unwanted hard edge of "exposing" clients, embarrassing them, or making them mad. Most people interested in clinical work would simply loathe to be called "pushy," and most new therapists— responsible and caring people—are worried about doing anything that might hurt their clients' feelings. However, without being confrontational or hostile in any way, it is often necessary to challenge clients by giving them honest feedback about the problematic ways they are affecting others or inviting them to take more personal responsibility by looking at their own participation in problems. Therapists are not responding effectively at the "separateness" end of this dialectic when they cannot do the following:

* challenge clients at times or disagree with them and see things differently,
* address interpersonal conflict directly and talk through the inevitable misunderstandings or ruptures that come up between them,
* encourage clients to look more realistically at their own contribution to problems with others, or
* set firm limits with acting-out clients and maintain clear therapeutic boundaries.

Therapists are encouraged to reflect on their own countertransference propensities in these ways and strive to expand their own experiential range during their clinical training. For example, with practice and rehearsal, new therapists can learn respectful ways to challenge clients, provide unwanted but necessary interpersonal feedback, and set limits:

THERAPIST: It's OK to be angry with me, and I welcome talking about that, however, cursing and swearing at me like that is not. I'd like to hear about your anger, without the swearing.

OR

THERAPIST: Tiffany, I'm not able to change my appointment time. It is already 5 o'clock and I have another commitment at 6. Let's talk for a minute about next week's appointment so we can be sure we can find an hour that works for both of us.

OR

THERAPIST: Gerald, this is the second session you've come to treatment intoxicated. Last session you said it was an unusual event, a retirement party at work. Let's talk together about what's happening.

In parallel, many therapists in training also struggle with issues on the relatedness side of this dialectic. In particular, new therapists often express concerns that they are "too sensitive" to their clients' feelings. For example, many caring and capable practicum students worry about overidentifying with their clients' feelings or problems:

* What if I feel my client's sadness too much, you know, and just start to cry myself?

- What if I take on the client's feelings and feel them too much—you know, be so affected that I can't be objective about things. Maybe I'm just too sensitive to be a good therapist?
- What if I can't stop thinking about their problems when I leave the session and go home? Some of these things really bother me.
- Her life is just like mine! Maybe I've got too many problems of my own to be a good therapist.

Therapists' goal is to find a new middle ground for themselves along this separateness–relatedness dialectic. To help with this, recall Sullivan's (1968) concept of the **participant-observer**. The therapist's intention is to be empathic and resonate with the client's feelings yet, at the same time, also remain separate by being clear that these are the client's feelings—not our own—that we are trying to respond to and understand. Learning to become a "participant" in the relationship—to listen with presence and empathically engage with the client, while remaining an "observer" who is simultaneously thinking about what things mean and how best to respond—may be the most challenging but important task in clinical training; we will continue to explore it in the chapters ahead.

The family dynamics presented in this chapter offer a parallel to the relationship between the therapist and the client. New and experienced therapists alike are continually seeking to develop their own capacities to respond at both ends of the separateness–relatedness continuum. Effective therapists, like Authoritative and Secure parents, have a broad experiential range that allows them to be emotionally attuned and responsive to clients, yet also able to set limits with clients, question their actions or challenge them, and tolerate clients' anger or disapproval at times. More specifically, therapists need to be able to respond to the emotional deprivation of clients who have been reared by highly Authoritarian or Disengaged parents, or caregivers with a Dismissive attachment style. At the same time, they also want to be able to set limits with the provocative, demanding, or testing behavior of clients who have been reared by Permissive parents or with other types of inconsistent parenting. Thus, the personal challenge for all therapists is to examine their own experiential range, which has been shaped in their own families of origin. Effective therapists are nondefensive and willing to explore the interpersonal spheres that are anxiety arousing for them, and work to broaden their own interpersonal range so that they can respond to the wide range of problems that clients present. To achieve this flexibility, therapists are encouraged to do their own **family-of-origin work**.

Family-of-origin work is an integral part of training for those who specialize in family therapy. In particular, Bowen (1978) has written extensively about how clinical trainees can systematically study, and sometimes change, problematic ways of interacting in their own families of origin. *By achieving greater understanding of their own family backgrounds, and perhaps greater empathy for all family members, therapists can make a significant contribution to their own clinical skills.* Readers are encouraged to read and embark on the great adventure of doing family-of-origin work (Boszormenyi-Nagy & Spark, 1973; Kerr & Bowen, 1988; Minuchin & Nichols, 1998). In particular, therapists can prepare a family genogram—a three-generational map of family rules, roles, myths, and structural family relationships

(McGoldrick et al., 2008). By learning about their families of origin in this way, therapists learn much about their own countertransference propensities, become more compassionate toward their clients' struggles to change, and better appreciate the profound influence of familial experience on adult life.

CLOSING

In this chapter, we have examined four basic dimensions of family functioning: attachment styles in adulthood, child rearing practices, structural family relations and the parental coalition, and the separateness–relatedness dialectic. These characteristics of family life will help therapists better understand many of the problems that clients present. For some trainees, it is disruptive to read this material because it violates familial rules and cultural prescriptions to question or examine what transpired in their families of origin. Also, many readers are parents or grandparents themselves. As they learn more about family functioning, and parenting practices in particular, guilt may be evoked as they recognize the limits of their own parenting skills, and the problematic consequences for their own children and loved ones that have resulted. Such readers may need to relinquish unrealistic expectations of themselves and forgive themselves for not knowing more about child rearing than they did at the time.

Suggested Readings

1. Clients make more progress in treatment when therapists can help them come to terms with the good news and the bad news in formative relationships with caretakers. However, it is especially challenging for both therapists and clients to do this when abuse has occurred. To help with this always complex issue, readers are encouraged to read "Ambivalent Feelings: Case Study of Sheila," in Chapter 6 of the Student Workbook.
2. "The Circle of Security Intervention" (Cooper et al., 2005), in L. Berlin et al. (Eds.), *Enhancing Early Attachments,* is profound reading. It describes an attachment-oriented intervention project that teaches at-risk parents how to provide children with comfort, protection, and security, while also supporting their exploration so healthy development is facilitated. Parents interact with their young children and then are taught how to "read" and respond to their children's cues in appropriate, attachment-facilitating ways. The parallels to therapist–client relationships are illuminating.
3. Therapists interested in learning more about working with the many Preoccupied and Dismissive clients they will be seeing in treatment, and in preparing case formulations and treatment plans from this relational/attachment perspective, are encouraged to read Daly & Mallincrockt (2009), "Experienced Therapists' Approach to Psychotherapy for Adults with Attachment Avoidance or Attachment Anxiety," *Journal of Counseling Psychology* 56), 549–63.
   This outstanding article illuminates the differing ways to conceptualize and approach working with clients who present with high attachment avoidance or attachment anxiety. The authors highlight the impact of each attachment style

on the therapist–client relationship and on the progression and outcome of treatment.

4. Basic information on child development and child psychopathology will help therapists understand their adult clients' personalities and problems. One outstanding text is *Psychopathology from Infancy through Adolescence: A Developmental Approach,* by Wenar and Kerig (2006); see especially Chapters 4–8.

5. Parentification is an essential concept to understand because so many clients and therapists have been scripted into this familial role. Further information about this important concept can be found in Chapter 9, *Helping Children Cope with Divorce,* in Teyber (2001).

6. Structural family relations provide an illuminating road map for understanding family functioning. A classic work in this area is Salvadore Minuchin's (1974) *Families and Family Therapy;* see especially Chapters 3 and 5.

7. Monica McGoldrick et al.'s (2008) textbook *Genograms: Assessment and Intervention,* provides a clear "how-to" approach for those interested in doing their own family-of-origin work. This engaging book provides illustrative genograms of multigenerational processes in notable families, such as those of Sigmund Freud, Bill Clinton, Woody Allen, and others. New therapists can learn much about themselves, and the countertransference propensities they will bring to their clinical work from doing this important, personal exploration with genograms of their own families of origin.

8. Cassidy, J. & Shaver, P. R. (Eds.) (2008) *Handbook of Attachment: Theory, Research and Clinical Applications* (2nd edition). New York: Guilford Press. This handbook summarizes and provides extremely useful information on attachment-informed therapy. Also, an outstanding collection of chapters addressing attachment and its clinical applications is found in Obegi, J. & Berant, E. (Eds.) (2009). *Attachment Theory and Research in Clinical Work with Adults.* New York: Guilford Press.

7 | INFLEXIBLE INTERPERSONAL COPING STRATEGIES

CONCEPTUAL OVERVIEW

In this chapter, an interpersonal model for conceptualizing clients' clinical presentation is delineated. The model highlights the role of clients' core conflicts and their *interpersonal strategies for coping with it*. Core conflicts are the central problems or key issues that pervade the client's life. The core conflict is the hub of the wheel that links together the different problems and concerns that clients present, and provides a focus for treatment.

The core conflict (also called the *generic conflict*) has arisen out of the client's repetitive interactions with significant others, especially parents or primary caregivers. Following the developmental issues presented in Chapter 6, this core conflict often arises from the combined effects of insecure attachments, problematic child-rearing practices, and faulty structural family relations. When these types of problematic interactions with caregivers were repetitive or ongoing, they give rise to pathogenic beliefs about oneself, faulty expectations of others, and a narrow or skewed view of what will usually occur in close relationships. To cope with the painful feelings, low self-esteem, and interpersonal threats that accompany these developmental experiences, *clients often develop a repetitive or fixed interpersonal style*. For example, the client is usually pleasing others or trying to be perfect, repeatedly taking charge or being in control, *routinely* taking care of others and meeting their needs, consistently remaining invisible or avoiding all conflict, and so forth. The key here is that this interpersonal coping strategy is not just used sometimes or flexibly employed in situations when it is actually needed. Instead, it is pervasively and rigidly overused—even in situations where it is not necessary or no longer adaptive. (See Millon & Grossman, 2007; and Zeidner & Endler, 1996, for a comprehensive overview of interpersonal coping styles and their implications for adaptive and maladaptive behavior.)

Defenses such as these interpersonal coping strategies are necessary for all people at times—they help us cope in difficult situations. Clients will feel understood and affirmed when their counselors can grasp how one of these interpersonal coping styles really was a necessary and adaptive way to cope at one point in time. That is, recalling the concept of "honoring the client's resistance," therapists can validate that these rigid or inflexible interpersonal coping styles once were an adaptive strength feature in the client's personality—they helped the client cope more effectively with reality-based problems. The problem for most clients, however, is *that these coping strategies are no longer necessary or adaptive in most current relationships.* In fact, rather than continuing to help, they now contribute to many of the symptoms and problems that clients present in treatment (Baumeister & Scher, 1988; Pezzarossa et al., 2002; and Wei & Ku, 2007). Thus, the purpose of this chapter is to clarify this as an overriding treatment goal. On the one hand, the counselor aims to help clients discriminate or sort through current relationships where they still need to respond in this pleasing or avoidant way (for example, with an authoritarian, dominating boss). And, on the other hand, discern other relationships (such as with a new friend) where the old coping style (continuing to take care of everything or always be in control) is no longer needed and only serves to create new problems.

To reach this treatment goal, the therapist's intention is to:

- identify the formative patterns and painful feelings that make up the client's core conflict,
- highlight the interpersonal coping strategy the client has used to cope with these developmental challenges,
- clarify how the client's interpersonal coping strategy is being expressed in current interactions with the therapist and others, and
- begin changing this pattern in the relationship with the therapist and others in the client's life.

As new therapists become more successful at formulating the client's core conflict and interpersonal adaptation to it, this conceptualization will help counselors recognize themes and find organizing patterns in the complex material clients present. As noted in Chapter 2, client conceptualization is an ongoing process that begins with the initial client contact. General hypotheses about clients' problems are formulated early in treatment, and these tentative working hypotheses are further refined or discarded as the therapist learns more about each client. However, as the client's conflicted emotions emerge, the key concerns that are central to the client's problems become clear. Having focused the client inward, the therapist can use the client's emotions that begin to emerge as guideposts to clarify this central conflict, and focus on how it is being played out with the therapist and others.

INTERPERSONAL MODEL FOR CONCEPTUALIZING CLIENTS

In this section, we explore an interpersonal model for conceptualizing the client's core conflict—how it originally developed, the client's interpersonal adaptation for coping with it, and how it is being expressed in current symptoms and problems. This model for conceptualizing the client's problems includes five components:

- The client's unmet developmental needs
- The original environmental block that engendered the problems
- The client's intrapsychic defenses against her own core conflict
- The client's interpersonal strategy to rise above or overcome the core conflict
- An interpersonal resolution of the core conflict

Adapted from Horney's (1970) interpersonal theory, these five components provide a useful model for conceptualizing client problems (see Figure 7.1). This model is an integrated system composed of several complex psychological processes. After discussing each component of this model, we will apply it to treatment by illustrating the model with three case studies. Let's begin by examining how enduring problems result when one or more of a child's basic developmental needs go unmet. Be patient—this is a lot of reading to digest before it is illustrated with case studies.

UNMET DEVELOPMENTAL NEEDS

This conceptual model begins with children's basic needs for secure ties. Figure 7.1 (this diagram starts at the bottom of the page and flows upward) shows children's developmental trajectory and highlights young children's emotional needs, in particular, children's need for a secure attachment. If parenting figures are consistently responsive to the children's bids for attention and affection, children will be able to freely express this need. Of course, as children develop, they benefit from learning to accept limitations on parents' ability to respond and to tolerate delays in parental response. If parents are consistently punitive or unresponsive, however, anxiety will soon become associated with the child's needs. Children are then forced to fashion a solution; the **compromise solution** they construct often alienates them from themselves and their authentic experience. Although many have written about this, none have captured it more succinctly than Carl Rogers (1951) did years ago in writing about **congruence.** Consider preschool children whose need for understanding and affection goes unmet from disengaged, authoritarian, or even permissive parents. When such children approach their caregiver for comfort when distressed, their need is often rebuffed:

DISENGAGED: What do you want now? You always need something—get out of here!

AUTHORITARIAN: What's wrong with you? Stop that crying right now, or I'll give you something to cry about!

PERMISSIVE: Just go ahead and do whatever you need, honey. I'm kind of busy now.

If these types of interchanges occur repeatedly, children will soon learn to anticipate rejection and feel anxious whenever emotional needs for comfort or soothing are aroused. Children are biologically organized to continue experiencing attachment needs, but their direct expression is blocked by the parents' unresponsiveness (see Part 1 at the bottom of Figure 7.1). The attachment-seeking child is then compelled to find some way to cope with this painful set of circumstances. Although clients will present with a wide range of problems that result from the varied developmental

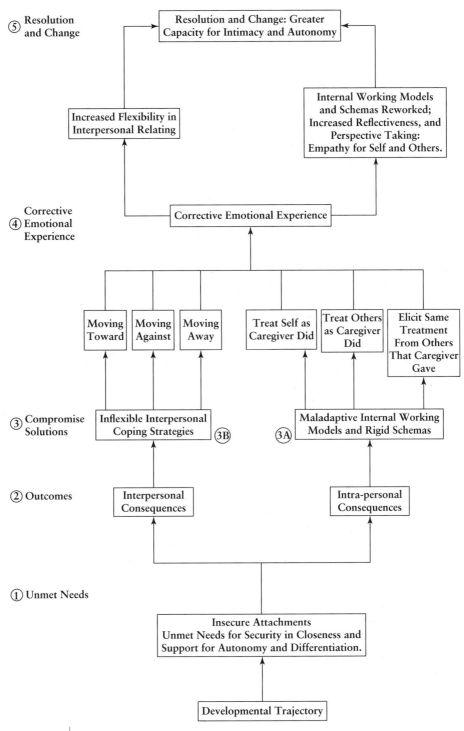

⑤ Resolution and Change

Resolution and Change: Greater Capacity for Intimacy and Autonomy

Increased Flexibility in Interpersonal Relating

Internal Working Models and Schemas Reworked; Increased Reflectiveness, and Perspective Taking: Empathy for Self and Others.

④ Corrective Emotional Experience

Corrective Emotional Experience

Moving Toward

Moving Against

Moving Away

Treat Self as Caregiver Did

Treat Others as Caregiver Did

Elicit Same Treatment From Others That Caregiver Gave

③ Compromise Solutions

Inflexible Interpersonal Coping Strategies   ③B

③A   Maladaptive Internal Working Models and Rigid Schemas

② Outcomes

Interpersonal Consequences

Intra-personal Consequences

① Unmet Needs

Insecure Attachments Unmet Needs for Security in Closeness and Support for Autonomy and Differentiation.

Developmental Trajectory

FIGURE 7.1   INTERPERSONAL MODEL OF THE CHANGE PROCESS

challenges presented in Chapter 6, enduring psychological problems often begin when basic childhood needs for secure attachments (both safe haven and secure base) are not met. Most core conflicts also involve a failure to provide children with clear communication and consistent emotional access, on the one hand, and support for autonomy and differentiation on the other. Children who grow up in a predictable environment with both warmth and consistent limits have the best developmental trajectory (Kobak et al., 2006; Sroufe et al., 2005). While children vary to some extent in their developmental needs, experience different kinds of family environments and developmental problems, and develop their own individualized coping strategies, most children experience anxiety when their emotional needs are blocked. Later, as adults, they also feel threatened when similar needs or feelings are activated in current relationships (for example, if the client has a problem or emotional need and has to ask others for help). To cope with this lack of safety, many clients begin to construct a **"compromise solution"** designed to manage the continuing anxiety evoked by these unmet needs and the unwanted responses they expect to receive from others now. A large body of empirical research finds that rigid self-defeating coping strategies that provide short-term benefits but long-term adjustment costs have been learned in problematic parent-child and insecure attachment relationships (Rubino et al., 2004; Schill & Williams, 1993; Wei & Ku, 2007).

Parts 2 and 3 of our model illustrate how both sides of this compromise solution works. On the one hand, clients turn against themselves to block or minimize the feeling or need that has become anxiety arousing (for example, "I hate myself. I have such a big mouth—I'm too much for everybody"). Simultaneously, clients implement different coping strategies to "rise above" the need and indirectly gratify it (for example, "If I just get a little thinner, then he won't leave me"). Part 4 will refer to the corrective emotional experiences that therapists provide to help clients resolve their core conflicts and the compromise solutions they've developed to cope with their early life experiences.

We will now explore both sides of the client's compromise solutions and learn about the complex psychological maneuvers that individuals adopt to cope with these core conflicts. In particular, we will see how clients' inflexible coping styles are replayed in the current relationship with the therapist, just as they are with others in their everyday lives.

COMPROMISE SOLUTIONS: INTRAPERSONAL AND INTERPERSONAL OUTCOMES

In Chapter 2, we proposed three domains for conceptualizing clients' problems: (1) maladaptive relational patterns, (2) rigid schemas with faulty beliefs about self and unrealistic expectations of others, and (3) core conflicted feelings. All three of these domains are intertwined, but the environmental block in Part 1 of Figure 7.1 principally evokes the core conflicted affect. As we will see, the faulty beliefs that arise from these core conflicts predominate in Part 3A, and the interpersonal strategies used to cope with them—which are the primary focus of this chapter—are highlighted in Part 3B.

Clients employ two psychological mechanisms in order to defend against the anxiety associated with expressing, or even experiencing, their blocked needs.

First, we will examine ways in which clients defend against their core conflicts by responding to themselves in the same way that others originally responded to them. These can be thought of as *cognitive schemas* that clients develop—that is, beliefs about themselves and others that, once adopted, regulate their behavior. Second, we will examine ways in which clients adopt a general *interpersonal coping style* to try to rise above their personal or emotional needs. In treatment, this style will often quickly become apparent in the way the client interacts with the therapist (for example, frequently pleasing, controlling, or withdrawing). Taken together, these two mechanisms of *blocking* and *rising above,* which represent attempts to minimize anxiety and provide a modicum of self-esteem, are clients' compromise solutions to their core conflict. Let's explore how this complex defensive system works.

INTRAPERSONAL OUTCOME OF THE UNMET NEED: BLOCKING THE UNMET NEED

With the limited skills and means available to them, insecurely attached children try urgently to get their caregivers to respond to their emotional needs but are unable to elicit the help they need. For example, this may reflect a general lack of parental responsiveness, parents responding on their own timetable rather than when the child initiates contact, a punitive response, inconsistent response, and so forth. To gain some active mastery over their helplessness and to ward off the anxiety aroused, *such children begin to block their own need in the same way that the caregivers in their social environment originally blocked it.* For example, children may deny their need or distress as insignificant and regard themselves as unimportant (for example, "I'm stupid"; "I don't matter") or may reject it and view themselves contemptuously as demanding ("needy"). Routinely, counselors will hear clients say the same hurtful responses toward themselves that others originally expressed toward them ("What's wrong with you!"). In turning against themselves, children are able to ward off the unwanted anxiety that would be engendered if they felt that their developmental needs were reasonable and legitimate but still viewed as unacceptable to those they need and love most—their caregivers. In addition to preserving attachment ties, this adaptation provides a sense of control and allows them to continue to view the world and their parents as reasonable and just. In this way, a certain kind of internal security or safety comes from viewing themselves as the problem, rather than the caregiver. This then becomes their schema for viewing themselves and others; by acting accordingly, they elicit responses consistent with the schema. For example, when they ask for something (a pay raise, a day off from work, help on a project) they do it in a way that conveys that they don't really expect that their request will be granted ("I don't matter"; "I don't expect to be given to"). Consequently, more often than not, their requests go unmet. This adaptation where children begin to do to themselves what was originally done to them has been referred to as identification or modeling.

Disruptive Internal Working Models and Rigid Schemas. In order to "overcome" or manage the anxiety associated with their unmet needs both for security and for support for autonomy and differentiation, clients often develop three disruptive internal working models and rigid schemas that cause problems for them.

1. First, clients **block** their own need internally and respond to themselves in the same hurtful ways that others have responded to them. That is, when the children of nonresponsive parents feel a need for reassurance or affection, they will block their need by feeling critical or contemptuous of themselves and their own need. As adults, their own emotional needs are still unacceptable, and they will often say to themselves the same critical things they heard from others years ago. Often they adopt the exact words and tone—but without recognizing the source of these internalized "tapes" as their own attachment figures who once said these words to them in this same tone of voice. Furthermore, *the judgmental, punitive, or rejecting affect that clients feel toward themselves as they replay these internalized tapes is the same affect that parenting figures originally expressed toward them years before* (for example, "What's *wrong* with you!").

2. Second, clients block their anxiety-arousing need by reenacting in current relationships the same scenario that was originally experienced in earlier, formative relationships. Thus, on an *interpersonal* level, clients will say and do to others what was originally done to them. For example, the child of strictly authoritarian or dismissive parents may grow up to be a parent who similarly disparages the same emotional needs in his or her own children—especially the child who most resembles the parent in gender, birth order, temperament, and so forth. Especially in the arena of parenting, clients who have not integrated or come to terms with their own conflicted feelings are prone to reenact the same maladaptive relational patterns that they experienced in their own childhoods. As parents, many clients feel helplessly dismayed as they watch themselves respond to their children in the same hurtful ways that their parents responded to them. Despite sincere pledges that "I will never do to my children what my parents did to me," variations of the old themes occur with clocklike regularity. Here again, in the interpersonal sphere, adult offspring gain some control over their own dilemma by repeating what was originally done to them rather than knowing and experiencing what they once felt.

3. Third, clients block the experience and expression of their own conflicted need by **eliciting** the same unsatisfying response from others in current relationships that they received in their childhood. For example, unless the children with problematic developmental experiences have other reparative relationships (with aunts and uncles, friends, teachers, and mentors) to help them resolve these concerns, they are likely, as adults, to select a marital partner who cannot respond well to their emotional needs. In the event that their spouse is capable of being affectionate or intimate, they may not be able to accept this emotional responsiveness. Why? The possibility of having a response to their old, unmet needs—even a well-intended, benevolent response—will often arouse shame or anxiety. That is, the spouse's affection is likely to arouse the pain of the original deprivation. Following their schemas, such clients also will anticipate that their spouse will reject their needs as others have done in the past—even though the spouse has not been tending to respond in this hurtful way.

Operating on schemas or internal working models that fit past relationships better than current ones, such clients may simultaneously elicit and reject nurturance from their spouse. This will certainly frustrate and confuse the spouse, especially if these mixed messages activate the spouse's own problematic relational templates, as they commonly do. Unresolvable marital conflict that results in emotional gridlock occurs when the spouses' schemas dovetail and each presents the other with a version of their old relational pattern (For example, think of a husband with a Dismissive attachment style who is married to a Preoccupied wife. His lack of responsiveness and communication will make her feel even more anxious and alone, while her need for emotional support and reassurance will be seen by him as more of the intrusive, demanding control he has learned to expect and abhor in relationships). In such cases, the likely outcome is enduring conflict, emotional disengagement from each other, or divorce unless counseling helps one or both partners recognize both patterns and change their mutual interaction.

To sum up, clients defend against their unmet needs or unacceptable feelings by:

- internally responding to themselves in the same rejecting or dismissive ways that attachment figures did,
- modeling the rejecting or unresponsive parent and responding to others in the same hurtful way, and
- repeatedly becoming involved with others who provide the same hurtful response to them that they originally received as children.

Although one of these three modes may be predominant for a particular client, many clients will employ all three mechanisms. In all three maneuvers, however, clients turn a passive experience that originally happened to them into an active experience over which they now have some control. Therapists can focus on how clients take up the same hurtful affect that was originally expressed toward them in childhood and actively turn it against themselves, or others, to block their own unmet need. This turning against the self (often expressed by clients as variations of "I can't") is only half of the conflict, however. Paradoxically, while defending against the anxiety of reexperiencing unmet needs or unacceptable feelings, clients simultaneously try to "rise above" it and have it met indirectly. This side of clients' compromise solution is shown in Part 3B of Figure 7.1.

INTERPERSONAL OUTCOME TO THE UNMET NEED: RISING ABOVE THE UNMET NEED

The most important component of this model for new therapists is recognizing the defensive interpersonal styles that clients adopt to cope with their developmental problems. Children's unmet needs and developmental problems do not just dissipate or go away as they grow up. These disruptive needs and feelings may be denied or sequestered away but they continue to be evoked in adult relationships, especially in their own marriages and parenting. Because these needs and feelings remain too anxiety arousing or unacceptable to be expressed or dealt with directly in current relationships, clients try to cope or **rise above** them by adopting various interpersonal coping styles. Horney (1970), Beck et al. (2003), Millon (1999), Benjamin (2003), Gottman & DeClair (2001), and many others have used related terms to describe the interpersonal coping styles that clients employ. To illustrate these interpersonal coping strategies, we will focus on Horney's three lifestyle adaptations: Moving Toward, Moving Away, and Moving Against others.

INFLEXIBLE INTERPERSONAL COPING STRATEGIES: MOVING TOWARD, MOVING AGAINST, AND MOVING AWAY

Clients adopt a fixed interpersonal style both to reduce the anxiety associated with having unacceptable needs and feelings revealed and to find some sense of value or self-worth. Let's first take a look at Horney's three coping styles—they can help therapists so much as they try to understand and respond to their clients.

1. Moving Toward: Some clients learned in their families of origin to cope with problematic familial interactions by **moving toward** or pleasing people. These clients learned they could earn some needed approval, and diminish the threat of further rejection, criticism, and so forth, by excessively *complying* with their parents and being unfailingly good (always helpful and nice, and even perfectly well behaved). These children are not learning to be appropriately well behaved; instead, in order to ward off anxiety and maintain their vulnerable and easily threatened self-esteem, they are giving up too much of themselves and their own identity. In this pervasively pleasing stance, clients have lost their own voice and the ability to "have their own mind." Generalizing this coping strategy they have adopted with caregivers, they carry over to others this pattern of relating in which they always try to please or accommodate others and avoid disagreeing, expressing their own interests or preferences, or asserting their own limits and boundaries. Thus, these clients have learned to defend against anxiety and win some approval by consistently moving toward others in a pleasing, servile way. As we started to see in Chapter 6, this coping strategy is commonly found in counselors—the caretaking career fits with their coping style.

2. Moving Against: Other clients have learned that aggressiveness and resistance to parental wishes, if pursued long enough, will ward off pain or insecurity. They may have needed permissive caretakers to act more strongly and take charge of the family effectively, which would have provided the safety they needed to feel secure without having to try and take control of everything themselves. These expansive and dominating individuals cope by **moving against** others. They want to be in complete control of themselves and their emotions at all times and, even more problematic, they dominate, intimidate, or try to exert too much control over others. They approach relationships competitively with the orienting attitude that they must win what they deserve, and they often assert themselves aggressively for this purpose. Their mind-set is that they should prevail in every situation or conflict. Thus, these clients adopt a repetitive interpersonal style of moving against others to protect themselves from the unwanted feelings evoked by needing someone, being in situations where they are not in charge or are unsure of what to do, or being close and intimate with someone.

The beginning therapist may dread moving-against clients because they are often critical or challenging with the therapist and try to control or compete with them. Difficult for the majority of experienced therapists as well, researchers find that many therapists have trouble with clients' anger toward them (Hill et al., 2008) and that these negative, angry, or provocative clients do frequently disrupt the treatment process (Matsakis, 1998). Practicum instructors and supervisors need to help new therapists prepare for these highly workable but initially challenging clients by role-playing and rehearsing responses.

3. Moving Away: In Horney's third interpersonal style, clients have learned that the best way to reduce the interpersonal threats they have grown up with and create some safety for themselves is by **moving away** from others through physical avoidance, emotional withdrawal, and self-sufficiency. They do everything for themselves and do not ask for help even when that might be appropriate. They move through life having "no need" of others—they often expect rejection or rebuff and attempt to take care of all their needs on their own. Like Millon (1981), Horney emphasized that their tendency to move away from people and avoid is often shaped by a heightened sensitivity to rejection and shame:

> "Change in an appointment, having to wait, failure to receive an immediate response, disagreement with their opinions, any non-compliance with their wishes, in short, any failure to fulfill their demands on their own terms is felt as a rebuff. And a rebuff not only throws them back on their basic anxiety, but it's also considered equivalent to humiliation" (p.116).

Because life is stressful for everyone and all people need defenses, healthy individuals also employ these three interpersonal styles at times. When they do not become exaggerated or used pervasively, there are adaptive strength features in them. However, they are not an identity or way of living for well-functioning people; rather, they are more flexible strategies used selectively to cope with a difficult person or problematic situation. With most clients who enter treatment, however, one style has become predominant or characteristic. It is overused, inflexibly, even in situations where this uniform mode is not adaptive or is not helping the situation any longer. The same coping style is also used unnecessarily—in safer situations where, realistically, this defensive coping strategy is not actually needed. As we will discuss later, *the therapist's role is to help clients learn to discriminate or assess more realistically when and with whom they need to cope in one of these ways and when they don't.*

Many have adopted or adapted Horney's three interpersonal coping styles. For example, Beck has used them in his cognitive behavioral work with personality disorders (Beck et al., 2003). John Gottman has utilized them in his illuminating research on marital interaction and how couples respond to their partners' bids for emotional engagement by moving toward, away, or against (Gottman & DeClair, 2001). Using closely related terminology, Jeffrey Young's Schema Therapy approach (2003) describes three similar coping styles that were adaptive and necessary in childhood—but imprisoning in adulthood when they are no longer needed but can't be dropped. Parallel to the Moving Toward coping style, Young describes the coping style of *Compliance:* the client feels helpless in the face of a more powerful figure and avoids further mistreatment by pleasing to avoid conflict or retaliation. Similar to the Moving Against client, Young describes the coping style of *Overcompensate.* This client makes others feel guilty, blamed, and intimidated; is often a workaholic who appears superior or perfect to others; and frequently counterattacks and controls. Closely following Horney's Moving Away client, Young describes a third coping style characterized by psychological withdrawal that he terms *Detached from own feelings and avoids others.* This client is cynical or aloof and avoids investing emotionally in people or activities, is excessively self-reliant, and isolates.

To review, we saw that clients turn against themselves in different ways to cope with problematic developmental experiences within their social or familial environment. In these turning-against-self defenses, clients respond to themselves and others in the same problematic ways that important others originally responded to them. We then discussed how clients try to cope with their developmental problems by adopting a fixed or **rigid interpersonal coping strategy**. These coping strategies included the interpersonal styles of moving toward, moving against, or moving away from others. Although there are personality strengths in these coping styles when they are used flexibly, they become maladaptive for clients and create problems because they are employed inflexibly in all types of situations.

With both sides of this compromise solution in mind, we are now ready to look more closely at how clients use their interpersonal coping style to try to **rise above** their core conflicts. As we will see, however, even though clients derive self-esteem and fashion an identity from their interpersonal coping style, it remains a defensive adaptation that avoids, rather than addresses, their basic conflicts and, in the long run, exacerbates rather than resolves their problems.

RISING ABOVE CORE CONFLICTS: A DEFENSE FRAMED AS A VIRTUE

When developmental problems are pervasive or severe, clients often adopt one of the three interpersonal styles: moving toward, moving against, or moving away. This **inflexible coping style** is a habitual way of relating to others that is also reflected in vocational, marital, and other defining life choices. There are many exceptions, of course, but individuals who move toward may find that careers in nursing or counseling are a good match for their caretaking skills and needs. Individuals who move against may find that careers in law, medicine, or money management are congruent with their interpersonal style of taking charge and needing to be in control. Individuals who move away may feel comfortable as researchers, artists, technical experts who work alone, or individuals in a meditative lifestyle. In their self-concept, many clients will privately frame this defensive coping style as a special attribute or **virtue** and use it to feel special and rise above their core conflict. That is, clients do not experience their interpersonal strategy as a defensive coping system to avoid anxiety and low self-esteem but as a virtuous way of relating to others that may make them feel special. However, this unrealistic sense of being special is compensatory and actually reflects their low self-esteem or shame-based sense of self. *Their self-worth is brittle and vulnerable—it relies too much on their ability to rise above their core conflicts by pleasing others, by achieving and succeeding, or by withdrawing and feeling cynically superior.* In these ways, clients frame their defensive interpersonal coping style as a superior quality or virtue. More contemporary therapists write about the "narcissistic" element in many different types of clients who are defending against a shame-based sense of self (Kohut, 1977; Stolorow et al., 1994). They describe what clients do to "restore" their sense of self-worth or power when they (too readily) feel shamed and overreact to feeling rejected or diminished by others or that they have failed in some way (see also Thomaes et al., 2007)—sensitive issues to work with, certainly, yet relieving for clients when counselors can take the risk and find respectful ways to broach these far-reaching issues with clients.

Thus, even though moving-toward clients are compliant and submissive, they do not experience themselves as servile. Instead, they often feel special—more caring than others who are not as compassionate. They accomplish this distortion by perceiving themselves as selflessly loving; always altruistic and sensitive to the needs of others or committed to ideals of peace or justice. In this way, the defensive interpersonal style becomes a way to cope with a lack of self-esteem—a virtuous quality that wins needed approval from others and wards off anxiety and shame.

Similarly, moving-against clients do not see themselves as angry, competitive, self-centered, or demanding, although they frequently behave in these ways. Instead, they see themselves as heroes or heroines—strong leaders entitled to direct others, who they often regard contemptuously as weak, dependent, or incompetent.

Finally, moving-away clients do not regard themselves as emotionally constricted, as avoidant and too easily hurt, or as being risk averse and narrowly limiting their lives. Instead, they prize their aloofness, perhaps as evidence that they are more refined, that they do not "need" others, or hold the illusion that they feel or experience things more deeply than others. They may be cynical and at some level feel better than the ordinary masses, whose involvement in everyday happenings may be seen as mundane and small-minded.

Thus, to lessen their anxiety, clients inflexibly employ an interpersonal coping style that allows them to feel special; gain a sense of identity and self-esteem; and rise above their unmet needs, unacceptable feelings, and pathogenic beliefs. Their ability to turn a defensive coping style into a virtue or talent that testifies to their specialness is what Horney terms a **neurotic pride system**. For these clients, however, their self-esteem is brittle and vulnerable. It is based on an unrealistic sense of being special and rising above rather than learning to value themselves and their own authentic feelings.

We now have the two sides of clients' core conflict. First, clients block their own needs and authentic feelings by responding to themselves in the same critical, dismissive, or rejecting ways that adult caregivers originally responded to them. Clients also come to feel toward themselves the same affect that, originally, the caregiver characteristically communicated toward them (for example, disappointment, resentment, indifference and so forth). This affect will usually be evidenced as low self-esteem when clients enter counseling, and clients are often seeking treatment to be rid of this intrapunitive feeling. Usually, it is relatively easy for therapists to recognize how clients are turning against themselves and to respond in more affirming ways.

Second, clients rise above their core conflicts and find ways to reframe their defensive strategy as evidence that they are special. This aspect of their compromise solution is usually more covert and is harder for therapists to address because the clients' neurotic pride is typically bound up with shame. Let's examine some other reasons why clients are reluctant to let go of their attempts to rise above their conflict.

Clients' interpersonal coping style has indeed granted some relief from the anxiety of unmet needs and unacceptable feelings by providing a sense of control over them. It has also allowed clients to earn some attention from others and respect for themselves. In this way, *clients' interpersonal coping style has become the primary source of their self-esteem and their principal avenue to succeeding in life*. Furthermore, clients have often developed impressive abilities in the service of their coping strategies. For example, many moving-toward clients have not only been able to make others like them but have also become genuinely skilled at caring for others. Likewise, moving-against clients have often been able to achieve a great deal of success and power, and may develop significant technical or other abilities along the way. Similarly, although moving-away clients have often been able to remain safely aloof, they may also have developed a rich inner life or been able to develop creative or artistic abilities.

One of the most important interventions therapists can make is to *help clients recognize the significant emotional price they pay for their interpersonal coping style*. Therapists do this *cost/benefit analysis* by recognizing their strengths and abilities but also highlighting, as concretely as possible, how their interpersonal adaptation is creating stress and limiting their lives now:

THERAPIST: You've just earned another award at work, where they have nicknamed you the "star," and you do so much for everybody at home. I'm so impressed by how capable you are and how much you are always doing for everyone. But I'm also concerned about how tired you look and how often you are sick. You often use the expression that you're "running on empty." I'm hoping that we can begin to listen to that feeling together here in counseling—I think it's telling both of us something important.

In this way, therapists want to affirm the genuine strengths that clients have developed through these coping strategies but simultaneously be compassionate toward the toll they take. Later in this chapter we will see how clients can retain the skills and strengths they have developed in these compensatory modes, but without utilizing them in the same inflexible or unidimensional ways as in the past. As before, therapists need to help clients honor their interpersonal coping styles and appreciate rather than pathologize them. They were absolutely necessary at one time and represented the best possible adaptation they could have made at that point in their development. Although they are still adaptive in some problematic relationships, they are overused now and are no longer necessary or helpful in many current situations. *An important treatment goal for many clients is to learn that more flexible or varied interpersonal coping styles will serve them better.*

To sum up, by turning their defensive interpersonal adaptation into a sense of being special, clients often gain significant secondary benefits and genuine personality strengths. It's no wonder, then, that some clients initially may be reluctant to explore this adaptation with the therapist. It has become the basis of their identity, the primary source of their self-esteem, and the only way they have had of defending against their seemingly unresolvable conflicts. *Once therapists fully grasp all that this defense has provided for clients, and help clients to appreciate both what it has accomplished and what it has cost, both therapists and clients will be less judgmental or impatient with the clients' symptoms.* Both will have greater empathy for the clients' struggles to change and will be able to work together collaboratively to develop more flexible coping strategies that will be more adaptive. Understandably, clients will not want to risk giving up what they have without having something better to replace it. As we will see, however, clients never really succeed in rising above their core conflicts through these compensatory adaptations. By acknowledging conflicts and addressing them more directly, clients can develop a wider experiential/interpersonal range that will be an important part of resolving many problems.

SHOULDS FOR THE SELF, ENTITLEMENT FROM OTHERS

Clients both gain and lose something by their compromise solutions. As a result of feeling special and virtuous, clients gain what Horney terms **false pride** and a certain type of **entitlement** from others. For example, self-effacing, moving-toward clients hold unspoken expectations of unwavering loyalty and approval from others, and they often demand constant reassurances. Expansive, moving-against clients claim leadership and control, and they expect deference from others. And, although it may look as if detached, moving-away clients place no demands on others, they silently hold expectations that others never criticize them or make demands upon them. Of course, the world doesn't go along very well with these "neurotic claims." They create ongoing conflict and leave clients exposed to continuing frustration, as others frequently do not accept such unrealistic demands and unspoken expectations.

Accompanying clients' unrealistic demands on others is a harsh and uncompromising set of "shoulds" that they place on themselves; these faulty beliefs reflect their impaired self-esteem and diminished sense of self. For example, moving-toward clients suffer under the self-imposed demands that they should be the perfect lover, teacher, spouse, parent, and so forth. They must always be caring and responsive to the needs of others and never feel angry, compete, or advocate on their own behalf. Moving-against clients demand of themselves that they should be able to overcome all difficulties and dominate. They must win and be on top, control their feelings and be "strong" at all times, and overcome personal insecurities and even physical illness simply by an act of will. Moving-away clients pay for their specialness by believing that they should be able to work tirelessly and always be productive. They demand that they should be able to endure anything without becoming upset and that they should never need help from anyone. As life and aging goes, such individuals will not be able to meet at all times these rigid demands they place on themselves. For example, inevitable failures occur:

- Moving toward: They cannot win someone's love or approval.
- Moving against: They do not succeed in an expansive new venture or taking command of an important situation.
- Moving away: They cannot remain aloof but are evaluated critically or forced to compete at work.

When such inevitable failures occur, intense anxiety, depression, and other symptoms often result and some clients will enter treatment in crisis. In particular, shame is often evoked when interpersonal coping styles fail—which clients may experience as a "defeat" or present as a "failure experience," and contributes to many different symptoms and problems that clients bring to treatment (Tangney & Dearing, 2002).

As we will see below, the "tyranny of the shoulds" is an exacting price to pay for the client's sense of specialness. One of the most helpful interventions therapists can offer is to be able to articulate, both accurately and empathically, how this coping strategy is central to the clients' basic sense of self—the basis of whatever self-worth they may possess, and *nothing less than the organizing theme in their life*. At the same time, therapists need to help clients *appreciate* these interpersonal coping strategies. They do in fact contain real strengths, and therapists can help clients recognize how once they really were necessary and adaptive. Overused and overgeneralized now, however, they are no longer necessary or adaptive in many current situations. Therapists help by clarifying the unnecessary price in symptoms and stress that the client is paying for this unidimensional coping strategy that no longer serves them well.

RESOLVING THE CORE CONFLICT

Part 4 in Figure 7.1 illustrates how the client's core conflict (key issue or primary problem) can be resolved. Ultimately, clients' compromise solutions to their core conflicts will not succeed. The client can block the developmental problem (turn against self) and try to rise above the core conflict by moving toward, away, or against. However, these compensatory maneuvers will not resolve the client's problems—no matter how successful the client becomes through these interpersonal coping strategies. As they repeatedly employ these interpersonal styles—even in situations where they are not needed or are not effective—they will engender problems in current relationships, and create personal stress that often contributes to health crises. Let's examine three reasons why the client's compromise solution will not succeed and then explore what can be done to help the client instead.

First, the "tyranny of the shoulds" places unrelenting demands on the client. These can be heard, for example, when clients enter treatment exclaiming, "I can't do all of this anymore! I can't be a perfect parent, a great spouse, and the most productive person in my office. I can't be everything to everybody anymore!" An individual's quality of life is diminished by these unrealistic expectations. They are depleting and fatiguing, and diminish the potential for personal and professional growth, even if these unrelenting demands they place on themselves (and on others) do not provoke a crisis that brings them to treatment. The struggle that comes from not having had a "secure base" in order to later feel secure in venturing into new relationships with flexibility is palpable in clients severely bound by "shoulds."

Second, clients' sense of entitlement and being special will usually fail as well. For example, pleasing clients will not be able to win the love or approval of everyone they need. Expansive clients will not always succeed or be able to make others defer to their demands, and detached clients will encounter criticism from others or situational pressures that force them to compete. Inevitably, clients' attempts to become special and rise above their conflict will lead to conflict at work and at home. Typically, clients enter counseling when situational life stressors, developmental transitions in adulthood, or aging/illness have caused their interpersonal coping strategies to fail.

Therapists help clients change when they grasp the pain that clients suffer when their attempts to be special and rise above their conflict fail; Kohut (1977) writes eloquently of the **narcissistic wound** that is so painful to clients' core sense of self. Even though the demands that clients place on themselves and others are unrealistic, it still can be excruciating when their interpersonal coping strategy fails. Tragically, a few individuals will even attempt suicide in response to such a failure, and others will contemplate it. Why are some clients driven to such an extreme response? In a word, shame. The interpersonal coping style that many clients have adopted is the basis of their identity and their primary source of self-worth, but it is a brittle substitute for the genuine self-esteem they lack. From these clients' point of view, they *are* their ability to please, to achieve, or to remain superior and aloof. Such clients overreact to seemingly insignificant events (for example, gaining three pounds over the holidays or losing their patience with a trying two-year-old). This overreaction occurs because their entire sense of self is threatened when their interpersonal coping strategy fails and their shame-based sense of self is revealed by their mistakes—which they frame as "failures."

When such students get an A minus rather than the A they demand of themselves, they may feel like a complete failure. Too often, the therapist's initial response to this disappointment is to challenge the client's overreaction to the grade and provide a more realistic appraisal of the situation. This well-intended, reality-based response misses the deeper meaning for this client. Such a reality check is certainly useful but, initially, a more productive response may be a bid for an empathic connection. That is, wondering aloud, the counselor may reflect that there seems to be no room for error for this person—it has to be perfect or, seemingly, everything is lost and the consequences seem so dire. In this way, the therapist is trying to empathically reflect how all-important this "failure" or disappointment seems to be from the client's point of view. Delving further, with a compassionate and curious attitude, the therapist invites the client to join and wonder aloud as well about how this grade seems to indicate something very important about the client and who they really are. Collaboratively exploring in this way, the counselor and client often find that this failure experience is viewed as "proof" or confirming evidence that something bad or unwanted about them is true. For example:

CLIENT: When I got passed over for the promotion, it proved to everyone that I couldn't make it.

OR

CLIENT: When I got frustrated and upset like that, I knew none of them would ever like me again.

Working collaboratively in this way, the therapist and client are trying to hone in and clarify, as precisely as they can, what this disappointing grade seems to say or mean about the client. For example, together they may find that it proves to the client that he is just an "imposter" who really is inadequate and can never do it right, or is letting someone down whose approval he "must" earn, and so forth. This pathogenic belief is true, in the client's mind, despite the client's Herculean efforts to rise above or disprove it by achieving A's. In other words, the therapist's intention is to discern more specifically what this grade really means or says to this client about herself. By exploring and *clarifying together* how completely this client's sense of self depends on his or her ability to achieve or please (which is a different intervention and interpersonal process than interpreting, challenging, or reframing), it reveals the anxious insecurity and emotional deprivation that originally led to this defensive coping strategy. Thus, therapists use this understanding to share their compassion both for the pain of the original problem and the burden of trying to rise above it that the client has suffered with for so long. By being able to help the client articulate this dilemma and holding a compassionate or respectful stance toward it, therapists behaviorally demonstrate that they grasp the real meaning of this prototypical experience that occurs over and over again for the client. This kind of empathic understanding often diminishes presenting symptoms of anxiety and depression. *It provides safety for clients—a secure base from which they can begin to question this coping strategy, appreciate how and why they learned it, start to observe themselves employing it in their daily lives, and begin to change this pattern of relating—first with the therapist and soon with others in their lives.*

In sum, by appreciating the broader life context of this "failure experience" when their rising above defense fails—an impact that has always seemed so irrational to others as well as to the clients themselves—therapists understand clients in a way that most clients have not experienced before. This compassionate understanding is a corrective emotional experience that helps clients free themselves from the tyranny of the shoulds and the unrealistic demands they place on themselves, and helps them be able to come to terms more realistically with who they are and how they choose to live.

Why is it such a crisis for so many clients when their interpersonal coping style fails? It triggers the unwanted feelings and faulty beliefs that were originally engendered by problematic familial interactions (for example, "I really am worthless" or "It's my fault they didn't want me"). One way that the therapist helps clients expand these narrow and repetitive coping styles, and discriminate how to use them more *flexibly* in situations where they actually are necessary or adaptive, is by helping clients understand the family roles and interactions that originally led to these interpersonal defenses of moving toward, moving against, or moving away. Usually, it is more effective to respond to these developmental problems *after* therapists have responded to the current crisis brought on by the clients' pathogenic beliefs that—if they are unable to successfully please, achieve, or remain aloof—they are not loveworthy, cannot protect themselves, are too needy or demanding for others, and so on.

Third, the most important reason the client's compromise solution cannot succeed is that the client can never rise above the original conflict. *No matter how saint-like or loving, how successful or accomplished, or how self-contained and independent, the client is still defending against the original conflict rather than resolving it.* These interpersonal maneuvers are defensive coping strategies. They are attempts to *avoid* the problem by blocking it, or to manage the problem by trying to rise above it, rather than address it directly. This is clearly illustrated when the client's increasing success in life is not met by good feelings of satisfaction or accomplishment but by increasing depression, cynicism, or unrelenting pressures to achieve and accomplish more.

Let's review this complex sequence. When the child's basic developmental needs are not met by caregivers, the individual will adopt a "compromise solution" that temporarily relieves the anxiety of having basic emotional needs unmet but does not resolve the reoccurring problem. On the one hand, clients try to block their own needs, just as they were originally blocked by significant others. Clients turn against themselves and take on the same self-punitive feelings and attitudes that attachment figures once held toward them. These clients lose their self-efficacy to feelings of helplessness and hopelessness.

On the other hand, clients try to rise above their core conflict by becoming special, turning their defensive lifestyle adaptation into a virtue, and obtaining some compensatory self-esteem. Clients will be highly invested in trying to overcome and rise above their core conflict; typically, to have to face it will feel shameful, frightening, or hopeless. Thus, Part 3B of Horney's model (Figure 7.1) indicates clients' exaggerated sense of being special and defiant attempts to rise above the conflict. For example, this client might vow to himself, "I will never need anything from anybody ever again." This rigid, defensive stance reflects an act of will to overcome these developmental problems and rise above them, rather than trying to resolve them by addressing them directly.

To resolve their core conflicts, the therapist works to help clients recognize and relinquish their compromise solutions. Attempts to block the original need and to rise above it are fragile solutions that cannot be sustained; failure experiences that precipitate crises and symptoms are inevitable. Thus, instead of avoiding their real problems in these two ways, the therapists' goal is to help clients approach them. Therapists help by allowing clients to experience their core conflicts in a more supportive relationship that does not meet the original need but does give clients the understanding and validation they need to live through the feelings that were so unwanted. The interpersonal solution provided by the therapist's more effective response also allows clients to integrate other reactive feelings, helps clients disconfirm pathogenic beliefs about themselves, and alters clients' faulty expectations about what may occur in relationships with others. Now, as adults, clients are able to understand and resolve the original conflict by reexperiencing and reevaluating with the therapist what was once so hurtful or problematic (for example, a client might say, "I'm really not being selfish or doing something wrong if I become more independent and successful, even though this makes my mom feel sad and withdraw sometimes"). In this way, what was once so unacceptable or threatening can now find compassion and understanding. Clients can integrate these difficult feelings and question these pathogenic beliefs and faulty expectations, rather than continue to defend against them or be ruled by them. When clients allow the therapist to help contain the conflicted feelings, old relational patterns are broken and new, more flexible schemas become possible. Therapists' compassion and validation comprise a new response that disconfirms these faulty beliefs and, in turn, allows clients to be more self-accepting and forgiving of themselves. Thus, clients can now integrate the previously unacceptable parts of themselves and can better accept both the good qualities and the limitations in other relationships that have been important but conflicted.

As clients receive a more effective response from the therapist than they have received from others in the past, they will not need to block their core conflict or attempt to rise above it anymore. The therapist's more accepting and affirming response to the client's original predicament alters the maladaptive relational pattern and alleviates the anxiety that was originally evoked. As this occurs with the therapist, clients are able to begin generalizing this experiential relearning with the therapist to others in their lives. With the therapist's help, that is, clients can begin responding in more adaptive ways that improve their relationships with others and expand the narrow coping style that they originally adopted to cope with their dilemma. In the chapters ahead, we will examine more specifically: (1) how the therapeutic relationship can be utilized to help clients resolve their core conflict; and especially, (2) how this emotional relearning with the therapist can be generalized to other relationships in clients' lives. For now, an extended case study of a moving-toward client, and two case summaries of the other coping styles, will help therapists use this conceptual model to guide their interventions.

CASE STUDY OF PETER: MOVING TOWARD OTHERS

DEVELOPMENTAL HISTORY AND PRECIPITATING CRISIS

Peter was an insecurely attached child, whose emotional needs went unmet in his authoritarian home. Further stressors accumulated when his wrangling parents divorced when he was six years old. Although this family change could have been an opportunity to improve parenting and establish calmer, more cohesive family relationships, it was not. Especially problematic, his parents could not work out a cooperative parenting relationship following the breakup. His father did not take an active parenting role and, by two years after the divorce, Peter saw his father infrequently and when they were together, their relationship was superficial.

Peter's mother was overburdened by the demands of raising three children on her own, working full-time, and trying to make some kind of personal life for herself. She was involved with several men in the years following the divorce but was never able to establish an enduring love relationship. Frustrated by the many demands and few pleasures in her life, Peter often became the target of her resentment. Impatient and irritable toward him, she criticized Peter when things went wrong in her life and often felt resentful of his needs. Although she tried her best to be fair to the children and give them a good home, she was not very responsive to Peter or affirming of him.

After his father left, Peter quickly learned that taking care of his mother was the best way to ward off her disapproval and to win whatever affection he could. Following the anxious/ambivalent attachment style, he was "preoccupied" with monitoring his mother's moods and concerns. By the age of ten, Peter had already adopted a pervasive interpersonal style of moving toward people. His teachers described him as an especially responsible and well-behaved boy who was "a pleasure to have in class—always so helpful." Let's turn the clock forward 15 years and see how these developmental challenges are being expressed in Peter's early adulthood.

At the age of 25, Peter is a graduate student in counseling. Becoming a therapist felt like a perfect career choice to Peter. He prided himself on his sensitivity and concern for others and took pleasure in being able to help those in need. And now that he was carrying his own client caseload, it was great to find that he enjoyed being a therapist as much as he thought he would. Best of all, Peter was finding that he was often able to help his new clients. At least, all of his clients seemed to like him and, in contrast to the experiences of some of his classmates, his clients kept returning to their counseling sessions each week. As his second-semester practicum got under way, Peter felt that he was on his way.

Later that semester, though, Peter had a setback. After presenting a videotaped recording of one of his therapy sessions in group supervision, Peter received some unexpected feedback. The practicum instructor told Peter that he was being "too nice" to his clients and that he seemed to need his clients' approval too much. The instructor went on to say that Peter seemed to be afraid to challenge his clients, reassured them too often, and tended to avoid potential misunderstandings or conflicts in the therapist–client relationship that needed to be addressed.

Peter was stunned. Although he had some awareness of his aversion to conflict, he did not truly understand what the instructor was talking about, and he felt hurt and confused by the criticism. It was important to Peter that his instructor like him and approve of his clinical work. Peter tried carefully to explain that the instructor did not understand the close relationship that Peter was developing with his clients or recognize all of the important issues that his clients had been revealing. The instructor responded that this was probably true, but Peter was missing the point. He repeated that Peter needed to think about his reluctance to address interpersonal conflict or say things that his clients might not want to hear. To make matters worse, two students in the practicum group chimed in and agreed with the instructor's comments. With that, Peter's anxiety became so high that he could no longer defend or explain himself, let alone try to understand or learn from their comments. Peter stopped arguing with them, looked down, and quietly nodded agreement throughout the rest of the supervision session.

Throughout the next few days, Peter felt just sick inside. He was so dismayed by the criticism that he couldn't think about anything else. For a while, he thought that he should drop out of the practicum group, but then he decided that, if he only tried hard enough, he could make the instructor see that his criticism was unfounded. Peter kept searching for a way to discount the feedback and stop the anxiety that was churning inside.

One week later, Peter found out that his girlfriend was having an affair with another student in the program. Although he tried to be understanding at first, he felt shocked and betrayed. He alternated between withdrawing and announcing that their relationship was over and desperately trying to win her back. Peter felt shattered. He became so anxious that he was unable to eat or sleep, let alone study. It felt as if a motor were racing inside of him—accelerating out of control. Peter began hyperventilating, experiencing heart palpitations, and having anxiety attacks. By the end of the week, he was unable to drive his car.

To make matters worse, Peter tried to keep all of this turmoil to himself. He thought he should remain "calm and together," and he was afraid that his supervisors would not want him to see clients if he was "so messed up" that he was having anxiety attacks himself. But despite his attempts to cover up his distress, his individual supervisor soon asked him what was wrong.

Although he never could have allowed himself to ask for it, Peter desperately wanted his individual supervisor's support, and he was greatly relieved to receive it. Peter explained how his practicum group was becoming one of the worst failure experiences of his life and how hard it was to accept what his girlfriend had done. The supervisor was supportive but also found a tactful way to say that the practicum instructor's comments fit with some of his own observations. Because he knew that his individual supervisor liked him, Peter was able to consider the feedback this time. The supervisor suggested that these were important issues for Peter to work with but that they could be dealt with better in his own therapy rather than in supervision. Peter agreed and began seeing his own therapist.

It helped that Peter's supervisor thought he could be a fine therapist, in spite of his anxiety attacks. However, Peter's anxiety remained paralyzing as he began his own treatment. Fortunately, Peter was assigned to a skilled and experienced therapist who soon conceptualized the predisposing vulnerability, interpersonal coping strategy, and situational stressors that precipitated his crisis. Peter's coping strategy of moving toward people had generally worked well enough for him up to this point. However, both of the crises that Peter had just experienced ran headlong into the heart of his core conflict and his interpersonal style for coping with it. As his ability to rise above his core conflict by pleasing others failed, the anxiety associated with his profoundly insecure childhood broke through, and anxiety attacks resulted. Let's try to understand more fully why Peter developed these symptoms.

PRECIPITATING CRISIS, MALADAPTIVE RELATIONAL PATTERNS, AND SYMPTOM DEVELOPMENT

Many people would have coped with the two stressful events that Peter experienced without developing such significant symptoms. As we have seen, however, we understand more when we apply the concept of client response specificity. We will use Peter to illustrate how symptoms develop when situational stressors evoke painful developmental experiences, confirm pathogenic beliefs, or cause interpersonal coping strategies to fail.

The first stressor for Peter was his practicum instructor's critical feedback. Criticism from a respected authority figure would be unsettling for most people, but generally they could cope with it. For someone like Peter, however, such disapproval carries far more weight. Because of Peter's strong need for approval, his history of receiving excessive criticism, and his lifestyle adaptation of trying to rise above these developmental conflicts by being helpful and nice in order to win approval, the practicum instructor's criticism felt huge.

Second, the practicum instructor specifically challenged Peter's interpersonal coping style of moving toward others. This confrontation not only aroused the anxiety of receiving criticism and feeling rejected but also undermined Peter's primary means of warding off these unwanted feelings (pleasing others and accommodating their needs). Thus, *the instructor's feedback triggered Peter's core conflict and, simultaneously, weakened his defenses against it.* In other words, Peter's developmental experiences made him acutely sensitive to criticism. He adopted the attitude toward himself communicated by his early caretakers—that he did not matter enough to be committed to (from his father) and was not good enough to be approved of and responded to (from his mother). In the most important relationship in his life, he learned to expect that his mother often would be frustrated, critical, and disappointed with him. His attempt to defend against the anxiety evoked by this set of circumstances was to try to be perfect—to please her and others and meet their needs, so that this self-critical schema could be refuted by their appreciation. This coping style was, of course, a formula for failure: It was impossible for Peter to be liked and appreciated at all times by everyone he met. There was no such thing as constructive criticism for Peter—he construed all disapproval as rejection, and he felt very uncomfortable whenever he could not win everyone's approval. When teaching his first undergraduate course, for example, Peter seemed only to notice the two or three students in the class who seemed disgruntled. He did not meaningfully register that the large majority of students enjoyed the class and thought he was doing a good job.

If not for the subsequent crisis with his girlfriend, Peter probably could have recovered from the first setback without developing symptoms. Most likely, he would have reconstituted his interpersonal coping style and been somewhat successful in winning the approval he needed from others in his life—different instructors, friends, and as it too often goes, perhaps even clients. As we will see, however, the subsequent stressor with his girlfriend also struck at the same core conflict. At that point, his moving-toward coping style toppled, and the intense anxiety associated with his insecure attachment history was evoked. More specifically, this was "shame-anxiety"—anxiety over the threat of having his unlovable and unworthy self exposed. This anxiety over having his shame-worthy self revealed, both to himself and to others, became too intense to be blocked; it broke through in anxiety attacks and symptom formation.

A partner's infidelity will be highly stressful for almost everyone. Here again, though, this particular stressor held far greater significance for Peter when viewed within the personal context of his developmental history and subjective worldview. Peter had to cope with far more than just the loss of trust with his girlfriend; he also suffered a blow to his identity and basic sense of self-worth. In Horney's neurotic pride system, Peter's coping style of moving toward entitled him to be special, so that others would love and prize him inordinately. Although Peter was not very aware of his own unrealistic claims, he expected his girlfriend to idealize him as more sensitive and loving than anyone else could be. Clutching to this defensive sense of being special, Peter couldn't believe that his girlfriend could actually be interested in someone else. Peter's neurotic pride system was painfully shattered when he found that he was not the commanding center of his girlfriend's life that he felt entitled to be.

Multiple stressors occurred for Peter in a short period of time. In and of themselves, these events would precipitate a crisis for many people and lead them to therapy. However, if these situational stressors do not tap into preexisting schemas, repeat familiar but unwanted relational scenarios, and confirm pathogenic beliefs, they will not usually provoke such strong symptoms. The client will often be able to recover in a relatively short time with crisis intervention or short-term supportive therapy. In contrast, when stressful life events tap squarely into a client's core conflict, a client such as Peter has to cope with far more than just the demands of the current situational stressors.

Thus, when Peter's girlfriend became sexually involved with another man, he had to cope with much more than just the pain of this betrayal. He had to cope with the even bigger developmental problems that were triggered: his strained attachment history; his interpersonal coping strategy of pleasing to ward off the criticism and rejection that he expected; and the wound to his self-esteem when his sense of being special was shattered. Clients such as Peter do not have a secure sense of self-esteem to fall back on in times of crisis. As unrealistic as it is, the sense of being special and the secondary gains that clients can earn with their interpersonal coping style is the only self-worth that some clients have been able to garner. It is understandable, then, that Peter was overwhelmed by anxiety attacks. Both of the situational stressors intensified Peter's original conflicts, and they both took away his coping strategy for defending against them. In this light, *we see that clients' presenting symptoms are not irrational.* Per client response specificity and subjective worldview, they make sense once they are understood in the broader context of the client's developmental history and the cognitive schemas, pathogenic beliefs, and inflexible coping styles that result from this history.

COURSE OF TREATMENT

Fortunately, Peter's therapist was both knowledgeable and kind. He was genuinely empathic to the pain that these situational crises brought on for Peter but also grasped what made Peter so vulnerable to them. The therapist recognized that the current crisis provided an opportunity to resolve the more important developmental problems that left Peter prone to reexperience crises such as these whenever others rejected or disapproved of him. In the months that followed, Peter was able to resolve the precipitating crisis. More significantly, he made progress in coming to terms with the impact of his developmental history and the problems that followed from it. As a result, Peter was able to expand his narrow moving-toward coping style and adopt a more flexible interpersonal repertoire. Here and in the next chapter, we will review the course of therapeutic events that allowed Peter to make these far-reaching changes.

Peter entered therapy in crisis, and the therapist was able to respond effectively to his distress. From the first session, Peter felt understood and cared about by the therapist. As his accurate empathy led to a strong working alliance, Peter's anxiety attacks stopped. The anxiety and distress that had been so disruptive for him steadily subsided as the therapist continued to provide an effective holding environment. However, as Peter began to feel more secure and function better in his everyday life, he began to be resistant to the feelings he had been sharing with the therapist and wondered about ending treatment.

The therapist observed that, as Peter felt understood and affirmed by him, Peter's unmet needs for acceptance and support were assuaged and the attendant anxiety diminished. As a result, Peter felt safer, his symptoms abated, and he began to function better. At the same time, however, he also began to reconstitute his interpersonal coping style of moving toward others. Rather than terminating therapy at that point—which might need to occur in crisis intervention or time-limited treatment—the therapist was able to help Peter go further and address the developmental conflicts that made him so vulnerable to situations where he faced disapproval or the threat of being left. The primary way that the therapist did so in the beginning was to help Peter identify his habitual coping style of pleasing and recognize how automatically and frequently it came into play. Thus, the therapist *named it*, so they could talk about it and begin to explore it together—compassionately and without judgment. This opened the door that changed everything. From these conversations, for example, Peter learned to focus inward to identify what he was feeling, and what was occurring with others, at the moment just before he felt compelled to employ this coping pattern and please. (Cognitive behaviorists use such "self-monitoring" techniques to increase self-awareness of the situations, thoughts, and emotions that trigger faulty coping strategies.) Usually, at such moments, Peter found that he was expecting others to be disapproving of him or to go away and cut off from him in some way. With help from his therapist, Peter became increasingly adept at recognizing this sequence and his propensity to truncate the anxiety aroused by engaging in pleasing behavior.

Specifically, as he listened to Peter's narratives, the therapist would highlight when instances of Peter's key relational themes were occurring. That is, the therapist would punctuate the interaction when Peter was (1) pleasing others, (2) avoiding anger and other interpersonal conflicts, and (3) expecting others to be critical or rejecting. In particular, the therapist used process comments to point out these three repetitive relational themes whenever he saw them occurring in their interaction together or with others. For example, when the therapist thought that Peter was trying to please him, too, he made this overt and inquired about this possibility:

THERAPIST: When we're talking, Peter, sometimes I find myself wondering if you might be thinking too much about what I or others might want to hear. What do you think—is there anything to that?

The therapist also highlighted whenever he thought he heard one of these three themes in Peter's relationships with others. For example:

THERAPIST: As I listen to you describe this interaction with them, it sounds like you are trying very hard to keep them from being angry at you. I feel like I've heard this before. Does it seem to you like this comes up a lot?

This feedback provided direction for their work together—a treatment focus— and served as the basis for developing mutually held treatment goals. For example, *the therapist encouraged Peter to risk not pleasing him and focus instead on just saying and doing what he wanted in their relationship.* Not having to be vigilant about what the therapist liked or wanted gave Peter the opportunity to "live inside his own skin" and "have his own mind" rather than be thinking about what the other person wanted to hear. This was a corrective emotional experience for Peter. He could enjoy this relationship where he could be cared about even when he was not trying to please the therapist or act in ways that, at some level, were designed to keep the therapist engaged with him and ensure that he would not leave him or be critical (such as frequently complimenting the therapist or telling him how helpful he is being).

The therapist continued to focus Peter inward. In particular, they explored the feelings that were evoked for Peter when, instead of employing his usual coping style of pleasing others, he risked new, more assertive responses that he had rehearsed with the therapist. Gradually, by focusing on what he was experiencing just before he employed his usual coping patterns, Peter began to clarify the emotional deprivation he had suffered as a child but had always been too ashamed to reveal to others or even acknowledge very fully to himself. Relatively quickly, the therapist's accepting presence allowed Peter to overcome the shame of having his pain revealed. Now, for the first time in his life, Peter could be fully seen by someone with all his strengths and problems and continue to be respected. Peter began to understand how he had coped with his childhood predicament by taking care of and pleasing his mother, and then others as well. As this significant progress occurred, however, Peter still continued to struggle with strong feelings of anxiety, shame, and sadness.

These contradictory feelings continued to wax and wane for Peter over the next few months. Peter was increasingly recognizing how unwanted and alone he had felt as a child and how ashamed of himself he had always been. (Poignantly, he recalled that the animal he felt most similar to during elementary school was a worm.) Accompanying these feelings, Peter recognized and began to challenge the false belief that he was not loved because he was somehow unworthy—that he did not matter enough to be important to others. As Peter continued to feel held by the therapist's understanding, his long-withheld feelings of sadness over this deprivation and invalidation could be expressed for the first time. Peter moved through the original environmental block as he received a far-reaching, corrective emotional experience. Repeatedly, the therapist's validating responses disconfirmed Peter's old relational expectations and were deeply comforting, offering him the affirmation for which he had always longed. Securing his own identity, this validation also provided him with a sense of being seen or known for the first time and accepted for who he really was.

Peter began to feel sadness and compassion for himself and mourn his childhood losses—rather than continue to blame himself for what occurred. In response to this primary affect of sadness, two reactive feelings followed closely behind. Peter found himself feeling angry: at his father for walking away from him, and at his mother for making him feel responsible for trying to earn her love and trying so hard to be "good" to ward off her disdain. As soon as this anger emerged, however, Peter became exceedingly anxious—afraid of being left on his own by the therapist, his girlfriend, his parents, and anyone else for being angry with them and "protesting" how they treated him. Each of these three feelings in Peter's affective constellation—the shame surrounding his unmet emotional needs, his anger over being rejected and dismissed so readily, and the anxiety of being left or emotionally cut off if he protested—was repeatedly aroused in treatment. With the therapist's help, Peter was gradually able to integrate and resolve these feelings. This occurred by being able to fully experience or know them himself, to risk sharing or revealing them to the therapist, and discovering that he could live through or contain them with the therapist's support. This was not a smooth or simple process, however. As the primary feeling of sadness or deprivation emerged, Peter's own shame and internal blocking defenses were activated. Let's examine how the blocking side of Peter's compromise solution was expressed in therapy.

Peter's two situational crises toppled his rising-above defense. He could no longer be special and win the approval he needed from either his practicum instructor or his girlfriend. As his interpersonal coping strategy failed, it exposed his shamefully unacceptable attachment needs (which Peter derided as "weak") and evoked intense anxiety. However, the other side of Peter's compromise solution was still operating: To defend against the painful feelings being evoked, Peter turned the same type of block that was originally imposed by the environment against himself.

In treatment, and throughout his life, Peter had tried resisting each feeling in his affective constellation. He did this by blocking both the subjective experience ("they don't want me") and the interpersonal expression (pleasing others to ward off disapproval) of his core conflict (excessive criticism and rejection). For example, as the sadness of his childhood deprivation emerged, Peter thought that the therapist would be critical and rejecting. He was afraid that the therapist would want him to "stop crying, grow up, and act like a man." This *transference reaction* again served to block his sadness until the therapist was able to draw it out more fully and clarify this misperception by talking it through with Peter. More important, Peter was able to experience the therapist's continuing respect for him in the face of his seemingly unacceptable emotional needs. Peter learned that in this relationship he did not have to be perfect, please or take care of the therapist to be cared about. Although his schemas led him to expect that the therapist would be irritated by his emotional needs and resent them, as his mother had done, Peter experienced a reparative new response of respect and empathic understanding.

Next, as Peter's anger toward his girlfriend, and then both of his parents, emerged, Peter felt guilty. He felt that he should be understanding of them, as his coping style demanded. He also felt ashamed of his inability to sustain this forgiving response. This reflected his moving-toward adaptation and served to block the expression of his profound need for more secure relational ties. The therapist was skillful in balancing Peter's protest or anger while still preserving what was good in their relationship. For example, he affirmed the validity of Peter's anger but also acknowledged the reality that Peter still loved his mother and had received many good things from her. With support for both sides of his ambivalence, Peter gradually felt safe enough to risk experiencing both the longing and the anger that had always been present but too anxiety arousing to acknowledge.

Why was it so threatening for Peter to experience his hurt and anger or share it with the therapist? While growing up in his family, Peter had learned that relationships were not resilient—there was no way to repair when the parent-child relationship was ruptured. His emotional needs in general, and any angry or sad feelings in particular, were simply unacceptable. He knew that, because of his father's weak commitment to him and his mother's readiness for such strong love withdrawal, he could easily lose the little emotional support he had. Aware that many different feelings could rupture his tenuous attachments, Peter kept what little he had by agreeing with them and believing that his needs were unimportant. In particular, his anger aroused strong separation anxieties. Realistically, he feared that he would be emotionally cut off if he expressed even disappointment, let alone the outrage he felt. Peter then tried to block these threatening feelings by telling himself that he should be calm and accepting—especially if he were going to be a therapist and help other people. The therapist knew that Peter was on his way when Peter stopped anxiously pleasing and pursuing his girlfriend and told her that he no longer wanted to try and work things out or get back together—it was time for him to move on and explore new relationships that would be a better match.

To sum up, the therapist provided a skillful balancing of the good news and the bad news in Peter's development. He was able to acknowledge the reality that Peter loved his mother and still felt loyal to her despite all that occurred, and that some of the qualities he valued most about himself were qualities derived from her.

However, these reality-based strengths were accompanied by the painful realities of Peter's emotional deprivation and invalidation. By exploring with Peter and then affirming what was legitimately good in his relationship with his mother, rather than simply blaming or rejecting her, the therapist helped Peter preserve important aspects of this tie. This gave Peter the security he needed to risk feeling his anger at both of his parents for what really was wrong in their relationship (to "protest," in Bowlby's attachment terms). Being able to protest with the therapist what had been so unfair and hurtful was not to reject his parents, hold a grudge, feel sorry for himself, or make his parents bad people. Instead, *it was to change the self-narrative of his life story and know that these things actually did happen, really were painful, and that he didn't cause or deserve them.* Rather than deny the reality of his circumstances any longer, being able to have the validity of his own feelings and perceptions for the first time led Peter to a stronger stance that he was able to generalize to other relationships. Peter became more confident and assertive, developed firmer boundaries with others and more clarity about who he was, and began to disconfirm the pathogenic belief that he would be criticized or rejected if he wasn't taking care of others. Peter came to understand that his mother's criticism and exasperation with him was unwarranted, had absolutely nothing to do with him, and was just the projection of her own hopelessness about ever being loved. Significantly, Peter was able to feel compassion toward her and appreciate the predicament she was in as she was coping with her own long-term depression, yet without denying the reality of the painful consequences this engendered for him.

As a result of these changes, Peter felt less compelled to make others like him, less threatened by interpersonal conflict, and more self-contained. Finally, Peter was also able to keep the best parts of his moving-toward style: He remained a genuinely caring and responsive person. However, this was no longer the unidimensional response pattern that it had been in the past. He no longer compulsively tried to get everyone to like him. He became increasingly more able to address interpersonal conflicts, although this would continue to be challenging for him; nevertheless, this growing ability was helpful in establishing more mutually rewarding relationships. Clearly, Peter had grown through this crisis and become a more resilient person. Therapy helped immunize him against his predisposing vulnerability to rejection, criticism, and loss. By experiencing a reparative response from the therapist and finding that some relationships could be different from those he had experienced and expected, Peter expanded his schemas of what can occur in relationships. In this way, the therapist helped Peter to stop carrying emotional baggage from his past into his present relationships and from reenacting the same maladaptive relational patterns and inflexible coping strategies.

Peter will certainly have further crises in his life, especially when circumstances again tap into his "old wound," but the changes he has made will help him respond to future problems in a more empowered, present-centered manner.

TWO CASE SUMMARIES

In the case study of Peter, we have seen how the interpersonal model for conceptualizing clients is used with a moving-toward client. Next are two case summaries that highlight salient features in the treatment of a moving-against client and a moving-away client.

CARLOS: MOVING AGAINST OTHERS

Carlos had been riding high, on his way to becoming a real estate "king," when his business failed. Buoyed by his initial success in a booming real estate market, Carlos had been woefully overextended when interest rates rose unexpectedly. His business went bankrupt, his large home and sports car were repossessed, and party friends deserted him as creditors and the IRS aggressively pursued him. As his dreams of wealth and power were dashed, Carlos became seriously depressed. Reluctantly, he entered therapy with an older Hispanic therapist who was widely respected in his community for helping migrant families and children. Even though Carlos felt that people should solve their own problems and that therapists were in this business just for the money, he decided to try it with the resolve that he was not going to let the therapist, or anyone else, take over and "tell him what to do."

After three months in treatment, despite his skepticism, his depression had in fact improved. Carlos could see how helpful the therapist had been and had actually told a friend that he didn't know where he'd be today if not for the therapist. The therapist had taken an active yet nondirective approach in working with Carlos. He had established early in the treatment process his respect for Carlos's success and empathy for the financial setback and ensuing psychological state Carlos was dealing with. For example, when Carlos told the therapist why he was in therapy, the therapist had reflected back that he understood Carlos's coming in as showing that Carlos wanted to "remain actively in charge of his own future" and that he wanted Carlos to identify what *he* wanted from coming in, and Carlos had resonated with that. On this particular day, however, Carlos was ready to quit. He was angry that he was still coming to therapy—he did not want to need the therapist, or anyone else, anymore. Carlos often felt frustrated and resentful like this. On this day, however, he was fed up with the therapist's fees in particular. As Carlos often did with others, he began the session by taunting the therapist:

CARLOS: Who do you think you are to charge so much? Do you really think you're worth all that?

The therapist acknowledged the feeling in Carlos's provocation:

THERAPIST: You're angry about my fees. Let's talk about them.

CARLOS: (*in a provocative and mocking tone*) What are you doing—trying to prove that you can gouge as much money as the psychiatrist next door? Did your mother want you to be a "real" doctor and make a lot of money or what?

Fortunately, the therapist was able to remain *nondefensive* and stay empathic to Carlos during this critical incident. The therapist did not retreat (flight) or counterattack (fight), which is what Carlos's provocations usually elicited from others. Nor did the therapist act on his own automatic response tendency—which was to try to explain and convince Carlos of his good intentions. Instead, the therapist tried to maintain a modulated stance, keep from overreacting, and offer an empathic bid as a way to change this familiar relational scenario that Carlos was trying to reenact again:

THERAPIST: Yes, you're angry with me. I do charge you for my time, and it's hard for you. And there are still problems in your life that haven't changed yet.

The therapist's accepting response only frustrated Carlos further, and he tried harder to embroil the therapist in conflict with him:

CARLOS: I hate your understanding; I hate coming here; I don't want to do this any-more! What do you do—be nice to people when they're falling apart so they become dependent on you? Then they can't leave you and you can keep taking all of their money. There's something wrong with you—you're sick!

THERAPIST: It seems right now that I am trying to make you dependent on me, so that I can take advantage of you financially?

CARLOS: *(loudly)* Yes, you idiot, I want to leave and get out of here, but I can't. I'll just get depressed again if I leave. I need you and you know it, and you use that to take the money I have left. I'm trapped here. You've got me and I hate it. I hate every-thing about this!

The intensity of Carlos's reactions told the therapist that, in addition to what-ever reality-based issues needed to be addressed regarding fees, Carlos's core con-flict was being activated right now. On the basis of other hypotheses that he had already formulated, the therapist reasoned that, for Carlos, having a need of some-one meant being used or exploited by them. Trying to articulate what Carlos was experiencing in their relationship at that moment, the therapist responded:

THERAPIST: Carlos, it sounds as if you don't experience the decision to remain in ther-apy as your own choice.

CARLOS: And what the hell does that mean?

THERAPIST: It seems like you feel trapped by your own need to be helped or under-stood. *(pauses)* I'm wondering if you have learned that to let yourself need some-body puts them in control of you and leaves you subject to their own self-serving needs.

CARLOS: Bingo! That's how the world is and that's what people do. It's about time you figured it out! I hate needing others—it's so weak, they just use you—like you do!

THERAPIST: Carlos, I'm sad that that is the only way relationships have gone for you. I can see why you feel that your only choice is to either resist by leaving the help you still want or to stay in treatment with me but feel exploited. What a sickening bind you've had to live in.

CARLOS: *(becomes quiet and nods in half agreement)*

THERAPIST: You have experienced something like this with others, Carlos, but I want you and me to have a different kind of relationship. I want to respond to your need for help, not to use your trust and vulnerability for my own gain. You've been hurt in this way too many times before.

Unable to provoke the therapist and elicit the fight-or-flight response that he had usually succeeded in getting, Carlos calmed down and was gradually able to accept what the therapist was saying. Carlos's transference reaction was that the therapist had manipulated Carlos's dependency in order to meet his own financial need. This faulty schema, that selectively filtered and biased his perceptions, slotted every rela-tionship into the same unwanted but familiar pattern. Although this pattern certainly distorted and disrupted Carlos's current relationships, developmentally, it accurately encapsulated his life story. Carlos had grown up in an orphanage—often feeling unwanted and alone. Sadly, when he was finally placed with foster parents during second grade, they took advantage of his need to finally belong to someone.

By the time he was nine years old, Carlos had become an exploited laborer on his foster parents' farm. To earn their approval and affection, he worked long days on the days he wasn't in school. At age 18, angry and very ready to leave home, Carlos had long realized that his foster parents were taking advantage of him and were not to be trusted. At some point during his early adolescence, Carlos's anxious efforts to please his foster parents had evolved into an angry defiance toward them and others. He had vowed to himself that he was never going to need anybody again. Attempting to disconfirm the shame of having felt used, Carlos resolved that he was going to be rich and powerful—strong enough that no one would ever be able to take advantage of him again. Carlos's feelings of exploitation and betrayal were further fueled by his racial and ethnic history. A Latino, Carlos was the child of undocumented migrant workers. According to the story he had been told—which he'd always suspected wasn't true—his babysitter had been instructed to leave him at the board and care facility if his parents were to be suddenly deported. Carlos was angry at their exploitation by farmers but also angry with his parents for never returning to claim him. These factors all contributed to Carlos's determination to never need anyone again. He was going to be rich so that no one would ever use or betray him; he would simply buy whatever he needed.

The dramatic confrontation with his therapist proved to be a corrective emotional experience for Carlos. Although his fear of being used was aroused in relation to the therapist, as it was whenever Carlos began to get close to anyone, his coping strategy of hostility and intimidation did not succeed in pushing the therapist away. The therapist was able to tolerate Carlos's antagonism and remain concerned about him. The therapist validated the betrayal in Carlos's familial and adoptive experiences, and how these original injustices were compounded by seeing the impact of racial exploitation in his community today.

As the therapist repeatedly provided a different response to his moving-against style than Carlos expected, the next feeling in his affective constellation emerged: His primary feelings of hurt and betrayal became the predominant issue in treatment. Just as the therapist was able to tolerate Carlos's provocative anger, he was also able to provide a holding environment to contain these painful feelings of vulnerability and loss that were so unacceptable to Carlos. Before long, they, too, became less threatening and Carlos began to have some compassion for his childhood predicament—just as the therapist did. Nevertheless, Carlos's shame about being helped and feeling sad, or concern that he was being used, continued to be activated in the therapeutic relationship. In this way, Carlos experienced the same fears and concerns over and over again with the therapist. However, each time these issues came up between them in some way, Carlos received a genuinely empathic and understanding response from the therapist that was different from what he expected and usually received from others.

In addition, it was helpful to Carlos when the therapist validated his perception that he had indeed been taken advantage of as a child, and that people of his culture were often devalued and exploited, and that indeed elements of devaluation and exploitation were present in some of his current relationships. However, the therapist patiently but repeatedly questioned whether others in his current life were *always* trying to exploit him. Meaningful changes started to take place in his life as he began to recognize that there were some others in his life who were trustworthy and interested in sharing a real friendship.

Honoring his interpersonal coping strategy, the therapist also observed that, as a child, Carlos had made the best adaptation that he could to his unsolvable predicament. The therapist pointed out that there was real strength in Carlos's attempt to rise above his circumstances by adopting an aggressive and defiant interpersonal style. This attempt to cope by rising above his abandonment and exploitation was no longer necessary or effective in many situations. However, it had indeed once served a necessary purpose—to protect him from the vulnerability he experienced as a child and was determined never to experience again. Eventually, Carlos understood what the therapist meant when he suggested that it took just as much courage for Carlos to face these heartbreaking feelings now, as an adult, as it originally did to defend against them as a child by compulsively striving to achieve power and control.

With the therapist's support, Carlos was gradually able to reexperience his old feelings of urgent need and simultaneous outrage at being used by his foster parents and, later, at being left by his biological parents. As he gradually let go of his combative and competitive stance with the therapist, and let himself undergo those painful feelings of his childhood that had always seemed disgusting and weak, Carlos began to change. First, his provocative, argumentative style gave way to a friendlier camaraderie with the therapist. Carlos began to disclose more easily and felt less competitive with the therapist. In particular, he became less concerned about needing the therapist or being taken advantage of by him.

These changes with the therapist began to carry over to other relationships as well. With the therapist's encouragement, Carlos began to establish male friendships for the first time; his need to dominate and always be right had precluded such friendships in the past. Carlos's relationships with women had always been short-lived, superficial sexual contacts, which often were manipulative. During counseling, Carlos realized that he felt less threatened in these casual, short-term relationships. If he became more committed in a love relationship, he again felt vulnerable to being left and being used. As Carlos stopped defending against his fear of being left and exploited as a child and began to feel compassion instead for his own childhood dilemma, his pattern of using women changed. At the time treatment ended, however, he was still unable to sustain an intimate love relationship on more egalitarian terms.

Other evidence of change came from Carlos's business associates, one of whom mentioned that he seemed "less driven" than in the past. His attempts to rise above his problems through achieving wealth and power were modulated and gave way to just an ambitious work ethic. When an established banker in town offered Carlos a good job as a loan officer and appraiser in his real estate division, Carlos felt ready to terminate therapy. All of his problems were certainly not solved, but Carlos now felt capable of successfully moving on in life on his own. It was important to Carlos to be reassured that he could return to therapy if he needed to in the future, and that doing this would not signify failure but rather strength in recognizing when he needed help. Assured that the therapist would welcome seeing him again in the future if he chose, Carlos terminated and did not recontact the therapist.

MAGGIE: MOVING AWAY FROM OTHERS

Six months after her adolescent daughter had been date-raped, Maggie was still in crisis—as if it all happened yesterday. Nightmares stole her sleep; migraine headaches persecuted her days. Feeling so helpless and inadequate to comfort her daughter, and enraged at the casual indifference of the lawyer and police, Maggie was afraid her life was spinning out of control. At work, her supervisor's evaluation said she was "irritable, sullen, and difficult for coworkers to interact with" and suggested that she seek counseling. Although she had always "hated" to ask for help with anything, Maggie contacted a therapist when she realized her job was in jeopardy. The therapist was responsive to Maggie's feelings of outrage, guilt, and helplessness evoked by her daughter's tragedy. To her surprise, Maggie felt understood by the therapist. Time and again, the therapist "got it"—understood what something really meant to her—and Maggie actually began looking forward to their meetings. She gradually came to trust the therapist a little more and slowly began to risk investing herself a bit in their relationship.

During one session, Maggie reported a dream from the previous night. In the dream, Maggie was alone in a vast desert night. No other people existed in this great, silent space. The desert night was black; no light shone from stars or moon. As she walked across the endless sand, a cool, dry wind began to move lightly across her face. Maggie lay down on the sand, closed her eyes, and silently slipped away into the darkness.

Maggie relayed that she had dreamt variations of this dream many times throughout her life. Because this was a recurrent dream, the therapist knew that it held much meaning and, in ways she couldn't fully understand yet, probably encapsulated her core conflict. Hoping that the dream could also provide an avenue for joining her in her central feeling of aloneness, the therapist tried to bridge Maggie's moving-away orientation:

THERAPIST: Can you close your eyes and find the dream again?

MAGGIE: Uh-huh. *(settles back and closes her eyes)*

THERAPIST: Describe what you see to me.

MAGGIE: I'm walking, but I'm tired of it; there isn't anyplace to go anyway. It's quiet and dark, empty. I'm alone. I can feel the breeze. It's sandy and the horizon is a long way off. Now I'm lying down on the sand. I close my eyes, and just slip away somehow, like going to sleep forever.

THERAPIST: I don't want you to be there alone. Will you let me join you?

MAGGIE: *(pauses, then speaks cautiously)* Well, thanks, but it feels sort of familiar, sort of safe, I guess, to be just by myself.

THERAPIST: You are familiar with being alone, that's the safe zone for you.

MAGGIE: Yeah, I guess it is…

THERAPIST: …is that what you want?

MAGGIE: *(very long pause, tears in her eyes)* No…actually, no…I really don't *want* to be alone *(looks at therapist)*. I do want to be with you, it's kind of comforting to talk with you, but it's kind of scary, too, sort of vulnerable.

THERAPIST: Uh-huh. It's kind of comforting to have me there and not be alone, but that vulnerable feeling is kind of scary too.

MAGGIE: Yeah, I guess I've just learned to do it all myself. Being alone, doing it on my own has just become the safe way. But no, I don't really want to be alone anymore.

THERAPIST: Good. Will you let me join you and maybe talk about the scary/vulnerable feelings if you want?

MAGGIE: Yes...but do you really want to join me?

THERAPIST: Yes, Maggie, I really do want to join you, and I don't want to do anything that would hurt or scare you if you invited me in.

MAGGIE: Well, maybe that would be nice, to not be alone anymore.

THERAPIST: Thank you for taking the risk to let me join you. Can we go back to the dream, but be together in it this time?

MAGGIE: All right.

THERAPIST: Good. Close your eyes and hold that same, familiar image. But you're not alone this time. I am walking toward you, and I reach my hand out toward yours. Will you take it?

MAGGIE: Yeah. It's nice...But I'm kind of scared, too...

THERAPIST: Mmm hmm. Now I'm in the dream with you—you're not alone there anymore. We're holding hands and walking together through the desert night. And it feels two ways—kind of nice, and kind of scary.

MAGGIE: Yeah, we're walking together. *(opens her eyes and looks at the therapist)* I'm not alone. It's better to have you with me. I think it's more nice than scary...but I sort of feel like crying when I say that...

This joining experience was a turning point in therapy. Maggie had grown up in a silent void, much like the setting in her dream. She had never known her father, and her mother was often "away" pursuing the next new boyfriend in her life. By 10 years of age, Maggie was regularly spending most of the weekend alone, fixing her own meals, and putting herself to bed while her mother was "out." Maggie felt unwanted and rejected by her mother, and not "safe" with the men her mother brought home—who "weren't always nice" and whom Maggie sometimes felt "scared" of. The dream reflected the emptiness of her childhood, the lack of comfort and protection she had experienced, and her lifestyle adaptation of moving away from others. However, the crisis with her daughter overwhelmed her coping strategy to withdraw and be self-sufficient and aloof. In letting the therapist join her in her dark, empty space, Maggie took the enormous personal risk of accepting the human contact she longed for yet had long since resolved to hold away. Of course, this single corrective emotional experience did not resolve her core conflict or change her interpersonal style for coping with it. However, in big and small ways, similar incidents of sharing continued to occur with the therapist.

Maggie's sense of aloneness was heightened by her consciousness of being "different" because she was bicultural, as her therapist, herself a bicultural woman, was well aware. The therapist's support and ability to articulate all the factors, familial and cultural, that contributed to Maggie's sense of isolation helped her begin to change in three ways.

First, Maggie's conflicted affective constellation emerged. The profound *sadness* resulting from her childhood neglect began to come to her—as we see at the end of the dialogue above. To ensure that the lack of response that Maggie had suffered as a child was not reenacted in their relationship, the therapist was careful to let Maggie know that her feelings were being heard this time. The therapist was skillful in communicating to Maggie that she was not in a dark, silent void anymore but rather in a caring and responsive relationship. Maggie became more comfortable sharing her loneliness with the therapist—and began to question the accompanying beliefs that she "didn't really matter very much" and that others "wouldn't be very interested in helping" if she had a problem and needed them. Feeling safe with the therapist allowed her to experience her deprivation and face it rather than dismiss it as she had always done before. She was able to register or know more realistically how painful this had actually been for her. This deep sharing and resonance from the therapist was relieving for her, but it evoked another threatening feeling—the intense *anger* she felt toward her mother for, in effect, walking away from her and for exposing her to all of these men that just kept coming in and out of their apartment. The therapist was affirming of her anger and validated her feelings of not being watched over and protected. (For example, Maggie would say, "I didn't deserve this—a mother shouldn't leave her daughter home alone to go chasing after the next stupid boyfriend or bring all those creeps into the house. The one she had around the longest was awful—he knew the bathroom door didn't lock and he used to always 'accidently' walk in when I was in there. He also used to talk about the size of my breasts and behind—he was a creep and my mom never stopped him from saying those things. I guess she preferred them to me. Even when they tried to touch me, she would say they were just 'teasing.' I hated it.") Feeling such anger toward her mother evoked the third feeling in her affective constellation—*anxiety*. Painful separation anxieties were aroused by this protest, as Maggie feared that being angry with her mother would only further push her away and leave her back where she started—feeling even more alone. The therapist affirmed each feeling in her affective constellation and, as she did, Maggie progressively became more comfortable with each feeling. In turn, Maggie also became a little more forthright, engaging, direct, and confident in her interactions with some others.

Second, as the therapist continued to understand and support Maggie through each feeling in her affective constellation, Maggie began to make important behavioral changes in how she responded to the therapist. Although Maggie had developed a strong working alliance with the therapist, she was still deeply reluctant to accept help from anyone and held a part of herself back. Interestingly, the therapist's bicultural identity had given her an *ascribed credibility*, based on perceived similarities of race and gender, which helped Maggie remain in treatment initially when she was so unsure of this process. Now, however, the therapist's *achieved credibility*—her skillfulness and the effectiveness of her responses—took center stage. As their working alliance continued, Maggie incrementally relinquished her coping style of moving away and gradually allowed the therapist to see her, know her, and help her. Maggie became less reserved toward the counselor and, instead, more expressive and responsive than she had been with others. She talked more freely about herself and found herself being curious about the therapist's own personal life. The therapist saw this as positive for this particular client and responded willingly at times. Maggie's nightmares and headaches already had been alleviated, but now a sense of humor and relaxed confidence were emerging as well.

Third, as Maggie talked about feeling "fuller" inside, the therapist observed that changes were occurring in other relationships as well. At work, her supervisor said he was pleased to observe that Maggie was "less irritable and sullen than before"—it was easier for others in the office to talk and work with her. When tensions surfaced with others, Maggie's initial reaction still was to withdraw and "go away inside." However, Maggie became better able to approach problems with others and talk through the interpersonal conflicts that inevitably come up in a busy workplace. For Maggie, it was especially helpful for her to describe a conflict she was having with a coworker or customer and have the therapist suggest new or different ways for her to respond. Together, they would role-play different responses that would keep Maggie involved in the situation, rather than withdraw or remain aloof as she had always done in the past. Maggie loved to take the part of the difficult customer or coworker and have the therapist role-play what Maggie could say or do in that situation. Maggie found this behavioral training invaluable. With success, she regularly found herself using almost the same words and phrases with others that the therapist had modeled.

Especially meaningful for the therapist to observe, Maggie became more accessible—more emotionally present—with her two adolescent children. For the first time, she talked more about her own interests and personal history with them. The nearly grown children welcomed this sharing and, in turn, began to say more about themselves and what was going on in their lives as well. Maggie was feeling closer to her children and, with coaching from the therapist, more capable of helping her daughter with the aftershocks of her assault.

In addition, Maggie began to change how she responded to her boyfriend. This had been a superficial relationship, like most others in her life had been. Even though they had known each other for more than two years, Maggie now invited more personal sharing and emotional closeness in their relationship and asked for more commitment from him. Specifically, she asked him to talk more about himself and to spend more time with her, which he was able to do. It was exciting but anxiety arousing to take each of these steps forward. Her old coping style of moving away was activated each time she took a step toward her boyfriend or others, especially when others did not respond in completely positive ways. Repetitiously, the therapist celebrated her successes in each of these arenas, and patiently helped her work through the setbacks and recover from the disappointments that regularly came along as well. As she continued to make progress and participate more fully in life, however, Maggie tentatively suggested that it might be time to end counseling.

Although things had been going well, Maggie became depressed as they started talking about ending. Even though she had brought up the topic of termination, she missed their next session ("I just completely forgot"), and arrived late to another. As before, the therapist continued to focus Maggie inward, and her profound feelings of being unwanted and alone, unprotected and vulnerable emerged again. The therapist suggested that they put off setting a termination date for a while and work further with this sadness. With more vividness and detail than before, Maggie recalled her childhood depression. Painful recollections returned—for example, the memory of being eight years old, sitting alone on the living room couch, and listening to the clock tick the empty afternoon away. Maggie sobbed as she recalled her aching wish for a mom she could come home to, which was only reenacted in her unfortunate marital choice. Following her early maladaptive schemas, Maggie had married a salesman whose job took him away from home for extended periods—and he had been preoccupied and emotionally unresponsive to her when he was home as well. Thus, in her first significant love relationship, Maggie chose someone who repeated, rather than resolved, her history of aloneness. However, touching the pain of her childhood neglect so directly, and sharing so fully how it was echoed in the disappointment of her marriage, relieved the depression that the suggestion of terminating had precipitated. Before long, Maggie again felt ready to end and, this time, successfully terminated.

CLOSING

In this chapter, we have studied an interpersonal model for conceptualizing clients. In an extended case study of Peter, a moving-toward client, we saw how this model can be applied to treatment. We also explored two case summaries that illustrate its application to moving-against (Carlos) and moving-away (Maggie) clients. The therapists in these three cases were helpful, in part, because they used their clients' rigid interpersonal coping styles as an orientating focus for treatment. These therapists were also helpful because they worked collaboratively and engaged the client in recognizing and expanding their coping strategies. Although new therapists often find these three coping styles relevant with their clients, they can intervene with them ineffectively. To help, Safran and Muran (2000) clarify that the therapist's premature attempts to identify a relational pattern (that is, force it) usually feels blaming to the client. They encourage collaboration and, again, as Kiesler (1996) and others guide, to share observations with "skillful tentativeness." Emphasizing the subjectivity of the therapist (for example, "My sense is…" "What occurs to me…" "I'm wondering if…") suggests a more *egalitarian* relationship. This allows clients to be freer to accept, reject, or modify the therapist's observations and make them their own. Similarly, once therapists identify these patterns, they can press clients to change the patterns too quickly. Instead, a better stance with most clients is to encourage a simple *awareness* of the relational patterns, rather than press clients to change their coping strategy prematurely. For example,

THERAPIST: Let's watch for this and see when it comes up. You know, just pay attention to moments when you find yourself accommodating or trying to please too much—and we'll start to track that together. As you become more aware of it, I think it will become easier for you to change it and respond differently if you choose.

Looking ahead, new therapists need further guidelines for conceptualizing clients and help to shape treatment plans that guide where they are going in treatment and what they are trying to accomplish with each client. To do this, further information about clients' current interpersonal functioning will be provided in Chapter 8. At this point, however, readers are prepared to utilize the guidelines for keeping process notes provided in Appendix A, and for writing case conceptualizations in

Appendix B. These guidelines have two overarching goals. First, they will help therapists formulate treatment plans by clarifying the maladaptive relational and cognitive
patterns that provide a focus for treatment. Second, they will help therapists recognize the process dimension and intervene more effectively by linking the current interaction with the counselor to the problems that clients are having with others.

The therapeutic relationship is the most important tool that therapists can use
to help clients resolve problems and change. In Chapters 9 and 10, we will discuss
further how the therapist can provide a corrective emotional experience and generalize this experience of change with the therapist to relationships outside therapy.

SUGGESTED READINGS

1. In the current managed-care milieu, it has become essential to write effective
   case formulations with clear treatment goals and programmatic intervention
   strategies. In Chapter 7 of the Student Workbook, an extensive case formulation of the character Hap Loman, from Arthur Miller's play *Death of a Salesman*, is provided that illustrates the format for writing case formulations
   provided in Appendix B.
2. Useful guidelines for conceptualizing clients are found in Chapters 5 and 7 of
   *Psychotherapy in a New Key: A Guide to Time-Limited Dynamic Psychother-
   apy*, by Strupp and Binder (1984). Hanna Levenson's (2003) article "Time-
   Limited Psychotherapy: An Integrationist Perspective" (*Journal of Psychotherapy Integration* 13(3/4), 300–333), is an excellent articulation of an integrated
   interpersonal approach with roots in attachment theory and object relations.
3. Readers are encouraged to examine Wiger's *The Psychotherapy Documentation Primer*, 2nd ed., (Hoboken, NJ: Wiley, 2005), a practical resource for all
   aspects of psychotherapy documentation.
4. Karen Horney's work is essential reading for interpersonally oriented therapists;
   it has been widely incorporated by psychodynamic and cognitive-behavioral
   theorists alike. Readers may be especially interested in her *Neurosis and Human
   Growth* (1970) or *Our Inner Conflicts* (1966), both of which stand the test of
   time. *Cognitive Therapy of the Personality Disorders*, by Beck et al. (2003),
   effectively links Horney's interpersonal coping styles to cognitively based interventions for clients with difficult-to-treat personality disorders, and John
   Gottman uses these coping styles to illuminate marital conflict (J. Gottman &
   J. DeClair, *The Relationship Cure*, NY: Three Rivers Press, 2001).

RELATIONAL THEMES AND REPARATIVE EXPERIENCES

CHAPTER 8

CONCEPTUAL OVERVIEW

This chapter explores further the interpersonal patterns that are causing problems in clients' lives, highlights different ways they can be played out in the therapeutic relationship, and suggests useful ways to respond to these reenactments. In particular, we are going to explore these maladaptive relational patterns to help therapists conceptualize what is really wrong for this client and better understand what is going on in the therapeutic relationship. We will learn how therapists can use their own feelings and personal reactions toward the client as one of the best ways to understand the problems this client is having with others. That is, clients often are adept at getting the therapist to feel or respond toward them along the same problematic lines that occur with others in their lives (for example, leading the therapist—like others in their lives—to feel overwhelmed, bored, or controlled by them; helpless or discouraged about being able to help; anxious about offending them or making a mistake; and so forth). In this way, *we will look very closely at how the problematic patterns that are disrupting clients' relationships with others can be brought into the therapeutic relationship and played out with the therapist along the process dimension.* Empowered by an understanding of what is going on between them, therapists can formulate better treatment plans and know where they are trying to go and what they are trying to do to help their client. Throughout, we will suggest different ways that therapists can intervene to change these problematic reenactments by making process comments, providing interpersonal feedback, and using other immediacy interventions to disconfirm rigid schemas and provide corrective relational experiences.

THREE WAYS CLIENTS REENACT THEIR PROBLEMS WITH OTHERS IN THE THERAPEUTIC RELATIONSHIP

The more serious and enduring problems that clients try to resolve in treatment usually originated in attachment and formative family relationships, and now are being played out in thematically similar ways in current relationships. Routinely, the same interpersonal patterns that are causing problems with others (such as being excessively controlling or competitive, responding passively or compliantly, being distant or withdrawing, and so forth) emerge in some form in the therapeutic relationship as well. As we have seen, clients do not merely talk with therapists about the problems they are having with others. In a deeply experiential or here-and-now way, they re-create relational themes with the therapist that parallel the problematic patterns they are playing out with others. The challenge for the therapist, then, is to find a way to alter or change this pattern that causes problems in clients' real-life relationships. Providing this experience of change with the therapist—as opposed to merely inter-preting or explaining it intellectually as so often occurs—greatly facilitates clients' ability to change these faulty patterns with their spouses, children, and important others. Thus, throughout the course of treatment, the therapist's aim is to:

1. identify these maladaptive relational patterns that keep recurring with others,
2. provide a new and better response that does not repeat the familiar scenario in their interaction, and
3. help generalize this experience of change in the therapeutic relationship to clients' interactions with others in their everyday lives.

Therapists practicing in varying interpersonal/relational, psychodynamic/inter-subjective, attachment, and cognitive/schema-oriented approaches are all working with three closely related but distinct ways in which clients bring their problems with others into the therapeutic relationship: eliciting maneuvers, testing behavior, and transference reactions.

Briefly, clients often employ *eliciting maneuvers* (Sullivan, 1968) to avoid anxiety and defend against their problems by getting others to respond in certain predictable ways. At other times, clients will employ a second interpersonal strat-egy, *testing behavior* (Weiss, 1993), to cope with problems. When "testing," clients behave in certain ways to assess whether the therapist is going to respond in the familiar but problematic ways they expect or in the more helpful ways they actually need. The third (and better-known) way that clients bring their conflicts into the therapeutic relationship is through *transference reactions or schema distortions*—clients' systematic misperceptions or cognitive distortions of the ther-apist. These three concepts illuminate much of what is going on in therapeutic relationships. In all three concepts, the basic assumption is that the client recreates certain problematic relationship patterns with the therapist as a way to *confirm or reject* behaviorally that the therapist (and others) are going to respond in the same unwanted way that important attachment figures once did (for example, be indif-ferent, judgmental, demanding) or in a new and more helpful (corrective) way (Gelso & Hayes, 1998; 2002). Let's look closely at these important interpersonal processes and the countertransference issues they often evoke in therapists.

ELICITING MANEUVERS

Many cognitive and interpersonally oriented therapists have described how clients develop fixed interpersonal styles to avoid anxiety and defend against the unwanted responses they expect from others (see, for example, Beck et al., 2003; Benjamin, 2003). Routinely, clients are inaccurate in their perceptions of others and restricted in their range of emotions. As we saw in the last chapter, they often respond to others inflexibly, using rigid coping styles of moving toward, moving away, and moving against. In part, clients systematically employ these interpersonal styles to (1) elicit desired responses from others that will avoid conflict and ward off anxiety and (2) preclude threatening or unwanted responses from others that will trigger their primary schemas and key conflicts. Thus an **eliciting maneuver** is an interpersonal strategy that wards off anxiety and brings about certain desired, safe responses. However, this maneuver keeps the client stuck, and change, personal growth, and interpersonal relationships remain stymied. To illustrate how eliciting maneuvers serve to defend against problems, we will return to the case study of Peter, the moving-toward client in Chapter 7.

Peter's moving-toward style tended to elicit approval and kindness from others. As a general way of life, Peter was helpful; he sympathized, agreed, and cooperated with people. For example, Peter tried to "understand" his girlfriend's infidelity and to win his instructor's approval. When Peter began to see a therapist, he continued to employ the same interpersonal style that he had used with others throughout his life. As many moving-toward clients do with their therapists, he tried to elicit his therapist's approval by being a good client who was quickly getting better—at least until the therapist began using process comments to question this compliance. For example, "Peter, sometimes I wonder if you are trying too hard to figure out what I want you to do. Any thoughts about that possibility?"

In contrast, Peter was rarely forthright, appropriately assertive or angry, or skeptical with anyone in his life. These responses, which a well-functioning person needs to employ at times, were not part of Peter's interpersonal repertoire. For example, he did not communicate how angry he was with his girlfriend or even have much awareness of how angry he actually was with both of his parents. Similarly, he did not set limits with his practicum instructor or classmates regarding how much critical feedback he could incorporate at one time. Thus, *his pleasing interpersonal coping style was designed to elicit approval from others and discourage angry or critical responses that would arouse intense anxiety for him.*

If Peter's therapist had merely responded automatically, without first considering what Peter's interpersonal style tended to elicit from others, Peter would not have changed much in therapy. That is, if the therapist had merely supported Peter and met his need for approval, Peter would have only reenacted in therapy the same rising-toward defense that he had used throughout his life. Shoring up these defensive coping styles is often the goal in some time-limited or supportive therapies and helps many clients in crisis regain their equilibrium. However, the therapist wanted to do more than this with Peter. Seeking more enduring changes and not merely symptom relief, the therapist tactfully used process comments to help make Peter's moving-toward style overt as a shared focus for discussion. Thus, instead of automatically responding to what Peter elicited and providing only approval, support, and reassurance (which he felt), the therapist focused on this eliciting maneuver as a part of Peter's problems that needed to be addressed in treatment. How did this intervention occur?

Over the course of several months, the therapist remained watchful for opportunities to tentatively wonder aloud—in a supportive and noncritical way—about instances where Peter's coping style of moving toward others might be operating with him and with others in Peter's life. Soon, working collaboratively on this, Peter and his therapist began to consider how on the one hand, this was serving to protect him from his problems, and on the other how Peter suffered as a result of this coping style. As they "named" this behavior pattern and began looking for it together, Peter increasingly reported instances where he had observed himself using this moving-toward style with others, and recognized how it elicited approval and support from others. He also began to see how this style discouraged the critical or rejecting responses that were so familiar from his childhood, and that were so painfully anxiety-arousing for him. As Peter made progress in treatment, he gradually became less preoccupied about winning approval from others and less worried about "fixing things" when small interactions with others held some conflict or just didn't go so smoothly. In other words, the treatment goal of expanding his interpersonal range was being met as he continued to report instances where he was feeling less self-conscious than before, speaking up more directly with others, and standing up for himself more assertively when the situation called for it. Peter was excited about these changes and described himself as feeling more "resilient."

Eliciting Maneuvers from the Client Trigger Countertransference Reactions in the Therapist.   This is where things get really interesting! The eliciting maneuvers used by moving-toward clients like Peter are familiar and not especially difficult for most new therapists to work with. In contrast, moving-against clients employ eliciting maneuvers that are often challenging for beginning and experienced therapists alike. In the first few minutes of the first counseling session, these clients often do something to take command of the relationship. Typically, moving-against clients readily find some way to intimidate or disempower therapists, making them feel insecure and one-down in the therapeutic relationship (or sometimes angry and competitive). For example, clients may insist on sitting in the therapist's chair, questioning therapists' adequacy, diminishing their credentials, or criticizing what the therapist has just said or done:

CLIENT: So, if you're just a trainee, what makes you think you know enough to be able to help me?

THERAPIST: Well, I've had some experience before I entered the program, and I have a supervisor to help me.

CLIENT: I see. Now, does this supervisor always tell you what to say, or do they let you newbies say what you think sometimes?

This provocative presentation elicits anxiety in many therapists—as it often succeeds in doing with others in the client's life. Many therapists respond by trying to be nice in order to stop the hostile challenges; simply continue along, hoping that ignoring the hostile subtext and acting as if it weren't occurring will stop it; or simply become quiet or withdraw emotionally. In contrast, some therapists will respond in kind and become competitive or punitive, and a few will even counter with their own overt hostility. Treatment will not progress when moving-against clients succeed in eliciting one of these defensive flight-or-fight responses from the therapist—as they commonly do with others. By reenacting this maladaptive relational pattern with the therapist, these clients have successfully neutralized the therapist's ability to help them (or to hurt them). In this way, they have defended against their anxieties associated with having problems, asking for help, relinquishing some control, getting closer to what's really wrong, and so forth. Thus, clients' eliciting maneuvers protect them from their core conflicts, but at the price of change.

Unfortunately, therapists tend to respond "reflexively" or automatically to the client's eliciting maneuvers with their own countertransference propensities—that is, with their own characteristic tendencies toward flight or fight (Millon & Grossman, 2007; Mueller & Aniskiewicz, 1986). What can therapists do instead? First, they can try to retain legitimate control of themselves internally by trying to understand what is occurring right now in their relationship with the client—rather than **personalizing** it so readily and overreacting. It certainly is challenging for new therapists to do this in the beginning, but with experience and help from a supervisor the goal is to begin formulating working hypotheses about:

- the impact this moving-against behavior might have on others in the clients' lives (such as making others feel like arguing with or withdrawing from them) and
- what conflicts or anxiety arousing situations the client might be avoiding by these maneuvers (feeling "weak" or ashamed, for example, of needing help or having a problem they can't solve on their own).

Second, on an interpersonal level, *therapists are trying to find ways to engage the moving-against client in some way in which the interpersonal patterns of domination, intimidation, competition, and so forth, are not reenacted.* Immediacy interventions offer us many possibilities.

- One option is for the therapist simply to withstand the client's criticism, without avoiding the challenge or becoming defensive. For example,

THERAPIST: *(calmly, straightforwardly)* So, your previous therapist was an experienced psychiatrist, and you're not sure that a younger social worker such as myself can help you. Let's talk together about that—tell me more about your concern.

- Another alternative is for the therapist to make a process comment and engage the client in talking together about the current interaction and what's going on between them.

THERAPIST: You're speaking to me in a harsh, loud voice right now, and you sound angry. What are your thoughts about what might be going on between us here?

- With some clients, the therapist may wish to use a self-involving statement or provide interpersonal feedback to begin exploring the effects of their interpersonal style on others and highlighting the relational patterns that ensue.

THERAPIST: Let me take a bit of a risk here and talk about the impact you have on me sometimes. Right now, I'm feeling criticized as you correct me again, as I've felt other times. That leaves me wondering how others in your life usually respond to you when you are critical like this—what usually happens next?

Some of these clients will be unaware of the distancing and disruptive impact their criticalness has on others, and this interpersonal feedback can provide a new opportunity to begin exploring this collaboratively. Other moving-against clients will say something that suggests that they "hit" first before someone can get them. To emphasize, however, the purpose here is not to win the battle for power and control and get one-up with the client. On a process level, this interaction would only continue, or recapitulate, the same cyclical pattern that recurs with others. Rather, this approach is intended to help the therapist find some way to engage with the client other than by automatically responding with counterhostility, control, or withdrawal to the eliciting maneuvers—as others usually do. It will also help clients develop increased awareness of when they employ their coping style, its impact on others, and most important for change to be achieved, the price they pay for this coping style. As clients hear interpersonal feedback from the therapist, become more aware of the impact they are having on others, and begin to reflect on when and why they act this way, they may begin to respond more flexibly and exercise more choice about responding this way.

- Still another possibility is for the therapist to explore how the client's eliciting maneuvers affect the client.

THERAPIST: (*nondefensively*) You have insisted on sitting in my chair, even though I asked you not to. Now that you are there, how does it feel?

Routinely, moving-against clients become aware of feeling alone, empty, or anxious at these moments, and the hollowness of their victory provides a new shared point of departure for the therapist and the client. Thus, as these four different types of responses illustrate, there is no simple formula for the "right" way for the therapist to respond, of course. Per client response specificity, one type of response will work well with one client and not at all with another. Therapists need to be flexible, assess how the client responds to each intervention, and modify their responses to find what works best for this particular client (Persons & Silberschatz, 1998). Thus, to respond effectively to these challenging clients, therapists' intentions are:

- to attend to their own subjective reactions and recognize what the client is eliciting in them (such as feelings of incompetence or competitiveness);
- to formulate working hypotheses about how others would typically respond and begin to identify the relational patterns that are occurring with others and causing problems in the client's life (for example, others tend to withdraw or argue);
- to try to find another way to respond (for example, via process comments, self-involving statements, or interpersonal feedback) *that does not go down the old familiar path and reenact the same problematic scenario that the client usually elicits from others* (for example, Therapist: "If your secretary was observing us through a one-way mirror right now, do you think she would agree that you are talking to me in the same contemptuous tone of voice that causes so much trouble for you at work?").

Although moving-against clients often intimidate new therapists, this doesn't usually last very long. With a little more experience and a supportive supervisor, new therapists learn not to charge so readily at this red flag the client is waving. With more exposure to moving-against clients, therapists learn to appreciate that the intensity and rigidity of clients' eliciting maneuvers, no matter how alienating initially, are commensurate with clients' degree of anxiety and conflict. As therapists better understand the purpose of these distancing maneuvers, they begin to find some compassion for the predicament such clients are living out, which is usually the very best way to step out of the old relational pattern being elicited. By flexibly trying out different approaches (it helps to practice by role playing with classmates and supervisors), it's really not so hard for therapists to find ways to engage these clients that do not reenact either the intimidated/withdrawal or competitive/hostile patterns that they usually succeed in eliciting from others.

As we have seen, the confrontational eliciting maneuvers of moving-against clients—such as Carlos in Chapter 7—place considerable stress on many new therapists. Surprisingly, however, the meeker, moving-away clients—such as Maggie in Chapter 7—often pose a greater challenge in the long run. Many who select counseling careers deeply enjoy and have strong needs for close personal relationships. When moving-away clients continue to maintain their characteristic stance of emotional distance, many therapists feel unimportant, become discouraged about the therapeutic relationship, and ultimately give up—disengaging as many others have done before (Robbins & Jolkovski, 1987). When this occurs, the clients' eliciting behaviors have successfully defended them against a meaningful relationship with the therapist and thereby allowed them to avoid the anxiety-arousing issues—which were reality-based threats in their history, evoked by risking genuine involvement. Thus, the client's conflicts, or the threats and dangers that involvement has brought in the past, will not be activated, but the therapist's only real vehicle for resolving these problems is closed as well. Rather than responding automatically to these clients' aloofness by giving up and, commonly, blaming the client (for example, Therapist: Maybe this client just doesn't really want to be in therapy or work hard enough to try and change right now), the therapist may use process comments to find other avenues for establishing an alliance:

THERAPIST: You're very quiet, Mary. Sometimes it almost feels as if you are hardly in the room with me. Would it be OK with you if we talk a little about what it's like for you to come here and take the risk to try and bring me in on your life?

CLIENT: I'm not sure what you mean. Am I doing something wrong?

THERAPIST: Oh no, you absolutely are not doing anything wrong at all. I guess I'm trying to say that sometimes I feel like I'm not reaching you—or getting close to what really matters in your life, even though I want to. Has anyone else ever said something like this to you—or maybe this is just me?

CLIENT: Well, kind of. The only really nice boyfriend I've ever had left because he said he never really felt very important to me, that he gave up trying to "pursue" me.

THERAPIST: Thanks for joining me. This is a sensitive thing for us to be talking about, but I think this is important. It sounds like other people feel held away from you in some way or that they have to work to reach you...or maybe that you can "go away" so readily or do without them so easily. I'm trying to understand how this goes in your life. Can you help me say it better?

CLIENT: I think I do withdraw from people easily. I know I'm too sensitive—my feelings get hurt so easily...over "nothing" they would say.

THERAPIST: I see, you go away when your feelings get hurt, and maybe others don't understand what's really going on for you. We'll have to watch for that here and make sure that doesn't happen between you and me, too. But right now, I'm appreciating that you're choosing not to go away from me, but taking the risk to stay present and engaged with me as we try to sort this through. This is feeling very different right now than how we usually are. I'm feeling more connected with you, and I like it, but I'm wondering what it's like for you?

CLIENT: I don't know...I'm not so sure. I just get afraid with other people. I know I shouldn't go away so easily, but it just feels easier...it's safer.

THERAPIST: It just "feels safer." And maybe it's not feeling very safe with me either right now.

CLIENT: Oh, you're very nice. But it is just easier to go away...it's just what I've always done.

THERAPIST: Well, I'm thinking that it's probably easier for you to "go away" because it really hasn't been safe for you in the past. Would you be willing to tell me a little more about how I might hurt you if you stay engaged with me like this, or what others have done in the past that you really did need to get away from? You lead and I'll follow.

To sum up, the therapist can attend to what clients systematically elicit in others and hypothesize how this interpersonal strategy is likely to affect others and cause problems in their lives. In doing this, the therapist attends to clients' developmental history, current life situation, and to the cultural context in which coping styles and eliciting maneuvers are embedded. Importantly, *therapists are also attending to their own feelings that are elicited by clients.* This self-awareness is one of the therapist's most important sources of information about clients and what their interpersonal style tends to elicit from others. To be able to learn from one's own reactions to the client, the therapist aims to be **nondefensive** enough to be able to observe what clients tend to elicit or "pull" in them (such as feeling controlled, ignored, pushed away, and so forth).

Attending to their own personal reactions in this way will often provide important information about clients' key concerns, coping strategies for defending against old problems, and how the relational themes being played out in the therapeutic relationship may be disrupting other relationships. Doing this is very useful in highlighting the *treatment focus*:

1. What has brought the client to this particular place in his life (developmental history);
2. What is causing problems in the client's life now and how might this be similar to what he has had to deal with developmentally;
3. What is the cultural context in which this all developed;
4. How are these issues potentially being replayed in the client–therapist relationship; and
5. How can I address and clarify what has been difficult for the client and provide an experience in our relationship that is different, reparative, and empowering?

We have thus seen that the therapist can generate working hypotheses to understand why clients characteristically elicit certain responses from others, such as reassurance for a moving-toward client like Peter, competition toward a moving-against client like Carlos, and disinterest from a moving-away client like Maggie. Easy to say but often hard to do, the therapist's intention is to keep from responding "reflexively" or automatically with the types of response clients usually elicit from others and, if possible, try and respond in a new and better way that does not reenact the old scenario that the client is pulling for. Over time, the therapist can begin to help clients identify their eliciting mechanisms, recognize the situations or types of interactions where they tend to employ these measures, and understand how these maneuvers have served to create safety in the past. It does take courage for therapists to do this, however, especially as they begin their clinical training. On the one hand, therapists are trying to tolerate their own unwanted feelings (irritation, impatience, boredom, uncertainty) long enough to sort through whether these unwanted reactions have more to do with their own lives and personal feelings (that is, countertransference) or, on the other hand, whether these feelings and reactions inform us about the client's eliciting maneuvers and help us understand what this client tends to evoke in other relationships as well (that is, eliciting maneuvers)? The key element here is for therapists to remain nondefensive toward their own reactions so that they can reflect upon them and learn more about what's going on in the therapeutic relationship. Let's explore this interface further and see how the client's eliciting maneuvers often tap in to the therapist's own personal issues—activating countertransference reactions that lead to reenactments and impede treatment.

Client-Induced versus Therapist-Induced Countertransference.   Let's distinguish two different types of countertransference. As we have seen, the client's eliciting maneuvers tend to evoke certain similar reactions in the therapist and others—which we will call **client-induced countertransference** because the client tends to induce these reactions in most people. In contrast, we are beginning to see that things quickly become far more complicated when the clients' eliciting behaviors happen to tap into the therapist's own personal issues or current situation—which we will call **therapist-induced countertransference** because it has a lot to do with the therapist personally. Thus, with therapist-induced countertransference, it's not just about what this client tends to elicit in others, but how the therapist's own life and personality have been brought into the equation as well. To illustrate, suppose that a hyper-activating client acts helpless and keeps escalating his distress, and the therapist tries hard to respond better and help but "fails" at each attempt. Ideally, the therapist would retain a certain objectivity, remain actively engaged in the work but not overly invested in the client's response, and keep exploring other ways to respond—for example, by making the process comment that whatever the therapist tries to do seems to disappoint the client or fails to help. However, what if this therapist grew up with critical caregivers and, no matter what she did, she could never quite do it right or please them? If the therapist's own personal concerns regarding performance demands or criticism are activated in this way, she may lose her effectiveness because the client's eliciting maneuvers have tapped into her own history and personal issues (in other words, "therapist-induced" countertransference). Continuing, the therapist who grew up being criticized excessively, and now is overreactive to this client's unspoken disapproval, may feel inadequate and withdraw—believing falsely that she is failing or incompetent as a therapist, or perhaps feeling resentful and becoming critical toward this client in turn. *When clients successfully elicit the same type of response from the therapist that they usually elicit from others, as is occurring here, the therapist and client are reenacting rather than resolving the problem.* Also termed "enactments" by intersubjective theorists and researchers (Aron, 1996; Renik, 1998), such situations replay unwanted patterns rather than provide a corrective emotional experience. We are into complex and challenging issues for new therapists here, but let's keep at it—all of this has much to do with treatment outcomes. As student therapists learn more about these ways of understanding what may be going on between the therapist and the client, they discern more specific ways to intervene and help their clients change.

Scenarios such as the one above where the client's eliciting maneuvers activate therapist-induced countertransference and lead to reenactments in the therapeutic relationship are commonplace—for beginning and experienced therapists alike, it's just part of the work. When this type of reenactment occurs, however, and clients' defensive eliciting maneuvers have succeeded in immobilizing or discouraging the therapist, clients often switch to the other side of their **ambivalence** and try to reengage the therapist, perhaps by reassuring the therapist about how helpful the sessions are. At some level, clients do not really want to succeed in immobilizing the therapist or ending the relationship by repeating the same scenarios that have caused problems with others in the past, and *they will often try to get the disengaged or discouraged therapist involved in the relationship again.* Premature and unnecessary terminations commonly occur, unfortunately, because the therapist (and perhaps the supervisor) often fail to hear the client's renewed bid for a relationship because the discouraged, frustrated, or disengaged therapist has given up on the relationship.

THERAPIST: This client is hopeless; he isn't ready to change. He's the King of Blame—he just wants to tell everybody else what's wrong with them—including me! No wonder his third secretary just quit and his teenaged son doesn't want anything to do with him.

Especially with the more troubled clients we work with, therapists want to keep both of these disengaging and subsequent reengaging maneuvers in mind.

In closing, the key issue for therapists here is to be mindful of tracking and trying to distinguish between their client's eliciting maneuvers and their own personal reactions toward the client—often referred to as *client-induced versus therapist-induced countertransference* (Springmann, 1986). How do therapists begin to sort through for themselves whether the client's eliciting maneuvers (client-induced), their own personal issues (therapist-induced), or, commonly, some combination of both are evoking their strong reactions? If this client makes others feel or react similarly (for example, others in the client's life are also bored, impatient, or intimidated, as the therapist is made to feel in the session), it is probably client-induced. If not, this reaction probably has more to do with the therapist's own personal life (therapist-induced countertransference). Of course, when reenactments and ruptures are being experienced, both may be occurring (Hill & Knox, 2009) and supervision is usually the best place to sort this through (Falender & Shrafransky, 2004). We will be exploring these two different conceptions of countertransference further, but first let's look at the related concept of testing behavior.

TESTING BEHAVIOR

We have seen how clients employ their inflexible coping style to systematically elicit responses from others that may seem self-defeating at first yet have served to effectively avoid anxiety or support self-esteem in the past. These eliciting maneuvers—or "security operations" as Sullivan (1953) first termed them—are interpersonal defenses.

Eliciting maneuvers are used defensively to avoid anxiety, but at other times clients use different interpersonal strategies to approach and try to resolve problems. Although this may sound contradictory, it reflects clients' ambivalent feelings about treatment—*the hope that things could change and get better versus the schema-driven expectation that the same unwanted outcomes will only occur again.* Clients often carry out this healthy, approach side of their ambivalence by means of **testing behavior** (Sampson, 2005; Silberschatz & Curtis, 1993; Weiss, 1993). In direct or in more covert ways, clients reconstruct interpersonal patterns with the therapist that follow those that have been causing problems with others in their lives. Clients "test" to ascertain whether the therapist will respond in the familiar but problematic way that others often have in the past or, as is hoped, whether the therapist can respond in a different and more satisfying way that changes the old predictable scenario; this will allow clients to expand schemas and expectations, and facilitate progress on their problems. There is nothing casual about this testing for clients; *they are vitally interested in assessing whether the therapist is going to confirm or disconfirm pathogenic beliefs.* Many clients have profoundly painful histories of exploitation, denigration, and rejection; for these clients, their trust in others has been betrayed. Their testing behaviors with the therapists often reflect their intense wishes for healthier, safer relationships, hoping that the therapist's better responses to them will disconfirm their fear of again being violated, disappointed, or hurt. *Through testing, in other words, clients are trying to assess safety/danger in the therapeutic relationship.*

Therapists respond effectively and **"pass"** **important tests** by *behaviorally* disconfirming the client's faulty beliefs and expectations—and this corrective or real-life experience (not just interpretations, explanations, or reframing) sets the change process in motion. Clients feel safer when the therapist does not respond in the familiar but problematic way that important others often have, and that they have now come to expect from the therapist as well. Immediately following this corrective emotional experience, clients often make visible progress in treatment. For example, within the next minute or two, the therapist may observe that clients act stronger or behaviorally improve in one of these ways:

- They express more confidence in themselves or in the therapist ("You know, I think I'm just going to go ahead and talk to my wife about this. I've gone around about it in my head for too long now. I'm going to bring this up with her and just put it on the table—it's not good for me to keep avoiding it like this.")
- They feel better as symptoms such as anxiety or depression diminish ("I feel more settled right now. Maybe she won't ever stop being so bossy, but that's just her—it really doesn't have much to do with me. Maybe she can't change, but I don't have to be so reactive and get so angry when she's demanding like that.")

- They take the risk of bringing up threatening new material with the therapist ("I've been ashamed to tell you about something that's happening: I've been having an affair for a long time. I know I should be talking to you about it, but I've been afraid to tell you. I guess I've been worried that you'll judge me—you know, you wouldn't respect me anymore if you knew about this.")
- They act stronger by being more honest or forthright with the therapist and discussing problems between them ("It's kind of hard to say this because I don't want to hurt your feelings or anything, but I've been thinking for a while that therapy hasn't been working very well for me. You're so quiet—I don't know what you're thinking a lot of the time. I think I need more feedback or something from you.")

Statements and behavioral changes such as these provide evidence that the new or reparative response from the therapist behaviorally disconfirmed the client's old schema or pathogenic belief (which often has been painful and engendered anxiety, shame, and guilt) and creates **interpersonal safety** for clients. Observing clients' reactions to their interventions, therapists will often see that this increasing safety in the therapeutic relationship allows clients to make progress in the next few moments and change how they are responding to the therapist and others in their lives. Later in this chapter, we'll return to this important concept of assessing the client's responses to the therapist's interventions but, for now, let's explore further this important concept of testing behavior.

In a healthy attempt to address and resolve problems, clients often actively (but without really being aware of it) elicit the same types of problems with the therapist that they are having with others. Routinely, clients are actively testing to see whether they can change these problematic relational patterns and disconfirm the pathogenic beliefs that stem from them. However, it is difficult for clients to find disconfirming evidence because repeated confirmation of the beliefs has made clients deaf to such evidence. Cognitive therapists have clarified the selective attention and biased filters that lead clients to keep slotting diverse experiences into the same narrow categories (Beck, 1995). However, the *power of immediate, in vivo relearning with the therapist allows clients to begin questioning, or sometimes even to relinquish, deeply held but faulty beliefs about themselves and expectations of others*. Thus, the therapist can **fail the client's test** by responding in a way that unwittingly repeats the hurtful relational patterns and confirms pathogenic beliefs— setting the client back. Alternatively, the therapist can pass the test and provide a corrective emotional experience by behaviorally demonstrating that, at least sometimes, some relationships can be different—which expands rigid schemas.

To illustrate, suppose a young adult client is being seen in a college counseling center for depression. Her therapist has formulated the working hypothesis that this client is ruled by the pathogenic belief that if she has her own mind and does what she wants with life, her mother will be hurt, feel sad, and withdraw. Thus, she believes that she deserves to feel bad whenever she **differentiates** in any way (does what she wants or meets her own need) because she has been selfish and hurt her mother. How can this hypothesis guide the therapist's interventions as he tries to pass the tests this client is apt to present? The therapist would likely fail her test, that is, he would be repeating the dysfunctional interpersonal pattern that the client lives out with others, if he expressed caution, worries, or doubts when this inhibited client said something about becoming more independent, speaking up, acting on her own behalf, or pursuing her own interests and goals rather than submerging herself to the needs of others. For example:

CLIENT: I don't think I want a career in business after all. I'm not liking these business classes I've been taking very much. The other students are fine, but I'm not really like them. I think I might like teaching better—kids have always been fun for me.

THERAPIST: Well, maybe education would be a better major for you but, gosh, that job market sure is tight right now, and it hasn't ever paid as well as it should.

CLIENT: (*sinking*) Yeah, I wasn't thinking about that. I guess I need to be more practical. Maybe teaching wouldn't work very well for me after all...

Per client response specificity, the counselor's reservations would be insignificant to many clients, and even helpful to some, but they are highly problematic for this particular client. In this reenactment, for example, the client's own choices and interests were disapproved of, which then exacerbated her presenting symptom of depression. In contrast, a helpful, counteractive response, in which the therapist would pass this test—and empower this cautious and compliant client—would involve *expressing interest in the client's own interests and supporting the client's initiative whenever it emerged*. For example:

CLIENT: I don't think I want a career in business after all. I'm not liking these business classes I've been taking very much. The other students are fine, but I'm not really like them. I think I might like teaching better—kids have always been fun for me.

THERAPIST: Kids are fun for you—I'm seeing that. I'd love to hear about your interest in teaching. Tell me more.

CLIENT: Well, I don't think my parents would like to see me be "just a teacher." I was a National Merit Scholar, you know. But I think I'd love having my own classroom, and I could have a more balanced life with summers off—you know, so I could have time to raise my own family.

THERAPIST: You sound excited about this, like it would give you the kind of lifestyle that you choose. You know, you really do sound different right now...your voice is more alive, has more feeling, as you talk about this possibility.

Testing behavior such as this pervades all counseling relationships—from the first session to the last. However, testing is especially likely to occur in the initial sessions, in situations where strong feelings are being evoked, and when the current interaction between the therapist and client parallels relational problems that the client is having with others. Once therapists have the concept of testing, many of these tests will be easy to recognize and pass, allowing clients to progress in treatment and make changes in their lives.

Let's consider a similar type of testing. Suppose a female client, who often has been acting in a passive and dependent way with others in her life, says to her male therapist, "Where should we start today?" Even though he has the best of intentions, the therapist will fail this client's test if he takes the bait and says, "Tell me about..." By telling this client where to begin, even though she has just asked for his direction, the therapist has *confirmed this client's pathogenic belief that she needs to take a subservient role with others and follow or comply with what they want.* The problematic belief that she ought not to assert herself, initiate, or take the lead is confirmed when the therapist tells this client what to talk about (recall that, per client response specificity, this same directive could be helpful to a different client). This client is likely to remain unassertive and dependent in her relationship with the therapist and, as a result, be unable to change this with others in her life. In contrast, the therapist is likely to pass this test and disconfirm this faulty belief if he repeatedly looks for ways to encourage her own voice and actively support her initiative whenever she does risk taking it. Thus, when this client asks him what to talk about or where to begin, the therapist can provide an experience of change by responding as follows:

- Well, maybe you could just sit quietly for a moment or two, think about the issues or concerns that feel most important to you right now, and bring me in on those.
- What would you like for us to begin with today?
- OK, I do have some possibilities, but maybe I could hear your thoughts first about the best way to use our time today.
- I'd like to talk about whatever you think would be most helpful. What might that be—what comes to mind?

Most compliant clients will not accept these invitations so easily, of course. Oftentimes, they will not accept the therapist's first invitation but continue testing to see whether the therapist really means it when he says he is interested in her choices and what seems most important to her:

CLIENT: (*vaguely*) Oh, I'm not sure, I can't really think of anything. What do you suggest?

Trying again to change the old relational pattern in their current interaction and engage the client in a new, more egalitarian way, the therapist might pass the test by giving a second bid for the client to take the lead:

THERAPIST: OK, let's just wait a minute then. Why don't you take a moment to check in and just be with yourself, and let's see what comes up for you.

CLIENT: (*pause*) Well, uh, OK…I guess I haven't been sleeping very well lately. I keep having this bad dream.

THERAPIST: The same bad dream keeps coming back. Let me join you there—tell me about it.

With each of these responses, the therapist's intention is to demonstrate that he genuinely wants to engage her in a more collaborative way where they can share control over the treatment process. Behaviorally, the therapist is demonstrating to her that he does not share her belief that she needs to remain dependent or submissive and be told what to do (while he is also discerning more clearly the treatment focus that, too often in other relationships, this is how things went). Repeated interactions of this sort—which consistently disconfirm her symptom-engendering belief that she isn't capable of acting independently, that others need her to depend on them, or that she is pushy or demanding if she says what she wants—will be reparative and bolster her self-efficacy and autonomy. As she continues to find it safe to act stronger with the therapist, she can begin to try this new way of relating with others in her life as well. She and the therapist then will begin to sort through others in her life with whom it is safe to act more assertively and independently, and others who undermine her confidence, try to control her, and do in fact want her to remain one-down.

Usually, it is relatively easy for new therapists to identify this type of testing around dependency/compliance issues and respond effectively. In contrast, tests that involve issues of power and control usually are far more challenging for new and experienced therapists alike. Clients with different developmental experiences—for example, those who could "run over" their caregiver, or who wielded too much power and influence in family matters, often test the strength or resolve of the therapist. When these confrontational or moving-against clients can successfully dominate, demean, or control the therapist, things quickly go downhill for therapist and client alike. Before these clients feel safe disclosing personal or vulnerable issues, or even begin to engage in a working alliance, they also need to **assess safety/danger** in the therapeutic relationship. For these clients, this means testing whether the therapist can handle them. For example, when clients are insulting, provocative, or push limits with the therapist, they often are testing whether they can manipulate or run over the therapist—as they could their caregiver. Or, is the therapist able to tolerate their disapproval or disdain without becoming defensive, set limits on acting out with the therapist, and sustain clear treatment boundaries—thereby offering the interpersonal safety they did not receive before? Let's consider a case example of this more challenging, provocative type of testing.

Graduate students in Carl's training program were supposed to videotape all therapy sessions. Early in treatment, however, Carl's client Lucy complained that the videotaping made her "self-conscious." She said it was already hard for her to talk about what was going on in her life and that the videotaping just made it "impossible" to talk about really important things. With each successive week, Lucy voiced her increasing exasperation with Carl and disappointment in treatment. She told Carl that he was not helping her and threatened to stop treatment if he didn't turn off the video recorder.

Carl was torn and truly did not know what to do. He knew the clinic rules and that his supervisor would want him to abide by them, yet he also felt he should be flexible enough to do what his client needed—and because he didn't want her to stop coming and lose another client. Without consulting his supervisor, he reached a compromise with Lucy and agreed to turn the videotape off for the last five minutes of each session. Lucy was appreciative and did indeed disclose new material to Carl. Carl felt badly about doing this without his supervisor's permission, but it also seemed like it was the only way to help his client and keep her in treatment.

Soon afterward, however, Carl realized that treatment was not going well. Even though she brought up some relevant new issues, Lucy became scattered and started jumping from topic to topic. Carl tried to pin her down and focus on one thing at a time—such as her ever-changing feelings or scattered thoughts, but she slipped away to a different issue each time. Lucy complained that she was "very upset" and needed more guidance from Carl, yet whatever he did failed to help. Realizing that he had made a mistake and that things were getting worse with his client rather than better, Carl decided to talk with his supervisor.

First, Carl and his supervisor spent some time discussing honestly what Carl's decision to turn off the tape meant for their relationship and their ability to work together collaboratively. With their relationship restored, Carl and his supervisor began exploring what had been going on between Lucy and Carl, and tried to understand what this interaction pattern might mean. To begin, they clarified how badly Carl had always felt about himself when others were critical of his performance. Carl relayed that he just "dreaded" to be in situations where, no matter what he did, he "couldn't do it right." Further clarifying their mutual interaction, the supervisor helped Carl recognize how this countertransference issue in his own life was activated by the critical and demanding stance that Lucy took with Carl—and with others in her life. Carl felt empowered by these new ways of thinking about what was going on in the client–therapist relationship and accepted his supervisor's firm limit that the videotape must remain on throughout every session. To his dismay, however, Lucy did indeed feel angry and betrayed when Carl reinstated this limit at the beginning of the next session. She accused Carl of being "untrustworthy" because he broke their agreement, told him that she didn't think he was going to become a very good therapist, and informed him that she could no longer talk with him about certain important topics. All of this was difficult for Carl, but the biggest threat came when she told him that she would have to think about whether she wanted to come back next week and keep working with him.

Carl felt "horrible" about her threats and accusations yet, with his supervisor's support, he tolerated his discomfort, tried to remain as nondefensive as he could, and stuck to the rules this time. Although she continued to blame, threaten, and complain, Lucy did return. In fact, despite her protests, she actually seemed calmer than before and began to focus more productively in therapy. Over the next few weeks, Lucy succeeded in making some important changes in her life. She began to share with Carl how "sickening" it had felt for her to be able to "boss my father around—just like my mother did." Following this disclosure, which Lucy had felt so ashamed about, she returned to the next session and said that she had apologized to her boyfriend for being "so demanding." Lucy told her boyfriend that she "didn't want to be this way anymore—with him or anybody else," and was trying to change it.

It is adaptive for clients like Lucy to test whether the therapist will confirm or disconfirm their pathogenic beliefs. Clients cannot progress to more vulnerable material or make changes in their lives if the current therapeutic relationship is not safer than past relationships have been. However, there is limited leverage in merely reassuring clients verbally about deeply held beliefs and expectations. Moreover, most clients are not consciously aware of testing therapists in these ways, and usually it is not helpful to point out testing behavior to them. Instead, as in all human relations, *it is what the therapist does that counts, rather than what the therapist says.* Therapists pass tests and provide a corrective emotional experience when they repeatedly respond in ways that behaviorally disconfirm pathogenic beliefs and alter maladaptive relational patterns. Fortunately, even though all therapists fail tests sometimes, they will have many opportunities to recover—just as Carl did. *It is especially important for new therapists to be reassured that they can readily recover from mistakes, failed tests, and reenactments.* Therapists can recover by acknowledging what went wrong and talking it through with the client, or simply by responding more effectively the next time this issue or theme comes up—often later in the same session or in the next session. In particular, therapists can recover from these failed tests that, like ruptures, regularly occur for even the most effective and experienced therapists if they:

- stay closely attuned for any **relational statements**—whether overt or subtle, that clients make about the therapist or their interaction together (Client: "You're probably not the kind of person who would approve of this, but...") and
- in each successive interaction, track how the client responds to what the therapist has just said or done (Therapist: "I just asked you to clarify that for me, and you became quiet. I wonder what went on right there for us—maybe something didn't feel quite right?").

When the therapist passes a client's test by responding in the way the client needs, the client experiences a new degree of safety in the therapeutic relationship (such as providing limits for Lucy, who held too much power over adults while growing up and did not experience appropriate limits and boundaries). An exciting process to participate in, therapists will often observe clients respond to this new safety by beginning to remember more clearly, feel more deeply, and share more fully. In the next few sessions, therapists will also observe clients taking risks, as Lucy did, and try out new and better ways of responding with others in their lives. Most clients will initiate these changes on their own:

CLIENT: Hey, guess what I did this week? I finally spoke up and told my boyfriend...

At this point, the therapist can support the client's willingness to risk engaging with others in this new and self-affirming way. This then leads to further discussion about discerning who in the client's life is capable of responding affirmingly to the changes the client has made and who is not. Thus, the client can safely process with the therapist how to generalize the new response style and work at consolidating a new set of beliefs about herself and others.

THERAPIST: You just did this differently with me—how very nice. You know, that makes me wonder if you couldn't speak up for yourself better and say more of what you want with your husband, too. What would it be like for you to try this with him, like you just did with me? What might the plusses and minuses be in doing that?

In conclusion, how can therapists clarify the interpersonal response that this particular client needs in order to pass a test and provide a reparative experience? To guide the responses they select for their clients, therapists can (1) consider client response specificity and (2) track the moment-to-moment interaction sequences and assess the client's positive or negative reactions to the therapist's interventions. Let's consider both carefully.

Client Response Specificity.   As discussed in Chapter 1, therapists need the *flexibility* to be able to respond differently to different clients. In counseling relationships, there are no cookbook formulas or "right" ways to intervene. One size does not fit all, that is—the same therapeutic response that passes a test for one client may reenact conflicts for another client with different schemas and expectations (Persons & Silberschatz, 1998). For example, in response to pressure from her client, Beverly agreed to reduce her fee substantially. Almost immediately, her client began wearing more expensive jewelry, delighted in telling Beverly about shopping trips, and described how wonderful it was to fly first class on her vacation. Realizing that the client was taking advantage of her, just as the client had done with her doting mother and indulgent husband, Beverly renegotiated a higher fee. Later in the session, the client brought up important new material about, first, feeling "pushy" and, later, "lonely" when she felt she had too much control over others. Working together, they entered a productive new phase of treatment where the client began to explore how this pattern of being demanding or dominating was being played out and disrupting other relationships in her life. Based on the client's willingness to begin looking at these unwanted parts of herself, Beverly concluded that, by being firm and not allowing the client to take advantage, she had passed the client's test and provided her with the corrective response she needed. This client was unaware of her own testing behavior with Beverly, however, and unable to articulate it in any way. Like most clients, the client did not feel safe enough to recognize or discuss this sensitive issue until she was *behaviorally convinced* that she could not manipulate Beverly as she could her mother, her husband, and her previous therapist (again, interpretations, explanations, cognitive reframing, and other verbal/cognitive responses are often useful but less potent than behavioral or experiential relearning).

Per client response specificity, in contrast, the same response of lowering the fee can be helpful for a different client. Chris was coping with a diminished income following a recent divorce and was having trouble making ends meet. Knowing that she grew up with demanding, authoritarian parents who gave her little, the therapist realized that offering Chris a more flexible fee arrangement could be a significant overture. This client would never have asked the therapist for a reduced fee (or asked anyone for help with anything) and was initially unwilling to accept any of the alternative fee arrangements that the therapist suggested. When the therapist asked about her reluctance to even consider his offer, Chris began to cry. As their discussion soon revealed, the therapist's generous response disconfirmed her pathogenic belief that she "didn't matter" and evoked the longing she had felt for her parents to "see her" and help when she had a problem. Chris had grown up under the demanding expectations that she had to do everything "perfectly" and should never need or expect any help or support; that was simply the family norm for her—what was expected. Clearly, changing the payment schedule with this client disconfirmed pathogenic beliefs about herself and faulty expectations about others. Within this session, treatment progressed into more intensive work about her emotional deprivation and resulting belief that she "did not matter much" to others. As these two examples illustrate, *the same response by the therapist will often have very different effects on different clients* (Silberschatz & Curtis, 1993).

Clients' cultural/ethnic backgrounds add complexity to response specificity. For example, self-disclosure of an issue that the therapist had great difficulty resolving in his own life could be viewed as unprofessional by some traditional Asian clients. However, the same disclosure with some African American clients may be welcome and convey that the therapist is human and less likely to be judgmental. In this way, racial, gender, religious, or other differences between therapist and client will affect how the client responds to the therapist's offer. Following our previous guidelines, however, the therapist's intention is to pay close attention to how the client responds to this intervention. For example, if the therapist decides that self-disclosure may be productive for this client, the therapist will closely observe how the client reacts to it—with a helpful new perspective or more involvement in the working alliance, for example, or with greater reserve and emotional distance from the therapist.

Client response specificity puts more demands on the therapist. It takes away the security of having a simple, rule-based approach to therapy. Manualized therapy approaches have been useful in terms of maintaining adherence to the treatment approach—especially in research studies, which was their original purpose. However, manualized treatments fail to honor client response specificity unless they specifically allow for variation based on client needs. The Motivational Interviewing literature has richly informed our understanding of how manualized approaches can be limiting when the client's need and lead is not taken fully into account and there is a press to strictly follow the manual. These MI researchers report that therapists in their treatment approach were more likely to bring about effective and lasting change when therapists applied the manual flexibly and tracked the client's response to interventions—attending and adjusting to how the client has just responded to what the therapist just tried to do (see Amhrein et al., 2003; Levant, 2005; Miller & Rose, 2009). As noted, client response specificity requires the therapist to remain close to the unique and personal experience of each individual client. However, within this highly "idiographic" approach, guidelines are available to help therapists discern more accurately the specific needs of differing clients. In particular, the therapist can learn to assess the specific relational experiences that each individual client needs by identifying the maladaptive interpersonal patterns that tend to recur in three arenas:

- current client–therapist interactions,
- relationships with important others in the client's current life, and
- the client's family of origin, where these these schemas and patterns were originally engendered.

We have begun to see that therapists learn how to pass tests and provide clients with the specific relational experiences they need by *assessing how clients react to different interventions from the therapist*. Let's go further with it.

Assessing Client Reactions.   How do therapists know whether they have passed or failed the client's test, or determine in general whether they have responded usefully to what the client just said or did? What guidelines exist to assess whether the therapist's response has been helpful or has reenacted some aspect of the client's conflict? If clinical trainees rely only on their supervisor's advice on how to intervene with their clients, they will remain dependent on guidance and insecure about their own clinical work. Similarly, new therapists will be prone to develop a false sense of security, and risk becoming rigid and dogmatic, if they settle on one theoretical approach too readily. New therapists do not want to readily adopt any single approach to treatment. Too often, new therapists "foreclose" on this aspect of professional identity development in order to keep from feeling incompetent, to avoid the anxieties and ambiguities inherent in clinical training, or to maintain ties with a mentor who is strongly identified with one brand of treatment as truth. Instead, *trainees want to try out different ways of responding and explore competing theories in order to sort through for themselves what works best for them and for this particular client*. Thus, even though supervisors and theories certainly are useful, beginning therapists develop more autonomy, and find their own identity as therapists, by learning how to **assess client responses** and determine for themselves the effectiveness of their interventions.

New therapists often suffer under the misconception that there is a single correct way to intervene with clients—and feel worried or insecure that they haven't done it right. A better approach is to observe how the client reacts to the therapist's response. In this approach, the client is determining the best way to intervene—based on how well the client can use what the therapist just said or did to make progress in treatment. With each successive turn of the conversation, therapists are trying to decide what is the best way to respond to this client right now—and choosing to respond by challenging, affirming, reframing, exploring, making links, providing feedback, and so forth. As soon as they respond, however, the therapist's intention is to *attend closely* and observe how the client seems to be responding to this intervention (with disinterest, agreement and further exploration, bringing forth new material, and so forth). As the therapist tracks the moment-to-moment interaction sequences—how the client is responding right now to what the therapist just said or did—clients inform the observing therapist of the responses they can utilize or benefit from. Clients do not usually verbalize this outcome and may not be conscious of it, but *they reliably guide the therapist by their behavioral response to how the therapist just intervened*. Let's look more closely at how this goes.

When the therapist responds effectively to a client's test and disconfirms pathogenic beliefs about self and faulty expectations of others, the client often feels safer and begins to act stronger with the therapist—behaving in new and more adaptive ways. This progress can be expressed in many different ways. For example, in the minutes after the therapist passes a test, clients may stop complaining about other people and begin talking more directly about their own thoughts and feelings. Or, they may look more honestly at their own contribution to the problem, as Beverly's client just demonstrated. Perhaps the client will relate a vignette with more clarity or less ambiguity than usual. The interaction with the therapist may acquire more immediacy or intensity and the dialogue becomes less intellectualized or superficially social. Clients will often engage the therapist more directly or be more present, perhaps by expressing warm feelings toward the therapist or by taking the risk to bring up a problem they are having with the therapist. Oftentimes, the client's affect will emerge or intensify. In the next minute or two, the client's feelings about whatever has happened—happy, sad, or angry—may be experienced more fully or shared more openly with the therapist. Therapists want to take note when clients act stronger in any of these ways. Yes, it is a pleasure to see the client is improving, but the client is also telling therapists what they have just done, and what they can do in the future, to make a difference.

Therapists also learn that they have responded effectively or passed tests when clients bring up relevant new material to explore. When clients initiate new topics in this way, they tend to become especially productive. In particular, clients may make useful connections, on their own, between their current behavior and formative relationships—without the therapist needing to make interpretations or guide the client back to developmental or familial relationships. The insights that clients generate in these circumstances are enlivening and hold real meaning for the client, and they are readily translated into new ways of thinking and acting in their everyday lives. In contrast, *it is not usually very productive for therapists to make historical interpretations or to lead clients back to try and make these familial connections.* Even when these developmental links are accurate, clients do not usually find them to be relevant to their current problems—just abstract ideas or distant possibilities that do not hold real meaning or lead to behavior change (for example, a client might exclaim in exasperation, "My mother! Why are you asking about my mother! I'm 31 years old—what the hell does my mother have to do with anything?!"). In this way, many clients will not benefit when therapists try to make familial/historical interpretations to help clients see the roots of their problems (for example, Therapist: "I think you're afraid to leave him and move out on your own now because you felt so alone as a girl—and you just don't want to feel that way again"). Rather, meaningful behavioral changes usually result when clients spontaneously make these developmental links on their own (the "aha" experience)—which is most likely to occur just after the therapist has passed a test or provided a corrective experience (for example, Client: "I think I've been afraid of moving out and living on my own because I don't want to feel all alone again—like I did when I was a girl").

In sum, the therapist has responded effectively or passed the client's test when the client responds by acting stronger in one of these ways. To illustrate, when Carl turned the videotape on for the full session, Lucy complained and even threatened to terminate. But by maintaining these appropriate professional boundaries, Carl behaviorally demonstrated that he was different from Lucy's father and was therefore safer. Carl could tolerate Lucy's disapproval and do what she needed rather than what she demanded. Although Lucy threatened and complained, she also began to work more productively with Carl and began changing with others in her life this way of relating that wasn't working.

If therapists observe how clients respond to their interventions and try to identify what they have just done to pass (or fail) the client's tests, they will be able to formulate treatment plans more effectively. These conceptualizations can then be used to shape subsequent interventions and better provide clients with the responses they need in order to change. For example, at other anxiety-arousing points in treatment, Lucy will again push the limits with Carl. If Carl has learned from their previous interaction, however, he will be better prepared to provide Lucy with the clear limits she needs and is covertly requesting.

What about the other side of the coin—how do therapists know when they have failed the client's test or unwittingly reenacted the client's conflict in some way? On having failed a test, the therapist will often observe that clients act weaker in some way—such as, in the next moment, becoming *more* compliant, defensive, confused, externalizing, distant, self-doubting, intrapunitive, and so forth. The client may also backtrack from healthy new behavior by undoing or retreating from recent successes, or become less of a partner in the working alliance. If therapists are closely assessing how the client is responding to what they have just said, they will observe that these reactions to failed tests often occur almost immediately. They are usually evident within the next minute or two and may continue into subsequent sessions. By tracking their interpersonal process in this moment to moment way, therapists are much better able to recognize when their current interaction may have just failed the client's test. In particular, therapists should consider the possibility that they have unwittingly reenacted some aspect of their clients' conflicts when:

- clients begin talking about others rather than themselves and the dialogue becomes more superficial;
- the therapeutic process becomes repetitive, intellectual, or loses any real focus; or
- clients become compliant, lose their initiative, or cannot find meaningful material to discuss.

We have just seen that when therapists pass tests (for example, overtly communicate their pleasure in the client's recent success experience so the client cannot misperceive the therapist as feeling envious or threatened as significant others have been), clients usually acknowledge the therapist's effectiveness by acting stronger and making progress in treatment. In parallel, when a therapist repeatedly fails a specific test (for example, by brushing aside the potential validity of a client's criticism of the therapist, just as the client's parent always needed to be right), *the client may inform the therapist that something is awry by recalling other relationships in which the same conflict was being enacted.* That is, the client may begin to tell the therapist vignettes about other people the client has known, or perhaps characters in books or movies, who are enacting the same problematic pattern. *Therapists always want to consider the possibility that, whenever the client is talking about another relationship, the client may be using this as a metaphor, analogue, or* **encoded reference** *to describe what is going on between the therapist and the client as well* (Kahn, 1997). As a sustained intention throughout each session, therapists are listening for the relational themes that characterize the vignettes clients choose to relay. For example, if clients repetitively share stories in which trust is betrayed, control battles are being enacted, others are not responsible and need to be taken care of, and so forth, they are often expressing how these same patterns are currently being activated or played out with the therapist. Yalom (2003) describes this as listening with "rabbit ears"—always considering the possibility that *whenever clients are talking about another person they may be making a covert reference to the therapist or what is going on between the therapist and client as well.* When this analogue occurs, the therapist can use a process comment to inquire about this possibility, make this relational theme overt, and bring it back into the immediacy of the therapeutic relationship—where something can be done to address and resolve the misunderstanding or repair the rupture in the working alliance. For example:

THERAPIST: Both of these people you have been telling me about failed to hear something important that others were trying to tell them. I'm wondering if something like that could be going on between you and me as well?

CLIENT: I don't have a clue what you're talking about.

THERAPIST: I'm wondering about the possibility that, like with these other two people, you might have told me something earlier that was important to you, but I didn't hear it very well. Do any possibilities come to mind?

CLIENT: Well, not quite, but since you mention it, I guess sometimes I think you do try to make things seem better than they really are. Like you need me to be happier than I really am or want to make my problems less than what they really are.

THERAPIST: OK, it seems like I'm trying to make things look better than they really are. Like I don't want to see how bad it really is for you sometimes? I can sure see why you wouldn't like that, and I'm glad you're taking the risk to talk with me about this. Help me understand it better though., I want to get it right—and change it. Are you saying I'm too optimistic, or maybe just looking at the world through rose-colored glasses? How would you say it?

CLIENT: (*laughing*) Well, I wouldn't say you put a yellow, happy-face sticker on my forehead when I come in the door, but sometimes you don't seem to want things to be as bad as they are.

THERAPIST: Was there a time today when I did that?

CLIENT: Uh-huh. I told you I was "terrified" when I woke up and they were gone. And you said, "Yeah, I can see how anxious you felt." Maybe you prefer that I only feel anxious, but I didn't. I felt terrified—that's what I said and that's what I meant, but I don't think that's what you wanted me to feel.

THERAPIST: OK, I'm seeing better what you mean now. Would you give me a redo and let me try that one again?! You weren't anxious; it was much more than that—you were terrified. (*pause*) Am I capturing it accurately this time—not making it less serious than it really was?

CLIENT: Thank you for hearing me, that does feel better. (*pauses*) You know, this makes me think how my dad was always so uncomfortable whenever I was upset. I don't think he knew what to do when something was really wrong, so he just tried to act like things were OK even when they weren't—like when my cousin got hurt....

To sum up, just as clients use eliciting behaviors to avoid and defend against their problems, they also use testing behavior to try to address and resolve conflict in the therapeutic relationship. By tracking the therapeutic process in the ways suggested here, therapists can assess whether they are providing clients with the corrective responses they need. The therapist's aim is to track their moment-to-moment, back-and-forth interactions and assess (or directly inquire about) how the client seems to be responding to what the therapist has just said or done. Although it often feels overwhelming to new therapists to try and keep track of so many things with their clients, most of this does become second nature with practice and a few years of experience. As therapists learn to track and assess client reactions in this way, they will be able to choose more confidently or know more easily how they want to respond next to what the client just said—and that's a good feeling to have.

TRANSFERENCE REACTIONS

The third way clients bring their problems with others into the therapeutic relationship is through **transference reactions**. Freud was writing about transference long ago, in 1910, and it remains one of the most salient but misunderstood concepts in therapy. Problematically, the field has used the same term to convey different meanings and, over the years, two competing definitions have evolved. Originally, transference meant displacing various positive and negative feelings and attitudes toward the therapist that clients have held toward important people in their past (see Cooper, 2001 for an array of theoretical perspectives on transference). This traditional psychoanalytic model of transference is based on the notion of the therapist as a "blank screen." Because the therapist is supposedly "neutral," whatever emotional reactions the client has toward the therapist are distortions that have been unrealistically *transferred* to the therapist—casting the therapist inaccurately into a role that better fits historical figures from the client's past. It is during vulnerable or distressing, affect-laden moments that clients do indeed tend to misperceive the therapist and others along old familiar lines. These persistent distortions of the therapist (and others) as wonderful, withholding, demanding, judgmental, and so forth, reflect the client's schemas or relational templates. The more troubled clients are, the more pervasive their transference distortions will be—causing significant problems in their close personal relationships, and distorting their reactions to the therapist as well (Gelso & Hayes, 1998).

In contrast, the contemporary definition of transference is broader and more useful. It refers to *all* of the feelings, perceptions, and reactions that the client has toward the therapist—both realistic and distorted. These reactions, which may be positive or negative, range on a continuum from fully accurate and reality-based perceptions of the therapist to distorted responses that are based on past relationships and have little to do with the reality of the therapist and how the therapist has actually responded. Thus, they do not necessarily represent distortions or pathology. Although transference distortions in the traditional sense certainly occur, the contemporary definition is emphasizing that many of the client's reactions toward the therapist are reality based and, most important, *almost all hold at least a kernel of truth*. For example, the client's affection for the therapist may be a genuine response to the very real caring the client feels in this relationship. Similarly, the client's anger or competitiveness may be based on the reality that the therapist sometimes has been late for sessions, emotionally unavailable, unable to grasp what is most important to the client, invested in the client making certain decisions or taking certain actions, and so forth. Therapists make a significant error when they too readily attribute the client's positive or negative feelings toward them to traditional transference distortions. Before doing so, we want to be nondefensive enough to be able to consider the potential validity or, more often, partial accuracy of those reactions. Maybe the client does have a valid complaint and something we have said or done has contributed to the client's (seeming) misperception or misunderstanding. With this more **contemporary approach to transference,** *there is much more receptiveness to the mutual interaction or reciprocal influence between therapist and client.* That is, transference is not just a one-way street that only says something about the client but, more often, reflects something about what is going on between the client and therapist—something that both share responsibility for and participate together in coconstructing. This is a more egalitarian approach that refutes the notion that the therapist can be a neutral or objective observer. Instead, both client and therapist contribute their own issues in this "shift from a one-person to a two-person psychology" (Hill & Knox, 2009). Related, attachment researchers find that therapists who themselves have Secure attachment styles are better able than insecure therapists (that is, Dismissing or Preoccupied) to remain nondefensive, have the reflective capacity to consider the client's differing point of view, and do not readily provide the automatic responses that the client characteristically tends to elicit or "pull" from others (Dozier et al., 1994; Dozier & Tyrell, 1998; Fonagy et al., 2002).

As we will see, therapists can use transference reactions both to conceptualize their clients and to help resolve problems within the therapeutic relationship. More specifically, contemporary short-term dynamic therapies have encouraged an important change in how therapists work with transference reactions. In decades past, transference reactions were used as a springboard to make interpretations about historical relationships and the genesis of conflicts with parents and early caregivers. Now they are used instead to *comprehend and change the current interaction between the therapist and the client* (Levenson, 1995; Mills et al., 1989). As noted earlier, therapists will often recognize important connections between current problems and developmental relationships, yet few clients will find these links meaningful without first addressing and resolving these distortions in the real-life relationship with the therapist. In this more contemporary approach, the therapist is working with transference in the immediacy of the current relationship (for example, Therapist: "What might I be thinking about or feeling toward you as you tell me this?"), rather than through the more distant or intellectualized approach of trying to interpret or lead the client back to historical relationships (for example, Therapist: "So, you're seeing me as being withholding here...could that perception fit your father better than me?"). Researchers find strong empirical support that this here-and-now approach is linked to better treatment outcomes (Andrusyna et al., 2006; Levy et al., 2006). In this contemporary or intersubjective approach to transference, the goal is to:

1. help clients feel safe to explore and discuss personal reactions toward the therapist;
2. work collaboratively to explore mutual contributions to problematic interactions or misperceptions in their relationship; and
3. work together to change interactions that are problematic or resolve perceptions that are faulty first with the therapist, and only then explore how these also might play out with others in their lives.

In most cases, clients' reactions toward the therapist accurately capture *aspects* of the interaction with the therapist. It is important for the therapist to acknowledge the potential validity of the client's perceptions. Indeed, without realizing it, we have usually said or done something that contributes to the misunderstanding, complaint, or rupture. Therapists want to enter into a genuine dialogue with the client to clarify what each was thinking, feeling, and intending during any misunderstanding or rupture of the working alliance. Addressing and resolving ruptures in this way, as soon as they emerge, is associated with treatment retention and treatment outcome (Muran et al., 2009). Thus, if therapists are willing to explore *nondefensively* what they may have done that activated the client's transference distortion or contributed to the misperception, this conveys to the client that the therapist doesn't always have to be right and is willing to risk a genuine relationship that goes two ways. Clients feel respected and empowered in this egalitarian relationship, and they often are able to become more authentic with the therapist and, in turn, with others in their lives.

Thus, in the interpersonal process approach, the therapist affirms the validity of many client reactions and tries to resolve actual distortions in the current relationship, rather than focusing back on the historical genesis of conflicts. Ironically, just after the therapist and client have addressed and resolved a significant misperception or misunderstanding between them, clients will often go on to make meaningful connections between formative and current relationships on their own. Before going further with how therapists can intervene with transference to help clients change, let's become better acquainted with the everyday expressions of transference.

Transference in Everyday Life. Transference reactions are commonplace—they simply pervade human relationships. To a greater or lesser degree, they occur for all people at times—including therapists, of course. We all systematically misperceive others as a result of overgeneralizing previous learning to the present situation— especially when we feel vulnerable or distressed. For example, imagine a male university professor who stands in front of a large lecture hall about to convene the first class meeting of Psychology 101. Before he has even begun to lecture, many students will be transferring expectations from significant male authority figures in their own lives onto the professor. Thus, in the first row of the lecture hall, 18-year-old Mary looks up to the professor admiringly. Just as she still somewhat idealizes her father, Mary respects the middle-aged professor, even though she actually knows little about him.

In contrast, Joe looks down on the professor disparagingly from high up in the last row of seats. Joe anticipates that the professor does not know as much as he thinks he does and that he will not be prepared very well for class. Joe expects that, just like his father who always had to be right and on top, this professor will probably try to "blow a lot of smoke" and get away with not knowing as much as he thinks he does in class. Joe has already decided that this know-it-all professor is not going to get away with much if Joe can help it. Let's hope that the professor is aware that people tend to transfer expectations onto authority figures who hold power and onto those who have high public visibility (movie *stars*, sports *heroes*, and political *leaders*). If the professor is not sensitized to the pervasiveness of transference reactions, his self-confidence as a teacher will take a roller-coaster ride. One day, Mary will follow him out of class and tell him what an interesting lecture he gave; the next day, Joe will drop his books loudly when the professor starts to speak and demand that he defend half of what he says.

Transference reactions such as Mary's and Joe's are everyday occurrences. The more conflicted or troubled one is, however, the more one distorts the current reality by inaccurately displacing reactions from the past onto the present. For example, Joe and Mary have strong initial reactions to the professor that are based on little real-life experience with him. If Joe and Mary function fairly well psychologically, their cognitive schemas will be flexible and varied, and they will become more realistic in their assessment of the professor as they have more exposure to his teaching. In contrast, the more thoroughly Mary idealizes her father and the more pervasively Joe disparages his father, the more likely it is that they will not be able to perceive the realistic strengths and limitations of the professor. Instead, their schemas will be more fixed and self-fulfilling, and they will maintain their initial expectations despite the fact that they are inaccurate or "overdetermined."

Transference reactions are also more likely in emotion-laden situations. For example, transference reactions are especially pronounced when two people are falling in love. Many married couples recall their courtship as the happiest time in their lives. During this time, each member may systematically misperceive the partner as someone who can fulfill unmet developmental needs—most commonly, perhaps, making one feel love-worthy or capable when one has not felt this way before. This romantic period is usually short-lived, however. The transference projections soon break down—they both realize that the other person cannot fulfill their unmet developmental needs, and their old schemas are reinstated. If the unmet needs and accompanying transference distortions are strong, the result is often dashed dreams or angry feelings of betrayal. In contrast, if the transference distortions of the partner have not been extreme, they can readily be shed, and a more realistic and enduring relationship can develop. Relinquishing these transference distortions of the new spouse is one of the major psychological tasks of courtship and early marriage.

Transference distortions may occur even more intensely in parenting. In the vignette below, tragically, a child "reminds" a parent of the child's father with whom she is now estranged and, in fact, has come to "detest."

> Ten-year-old Adam was brought into treatment at the request of his teacher, who reported that Adam was "sullen and withdrawn" and often spent his class time drawing pictures of angry/muscular men, swords and other weapons, and blood. The teacher became increasingly concerned when he began getting into fights with others at school, and made the referral when she noticed that he had cuts on his arms. In evaluating Adam, the therapist ascertained that Adam's mother had become increasingly negative and hostile toward him in the past year (for example, setting up his bedroom out in the garage). She had been telling Adam that she "hated" his worthless father and that Adam was "just like his father" and that "the apple doesn't fall far from the tree."
> Adam reported feeling acutely suicidal and had to be hospitalized (where he did, in fact, make an attempt—trying to electrocute himself by inserting a wire paper clip he had been using to "scrape himself" into an electrical outlet). Adam's mother argued against any need for treatment and had been skeptical about his need for hospitalization. She confirmed Adam's belief that "my mother hates me" when she said to the therapist in front of Adam: "Are you sure he isn't just trying to manipulate you—just trying to get attention like his father always did?"

In treatment, the therapist can systematically utilize the client's transference reactions to better understand the client's problems with others. One of the most effective ways to discern the client's schemas and faulty expectations is to *check in regularly and ask about how the client might be perceiving or reacting to the therapist or what is going on between them right now*. Routinely, aspects of the same patterns and themes that cause problems with others will be activated toward the therapist or played out in the therapeutic relationship. If the therapist tracks these transference reactions and successfully engages the client in a dialogue about them, they clarify faulty schemas to the client and help change the problematic relational patterns that are being reenacted with the therapist. As this important relearning occurs with the therapist, the client is empowered to begin changing this pattern with others.

Using Transference Reactions to Conceptualize Clients.   Attachment histo-ries provide one useful lens for understanding transference reactions and helping us conceptualize our clients. As Bowlby noted (1973; 1988), repetitive interactions with early caregivers shape children's expectations (and subsequent transference distortions) of others. Researchers have shown that the attachment patterns individuals develop impact both transference and countertransference reactions (Dozier & Tyrrell, 1998; Mallinckrodt et al., 1995; Martin et al., 2007; Woodhouse et al., 2003). For example, researchers find that clients with Secure attachment styles often are able to enter treatment more easily, and they have to capacity to explore their issues with more depth and intensity than clients with insecure attachment styles. In contrast, Preoccupied clients are often concerned about the dependability (and willingness) of others to meet their needs—and this will be quickly evident in their relationship with their therapist. At first, they may be easy to engage—perhaps overly so. Often-times, Preoccupied clients will be too eager to please the therapist, readily self-disclosing and emotionally expressive from the beginning—sometimes "a bit too much" or potentially even "overwhelming" for some therapists. Despite this quick and intense beginning, however, the therapeutic alliance and treatment often doesn't progress so well over time (Smith et al., 2010). Preoccupied clients may not accept, or even actively "resist," the support they are seeking. Also, they may become frus-trated or disappointed with the therapist because their high level of need cannot be fully met. At the same time, problematically, they are less likely to talk about this or other problems with the therapist and address their disappointment directly *for fear of losing the relationship with the therapist.* Thus it becomes especially important that therapists ask these clients directly about any negative feelings toward them, and to respond nondefensively. This is one way to help these clients discover that this is a safe relationship that challenges their internal working models because, unlike others in the past, the therapist is not "overwhelmed" by them—or too caught up in his own emotional problems or personal dramas to be able to "see" and respond to them in a consistent and dependable way.

In sharp contrast, Dismissive clients will find it much harder to acknowledge a problem, ask for help, and enter treatment in the first place. They are hesitant to talk about feelings or personal problems, and shy away from real engagement with the therapist. Their expectation, or transference distortion, tends to involve thinking the therapist "will not really want to hear about" their emotional needs or problems (and actually has disdain for them as being "weak"), will be intrusive and controlling—trying to "take over" if they let themselves get close during a ses-sion or need anything for a few moments, and expect that the therapist, too, will easily become exasperated or frustrated with them and be critical. Because opening up to others was so threatening developmentally and makes them feel vulnerable, it becomes important to respect their need for a slower pace in the beginning of treat-ment, and to be alert to the likelihood that the Dismissive client will have difficulty remaining engaged and attending sessions following significant disclosures, sharing strong feelings, or feeling closer to the therapist.

It's important to note that some therapists are uninterested in the client's trans-ference reactions because they do not see them occurring in therapy. Therapists will not see the client's transference reactions, and the invaluable diagnostic/assessment information they provide about the client's schemas and expectations, unless they actively intervene to draw them out. How can the therapist highlight or identify the client's transference reactions? Let's look at two ways of making the client's transference distortions overt so that the client can join with the therapist collaboratively to understand and change them.

First, *the therapist is looking for tactful but direct ways to inquire about the client's feelings and reactions toward the therapist.* As we saw in Chapter 3 when discussing resistance, it is important for both therapist and client to learn about the client's reactions to being a client and seeking help. Now, further along in treatment, it is also important to learn about the client's perceptions and reactions toward the therapist. This is especially important when:

- conflicts or misunderstandings arise between the therapist and the client, and
- the client has just shared strong emotions or self-disclosed.

It is during these two types of sensitive interactions that *the client is most likely to misperceive the therapist as responding in the same problematic or unwanted way that others have in the past.* For example, if clients disclose something they feel embarrassed about or enter a strong feeling, they will often be certain (erroneously) that the therapist feels, for example, burdened or depressed by their sad feelings. Or, they may inaccurately believe that the therapist doesn't respect them anymore now that they have disclosed this information about themselves. Clients believe the therapist is responding in this unwanted way because this is how important others have responded in the past—even though the therapist doesn't think or feel anything like this and has not communicated disapproval in any way. It is difficult for most graduate student therapists to grasp this in the beginning. Because the interaction or sharing seemed so positive to them, and to be so well-received by the client, new therapists find it hard to believe that it could have been misperceived in some problematic way by the client. However, if the therapist doesn't check in with the client and ask how it has been for them to share this significant feeling or disclose this important issue with the therapist, many clients may miss, reschedule, or come late to the next session, or return but keep the next session superficial. Thus therapists need to ask about or clarify potential transference distortions with the client and, if necessary, *make it unambiguously clear that they are not thinking or feeling the same unwanted things that important others have in the past*:

THERAPIST: That was a lot to share—there's so much sadness there.

CLIENT: Yeah, I guess I was crying pretty hard.

THERAPIST: Yes, and I'm honored that you choose to share such important feelings with me. May I check in with you for a minute about this—about how it feels right now to have shared so much and cried so hard with me?

CLIENT: Well, you're always very nice, but this has to be pretty tiring for you, too.

THERAPIST: Tiring? No, it's not tiring at all. Actually, it's enlivening for me—I feel close to you right now—that we've shared something that's meaningful for both of us.

CLIENT: Really? I don't wear you out with all this?

THERAPIST: No, I don't feel worn out or tired in any way. Actually, it's just the opposite; I enjoy being able to respond to you like this.

CLIENT: Boy, this is really different for me, that it's OK for you if I feel sad or cry sometimes. I think my mom needed me to be happy all the time, so I acted that way, even though I wasn't. When looking at pictures of myself, other people often say I look pretty or have a nice smile, but I see a sad face...

In this way, therapists want to **debrief** with the client after sensitive interactions, and check whether the client is distorting or accurately perceiving the therapist and, if transference distortions are in play, **differentiate** themselves from significant others and the client's internal working models, (for example, Therapist: "Maybe you're telling me that others in your life have felt burdened by you, or that your needs made people "tired," but I think I'm different from some important people in your life. I'm not feeling "tired" or burdened or anything like that at all"). Additionally, at other less significant times as well, therapists will want to check in with clients and explore their reactions toward the therapist. Routinely, *the following type of questions will open some of the most important, and unexpected, information that clients share during treatment—revealing perceptions and reactions that are central to the client's problems but that would not become accessible if the therapist did not inquire in these ways:*

- As you were driving to our session today, how did you feel about coming to see me?
- When you find yourself thinking about therapy; what kind of thoughts do you have about me and our work together?
- I'm wondering what you might be thinking is going on for me right now as we talk about this sensitive issue?
- I wanted to check in and get your thoughts about our work together and how you and I are doing together. You know, what you like about our time together, and anything about our interaction that may not feel so good.

Therapists can also explore what the client *projects* onto the therapist—what the client believes the therapist is thinking, expecting, or feeling toward the client at that moment:

- What do you think I am feeling toward you as you tell me that?
- How do you think I am going to respond if you do that?
- What do you think I expect you to do in that situation?

Oftentimes, it is uncomfortable for the beginning therapist to break long-held familial and cultural norms and invite clients to talk about "you and me" so directly. Doing so, however, often brings up key issues and concerns that are central to the client's problems and the therapeutic relationship, even though the therapist had no idea this was going on for the client. Although it may not seem this way in the beginning, therapists can learn to become more safely forthright with their clients than they have been able to be with others in their personal life.

The distinction between everyday social interaction and clinical intervention is significant here. Because many therapists have a moving-toward interpersonal style, it is especially difficult for them to become more forthright in this way and risk the client's disapproval. By role-playing and rehearsing this interpersonal skill with classmates and supervisors, and by viewing themselves on videotape, new therapists can practice being more direct or straightforward with clients without crossing the line and becoming demanding or intrusive—which we don't want or need to do. If trainees practice this intervention with colleagues and take the risk to pose such questions, clients' transference reactions will be made overt, and often the therapist will be taken right to the core of the client's concerns. The new issues that are revealed are often rich in meaning, and routinely surprise the therapist as the client replies:

- I think you are feeling disappointed in me.
- I think you'll act nice but really wish inside that I would just stop coming to see you.
- You're probably thinking that it's wrong of me to even be thinking about getting a divorce—that it's selfish for me to do that to my children.

As schemas and misperceptions such as these are made overt, therapists can clarify their actual responses to the client, which in itself often provides the client with a reparative experience. When transference reactions are revealed in this way, therapists have the opportunity to respond to the client with an immediacy and authenticity that is not often found in other relationships. For example, the therapist might clarify the transference distortions just listed by replying as follows:

- No, I'm not feeling disappointed in you at all. Actually, I'm very pleased with how well you have been doing. I'm wondering where that feeling of "disappointment" comes from, though. Did I do something today that made you think I was disappointed in you, or maybe this comes up with others, too?
- I want you to stop coming? Oh, no, that's not going on for me at all—I enjoy working with you very much. But I'm so glad you're telling me this—so we can work it through. Do you know when you started thinking this?
- No, I wasn't thinking that you are being selfish. I was thinking about how hard you are trying to do the right thing—to find a balance between your needs and what you want to do for others. Tell me more about "selfish." Is that an important or familiar word for you?

To emphasize, most beginning therapists find it difficult to consider or explore the possibility that their client is having significant feelings and faulty perceptions toward them that are not based on the reality of how they have responded to the client. With experience, they often become more familiar with clients' distortions and may be better prepared to explore and clarify them. That is, beginning therapists usually find it hard to accept that their clients have deeply held but unspoken beliefs that the therapist really doesn't like them, feels overwhelmed or bored by them, believes they are needy, dependent, or frustrating, and so forth. Even though the therapist may never have responded to the client in a way that would reflect any of these problematic perceptions, the discerning therapist still works under the assumption that such client reactions are likely to be present yet covert. If these distortions are not made overt and rectified, the potential for change is limited. For some trainees, this may be the critical point: They can accept intellectually that the client may hold such faulty expectations of others. However, it is often too threatening—evoking too much guilt over making a mistake or having done something wrong—to explore the possibility that the client may have such unwanted reactions toward the therapist.

To review, exploring clients' reactions toward the therapist reveals three important types of information. First, just as adopting an internal focus will reveal issues that would not have become accessible otherwise (Chapter 4), exploring clients' reactions toward the therapist will also uncover important new issues. As we have seen, the faulty perceptions and expectations that are revealed are often central to the clients' presenting problems and now can be addressed and resolved in the immediacy of the real-life relationship between the therapist and client.

Second, exploring transference will also highlight how aspects of the same problems that clients are having with others—and that they originally learned in formative family relationships—are being reenacted with the therapist. Unless the clients' reactions toward the therapist are directly invited, these distortions and reenactments are not likely to be identified or resolved. Thus, transference reactions reveal clients' key concerns and, as we will explore further, also provide an opportunity to resolve them—first, in the here-and-now, real-life interaction with the therapist, and then with others.

Third, as paradoxical as it sounds at first, transference reactions ultimately provide evidence to clients that their feelings and perceptions really are trustworthy. We're back to our theme of self-efficacy. Because clients' expectations and emotional reactions may not accurately fit the current situation in which they are activated, they often do not make sense at first to clients or others. As a result, many clients experience their emotional responses toward spouses, children, employers, and others as "irrational," "stupid," or "crazy." Consequently, they do not trust the validity of their own experience and, as a result, their self-efficacy is undermined. This also occurs in the therapeutic relationship when clients have emotional reactions toward the therapist that seem inappropriate or exaggerated. However, clients' emotional reactions are not irrational. Though they may not fully fit the present circumstances, they do make sense when they can be understood within their original context. Real-life events have legitimately caused clients to feel and expect what they do. When therapists help clients trace feelings and expectations back to where they learned them, *they will prove to have been an appropriate reaction to someone else in another time and place* (by asking, for example, "When is the first time you can remember feeling this way? Who were you with and what was happening?"). Once the therapist and client first have been able to clarify and resolve the transference distortion in their current interaction, this developmental link can often be made readily. In contrast, the developmental context will not come alive or be meaningful to the client if it is not first resolved in the current interaction with the therapist. In this way, transference reactions, like resistance, ultimately reveal that the client's feelings make sense.

To sum up, clients do transfer emotional reactions from past formative relationships onto the current relationship with the therapist. However, therapists begin by clarifying these misperceptions in the current, here-and-now relationship with the therapist. As they are sorted through in the therapeutic relationship, the client can use this real-life experience of change to begin changing these same patterns and faulty expectations with others:

CLIENT: He doesn't want to listen to me, maybe I am just too much—too needy and too demanding.

THERAPIST: Have you ever felt that way with me—that I find you "too needy and demanding"?

CLIENT: Well, sure, I think that a lot, but you can't say anything because you're a therapist. You know, you have to be nice—that's your job.

THERAPIST: Well, I don't think that being nice is my job—but giving you honest feedback is. If I thought you were being demanding or doing something else that was causing problems in your life, I'd tell you. It wouldn't be easy for me to say that, but I would take the risk and try to talk with you about it in a constructive way. So, no, I don't find you needy or demanding with me. I like it when you speak up for yourself—I think it's a strong and healthy part of you.

CLIENT: I'm glad you're saying that. I think I go along with other people too much, and don't say what I want, because I'm so afraid of seeming, you know, "demanding." I guess it's easier for me to say what I want in here with you than it is in the rest of my life.

THERAPIST: I'm glad it's easier with me—but what would you like to be able to say the next time you feel like your boyfriend isn't listening to you or taking you seriously?

CLIENT: I don't know—I'm just not good at this, but I guess I do have a right to speak up.

THERAPIST: Yes, you really do. Here's a suggestion: What if, next time, you just said to your boyfriend, "I don't think you're really listening to me right now"? How would it be for you to try something like that? I'm thinking it's a good way to see if he can meet your needs in this reasonable way or if this relationship can't give you what you want—as you've been wondering about.

CLIENT: You know, I can say that to him. And you're right, if he can't really listen to me when I ask him to, I probably shouldn't be thinking about moving in with him. It's just not a good match.

THERAPIST: Yeah, if he can generally respond well when you ask him for what you need, green light; if he usually doesn't hear you and can't change and listen when you ask for it, red light.

CLIENT: This makes sense. I'm going to start checking this out with him.

This type of transference clarification gives clients a real-life experience of change; they learn that their inaccurate expectations or problematic schemas do not fit in this relationship. As they have a real-life experience of change with the therapist, they find that at least some relationships, sometimes, can be another way. This experiential relearning, which comes from clarifying the misperception with the therapist, allows many clients to make two important steps. First, they often are able to make meaningful links to past or formative relationships that shed light on current patterns and problems. Second, clarifying the transference distortion and finding that the therapist is not responding in the familiar but unwanted way, clients are empowered to try out these new ways of responding with others that have succeeded with the therapist. Together, the therapist and client then begin the essential process of sorting through who responds well to these changes (for example, the client's husband) and who in their life doesn't (the client's father).

OPTIMUM INTERPERSONAL BALANCE

In this chapter, we are trying to understand the patterns and purposes in clients' interpersonal functioning. In particular, we are exploring different ways that clients tend to play out with the therapist related relational patterns and themes that cause problems in other relationships. However, as we consider the client's eliciting maneuvers, testing behavior, and transference reactions, it draws us further into the closely related issue of countertransference as therapists get pulled into "enactments" with their clients. By striving to maintain an **optimal interpersonal distance** with clients, therapists can manage their own "client-induced" and "therapist-induced" countertransference reactions more effectively and be better prepared to provide clients with a corrective response that does not reenact familiar but unwanted relational scenarios. We have already addressed two important constructs in the attachment literature: safe haven and secure base. As a "safe haven," the therapist uses empathic understanding and emotional availability to make it safe for clients to share their distress and feel responded to. Additionally, as a "secure base," the therapist has the task of facilitating autonomy and differentiation while sustaining connection—such as the task of respectfully challenging the clients' schemas or questioning the current utility of their interpersonal coping strategies. Thus, the therapist has to monitor and balance these two different trajectories—trying to balance the separateness–relatedness dialectic and avoid two common types of countertransference—enmeshment and disengagement.

To provide therapeutic relationships that facilitate change, therapists want to establish and maintain an effective degree of involvement with their clients. On the one hand, if the therapist becomes overly invested in the client's choices or change, this will be an unwanted reenactment for many clients (**enmeshment**). Especially for clients who grew up with too much parental control or intrusiveness, or too little support for pursuing their own interests and goals, their fear and expectation that they will again be controlled by the therapist (and others in their life they get close to) will be confirmed. And, even for different clients who do not share this developmental background, it will not be safe for them to risk letting an overinvolved therapist influence them or help them change. On the other hand, if the therapist is emotionally removed or too personally distant (**disengagement**), a corrective emotional experience cannot occur either. Although this will be especially problematic for clients who had authoritarian, rejecting, neglectful, Dismissive, or otherwise distant parents, it will also diminish the potential for treatment gains for most clients because the relationship is too insignificant to effect change. That is, even if the disengaged therapist responds in new, healthier ways that counteract what clients have found in the past, so what? *It doesn't have much impact and doesn't facilitate change because the relationship is not very meaningful to the client.* Thus, throughout the course of treatment, the therapist's intention is to monitor the balance of separateness versus relatedness in the therapeutic relationship and maintain an optimum interpersonal balance.

At times, all therapists will have difficulty maintaining an optimum interpersonal balance with some clients—whether because of client-induced or therapist-induced countertransference. To manage these "pulls," the therapist's aim is to be both a genuine participant in the relationship and an objective observer of it.

Sullivan (1968) captures this double role in his enduring term **participant-observer.** However, combining genuine empathic engagement with an "observing ego" or objective viewpoint is no easy task for beginning (or experienced) therapists. As in the separateness–relatedness dialectic, it is a challenging paradox indeed to be close (provide empathic understanding) and separate (maintain clear boundaries) at the same time (Jordan et al., 1991).

Without question, this balancing act is one of the most maddening challenges facing many new therapists. Perhaps the best analogy is being a 16-year-old student-driver in a car with manual (stick shift) transmission. Simultaneously, the new driver is trying to press down on the clutch with her left foot, brake or accelerate with her right foot, steer the car with her left hand, and shift the gearshift with her right hand—all the while she is trying to look ahead down the road, see where she is going and attend to what is rapidly approaching, and listen to instructions from an impatient driving instructor (perhaps the varsity football coach) sitting next to her and barking instructions! In just a short time, of course, the new driver will be able to do this almost automatically—and even change the playlist on the MP3 player and carry on a meaningful conversation at the same time. Similarly, new therapists need to be patient with themselves. With practice and experience, it will become much easier to maintain this self-reflective capacity and simultaneously be both a participant in and an observer of what's going on right now in the therapeutic interaction.

Let's look more closely at both sides of this particular form of countertransference—enmeshment and disengagement—and link it to our two differing types of countertransference: client-induced and therapist-induced.

ENMESHMENT

At times, all therapists will become overidentified—"enmeshed," in family systems terms—with certain clients. For example, if this client is well practiced and skillful at getting therapists and others to rescue, direct, or take responsibility for them, this is an instance of *client-induced countertransference.* That is, it probably has more to do with what the client elicits in others than with the therapist's own personal issues. With the help of a supportive supervisor, the therapist can come to recognize that the client is making the therapist feel responsible for him—just as he does with others in his life. It is relatively easy to begin talking with the client about this relational pattern and changing how it is potentially being reenacted in the therapeutic relationship. However, this becomes a more complicated issue when the client's attempts to make the therapist take responsibility for them, satisfy unmet developmental needs, or make decisions for them taps into the therapist's own personal issues (for example, this therapist grew up in a parentified role, or has an anxiously Preoccupied attachment style with a very high need for approval). With this type of *therapist-induced countertransference*, the therapist loses his objectivity and neutrality with the client because his own personal concerns have been activated; that is, the parentified therapist may continue in a caretaking, dependency fostering role with the client, or the Preoccupied therapist may have difficulty maintaining clear boundaries or challenging the client because he can't risk the client's disapproval (Sauer et al., 2003). Or, to take another common example, the therapist becomes invested in the client choosing to stay unhappily married—as the therapist experiences how hurt he had been by his own divorce, his parent's divorce, or begins to feel the pain of other losses he has had difficulty coming to terms with.

In these ways, certain therapists, because of their own developmental history, will have a tendency to repeatedly become *overinvolved* with many of their clients. How can therapists recognize when this traditional or therapist-induced type of countertransference is occurring? When therapists find they are frustrated, angry, or critical of a client for not changing, they have become overly invested and usually are trying to manage a personal issue through the client. Similarly, therapists often are enmeshed when they have dreams about a client, often think about a client outside the therapy session, feel depressed (rather than appropriately concerned) when a client is not changing, or become envious or elated over positive changes in the client's life. Further, enmeshment is usually occurring when therapists describe the client as being "just like me" and cannot recognize the many differences that, in reality, always exist between therapist and client. This overinvolvement, occurs, for example, when a Preoccupied therapist loses clear boundaries with a Preoccupied client who is pressing for intense engagement and fusion—which is very different than genuine care and liking for the client which, of course, is helpful and appropriate when it is modulated and without idealization. When enmeshment occurs, therapists usually need the client to change in order to meet their own needs—for example, to shore up their own feelings of adequacy as therapists or to manage personal problems or feelings of their own that are similar to the client's. Recalling the concept of change from the inside out or an internal locus for change, enmeshed or Preoccupied therapists are trying to solve an internal problem of their own, externally, through the client.

Treatment progress and change usually slow to a halt at the point when therapists become overinvolved with a client. When they become enmeshed, therapists lose sight of the fact that clients have their own subjective worldview, shaped by many factors that actually differ from those of the therapist. They tend to become controlling of the client and to project their own problems and solutions onto the client. As we have seen, this response routinely reenacts the client's developmental predicament in the therapeutic relationship. For example, many clients have grown up in families with overly controlling or intrusive parents. When the therapist's experiential range is too limited to support the client's own autonomy and differentiation, the therapist is responding along the same problematic lines as caregivers did in formative relationships. This reenactment with the therapist will hinder clients as they try to resolve their conflicts around intimacy and control that originally brought them into treatment. Thus, one reason for paying attention to maintaining an optimal interpersonal balance, and always being receptive to the possibility of enmeshment as a form of client-induced or therapist-induced countertransference, is to safeguard against responding to clients in ways that originally lead to their problems.

Further, when therapists become too close or overidentified, they lose their ability to be a participant-observer. That is, therapists cannot think objectively about the client or sufficiently step out of the interpersonal process they are enacting with their clients to consider what their interaction may mean. In parallel, their heightened emotional reactivity to the client will cloud their understanding and diminish their ability to provide a holding environment and remain a "steady presence" to contain the client's feelings and distress (See Fonagy et al.'s (2008) **reflective capacity**—the ability to "disembed" or step out of affectively charged interactions sufficiently to reflect on their own and the client's thoughts, feelings, and behaviors). And, instead of accurate empathy, enmeshed therapists will offer clients only a global, undifferentiated sympathy (as in, Therapist: "Oh, I totally get it!") that does not consistently capture the specific, personal meaning that this particular experience holds for the client. Most important of all, perhaps, therapists will fail to discriminate the real differences between themselves and the client that do exist and will begin to see the client as being just like themselves. Although this overidentification may feel like closeness or deep understanding, and can be reassuring to some clients at first, initial positive changes do not sustain well in an enmeshed therapeutic relationship.

DISENGAGEMENT

Just as many therapists will become enmeshed with clients at times, many therapists also will hold themselves too far away from certain clients at times. Per client-induced countertransference, therapists will often observe themselves initially distancing from moving-against and other "negative" clients who are characteristically hostile, critical, competitive, distrustful, controlling, manipulative, and so forth (Martin et al., 2007). The therapist is distancing herself just as others in the client's life often distance from them. Similarly, it is common for clients with a Dismissive attachment style to evoke feelings in the therapist of boredom, detachment, or being unimportant and unneeded—feelings that the client elicits in others in his or her life, and feelings that also mirror the Dismissive client's own childhood experiences (Daly & Mallinckrodt, 2009; Slade, 1999). As therapists recognize what their reactions are teaching them about this client, they can begin looking for opportunities to broach this topic and begin talking collaboratively about it.

Per therapist-induced countertransference, therapists also tend to disengage or distance from the client when the client arouses the therapist's own personal conflicts. For example, the therapist may disengage from the client as the client talks about the impending death of her aging parent. In this situation, the therapist may be activated by her own sadness about losses in her own life, by unresolved feelings about not being able to say good-bye to her own deceased parent, and so forth. The purpose of supervision is to help therapists recognize their own countertransference reactions, find a more appropriate way to manage their own personal feelings (for example, talk with a supervisor or colleague about them), and empower the therapist to be able to reengage the client with their own boundaries better in place (Falender & Schafranske, 2004).

Most important, however, some therapists will consistently be too removed from their clients because of their own enduring attachment and coping styles. This distance renders treatment ineffective for many clients. When therapists cannot be emotionally available or responsive to their clients, the working alliance is lost or, more likely, does not ever develop. Treatment loses its intensity and often becomes intellectualized when the therapist cannot be a "participant" in the real relationship with the client (Marmorash et al., 2009). Therapists need to be able to risk affecting and being affected by the client—nothing ventured, nothing gained.

Therapists also lose their intuition and creativity when they distance themselves from the client. When this detachment occurs, the client does not have the supportive relationship or holding environment necessary to explore affect-laden problems. Counseling is more than an intellectual discourse, and little enduring change will occur in an emotionally disengaged relationship. As emphasized above, *without the impetus or real meaning provided by genuine personal involvement, a corrective emotional experience will not occur.*

Finally, a therapeutic relationship in which the therapist is too distant or uninvolved will also reenact developmental problems for many clients. For example, for clients whose aloof, Dismissive, self-centered, or authoritarian parents could not meet their childhood needs for emotional support, a distant therapist might be frustrating but ultimately "safe." That is, their anxieties over their insecure attachments and unmet needs—and the resulting problems associated with needing or getting close to others—will not be activated in this therapeutic relationship (as in a therapeutic match between a Dismissive therapist and a Dismissive client, where both may be too comfortable with distance to engage each other or the client's problems in a meaningful way). Unfortunately, just as the client's conflicts are not aroused in a disengaged relationship, the opportunity to resolve these conflicts is lost as well. Thus, when therapists defend themselves against their own anxieties that are aroused by an authentic relationship and genuine emotional involvement with clients, many clients will not be able to change. Novice therapists who are consistently unwilling to risk genuine involvement with their clients are less likely to be effective (and to benefit from supervision) than those who are willing to risk a real relationship but tend to become overinvolved at times.

Optimum Middle Ground of Effective Involvement

To be an effective participant-observer, the therapist is trying to maintain an optimum middle ground of involvement. For two distinct reasons, maintaining this middle ground of involvement is not easy. First, let's explore further the therapist's side of this—therapist-induced countertransference reactions that get in the way. Second, we will explore further client-induced countertransference and see what clients sometimes do to make it harder for therapists to maintain an optimum interpersonal distance.

Therapist-Induced Countertransference: Learning How to Manage Our Own Reactivity.   Therapists are continually confronted with personal issues in their own lives by the sensitive, affect-laden material that clients present. For every therapist, it's just part of the work—it comes with the territory. As a basic element of their training, all therapists are encouraged to pay attention to their own countertransference propensities and become more aware of how they tend to respond when clients activate their own anxieties or personal issues. For example, when the client talks about an issue that is sensitive or difficult for you, what is your initial response tendency? Some therapists initially withdraw from the client and distance themselves by intellectualizing—for example, by interpreting the client's affect before the client fully experiences it or ineffectively "explaining" or interpreting what something means. Others become controlling and overly prescriptive, telling clients what they should do in a particular situation, usually without fully listening to what the client thinks is wrong. Whereas some therapists begin to talk too much, many student therapists become too quiet and unresponsive, leaving clients feeling that they aren't getting any help or don't have a real partner with them in the counseling room. With only a little effort and some nondefensiveness, however, therapists can begin to identify their own countertransference propensities and learn how to manage them responsibly.

Becoming familiar with our own response tendencies becomes even more important with moving-against and other difficult clients who respond to the therapist with negativity (Binder & Strupp, 1997). Through eliciting maneuvers, testing behavior, and transference distortions, some clients will respond in negative ways to therapists—critically and competitively, with anger or blame, contemptuously or manipulatively, and so forth. Although many clients are going to respond in these ways at times, too many therapists have not been given the practice and preparation they need in order to respond nondefensively to these challenging situations that are so unwanted by most therapists—whether novice or experienced. Therapists-in-training need to role-play or rehearse alternative responses with supervisors that offer more neutrality, and therefore are less likely to reenact the familiar but problematic responses that this client's negativity usually elicits from others. New therapists also are encouraged to anticipate and explore how their own personal issues and response tendencies will be triggered by different kinds of challenging, unwanted responses from the client—such as power struggles, sexual innuendos, and so forth.

Whereas much is often said about "nice" things such as empathy, genuineness and warmth, or a working alliance, less training is usually provided to help beginning therapists respond to a demanding, passive-aggressive, or demeaning client (the one who comments, for example, "Looks like you've you been gaining some weight. What's your husband say about that?"). In this regard, researchers find that therapists do not register or respond to client's negative reactions toward them very well—*therapists tend to deny or avoid the unwanted remark*. It seems that it is much easier for therapists to miss or gloss over negative/hostile responses from clients than positive responses (Hill et al., 1992). However, we don't want to tune out or miss these important messages from the client. Why? The negativity coming toward us is often at the heart of what is causing problems in the client's relationships with others. We need to make the most of the time we have with clients, and addressing clients' problems with others—especially as they are occurring right now with us—is usually the best way to do this.

Thus, our intention is to approach negativity nondefensively, rather than avoid it and act as if something significant was not being said—even though this may be hard for us to do. Our goal is to work with it in an up-front way that helps clients learn about the problematic impact they are having on others and better ways to negotiate conflict and manage feelings in relationships:

CLIENT: Your office looks a little tired. This furniture and brown color went out back in the 1970s.

THERAPIST: (*calmly*) Things look a little out-of-date to you in here.

CLIENT: Yeah, you should do something to change it.

THERAPIST: No, I don't think I want to make any changes; I'm comfortable here. But maybe it would be helpful for us to talk about this for a minute. You've said things to me before like that, and I'm wondering how others usually respond when you criticize them or tell them what they should do?

CLIENT: Geez, Doctor, relax. I'm just trying to be helpful. You shouldn't be so defensive.

THERAPIST: No, I'm not feeling defensive right now, but I am still wondering how others usually respond when you advise them like that.

CLIENT: Well, I don't know, I guess mostly they just do it.

THERAPIST: Uh-huh, I can see that—you do have good taste and express strong opinions in an authoritative voice. But I'm wondering how it affects your relation- ships. Maybe you're right and this is just me here, but I'm thinking that most people wouldn't like too much of this. What do you think?

CLIENT: (*pause*) Well, no, maybe some people at work, or at my church, don't really like me very much. Maybe they'd agree with you and say I'm "bossy."

THERAPIST: Well, I'm not saying you're "bossy," but I am wondering how this goes with others. So, yeah, I can imagine some people feeling intimidated and wanting to get away from you, and others feeling resentful and wanting to argue with you. What it's like for you to be seen as "bossy" in this way—how does that feel?

CLIENT: Well, when you say it like that, it makes me feel kinda bad—almost like crying...

THERAPIST: Crying. Let's sit together with that feeling and just let it be there.

CLIENT: (*slowly*) I'm taking over...just like my mother always did. Giving advice and opinions that others haven't asked for and don't want. Sometimes I wonder if I'm going to ruin my marriage—just like she ruined hers...

This therapist was effective because she remained nondefensive in the face of the client's criticism and control. She *approached* this sensitive issue—she took the risk of naming it and asking about it—rather than just going forward again and pretending as if something important wasn't happening between them. In contrast, other therapists often respond ineffectively to such negativity with their own personal or countertransference reactions:

CLIENT: Your office looks a little tired. This furniture and brown color went out back in the 1970s.

THERAPISTS: (*Moving toward and complying*) Oh, I'm sorry. You know, you're probably right. Maybe it is time for me to update things a bit in here.

(*Moving away and avoiding*) OK, well uh, what do you think we should work on today?

(*Moving against and chiding*) Maybe you should spend more time thinking about your own problems and less about what everybody else should do.

With practice, and by observing instructors role-play or model a range of effective responses, therapists can learn better how to *remain nondefensive and approach the negativity*—the key to responding effectively. This approach empowers the therapist to intervene effectively by providing interpersonal feedback, making process comments, or simply adopting a neutral stance to question, highlight, or explore this negativity. Although still poorly integrated into clinical training, psychotherapy researchers have been trying for decades to tell practitioners that negativity will occur with many of their clients and that therapists need to learn how to deal with it more effectively. It is an important part of the treatment process that can be especially difficult for new therapists. Unfortunately, however, even experienced, highly trained therapists tend to respond with the same fight-or-flight reactions that clients elicit from others and that cause problems in their lives (Safran & Muran, 2003).

To help therapists better maintain an optimum interpersonal distance with clients, let's look further at what therapists can do to manage their countertransference reactions to clients' negativity. Angus and Kagan (2007) emphasize that it is critical for therapists—early in their training—to begin tracking their response tendencies and countertransference propensities, and they provide guidelines to help with engaging in this process.

One useful exercise for therapists is to write down their initial reaction propensities in different anxiety-arousing situations—for instance, what you are most likely to say or do first when a client is demeaning or critical of you. Or, with a partner, trainees can role-play their responses to different clients in challenging situations. For example, one person can role-play being a competitive, intimidating, or insulting client, perhaps someone who is questioning whether the therapist is really capable enough to be of help. Taking turns, the other trainee practices responding to this provocative client. In small discussion groups of three or four, trainees can provide each other with feedback, or watch themselves being replayed on videotape, and share their observations of the other therapists' responses while role-playing these challenging situations. With practice and guidance from supervisors, new therapists can learn to approach these unwanted situations, find constructive ways to make them overt, and begin to explore them collaboratively with clients, and turn them into corrective relearning experiences for clients.

Client-Induced Countertransference: Eliciting Overinvolvement and Underinvolvement.   In addition to therapist-induced countertransference reactions, the balance between overinvolvement and underinvolvement can be difficult to maintain because *clients often try to shift or move the therapist along this continuum of involvement as part of their interpersonal coping strategy.* This is the client-induced countertransference we have been discussing. When the therapist has been responding effectively, some clients may become threatened by the anxiety-arousing material that is coming up, by succeeding and getting better, or by the alliance strengthening and the therapist becoming more important to the client. To defend against these anxieties or other difficult emotions that can be activated by the therapist's effectiveness, or to test the therapist to further assess safety/danger in the therapeutic relationship, some clients will try to engender either enmeshment or disengagement in the therapeutic relationship. For example, some clients may try to bore the therapist by talking about issues they are not really concerned about (which can be more problematic for a therapist with a Preoccupied attachment style who needs engagement and intensity). In contrast, others may try to elicit too much involvement from the therapist by exaggerating their distress and escalating their demands for help (which may present more of a challenge for therapists with a Dismissive attachment style whose propensity is to minimize strong emotions and bids for intense engagement). In these and other ways, the client's eliciting behaviors may press the therapist away from an effective middle ground of involvement.

If therapists can maintain this **middle ground of involvement**—remain both a participant and an observer in the counseling relationship—they can provide a more even therapy with no stormy ups and downs or lengthy impasses. It is also safer for the client to be vulnerable and trust when the therapist is dependably available without swinging to either side and becoming overreactive or unresponsive. By offering this steady presence in the face of the client's eliciting maneuvers, the therapist is providing a secure base for the client. More specifically, when therapists maintain an optimum interpersonal balance, the client does not have to be concerned about disappointing or letting down an overinvolved therapist or eliciting the responsiveness or emotional engagement of a distant technician. The attachment researchers have appreciated best the importance of the caregiver's *consistent* responsiveness in developing security and trust.

The essential balance is for therapists to maintain their own personal boundaries (separateness) while still being responsive to the client (relatedness). That is, therapists are striving to enter the client's subjective worldview—to be fully present and connected—yet still maintain appropriate self-boundaries and self/other differentiation. Regardless of the therapist's theoretical orientation, treatment is more likely to succeed when the therapist can maintain this optimum interpersonal involvement. Conversely, treatment usually reaches an impasse—or terminates early before the client's work is complete—when the therapist and client become enmeshed or when they do not connect sufficiently to matter to each other and mutually invest in the relationship sufficiently to establish an alliance.

How can we help therapists recognize the interpersonal process they are enacting with their clients and assess their balance of involvement with clients? Therapists can do this by considering the following questions after every session:

1. What are my feelings and personal reactions toward this client?
2. How might my reactions parallel those of significant others in the client's life?
3. Are my feelings typical of my reactions to others, or are they more confined to this particular client?

By writing answers to these questions after each session, therapists ensure that they are attending to the process dimension. Asking themselves such questions will also help therapists distinguish their own personal reactions toward the client (therapist-induced countertransference) from the client's eliciting behaviors and interpersonal coping style that affect many people (client-induced countertransference). Building upon these three questions, therapists are encouraged to utilize the guidelines for writing process notes in Appendix A after each session.

As noted, all therapists will sometimes become overinvolved with a client. Indeed, many theorists have described the therapeutic process as therapists' ability to become immersed in clients' modes of relating and then to work their way out of these enactments (Gill & Muslin, 1976; Levenson, 1982; Mitchell, 1988). However, *new therapists are especially prone to become overidentified with their clients and to misperceive their clients' problems as being the same as their own*—it's par for the course. This is one of the most common ways in which clients' patterns get replayed in the therapeutic process, however. New therapists often feel discouraged when their supervisors accurately point out that, here again, they are reenacting aspects of the client's problems in their interpersonal process. New therapists need to be patient with themselves and accept that it is difficult to maintain this effective middle ground of involvement that safeguards against reenacting the client's maladaptive relational patterns. By examining their own experiential range, anticipating their own countertransference propensities, and tracking the separateness–relatedness dialectic in the therapeutic relationship, new therapists will become increasingly able to maintain an optimum interpersonal balance.

When therapists realize that they have become over- or underinvolved with a client and find that they cannot realign the relationship on their own, it is best to consult with a colleague or supervisor. One of the most productive uses of supervision is to help therapists regain control of their own emotional reactions so that they can reestablish an effective degree of involvement with a client who has successfully pushed them away or drawn them in too much. Maintaining this optimum interpersonal distance is one of the best ways to provide a reparative relationship and keep from reenacting the familiar but problematic scenarios that occur with others.

ENDURING PROBLEMS TEND TO BE PARADOXICAL AND AMBIVALENT

THE TWO SIDES OF CLIENTS' CONFLICTS

Therapists can better understand their clients' problems when they recognize the paradoxical and **ambivalent** nature of many conflicts. Usually, there are two sides to the client's central conflict. Before problems can be resolved, therapists need to respond to *both sides* of their dilemma—with interest, curiosity, and an empathic exploratory stance for both sides rather than overtly arguing and pressuring clients, or subtly leading and cajoling them, toward one side or the other (Miller & Rollnik, 2002). Often, because of their own countertransference, therapists may fail to recognize both sides of their client's conflict or, more commonly, because they believe they know what is "right" and know what the client "should" do, therapists press the client toward only one side of their ambivalence about a decision (e.g., Ineffective Therapist: just leave the boyfriend; you need to stay in school, you should apply for that position, you shouldn't buy him that car, it's a mistake to get married so young). Therapists are more helpful to their clients when they can articulate and highlight more clearly, and then help clients explore and sort through for themselves, *the two opposing feelings, competing needs, or contradictory beliefs that make up the double-binding conflict.* Colloquial expressions convey the two-sided nature of conflict. We often say that someone is "stuck between a rock and a hard place," that someone has grabbed "both horns of the dilemma," or that a difficult relationship is "like a double-edged sword." These expressions capture the push-pull nature of conflict—that you "can't live with them and you can't live without them" or are "damned if you do and damned if you don't."

Therapists working within different theoretical perspectives have used differing terms to describe this. Long ago, for example, the behavioral therapists Dollard and Miller (1950) highlighted how clients were often struggling with "approach-approach" or "avoid-avoid" conflicts. Communication theorists and early family therapists highlighted double-bind interactions and paradoxical communications (Bateson, 1972).

Contemporary attachment researchers reveal the immobilizing plight of children who are attached to a caretaker who is sometimes helpful but, at other times, also frightens or hurts them (Hesse & Main, 1999). Recalling the "Category D" attachment style, imagine a young child who is growing up with a highly inconsistent or unpredictable caregiver. This attachment figure may have borderline or narcissistic features, struggle with bipolar depression, or be preoccupied with strong unresolved grief reactions. They also may be dissociative at times as a result of the caregiver's own unresolved physical or sexual abuse in childhood. In a truly maddening paradox, this child sometimes receives genuine comfort and understanding from this caregiver. At other times, however, this same benevolent caretaker who meets their emotional needs also becomes the source of terror and shame for the child. This occurs when the caretaker acts in strange or disturbing ways that the child cannot understand (for example, observing a drug-induced parent talking animatedly to imagined others with exaggerated hand and facial expressions), is enraged or threatening toward the child ("I wish you were never born! You're ruining my life. I hate you"), or cowers on the floor, weeping or terrified, in response to only their inner experience and no visible event to help the child make sense of their parent's experience or disturbing behavior. This child has an unsolvable dilemma: The source of comfort is simultaneously the source of danger. Aptly calling it "fear without resolution," the person who helps me is also the person who hurts me. To the child, seemingly, there is no solution to this double bind—"I can't walk away from this caregiver who I depend on yet, at the same time, I can't be close and need this person who frightens or hurts me so much." In this regard, clients with the most intractable symptoms, and who tend to internalize or repeat their caregivers' own problematic parenting or symptomatic behavior, grew up with these highly ambivalent relationships (Benjamin, 2003). Indeed, life is easier when caregivers are generally good or generally bad—when there is too much good and bad news mixed together, problems become more complex and harder to resolve (for example, a client is molested by her stepfather who, in other ways, is the most responsive and helpful adult in her life). In sum, theorists from many theoretical perspectives all have tried to help therapists recognize and respond to the double-binding, paradoxical, or two-sided nature of the conflicts that our clients present.

There are many different facets to this issue of the two-sided nature of conflict. One important aspect is how therapists often fail to appreciate that many clients have concerns both about being hurt by the therapist as they have been hurt by others, and about hurting others (Weiss, 1993). The counseling field has well-developed concepts regarding clients' expectations about again being hurt, misunderstood, betrayed, and so forth, by the therapist—as they have been in other important relationships (that is, transference). In contrast, far less attention has been paid to some clients' worry and guilt about hurting the therapist and others (and their internalized attachment figures) by doing well in life. This commonly occurs, for example, when these clients do better than their parents and surpass their caregiver in some arena such as having greater financial success, enjoying a happier marriage, or simply living longer or better than their caregivers. It may also occur when clients succeed at reaching a sought-after goal, appropriately stand up for themselves or advocate on their own behalf, pursue their own interests or career choices, and so forth. Thus, when these clients make progress or improve in treatment (for example, by graduating from college, earning a promotion, or having a relationship with someone who treats them well), they may retreat from these success experiences. In various ways, they may sabotage themselves, be unable to sustain progress toward their own chosen goals or, most commonly, be unable to enjoy or feel good about the success they achieve. Therapists will observe that some clients will become depressed as a result of separation guilt or survivor guilt. Even more will become anxious as success threatens their attachment ties to internalized caregivers who did not support their individuation, or it deviates from their prescribed familial role (for example, to take care of others and submerge their needs rather than pursue their own interests or goals). Clearly, the therapist doesn't want to sit back and relax as their clients begin to make progress in treatment and improve their lives—this is often where the real work begins!

To illustrate the two-sided or double-binding nature of many clients' conflicts, let's return to a previous case study. As we will see, once the therapist is able to recognize the two-sided structure of the conflict and respond empathically to both sides of the client's ambivalent feelings, the client's seemingly irrational or self-defeating behavior makes sense and becomes resolvable.

Recall Anna, the young adult client who was discussing her dream with her therapist at the end of Chapter 4. Anna wanted to end her chronic depression, become more involved with friends, and find a career path that she could pursue. Although she had expressed these wishes for many years, she could not follow through and realize them. She rarely initiated activities with peers and often declined the invitations she received from others. She repeatedly enrolled in college but would eventually "lose interest" and drop out after one or two semesters. Without recognizing the two-sided conflict that plagued Anna, her self-defeating behavior seems irrational.

Anna is like many clients who grew up with a Preoccupied (Anxious/Ambivalent) attachment figure who could not support her differentiation and autonomy. On the one hand, Anna wanted to grow up and have her own independent life with her own friends, marriage and family, and career. At the same time, however, she felt a responsibility to remain close to home and support her mother who acted as a guilt-inducing martyr—often having medical crises whenever Anna started to establish her own goals or pursue her own interests. The recurrent scenario was that whenever Anna began to succeed with friends, in school, or with life in general, her mother would look hurt, withdraw physically and emotionally, and induce guilt (for example, "I saw the doctor yesterday and he said my heart isn't good..."). Unable to differentiate from this tie that binds, Anna continued this family drama by repeatedly failing in different arenas of her life, and in one way or another, having to move back home and live dependently again. Upon her failure and return home, however, her mother would become less of a wounded martyr and become more available and responsive to her.

Having to sacrifice their own independence in order to maintain their attachment ties in this way, clients like Anna become depressed over their inability to individuate and have their own lives, and lose any sense of agency or self-efficacy from such binding guilt.

After a few months, however, the same cycle will re-play again. Fighting the depression that came from her compliance, Anna will start to act on her own healthy wish for more autonomy, seek peers and friends, and re-enroll in school. However, at this point, her mother again will withdraw and act hurt, and Anna will not be able to sustain her effort to individuate and launch her own adult life at her mother's expense. In a few months, she will still be feeling guilty about hurting her mother, and anxiously insecure about her feeling of being "alone." To abate her separation guilt and her separation anxiety, Anna will again return home, feeling hopeless and a failure. Completing the scenario, her mother will become less self-centered and more comforting and supportive of Anna in her dependency and social failure. This type of cyclical interpersonal pattern may be reenacted for years, or a lifetime, unless both sides of the client's conflict can be named, made overt as a shared treatment focus, and worked through in therapy.

Without understanding these binding family dynamics (see Haley, 1980; Mallinckrodt & Daly, 2009), the therapist and others might erroneously conclude that Anna is just immature and dependent, insufficiently motivated to do the hard work necessary to change, receiving too many secondary gains from being home and depressed, masochistic, has a chemical imbalance or untreatable Axis 11 Avoidant Personality Disorder, and so forth. However, once we understand the cyclical relational pattern that immobilizes her (that is, if Anna meets her own need and tries to emancipate she believes she is a bad daughter who is selfishly hurting her mother), we see that such pejorative labels are unwarranted. Clients' contradictory or self-defeating behaviors do make sense, but not until we understand the two-sided, push-pull, or double-binding nature of their conflicts.

EXPLORING AMBIVALENT FEELINGS

Understanding that many clients' problems are a two-sided or double-binding conflict, *therapists will be more effective when they can help clients explore and sort through both sides of their ambivalence.* That is, therapists want to take a neutral, exploratory stance and invite clients to expand and elaborate their thoughts and feelings on *both sides* of their conflict. For example:

THERAPIST: It sounds like one part of you wants to get married but another part of you doesn't. Tell me about both sides—the part of you that wants to marry him and the part of you that doesn't. I have no investment in what you decide; I'd just like to help you sort through the pros and cons on each side.

In this way, the therapist welcomes and invites both sides of the client's ambivalence. The therapist's intention is to respond affirmingly to the opposing or contradictory feelings that accompany both sides of the conflict. As clients find that they have the support of the therapist to explore all of their feelings and concerns about a particular decision, they are able to clarify their own preferences, take responsibility for their own choices, and act on their own decisions. In contrast, therapists will not be effective when they only respond to one half of the client's ambivalence, do not encourage the client to freely explore both sides of the issue, or jump in and subtly advocate or overtly tell the client what to do. In their Motivational Interviewing work with problem drinkers, Rollnick and Miller have found strong empirical support for listening for and then supporting the client's own intrinsic motivation to change rather than trying to instill it or "motivating" clients to make a particular decision, such as to stop drinking (2002). When the therapist cannot "hear" both sides of the client's ambivalence, or presses clients to act or choose the way the therapist wants, the therapist's own countertransference issues are usually in play, as the following vignette illustrates.

Marie, a depressed, 25-year-old graduate student, entered time-limited (20 session) treatment at the Student Counseling Center on campus. During the first few sessions, she explored how angry she was at her mother. Her mother expected Marie to be perfect, insisted that Marie never had any problems, and always needed Marie to be "happy." In counseling, Marie began to realize how much she resented her mother for denying so many of her true feelings and for having so many expectations of how she "should" be. Recalling similar issues in her own childhood, Marie's therapist resonated with her anger and actively supported her indignation.

At first, it was liberating for Marie to have her long-suppressed anger affirmed. She was both excited and relieved to be realizing that she no longer needed to keep fulfilling her mother's unrealistic expectations. This good feeling was short-lived, however. In the next few sessions, Marie's newfound freedom gave way to a growing despondency and return to her long-standing dysthymia. The therapist thought that Marie was worried about becoming the "bad child" in the family for defying her mother's expectations, feeling disloyal to her family for complaining about her mother to the therapist, and feeling guilty about being angry with her mother.

While asking about and looking for opportunities to explore these reasonable possibilities, the therapist continued to draw out and encourage Marie's anger. However, moving in a different direction than the therapist, Marie began to share fond memories about her mother and recalled special times they had spent together baking and making cookies after school. The therapist was not very responsive to Marie's positive feelings toward her mother, however. The therapist thought that Marie's fond recollections were part of denying her anger toward her mother, avoiding the reality-based conflicts in their relationship, and was concerned that Marie would continue to be ruled by her mother's problematic expectations.

As often happens in treatment, Marie's therapist only recognized one side of her ambivalence and failed to respond to the feelings on both sides of her conflict. Yes, Marie was angry with her mother, and these feelings needed to be explored and affirmed. At the same time, however, Marie knew that her mother had offered her many fine things as a parent and she did not want to forgo the good things they had shared. Marie's conflict, in so many words, was that she wanted to keep the good things she had experienced with her mother without having to take on or adopt the problematic aspects of their relationship as well. Thus, on the one hand, Marie wanted to reject the unrealistic demands to be perfect and happy that her mother had placed on her. Yet on the other, Marie became depressed when it seemed that the only way to do this was to give up her identification with her mother altogether. This had been the most important relationship in her life and, understandably, Marie did not want to risk losing the good parts of this relationship as well.

Realizing that time was short and treatment was stagnating, her therapist sought consultation from a colleague. As an outside observer, it was easy for the colleague to recognize both sides of Marie's ambivalence, and encouraged the therapist to *give Marie support for simultaneously having contradictory feelings of appreciation and anger toward her mother*. At first, the therapist was reluctant and began arguing with the colleague. Soon, however, she began to consider the possibility that her own countertransference might be operating and that, because of her own history, it was easier to support Marie's anger toward her mother than her appreciation for her. As a result of this helpful consultation, the therapist watched at their next session for an opportunity to invite and affirm the other side of Marie's feelings as well:

THERAPIST: It sounds like you feel two different ways toward your mother at the same time, and that makes sense to me. You're angry with her for expecting you to be perfect and happy all the time, but you also treasure the many fond times you've had together. Tell me more about your angry and your loving feelings toward her—bring me in on both sides of this.

This new approach had an immediate impact on Marie. As the therapist encouraged Marie to explore both sides of her feelings and, in effect, gave her permission to love her mother and be angry with her at the same time, she felt enlivened and began to make real progress. Marie began to clarify what qualities of her mother she wanted to keep and make a part of herself (such as her mother's easy laughter and genuine warmth at times) and what aspects of her mother she wanted to discard (her mother's preoccupation with appearances and what other people might think). This process of **differentiation** also allowed Marie to clarify which issues she wanted to bring up and try to talk about and hopefully change in her relationship with her mother, and which issues she preferred to let be and simply accommodate herself to.

Marie continued to make progress in treatment as the therapist stayed this course and continued to provide support for both sides of her ambivalence. At times, for example, the therapist would observe, "It really made you mad when she did that." However, this would soon be followed by the reflection, "I can see how important you two are to each other, and how much you love her—even though these problems get in the way sometimes." Therapy was successfully terminated at the twentieth session. Marie was no longer depressed and had better integrated aspects of her own identity as a woman. In the months following termination, Marie was continuing to initiate a rapprochement with her mother, was sorting through what could change and be better now in their relationship and what couldn't, and was exploring realistically how much authenticity she could have, including the closeness they could share now at this different point in their lives.

To summarize, therapists help clients change when they recognize the two-sided structure of their clients' conflicts, and help clients explore and integrate both sides of their ambivalent feelings. Trained to look for pathology, therapists tend to miss personal strengths or positive aspects of conflicted relationships. However, clients will make far more progress when the therapist can actively encourage clients to keep the good parts of the parental (or spousal) relationship while they are realistically examining the problematic parts. In this regard, it usually is very helpful if *the therapist can tell clients explicitly that they are not being disloyal if they talk about problems with a parent or spouse, and that the therapist knows that their complaints do not reflect all of their feelings about this person.* In this regard, many clients will not be able to engage deeply in treatment and work productively until the therapist recognizes and supports both sides of their ambivalence. For so many clients, a primary treatment goal will be to integrate their positive and negative feelings. This means they can be disappointed or angry with someone they love while still remaining connected to other, more rewarding aspects of the person which, in turn, will also allow them to still feel worthwhile even though they are disappointed with themselves about something.

Finally, just as therapists in individual practice are looking for opportunities to affirm both sides of the client's conflict or ambivalence, couple therapists are trying to support both partners equally in an interpersonal dispute. The primary task in marital therapy is to resist the couple's eliciting maneuvers and keep from taking sides by participating in the good guy/bad guy, overadequate/inadequate, healthy/neurotic, or other polarizations that couples so commonly present. When couple counseling fails, one member of the couple usually stops treatment because she or he has (often accurately) come to believe that the therapist has lost neutrality and taken sides against them in the marital conflict.

CLOSING

In this chapter, we examined three dimensions of clients' current interpersonal functioning: (1) how clients' problems with others are brought into the therapeutic relationship through eliciting maneuvers, testing behaviors, and transference reactions; (2) how therapists can establish an optimum degree of interpersonal relatedness and enact a corrective balance of separateness—relatedness; and (3) how therapists can understand the two-sided structure of many clients' conflicts and support clients in exploring both sides of their ambivalent feelings. These interpersonal factors, together with the developmental/familial perspective presented in Chapter 6 and the model of interpersonal coping strategies presented in Chapter 7, will help therapists better understand and intervene with the problems clients present.

With these conceptual guidelines in mind, we now return more fully to the treatment process. The next chapter explores how therapists can use the interpersonal process they enact with clients to identify maladaptive relational patterns, begin resolving clients' problems in the current interaction with the therapist, and then help clients generalize this new behavior with the therapist to other important relationships in their everyday lives. Working in the here-and-now with the process dimension is essential to all forms of interpersonally oriented therapy. However, if therapists are to use process-oriented interventions and the therapeutic relationship to effect change, they need to be able to talk forthrightly with clients about their current interaction and what's going on between them. For most new therapists, however, it often feels uncomfortable to meta-communicate in this way and "talk about you and me." These immediacy interventions are powerful yet challenging for most new therapists to adopt. Thus in Chapter 9, we explore therapists' concerns about working with the process dimension and provide guidelines to help therapists intervene more effectively within the therapeutic relationship.

SUGGESTED READINGS

1. In a managed care milieu, writing effective treatment plans becomes an essential skill for clinical trainees. Illustrating the guidelines for writing case conceptualizations that are provided in Appendix B. Chapter 8 of the Student Workbook provides an in-depth case formulation of "Linda" in Arthur Miller's play, *Death of a Salesman*.
2. The outstanding *Attachment Theory and Research in Clinical Work with Adults*, J. Obegi and E. Berant, eds. (New York: Guilford, 2009), includes contributions from leading attachment clinicians and researchers. A must-read for attachment-oriented clinicians, it provides useful guidelines for understanding and intervening with adults who have experienced significant developmental disruptions and highlights the importance of the therapist–client relationship in facilitating change.
3. Readers are encouraged to learn more about clients' testing behavior by reading Chapter 4 of *How Psychotherapy Works* by Weiss (1993) or Chapters 1 and 2 of *Transformative Relationships*, G. Silberschatz, ed. (New York: Rouledge, 2005).
4. Chapter 2 of Jay Haley's (1980) *Leaving Home* (New York: McGraw-Hill, 1980) elucidates the family dynamics and structural family relationships that impede individuation and emancipation for many young adult clients.
5. Family systems concepts of "cohesion and adaptability" can also be used to describe an optimal degree of interpersonal relatedness. Interested readers may examine *Families: What Makes Them Work* by Olson et al. (Newbury Park, CA: Sage, 1983).

RESOLUTION AND CHANGE

PART

9 | AN INTERPERSONAL SOLUTION

CONCEPTUAL OVERVIEW

The three previous chapters helped therapists conceptualize their clients' problems and formulate treatment plans; we now focus further on the process dimension and how change occurs. To begin, therapists need to empower clients by helping them develop new narratives for their lives that allow them to make sense of what has occurred to shape who they have become and the problems they are having (Crits-Christoph et al., 1999). Unless they develop self-narratives that are more realistic and fit closer with the good and bad news they have actually experienced in their lives, and are defined less by family myths and roles that do not fit the facts so accurately (such as, "I'm too needy and deserve to be left; I'm bad and will continue to be rejected; I'm selfish and hurting others if I succeed or follow my own interests," and so forth), it will be harder for them to guide where they are going in the future or to sustain changes after treatment has stopped. In this way, seeking a more realistic assessment of the formative experiences and interaction patterns that clients have experienced helps both clients and the therapist understand how they got here—where faulty schemas developed and why current relationships are being framed and recreated in the way they are (Luborsky et al., 1997). However, a consistent or primary focus on historical relationships and what happened in the past will not produce change and, too often, only serves to help clients avoid the anxiety of addressing current problems.

Looking instead for more than just insight or understanding of the past, change occurs when clients experience a behavioral or in vivo resolution of their problems in the current relationship with the therapist, and this experiential relearning is generalized to clients' everyday lives and current relationships. Said differently, we are interested in exploring developmental and familial relationships, but more to illuminate clients' faulty beliefs, expectations, and outdated coping strategies than to gain insight or catharsis. This clearer understanding of the problematic ways clients have been responded to in important relationships throughout their lives helps therapists formulate a treatment focus—clarifying the reparative experiences they need with the therapist in order to change (Levenson & Strupp, 1997). In this way, understanding formative relationships may be a relevant aspect of change for some yet in itself is usually insufficient. Instead, we are focusing on new experiences with the therapist that behaviorally disconfirm or counteract problematic messages learned in close relationships.

If the primary focus of treatment and fulcrum of change is in the current interaction between the therapist and the client, rather than a psychodynamic exploration of the past, what can the therapist do to help clients change? Does the therapist need to take responsibility for finding solutions and telling clients what choices and decisions to make? No, therapists can't just prescribe solutions or direct clients regarding what to do—telling clients how to live their lives only fosters their dependency and undermines their self-efficacy (Bandura, 2006; Barber et al., 2009). Researchers find that clients are more likely to consider, verbalize, and ultimately make changes when they experience the therapist as empathic and as facilitating their own reasons for making changes, rather than directing or telling them what to do (Miller, Benefield, & Tonigan, 1993; Miller & Rose, 2009). That is, the therapist helps by: attending closely to clients' own interests, motivations, and goals; actively supporting their initiative and autonomy; and by an empathic or attuned listening that helps to discern clients' key concerns and clarify what their problems really are about. This combination of providing both empathic understanding and support for the client's own initiative (in attachment terms, providing both a safe haven and a secure base) is key to clients' ability to disclose more deeply, engage in greater self-reflection, and from this construct a more coherent self-narrative with a greater sense of personal agency (Angus et al., 2004; Angus & Hartke, 2006; Bandura, 2006). Finally, this empathic attunement (relatedness) coupled with support for differentiation (separateness) also allows therapists to provide clients with a new and reparative response to the problematic patterns that they have experienced with important others in the past and come to expect from others. Change occurs as this new way of relating with the therapist expands clients' cognitive schemas, alters their beliefs about themselves and expectations of others, and allows them to increase their interpersonal range with others.

We have emphasized that change often occurs as clients, first, find new responses to their problems in their relationship with the therapist, and then work through similar issues with others beyond the therapy room. To highlight the power of this experiential relearning over just cognitively interpreting, reframing, or explaining what something means intellectually, the attachment researchers put it best: You do not get an empathic child by teaching or admonishing the child to be empathic; you get an empathic child by being empathic with the child (Karen, 1998). The therapist's intention is to make this experience of change an ongoing pattern of interaction in the therapeutic relationship—one trial learning is not sufficient. As this occurs, the focus of treatment shifts to helping the client *generalize* this in vivo relearning to other arenas in the client's life where, in parallel, similar themes are being played out and contributing to the client's presenting symptoms. Conversely, if the therapist and client merely talk about important issues and behavioral options, but the therapist does not actively help the client link these new ways of relating with the therapist to other relationships where the same kinds of interactions are causing problems, change will not occur. Thus, the purpose of this chapter is to further clarify the process dimension and illustrate how therapists working within different theoretical orientations can use the therapeutic relationship to help clients change (Goldfried, 2004; Levenson, 1995; Sampson, 2005).

RESOLVING PROBLEMS THROUGH THE INTERPERSONAL PROCESS

In this section, we explore how the interpersonal process that therapists enact with clients can be used to resolve clients' problems. First, we will review the course of therapy up to this point in treatment and highlight how clients' problems are brought into the treatment setting and reenacted with the therapist. This review serves to introduce the next stage of treatment—providing a corrective emotional experience and helping clients apply this relearning to other relationships. Following this overview, we will look at five case vignettes that illustrate implementation of these constructs.

RESPONDING TO CLIENTS' CONFLICTS IN THE THERAPEUTIC RELATIONSHIP

Ongoing, problematic interaction patterns with caregivers give rise to developmental conflicts that clients have not been able to resolve on their own. Symptoms and problems develop as clients try to cope with difficult feelings and faulty beliefs by adopting fixed interpersonal coping strategies. Although these coping strategies once were necessary and adaptive, they are no longer necessary and create problems in many current relationships. The therapist's first task is to establish a collaborative working alliance with the client by making accurately empathic connections. As documented consistently in the empirical literature, therapist–client alliance is a highly significant predictor of treatment outcome (Barber et al., 2009), and the therapist's empathic attunement facilitates clients' self-reflection and sense of agency (Bandura, 2006) as well as their motivation for engagement in treatment (Westin et al., 2004). Once an alliance is established, client and therapist then work together to identify and change the maladaptive relational patterns that are occurring with others and in the therapeutic relationship, as well as the conflicted feelings and pathogenic beliefs that accompany them. As this work proceeds, the therapist's intention is to watch for and recognize the **pull** from the client to reenact these relational themes in the therapeutic relationship. The therapist's goal, then, is to respond differently than others usually have and change the familiar but problematic scenario—providing a different type of relationship that resolves, rather than reenacts, the client's cyclical relational themes (Beebe et al., 2005; Goldfried, 2004; Sampson, 2005). As therapists respond in ways that counteract the problematic expectations learned in the past, clients experience new ways of relating and begin to question or evaluate the costs and gains of their habitual coping styles. In this process, clients are empowered to choose more flexibly how they want to respond in current relationships, and to develop more realistic and affirming self-concepts.

The therapist's primary role is not to give advice, explain, reassure, interpret, self-disclose, or focus on the behavior or motives of others, although each of these responses will be effective at times. Instead, the therapist encourages the client to take ownership of the treatment process by setting the direction for therapy. The therapist does this by helping the client identify the issues or concerns that the client feels are most pressing or important right now. The Motivational Interviewing literature documents the importance of engaging clients in this collaborative process that is empathic and, in particular, actively supports the clients' own interests, motivation, and initiative. These researchers find that clients engage more actively, become more committed to treatment, share more deeply, and rate their sessions as deeper, more valuable, powerful, and special when therapists refrain from being too directive and, instead, foster the clients' sense of personal responsibility and self-efficacy by joining them in their own interests and concerns (Pesale & Hilsenroth, 2009; Villanueva et al., 2007). Supporting and encouraging clients' leads in this way and following their interests is not a passive or patient waiting game, as is often misunderstood. The therapist is active but not directive in shaping a *treatment focus* by listening for and highlighting:

- maladaptive relational patterns or problematic interpersonal scenarios that keep occurring,
- pathogenic beliefs about self and faulty expectations of others, and
- affective themes that recur throughout the material the client presents.

The therapist continues to focus the client inward, reflects the core affective messages, links the client's maladaptive relational patterns with others to her own current interaction, and maintains a collaborative, working alliance by ensuring that clients are active participants who feel *ownership* of the change process. As the client's faulty beliefs, expectations, and coping styles recur throughout the narratives they relate, themes emerge that clarify the treatment focus and provide direction for the ongoing course of therapy (Barber et al., 2009). For example, the therapist will be able to do the following:

- Point out when and how clients employ their interpersonal coping strategy to rise above their conflicts.
- Help clients become aware of how they block their own needs and feelings by responding to themselves in the same problematic way that others originally responded to them.
- Help clients explore why they become anxious at a particular time, what types of responses they tend to elicit from the therapist and others to manage this anxiety, and how they systematically avoid certain interpersonal modes or feelings.

Although this conceptual awareness of core conflicts and interpersonal coping syles will be an important part of helping clients change, something more is needed. At the same time as the therapist and client explore the content of the client's concerns, they also tend to play out these same relational themes in their interpersonal process (Henry & Strupp, 1994). As we have seen, the client and therapist do not just talk about issues in therapy; they actually relive them in the therapeutic relationship. This occurs in three ways.

Transference Reactions.   First, transference reactions bring the client's core conflicts into the current relationship with the therapist (Connolly et al., 1999). As the client's conflicts begin to emerge in therapy, the client will become increasingly concerned that the therapist has been responding—or is going to respond—in the same unwanted ways that significant others have in the past. These fears and misperceptions follow from the client's cognitive schemas, and *they are most likely to occur when strong feelings have been evoked and the client feels distressed or vulnerable.* In some cases, treatment will reach an impasse until the therapist can help the client differentiate the therapist's actual intentions from the problematic responses that the client expects and has in fact received in other important relationships:

THERAPIST: As you tell me about this, what do you think I might be feeling about you?

CLIENT: Well, I guess I'm concerned that you might be a little disappointed in me.

THERAPIST: Disappointed? Oh, no, not at all. Actually, I was thinking…

Eliciting Behaviors.   Second, clients will systematically **elicit** responses from the therapist that pull the therapist into reenacting their old scenarios or confirming their problematic expectations of others. When therapy stalls or ends prematurely, clients have usually reenacted their generic conflict in this way with the therapist. This unwanted reenactment occurs when clients elicit responses from the therapist that are thematically similar to those they have received from others in the past—responses that clients perceive, once again, as rejecting, idealizing, competitive, critical, and so forth. Clients elicit these familiar yet problematic responses from the therapist for different reasons—sometimes defensively, to avoid the internal aspects of their conflicts, and sometimes more adaptively, by testing whether they can obtain a better response to their familiar relational scenario than they have come to expect from others:

THERAPIST: How do others usually respond when you talk to them like you're talking to me right now?

CLIENT: They probably don't like it much—you know, everybody just gets pissed with me.

THERAPIST: I can see how that happens, and I have some ideas about how we could try to make that different here in therapy—in our relationship. I think if we could start talking about what's going on between us and change how this goes in here, that would be a very helpful step to start changing this with others in your life. How does that sound to you?

Testing Behaviors.   Third, the manner or way in which the therapist responds to clients—the therapeutic process—may unwittingly reenact their conflicts. As we have emphasized, clients' relationships with their therapists need to provide different or more helpful responses to their conflicts than they have received in the past. However, it is not always easy to provide this corrective emotional experience. Clients will expect, and often successfully elicit, the same problematic responses from the therapist that they have received from others in the past. The task for therapists, then, is *to identify how their interpersonal process with the client may be reenacting some aspect of the client's problem with others and, when this is occurring, to use a process comment to make this interaction overt as a topic for discussion.* The therapist then works together with clients (and with the help of a supervisor when the clients' issues have evoked the therapist's own matching or **reciprocal conflicts**) to establish a different pattern of interaction that does not repeat the old relational scenario. As we have seen, however, this idea is far easier to say than to put into practice:

THERAPIST: I just disagreed with you—I'm wondering how that was for you?

CLIENT: Well, kind of different, I guess. I'm not used to people speaking up and disagreeing with me much.

THERAPIST: Yes, I can see why they don't—you are so forceful when you speak. I hear you that you want to "communicate better" with your 15-year-old daughter, but it makes sense to me when she says that she feels "dominated" by you and reluctant to disagree or speak up. I'm a 45-year-old, educated professional, and you're a challenge even for me.

CLIENT: We need to work on this, don't we? I think this is what my wife has been trying to tell me for a long time.

Thus, if it is to lead to change, the therapeutic process needs to enact a resolution of clients' conflicts rather than a repetition of them. When this occurs, clients have received far more than just an explanation for their problems. They have experienced a meaningful relationship in which their old conflicts have been aroused, but this time they have found a better outcome than what usually occurs. They have been able to reveal themselves, to speak up or act more boldly, to ask for help or express a need, and so forth, without receiving the unwanted response they have come to expect (for example, being ignored, blamed, aggrandized, criticized, diminished, and so forth). When this occurs in the context of a meaningful relationship, such in vivo or experiential relearning is powerful indeed.

Experiencing an interaction with the therapist that is incompatible with their early maladaptive schemas does not make up for the deprivations or disappointments that have shaped the client's life, of course. However, it does behaviorally demonstrate that change can occur—that at least some relationships, sometimes, can play out in a different and better way. This corrective emotional experience begins to expand clients' schemas for what can occur between people in relationships and broaden their interpersonal and affective range. For example, they can voice their opinion with some others without being ridiculed or interrupted; they don't have to fear being left if they disagree in some relationships; they don't have to be considered arrogant if they feel good about an accomplishment, and so forth. At this critical juncture in treatment, when the client is having a meaningful experience of change with the therapist and the therapeutic process is providing a resolution of the client's conflicts, two important things occur. First, *intervention techniques from varying theoretical orientations all become more effective.* That is, cognitive, interpretive, self-monitoring, educational, skill development, and other interventions all can be used more productively by clients because the useful new content in these interventions is congruent with the corrective process they are enacting. Second, as clients have repeated experiences of change with the therapist, the therapist can help clients begin to apply this in vivo relearning to other arenas in their lives where similar conflicts are being enacted. This next period of therapy—transferring this experiential relearning that has occurred with the therapist to other relationships—is called the *working-through* phase of treatment and will be the focus of the next chapter.

With this overview of the change process in mind, let's now examine more closely how therapists can use the interpersonal process they enact with their clients to effect change.

USING THE PROCESS DIMENSION TO FACILITATE CHANGE

To effect change, the therapist is trying to keep from replaying in treatment what has gone wrong in other important relationships. Although such reenactments are inevitable at times, we do not want them to come to characterize the ongoing interaction with the therapist—we want instead to identify and highlight them, and change them by talking them through with the client. To alter clients' schemas and expand their interpersonal range, clients need to have the real-life experience of a relationship that does not go down the same old problematic lines they have known before and come to expect (Daly & Mallinckrodt, 2009). This means that *therapists are:*

1. listening keenly to identify the core messages that clients learned in childhood; and
2. sustaining a focus on providing unique responses that differ from and disconfirm these problematic expectations.

Providing this in vivo or experiential relearning is the core component of change in the interpersonal process approach.

Within this context, interventions from other theories can be helpfully integrated. Following Rogers (1975), it is essential for the therapist to listen empathically and work collaboratively. Also, it is helpful with some clients to suggest psychodynamic interpretations and facilitate client insight about the sources or development of their problems (Hill et al., 2007). Behavioral and cognitive therapists significantly help clients by training new relaxation, desensitization, and assertiveness skills; cognitively reframing situations; and suggesting new, more adaptive behavioral responses for clients to try between sessions. Each of these and other interventions will certainly be useful, but each can become more effective when the therapist is also attending to the interpersonal process that is being enacted with the client.

Clients benefit when they live out with the therapist a relationship in which their core conflicts are activated and come into play with the therapist but, in contrast to what usually transpires with others, the therapist does not respond in the familiar but unwanted ways clients expect. This recommendation sounds easy but is often so difficult to implement. Why? Therapists often fail to recognize how they are metaphorically or thematically reenacting aspects of their clients' conflicts in their interpersonal process. Clients do not just talk about their problems in the abstract; they reenact the same relational themes with the therapist—in the three ways described earlier. When treatment fails, the working alliance usually has been ruptured because the treatment process has repeated some aspect of the same interpersonal problems the client has been struggling with in other relationships— the problems that originally brought the client to treatment. In particular, *this often occurs when the client's eliciting maneuvers, testing behavior, or transference reactions have tapped into the therapist's own personal issues and created a mutual or shared conflict for both of them.* For example, perhaps the client's need to control others brings up the therapist's own concerns about being controlled by others, or the client's criticalness evokes the therapist's own shame-proneness and defensive need to be right, and so forth. Appreciation of the complexity added by this hand-in-glove or dovetailing between the therapist's and the client's personal dynamics leads us to see why it is highly challenging to enact a solution to the client's old scenario rather than a repetition of it. To help, we look below at five different case examples to illustrate how the process dimension can be used to facilitate change.

Example 1: Reenacting the Problem—An Inability to Recognize the Interpersonal Process. Therapists-in-training usually find it easy to understand the process dimension conceptually, and relatively easy in group supervision to observe when a client's problematic relational patterns are being reenacted with another therapist. It is, however, far more difficult for therapists to recognize how the interpersonal process they are enacting with their own clients may be reenacting the conflict rather than resolving it.

An alcoholic client began treatment by expressing how much pain he was in as a result of his drinking problem. His wife was threatening to leave him and he was on probation at work. His 10-year-old son had recently seen him in an embarrassing situation while intoxicated and had asked him not to drink anymore. The client was distraught, and even though the structure of his life was collapsing he could not stop drinking. During their first session, the distressed client said, "Nothing ever works out for me. Every relationship I have seems to go bad and fall apart in the end." The therapist was moved by the client's plight but, unfortunately, lost the clarity of her boundaries and became overly invested in having him change. The therapist found it difficult to listen to how hopeless he sounded, and she began reassuring him that things would get better. The therapist disclosed that she had had a drinking problem of her own years ago and that she knew what he needed to do in order to stop drinking. The therapist and client readily established a friendly rapport in which the client successfully elicited a great deal of support and encouragement from the responsive therapist. The client enjoyed this support greatly and sincerely tried to follow the therapist's good advice.

Things went better for about a month, but then the client "fell off the wagon" and began drinking again. When he started arriving late for his therapy sessions, and then missed two of them, the therapist began to feel "let down." It was the final straw for the therapist when the client arrived for therapy smelling of alcohol. The therapist felt angry and betrayed because she had extended herself to him more than she usually did with other clients. She felt that he was letting them both down and told him so. Unwittingly, the therapist became punitive toward the client in this way, and she induced guilt over the impact of his "irresponsible" drinking on his family. The client was contrite and tried, unsuccessfully, to elicit her sympathy and support again. The client did not show up for his next appointment, however, and did not return to therapy.

This interaction reenacted the client's conflict in two ways. First, the client initially elicited sympathy from the therapist, but the therapist's support eventually turned to criticism and control. This confirmed the client's pathogenic belief that nothing ever works out and relationships all go bad in the end—the specific relational pattern that he had forewarned the therapist about during their initial session.

Second, the therapist benevolently tried to rescue this client, who was indeed behaving as a victim. The caretaking, advice-giving response from the therapist inadvertently defined the client as dependent and incapable of managing his own life. This response paralleled a similarly belittling attitude that the client had received from his indulgent but undermining parent, who could not serve as a secure base and support his exploration and autonomy. The client's cyclical pattern of establishing rescuer/victim relationships was reenacted in treatment and prevented the client from addressing his real concerns over shame-based feelings of inadequacy, which he defensively held at bay, in part, through drinking.

In reviewing this case several weeks later, the therapist reported, "The client just wasn't ready to stop drinking yet. He's going to have to get worse and really hit bottom before he'll be able to stop denying there's a problem and do something about it." The therapist failed to recognize that the interpersonal process they had enacted in treatment unwittingly repeated the client's familiar but problematic interpersonal scenario and prevented change from occurring. As is often the case when treatment fails, the therapist attended only to the content of what they talked about (his drinking) and not as well to how their process might be replaying aspects of the client's interpersonal problems (such as finding judgment and blame when he had a failure experience, or successfully getting others to rescue him when he acted like a victim). Similarly, this case illustrates the empirical support in the Motivational Interviewing literature for engaging clients in a collaborative process to facilitate their commitment and sense of personal responsibility, and the need to support their sense of self-efficacy in order to enhance treatment efficacy (Miller & Rose, 2009; Villanueva et al., 2007).

All therapists, beginning and experienced alike, will reenact aspects of the client's patterns and problems at times without realizing that this parallel is occurring. From an emotionally neutral vantage point—that of a therapist's supervisor or colleague looking at the session on videotape—it is relatively easy to see how the old pattern is being reenacted in the interpersonal process. In humbling contrast, it can be exasperatingly difficult for therapists to see their own interpersonal process while engaged in an intense, affect-laden relationship. This is complex work, and it is challenging to perceive and assess at both the content level and the process level simultaneously. In the first year or two of their clinical training, it is especially difficult for new therapists to attend both to the content of what is being discussed and to the process that is being enacted as the discussion proceeds. With more experience and guidance from a supportive supervisor, trainees will become more effective at observing the process level. Nevertheless, *trainees will find that using or applying this new understanding will continue to be a challenge for a while.* That is, trainees often have difficulty making process comments and intervening in the moment even though they are starting to recognize it with their clients. For example:

THERAPIST: (*internal dialogue*) OK, I can see that he's talking to me just like he does to everybody else. Great, I'm getting just as bored and frustrated as they do, but I sure don't want to bring this up. How can I possibly talk with him about what's going on without hurting his feelings?! I don't want to make him feel bad, or get mad—he might not come back, and then I'll have lost two clients....

Clearly, therapists in training are facing a complex challenge in their professional development here, and they need to be patient with themselves. The ability first to recognize and then to intervene on the process level is often acquired gradually over a period of a year or two, or more. In order to adopt a process-oriented approach and be comfortable using immediacy interventions, student therapists will benefit from further reading on this topic, rehearsal/practice with classmates, demonstration/role modeling from practicum instructors, and the guidance of a supportive supervisor. It is not easy in the beginning to work this way and, realistically, student therapists need help (see the suggestions at the end of this chapter—especially the practical guidelines and exercises suggested by Angus and Kagan, 2007).

Example 2: Process Comments Keep Therapists from Repeating Unwanted Patterns.
In the previous example, the therapeutic process reenacted the client's old scenario. In the next illustration, a similar dynamic begins, but this time the therapist recognizes the interpersonal process being enacted with the client and uses a process comment to realign their problematic interaction.

Many clients enter treatment because of depression. They feel sad, believe they are bad, and experience themselves as helpless to change their circumstances. In many cases, these clients communicate their very real suffering to the therapist with emotional pleas for help. With some clients, however, whatever the therapist (or anyone else) tries does not work (recall that many Anxious/Ambivalent children and Preoccupied adults have a "resistant" component in their attachment styles. The therapist's efforts to meet the client's request for help may be met with some version of "Yes, but..." A problematic cycle begins: This client, who may be presenting with both anxiety and depression—perhaps an agitated depression—feels increasingly distressed and intensifies his plea for help. For example, such clients might even exclaim that they "can't go on living" if things don't change. In response, the concerned therapist becomes more anxious, more active, and tries even harder to find some way to help. However, nothing the therapist does has any impact or provides any relief for this hyperactivating client.

As the client rejects the help that he actively has been eliciting, the therapist's own personal need to be helpful is frustrated. In a developmentally appropriate way, for example, the new therapist's own tenuous confidence or fledgling sense of adequacy as a helper may be threatened by this client's response—especially if the student therapist isn't feeling supported by his or her supervisor. Or, more dynamically, other therapists who were parentified in their family of origin may feel guilty for failing to meet the emotional needs of others. When the client's eliciting behavior has tapped into the therapist's own personal issues in one of these ways, some therapists will work harder to please; others will withdraw and emotionally disengage; and a few will become punitive and critical toward the client. (Think about your own response tendencies for a moment, and consider how you would likely respond to this **mixed message** from the client.) When any of these occur, however, the client does not experience a reparative relationship with the therapist.

Treatment will stall at an impasse or terminate prematurely when the therapist remains **embedded** and continues to please or be critical toward or disengaged from the client. The depressed client's conflict is reenacted as the client again feels dependent on a relationship with someone who is critical, too easily controlled to be trustworthy, inconsistent or emotionally unavailable, or threatens to terminate the relationship. What is the alternative? In successful treatment, the therapist is able to make a process comment (**metacommunicate**) and begin a dialogue with the client about what might be going on in their present interaction. Rather than focus solely on the content of what they are talking about (for example, depression), *the therapist can wonder aloud or tentatively inquire about how they seem to be responding to each other*. This is often the best way to effectively change or alter the reenactment:

THERAPIST: Let's talk about what is going on between us right now and see if we can understand what's happening in our relationship—it might have something to do with your depression. It seems to me that you keep asking me for help, but you also keep saying "Yes, but…" Maybe I get frustrated then, feeling that you don't allow me to help. What do you see going on between us?

With a process comment of this type, the therapist is inviting the client to be a collaborator and join in trying to understand their mutual interaction. The therapist is also providing interpersonal feedback and helping the client look at how his depression is expressed to and experienced by others. (This client recently said, "My wife used to be supportive, but it seems like she's giving up on me and kinda going away.") As we will see, the process comment also allows the therapist to break the cycle, at least for the moment, of the client's escalating need and the therapist's increasing frustration.

How does this process comment help the client to change? In this moment, the client is experiencing a relationship with the therapist that provides a new and different response to his old relational pattern. The therapist has remained engaged—in this moment by inviting the client to jointly explore the impasse. The therapist's consistent availability, in the face of consistent rebuffs, differs from the sequence he has often experienced in other relationships: initial attempts to help, followed by subsequent frustration for the helpers who become critical and blaming, and ultimately ending in others withdrawing from him. It signals a sustained interest in the client and commitment to being a collaborative partner in understanding what is happening. This is indeed a new and different experience for someone who typically expects, elicits, and then experiences abandonment or rejection at this juncture. The client has not experienced this consistent engagement (without the therapist/others overreacting on the one hand or giving up on the other), or the holding environment it provides, in past relationships. If this corrective stance toward the client continues and is repeated in many other ways—large and small—throughout treatment, the client is living out a resolution of his core conflict in the relationship with the therapist. When clients have this real-life experience of change, rather than passively or intellectually hearing advice, reassurance, or interpretations, they are empowered to begin making similar changes with others in their lives.

A key concept in the interpersonal process approach is that the capacity for change increases as the therapeutic relationship becomes more meaningful—that is, as the therapist becomes someone who matters to the client (see important discussions of the "real relationship" in psychotherapy by Gelso, 2002 and Gelso & Samstag, 2008). When an interpersonal solution is enacted in a valued relationship with the therapist, the therapeutic alliance is strengthened and change is facilitated in two ways. First, this new type of relationship provides interpersonal safety for the client (Weiss, 2005). The interpersonal safety of this new context permits clients to come to terms with the core conflicted feelings and pathogenic beliefs that accompany their maladaptive relational patterns (Westin et al., 2004). In other words, clients have the supportive relationship necessary to experience and integrate those feelings and situations that previously needed to be disavowed or split off as they were too threatening to be dealt with. Second, the experience of change with the therapist *demonstrates* to clients, rather than merely tells them, that current relationships can be different than they have learned to expect. It does this by expanding their cognitive schemas for relationships, by altering their core beliefs about the possibilities that relationships hold for them—for example, "I do matter and can be cared about"; "I can say what I want and not be left"; "I do have the right to set limits and stop letting my children run over me." Next, with the therapist's assistance, the client can begin to transfer this relearning and establish better relationships with others along these more flexible and self-affirming lines. As an illustration, let's consider the depressed client's response to the therapist's process comment:

CLIENT: Yeah, I do keep saying "Yes, but..." to you—like my wife always says I say to her. So this is all just pointless. If I'm doing this with you, too, then there's no way I'm going to get better. Maybe we should just forget the whole thing and stop now.

THERAPIST: Oh, no—that's not what I'm suggesting at all. It makes sense that you are having the same problems in here with me that you have with other people. In a way, it is a problem, but in another way, it gives us the opportunity to resolve your problem right here in our relationship.

CLIENT: How?

THERAPIST: If you and I can work out a better way of doing this in our relationship—you know, find a way to keep this old pattern from repeating in our relationship—I think it will go a long way toward helping you change this with your wife and others.

CLIENT: How can we do that now if we haven't been able to up to this point?

THERAPIST: Well, I think it's very possible. In fact, I think we are breaking that pattern right now, just by talking about the way we interact together. Tell me, how is it to be talking with me about our relationship and the way we respond to each other?

CLIENT: It's different, but I like it.

THERAPIST: I like it, too. I feel like I'm working with you. I don't feel pushed away now.

In this example, the therapist has used a process comment to effectively alter the current interaction with the client. For the moment, the process comment has kept their interaction from repeating the client's familiar predicament. Some variation on the old "Yes, but..." pattern will probably soon reappear, and the therapist will have to make another process comment and work through a similar cycle again. However, with the process comment, the therapist is temporarily providing the client with a different response that does not fit very well in the client's old scenario. If the therapist continues to find ways to provide these types of *counteractive responses* throughout treatment, the client will experience a reparative relationship and, as we will focus on in the next chapter, begin changing this pattern in other relationships as well. Below, the next example illustrates a macro perspective that tracks the process dimension over the course of treatment.

Example 3: The Process Dimension as a Means to Bridge Differing Theoretical Orientations.
There are many ways to help clients change, and therapists working from different theoretical orientations can all help clients change. Every theoretical approach clarifies some aspects of clients' problems, and every theoretical approach has limitations. Of great interest, researchers from an integrative perspective find that *"master" therapists representing different approaches tend to act more like each other than like the less skillful adherents of their own treatment models* (Goldfried et al., 1998). Continuing further, researchers consistently find that within each theoretical orientation, the primary component of change is not the theory per se (for example, client-centered versus psychodynamic) but how effectively individual therapists apply it in their therapeutic relationships. That is, *some cognitive behavioral therapists are far more effective than other cognitive behavioral therapists and account for most of the combined treatment effect* in psychotherapy outcome studies (Luborsky et al., 1997; Norcross, 2002; Teyber & McClure, 2000; Walborn, 1996). Decades ago, Kiesler (1966) identified this issue and called it the **uniformity myth**—as if all therapists were equally effective and differences in treatment outcome were due to theoretical orientation rather than more and less effective therapists within each brand of treatment. Kiesler and many others since have encouraged psychotherapy outcome researchers to study **within-group differences** (that is, more and less effective psychodynamic therapists, and more and less effective cognitive-behavioral therapists), where large differences in treatment effectiveness are consistently found. This is far more informative than continuing to focus on **between-group** differences (for example, cognitive-behavioral versus interpersonal-dynamic treatment approaches), where only very small or no differences in treatment effectiveness are found (Bergin, 1997; Blatt et al., 1996; Garfield, 1997). Summarizing this research literature, Lambert & Ogles (2004) argue that instead of continuing to try to identify empirically supported treatments we should be trying to identify empirically supported therapists. In sum, regardless of theoretical orientation or treatment approach, change is likely to occur when therapists can establish a strong working alliance through empathic understanding and repair this alliance when it is ruptured (Angus & Kagan, 2007). Working within any theoretical modality or treatment length, therapists will find that change is also more likely to occur when the therapeutic relationship provides a resolution of the client's conflicts rather than a repetition of them.

After two years in analytically oriented therapy, Rachel still could not take charge of her life. She was always complying with her husband's demands, and it was impossible for her to get her children to do what she asked. She also complained of her stultifying daily routine as a homemaker, yet she was never able to do anything to improve it. At her husband's insistence, she would periodically enroll in a class or interview for an office job but she never followed through on any of his suggestions.

In therapy, Rachel had spent many hours exploring her childhood. Her therapist was skillful in seeing unifying themes in the recollections that she shared with him. For example, her therapist astutely observed how she was subtly discouraged from initiating activities on her own as a child. Additionally, she was not allowed to feel good about her success experiences. Any time she expressed an enthusiastic interest, tried out a new venture, or achieved something she was proud of as a child, her parents did not notice it—or, if they did notice, they somehow didn't seem very happy about it.

Rachel's therapist once explained to her, in a sensitive and nonjudgmental way, that she had a passive-dependent personality style. Rather than feeling labeled or put down, Rachel was impressed that her therapist seemed to understand her so well. He knew so many things about her without her even having to tell him. Although her problems had not changed much yet, she still believed her therapist would cure her. He was so bright and insightful, and he seemed genuinely to care about her. He didn't like to tell her what to do, but when things became too much, he could usually help by explaining what the problem really meant. It was comforting to be with him, and Rachel did not know what she would do without him.

It was Rachel's husband, Frank, who finally ran out of patience with the slow course of treatment. After two years, he was fed up with the unending therapy bills and his wife's unremitting discontent. His wife's helpless dissatisfaction felt like an unspoken but unending demand for him to love her more, give her more, or somehow fill up her life. He was tired of these subtle, nagging demands and he wanted a change.

A man of action, Frank obtained the name of a behaviorally oriented therapist on the faculty of a nearby college. A friend told Frank that this therapist was a "problem-solving realist" who could make things happen quickly. This sounded like just the right approach to him. Frank insisted that Rachel stop treatment with her present therapist and begin with the new therapist. Initially, Rachel was sickened at the thought of leaving her therapist, but she sensed that Frank was truly at the end of his rope. Although she still believed in her therapist and felt loyal to him, she was afraid that Frank might actually leave her if she did not go along with this demand.

After her first two sessions with the new therapist, Rachel was surprised to think that perhaps Frank might have been right after all. The new therapist wasted no time in taking charge of the situation. It was actually encouraging to have the therapist outline a treatment plan with steps for her to follow. The therapist discussed a list of specific treatment goals with her and they planned a set of graduated assignments for meeting these goals on a scheduled timetable.

In their first hour together, the therapist had Rachel role-play how she responded to her children when they disobeyed her. Then, with the therapist serving as a model and coach, they rehearsed more assertive responses that Rachel could try with her children. The therapist also had Rachel enroll in an assertiveness-training class that the therapist was running for some other clients. Each week, Rachel was also to complete a homework assignment. For the second week, she was to call one new person she might like to get to know better and ask her to lunch. She was supposed to report back to the therapist on this assignment at the beginning of their next appointment.

Frank was encouraged by this practical, problem-solving approach to his wife's problems. He began to think that something might change after all. Rachel was surprised to find that she actually felt hopeful, too. She felt reassured by her new therapist's goal-oriented, problem-solving approach. In fact, Rachel became determined to make this therapy work, even though she had not wanted anything to do with it initially. She promised herself that she was going to try to do everything the therapist asked of her.

Therapy progressed well for the first few weeks, but things soon started to slow again. Without really knowing why, Rachel began to find it hard to muster the energy to attend the assertiveness class. She knew her therapist would be disappointed in her but she just couldn't seem to help it. Although she felt guilty and confused about it, Rachel began to come late to her therapy sessions. Over the next month, she began to miss them altogether and soon faded out of treatment.

Both of these therapists failed to have a significant impact on Rachel's problems, even though the treatment approaches they used were seemingly very different. Theoretically, the analytically oriented therapist might attribute the unsuccessful outcome to the great difficulty in restructuring the passive dependency needs of a basic oral character. The behaviorist might note that Rachel was not sufficiently motivated to change because she was receiving too many secondary gains from her help-seeking behavior. If we look at the interpersonal process that transpired between both therapists and the client, however, a different picture emerges. These two therapists actually responded to Rachel in a very similar—and problematic—way.

Let's look carefully at their interpersonal process. With both therapists, Rachel reenacted the same maladaptive relational pattern that she had with her children and with her husband. Her presenting problem was that her children ran over her and she could not make them listen to her. Her disturbing degree of compliance with her take-charge husband was a profound example of the same problem. Therapy failed because her pattern of being passive and compliant was reenacted with both therapists. Unwittingly, both therapists provided a **hierarchical relationship** in which she remained the passive helpee led by an all-knowing helper. This interpersonal process reinforced her pathogenic belief that the source of strength and the ability to solve problems did not reside in her but in the therapist or others. She was not a collaborative partner in her own treatment, so agency, motivation and commitment to change were difficult to sustain (Horvath & Bedi, 2002).

In order for Rachel to change, she needs to experience a therapeutic relationship that behaviorally affirms her own efficacy. This would be a relationship in which she is actively encouraged to initiate what she wants to explore in treatment, invited to set personal limits with and share control with the therapist, and is more overtly supported by the therapist in making her own decisions and choices. The client's symptoms will visibly improve in the context of a new interpersonal process in which the therapist:

- encourages her to act more independently *in their relationship,* and
- joins with her in collaboratively exploring the anxiety and guilt that arises each time she tries to act more assertively or independently with the therapist and with others in her life.

This experience of change with the therapist behaviorally shows her that her own strengths are valued and can be used, in conjunction with the therapist's skill and understanding, in a more productive relationship.

Unfortunately, both of Rachel's therapists were comfortable with the hierarchical helper-helpee relational process and did not bring it up as a focus for treatment. Neither therapist offered the meta-communication that, in certain ways, their current interaction was reenacting the same pattern that was problematic for Rachel in other relationships. Such a process comment, tactfully and tentatively shared, would have allowed Rachel to begin looking at this symptomatic pattern in a supportive environment. More importantly, it would have given her permission to change her dependent coping style *with the therapist* and to begin expressing her own feelings, interests, and authentic voice *in their relationship*. The sequence is often important: If Rachel can experience it first in her relationship with the therapist she can then feel more empowered to generalize and begin to do this in her relationships with others. The new experience with the therapist delivers two very important messages: (1) that some relationships can be different (that is, Rachel will not be abandoned, ridiculed, or ignored if she asserts herself and has a "voice" with the therapist—and perhaps some others as well); and (2) she is already succeeding and experiencing empowerment, now, by acting as more of an agent in her own life and future trajectory than she has before (Bandura, 2006).

What do these changes in the ways she interacts with the therapist have to do with solving the real-life problems with others that brought her to treatment? As long as the therapists were telling her what to do, Rachel could not set limits with her children, take a more assertive stance with her husband, become more aware of her own genuine interests and goals, or follow through and act on what she wanted to do. That is, the highly relevant content of what she discussed with both therapists (issues about autonomy, assertiveness, and her own identity) was not matched by the interpersonal process they enacted. *Unless Rachel has the actual experience of behaving as an active, equal participant in her relationship with the therapist, she will not be able to adopt this stronger stance in other areas of her life.* In other words, the process must be congruent with the content.

To enact a more egalitarian or collaborative relationship with someone like Rachel is not a simple task for the therapist, however. It requires therapeutic skill and thoughtful monitoring of the process dimension. In response to her life experiences, Rachel has become accomplished at getting her therapists, her husband, her children, and others to lead, direct, and control her. Both the behaviorally and analytically oriented therapists could have successfully used their differing theories and techniques to help Rachel if they had enacted a different process in their relationship. If they had encouraged her to initiate more in the session, and then focused together on exploring her reluctance to lead, Rachel's core conflict would have emerged quite overtly in the therapeutic relationship. That is, as soon as either of the therapists invited her to follow her own agenda and bring up whatever she felt was most important to talk about, encouraged her to disagree with or express any dissatisfaction she may have with the therapist, or celebrated whenever she was acting stronger or behaving competently in the therapeutic relationship, Rachel would have become anxious. At that moment, either therapist could have focused her inward on this anxiety so that they could begin exploring together what the threat or danger was for her to step out of her compliant and help-seeking mode (such as the belief that others would leave her if she wasn't always pleasing and "nice"; that she would be acting like her dominating mother—whom she was very afraid of being like; that she was being selfish or demanding; and so forth).

Simultaneously, either therapist could have given her permission to act more assertively within the therapeutic relationship and responded to her in ways that facilitated this new behavior in their interaction together. The therapist could do this in two ways. First, the therapist could watch for and affirm this effective new stance whenever it emerged in their relationship or with others. Second, the therapist could respond to instances in which Rachel "undid" herself by retreating to the safer, nonassertive mode just after she had risked acting in a stronger way with the therapist. Given Rachel's life experiences and interpersonal coping style, she was likely to become apologetic or confused, or to act dependently and ask for direction, soon after disagreeing with the therapist, making an insightful connection on her own, or redirecting the session more toward her own interests or concerns at that moment. If therapists have formulated working hypotheses about these propensities, they can track this potential reenactment in their interpersonal process. If so, they will be prepared to help their client identify this problematic pattern *as it is occurring*, and then explore the threat or danger she feels when she has just retreated from her new, stronger stance with the therapist:

THERAPIST: What do you think I might be feeling or thinking about you, and how are you afraid I might respond, after you have just acted more assertively by disagreeing with me like that?

With both of her therapists, however, Rachel only continued to defend against her anxiety over being more assertive, independent, and successful by retreating to her dependent, help-seeking role and, in their interpersonal process, merely reenacting this familiar pattern.

In most therapeutic relationships, the therapist and client will temporarily reenact the client's core conflict in their interpersonal process—it just happens. However, in successful therapy of every theoretical orientation, the therapist and client do not *continue* to reenact the maladaptive relational pattern in an ongoing way. Instead, they are able to recognize this recurrent pattern (and perhaps encourage the client to give this pattern a name to help them talk about it more easily, such as, for Rachel, "going along") and work out a different type of relationship that changes the course of the old familiar scenario (in Intersubjective terms, **deconstruct** or **disembed** from this enactment). Once clients find that their conflicts can be activated or come into play with the therapist, but do not have to result in the same hurtful or frustrating outcomes they have come to expect, their schemas and expectations expand and become more flexible. At this pivotal point in treatment, it is relatively easy for therapists to move to the next phase of treatment (that is, the Working Through phase) and help clients generalize this experiential relearning to other relationships in their lives.

Example 4: Resolving Conflicts by Working with the Process Dimension.
Excessive performance demands often leave trainees feeling pressure to do something to make their clients change. Unfortunately, these internal pressures on the therapist are often translated into interpersonal pressures on the client to change, oftentimes before either the therapist or the client know what's really wrong, let alone what they want to do about it. This is often evidenced by a premature emphasis on intervention techniques and usually at the expense of too little exploration and understanding. Unless the therapist has conceptualized what has gone awry for the client in other relationships and considered how these relational themes could be reenacted in their therapeutic process, intervention techniques will often fail. In contrast, if the therapist has been **assessment oriented** and asking

herself in an ongoing way, *What experiences has this client had to bring him to this place in his life,* and has been able to generate two or three good working hypotheses about the interaction between the therapist and client—and the specific relational experiences that this client needs in order to change—it is usually easy to find effective ways to intervene. Thus, therapists are encouraged to ask themselves repeatedly, "What does this mean?" rather than, "What should I do?" As we will continue to see, the second question is usually answered by the first.

Therapists can employ a wide range of techniques from different theoretical modalities (Wachtel, 2008; 1997). The key, then, is to observe closely clients' responses to what they have just said or done (Curtis & Silberschatz, 1997; Hill, 2009; Weiss, 1993). That is, the therapist evaluates the effectiveness of each successive intervention in terms of the client's ability to utilize this type of response to make progress in treatment. Based on this behavioral feedback from the client, *therapists need the personal flexibility to modify their interventions* to provide the responses that work best for this particular client (Binder, 2004; Teyber & McClure, 2000). Unfortunately, researchers find that many therapists do not flexibly alter their interventions to match the client's needs, but dogmatically stick to the same approach whether the client is finding it helpful or not (Najavits & Strupp, 1994).

To illustrate the need to use our understanding to guide our interventions, we examine two critical incidents in the treatment of an incest survivor. In these incidents, the therapist uses intervention techniques that have a highly significant impact on the client: validating the resistance and role-playing. It is not the intervention techniques in themselves that facilitate change, however. Both interventions are effective because they follow from the therapist's understanding of what is transpiring in the current interaction, and because the therapist then is able to work with these issues in the immediacy of the client–therapist relationship.

Early in treatment, Sandy had confided to her male therapist that she was an incest survivor. Grasping this profound betrayal, he anticipated that trust was likely to become a central issue in their relationship. Therapy had gotten off to a good start but, before long, progress began to slow as the material that Sandy presented became repetitious. About this time, Sandy recounted two different narratives in which the relational theme was feeling unsafe with men. Based on this and other material they had been talking about, the therapist hypothesized that, without being aware of it, Sandy was using these vignettes to broach the topic of trust in their relationship. As their relationship was becoming more important to Sandy, it seemed that her deep concerns about safety and betrayal were now being activated with him.

Responding to these **embedded messages** about their relationship, the therapist began to talk with Sandy about trust between them and asked about the different thoughts and feelings she was having toward him. Sandy genuinely liked the therapist and was finding him helpful. However, when the therapist explored the trust issue and asked specifically whether she felt safe with him, her affirmative response sounded half-hearted and unconvincing. It soon became clear to both of them that Sandy was emotionally removing herself from the therapist as they talked more directly about safety in their relationship.

The therapist responded to Sandy's concern by affirming both sides of her ambivalent feelings and, using immediacy, working with her concerns directly in terms of their relationship.

THERAPIST: I know that one part of you likes and trusts me, but it makes sense to me that another part of you doesn't feel safe. I think that both sides of your feelings toward me are valid and important for us to work with.

SANDY: (*sheepishly nodding and gesturing vaguely to indicate that this was true*)

Because they had been talking about issues of trust and betrayal for some time and this groundwork had been prepared, the therapist thought that this might be an opportunity to try to go further with these issues. Rather than trying to talk her out of these concerns or convince her of his trustworthiness, the therapist validated Sandy's distrust and actually went on to articulate or develop it more fully:

THERAPIST: If I violated your trust in some way after you had taken the risk to ask me to help you, it would be very bad for you. If I tried to approach you sexually or foster any other kind of relationship between us, it would hurt you very much. Maybe it would even hurt you so much that you might not be able to risk trusting or asking for help again.

SANDY: (*tearing, looks at the therapist and slowly nods in agreement*)

The therapist continued to elaborate Sandy's concerns about mistrust and exploitation from men in general within the immediacy of their relationship:

THERAPIST: If I took advantage of our relationship in some way, I think you would feel hopelessly betrayed. I think you might become very depressed again, enter into other relationships that would be hurtful or exploitative, and may even start thinking again that you do not want to be alive.

As the therapist spoke and made the concerns that they had been discussing with others overt in terms of their relationship, he observed that her entire demeanor changed. She remained tearful but became alert and present, solemnly nodding agreement with what he was saying. As they continued to talk together about this, the therapist went on to describe his own attitude, and reframe her distrust as a strength:

THERAPIST: Yes, I can see how much it would hurt you if I betrayed your trust, and no part of me wants you to have that wounding experience again. In fact, I respect the cautious part of you that isn't sure about trusting me. That distrustful part of you is your ally—it's a strength of yours. We need her—because she is committed to not letting you get hurt again. That's why it's so important that we go at your speed in here, that you have control over what we talk about, and that you can say no to me and know that I will honor your limits.

After the therapist responded so affirmingly to Sandy's resistance, and worked with this previously unspoken but dreaded expectation of betrayal in terms of their relationship, important changes occurred. Whereas in the past Sandy had only alluded vaguely to her childhood abuse, she now chose to share it more explicitly. Over the next weeks, she recounted in painful detail and with strong emotions how she had been molested over a period of years by her stepbrother, who was 12 years older. When she first went to her mother for help, her mother was not supportive, denied that the abuse was occurring, and told Sandy to "never talk about anything like that again." Sandy did not even consider seeking protection from her stepfather, who had always been distant and unresponsive. After failing to receive help, Sandy recalled sitting alone on the floor of her closet for long periods with the door closed, feeling afraid, ashamed, and "out of it." Sandy said that this was the point in her life when she resolved that she was "always going to just be alone." In response to this deep sharing, the therapist was compassionate, validated her experience, and began to help her with the many significant connections she began to make between this familial tragedy and the symptoms and problems she was struggling with in her current life.

The therapist had already learned in their work together that role playing was an effective intervention that Sandy enjoyed and found useful. During one of these sessions, the therapist talked with Sandy about using a role-playing technique to provide a different response to the abuse she had suffered, and Sandy welcomed the suggestion.

THERAPIST: I wish someone could have been there to stop him and protect you. No one was there for you but, if I had been there, I would have walked into your bedroom when he was there, turned on the light, and in a loud voice commanded, "Stop it! Get away from her and leave her alone right now! I see what you're doing and it's not fair. You're hurting her, and I won't allow it."

After speaking this forcefully, as if he were actually saying it to the perpetrator, the therapist rolled Sandy's coat into a ball. Using it as a little Sandy doll, he spoke to it reassuringly:

THERAPIST: You're safe now, and he's gone for good. I'm going to call the police now and help protect you so you'll never have to worry about him hurting you again.

As the therapist metaphorically gave Sandy the protective response that she desperately longed for as a girl but did not get, the full intensity of her isolation and shame was evoked. As these important feelings ran their course, Sandy became more composed and said, "I'm not all alone anymore. I'm going to be all right." The therapist, still holding the rolled-up jacket tenderly, asked Sandy if she could join in and help him take care of this little girl who still needed help. Sandy's affirmative response was almost joyous, and the therapist carefully tucked it in her arms.

THERAPIST: This is the little girl you were. She needs you to open up your heart to her and give her a home. You need to take care of her—and not push her away anymore like they did then—and you have done since. You need to hold her, talk to her, and listen to what she tells you. I want you to join me in taking care of this part of you, so this little girl is no longer alone behind the closet door.

Sandy readily accepted this responsibility and later bought a doll that she used to represent the part of her that needed to be cared for but had not been protected. She had always thought of herself as "unattractive" but went to considerable effort to find and purchase a doll for herself that she felt was pretty.

Sandy entered therapy almost 25 years after the abuse, anxious, unassertive, and almost incapable of leaving home alone. Just weeks after this session, however, she got her driver's license renewed and began driving again for the first time in several years. She also got a job as a waitress—her first paid employment in six years. Although she had always foiled any type of success for herself, she enrolled in a local community college and began receiving A's in many of her classes. Whereas Sandy had felt helpless, had characteristically acted as a victim, and had been repeatedly taken advantage of by others, she increasingly became more assertive and expansive. All of Sandy's problems did not go away, of course, but long-standing symptoms were resolved and her life turned forward on a new trajectory.

What allowed Sandy to get stronger in these significant ways? Several factors made this role-play intervention effective. By using this technique, the therapist brought Sandy's conflict into the therapeutic relationship and provided her with the reparative experiences she needed. In sharp contrast to what occurred in her family, this time Sandy experienced protection, validation, appropriate boundaries, and a supportive holding environment. This corrective experience with the therapist disconfirmed her pathogenic belief that she did not matter enough to be protected or looked after. The therapist's compassionate response also helped Sandy shift from an identification with her parents' rejection and abandonment of her—and her own resulting shame and self-hatred—to an identification with the therapist and his compassion for her. Eventually, this healthy new identification led her to be able to care better for herself and believe that she did matter and was not alone with her problems for the first time in her life. These far-reaching consequences were set in motion by the therapist's affirming response to Sandy's vulnerability. Even though it was enacted in role-play, the therapist provided Sandy with a corrective emotional experience that played an important part in allowing her to become stronger and more capable of protecting herself.

Example 5: Recognizing the Process Dimension in All Interpersonal Relations.
As we have seen, the most likely reason for counseling to fail is that the therapist and client unwittingly reenact aspects of the client's problems in their interpersonal process. At first, therapists in training often find it challenging to track the process dimension, and to distinguish the process that is being enacted from the content that is discussed. People find reparative experiences and change in many different kinds of relationships—not just therapy. So, for our last example, let's look at how the process dimension operates outside the therapy setting—at a sporting event.

In a championship basketball game, the defending champions confidently face an underdog team made up of talented but younger players who are far less experienced. The defending champions are heavily favored to win.

Late in the fourth quarter, the game is tied. The fans, on their feet, are cheering at the prospect of an upset. However, the coach of the young team can feel that the game is about to slip away. The screaming fans, the pressure of time, and the unrelenting play of their opponent is rattling his players. To help compose his less experienced players, the coach calls a two-minute time-out. His team has played well up to this point but the seasoned coach knows that, in these pressure-packed closing minutes, the other team's experience is likely to prevail.

Crouched on the sideline, the coach positions his five players in a semicircle in front of him. With the roaring crowd only a few feet away, the coach quietly makes eye contact with the first player on his left. Without speaking, he calmly holds the first player's gaze for about 10 seconds (a very long time amid such pandemonium). Having compelled the first player's attention, the coach turns to the second player and maintains eye contact with him for another 10 seconds or so, and so on with each of the five players. Each time, as he turns to engage the next player, the previous player's attention remains riveted on the coach. Finally, after making contact with each player in this way, the coach holds up a chalkboard with one word written on it: *POISE*.

For the remaining seconds of the time-out, the players and coach stand together silently. The referee blows his whistle, the players return to the court, and the underdogs defeat the defending champions. The players did not make mental errors in the closing minutes of the game. They did not turn the ball over to the other team, they made clutch free throws in the final seconds, and they smoothly executed the plays they had practiced all year. In a word, they played with poise.

This accomplished coach astutely identified the key issue that could defeat his team in the final minutes: being intimidated by a more confident and experienced team, losing their composure, and making mental mistakes that would cost them the game. Having recognized the central issue, however, the coach did not just *tell* his team that they had to remain poised. Instead, he gave them the *experience* of composure during the time-out. The players were able to finish the game with such composure because the coach used the relationship he had developed with his players to give them the experience of composure during the final tense minutes. The players were able to generalize this experience with the coach during the time-out to their performance on the court and play with poise.

Similarly, it is far more effective if the therapist–client interaction provides an experience of change than if the therapist merely tells the client what to do or explains what something means. It can be challenging to provide a relationship that enacts a resolution of clients' conflicts, however, and there are no simple formulas that tell the therapist how to do so. Even a master clinician such as Aaron Beck, who writes treatment manuals himself, does not adhere strictly to the protocol but applies it flexibly to adapt to difficult situations and different clients (Newman, 1998). So, even though therapists may wish there were a cookbook formula to tell them what to do sometimes—especially as they begin their training and new therapists have to cope with so much ambiguity—every client and every therapeutic relationship really is unique. In the Motivational Interviewing literature, Miller and Rollnick (2002) report that they improved their treatment outcomes when they *stopped* having their therapists follow a treatment manual with a standard protocol and, instead, more closely follow the client's interests, initiative, and lead. The therapist needs to be willing to take the personal risk of entering into a relationship that matters with the client, collaboratively assess what's gone wrong for this particular client to lead them to this point in their lives, and have the flexibility to provide the new relational experiences that allow this client to change.

Providing a Corrective Emotional Experience

As we saw in Chapter 8, the first step in working with the process dimension is to attend to clients' maladaptive relational patterns. The therapist is trying to identify how the same problematic scenarios that are causing problems with others in the client's life are starting to be played out or reenacted in the therapist–client interaction as well. As the therapist and client begin to identify these faulty patterns with others, they can start to highlight or punctuate when similar patterns are beginning to occur between them. Working collaboratively, they begin to discuss what's going on between them and find new ways to interact that resolve, rather than reenact, these patterns. In this way, the therapist is trying to give clients the real-life experience that clients do not have to be controlled, judged, ignored, idealized, and so forth, in this relationship as they often have been in the past. As clients live out this experience of change with the therapist, *the therapist can then help clients generalize or transfer this in vivo relearning beyond the therapy setting.* That is, based on this real-life experience of change with the therapist—not just an interpretation or explanation—clients' cognitive schemas for relationships change and their interpersonal range expands.

At this point in treatment, it is relatively easy for the therapist to help the client begin to sort through relationships with others more realistically. That is, by trying out new ways of responding to others in the client's life, the therapist and client together can begin to *discern* others with whom certain problematic patterns and unwanted responses continue to occur, and others who are able to respond in more affirming, accepting, or rewarding ways. For example, the client may test the waters again only to find out that his father still has to be "right" all the time, just as he did years ago. In contrast, he finds that his mother is less rigid than his father and can now listen or hear him better than when he was growing up. At times, however, she still tries to demand that he just comply and do everything her way. In this reassessment, he also clarifies that his girlfriend really is different than his parents and can be very cooperative, but he needs to speak up more clearly and better communicate to her what he does and doesn't want.

The key point here is that *issues that have caused problems for the client in other relationships can be made overt and talked about—especially when they begin to occur in the therapeutic relationship—and resolved differently with the therapist.* The best way to begin working with the interpersonal process and providing this corrective emotional experience is by asking clients, in an open-ended way, how the problem that they are talking about in relation to others could be occurring in their relationship as well:

THERAPIST: I'm wondering if this battle for control that you have been describing with your partner ever goes on between us here in counseling. Does it ever seem like I am trying to control you or that you are controlling me?

OR

THERAPIST: It sounds like others have often criticized or "judged" you, and it really hasn't felt safe for you to talk about important things. I'm wondering if you have ever felt criticized by me or been afraid that I will judge you too?

By bringing the client's conflicts with others directly into the therapeutic relationship in this way, the therapist creates the opportunity to disconfirm pathogenic beliefs and provide a reparative experience. In order to change with others, this client first needs to find that the therapist is neither trying to control nor willing to be controlled by the client, as often occurs for the client in other important relationships. If clients have this real-life experience of change, they can resolve this conflict, find that sometimes control can be shared in close relationships, and learn that some relationships can be different from those in the past. This immediate, real-life experience of change is a powerful way to intervene. As we have emphasized, however, many therapists find it challenging to make this process comment or link the client's problems with others to how the therapist and client may be interacting together right now. Thus, the purpose of the next section is to help therapists make process-oriented interventions and take the risk of using their real-life relationships with clients to facilitate change.

INTERPERSONAL PROCESS INTERVENTIONS

The interventions we've discussed have repeatedly invited the therapist to work on the client's problems within the *immediacy* of the therapist–client relationship. The following examples review how process-oriented interventions link clients' problems with others to the here-and-now relationship between the client and therapist. At the beginning of treatment, for example, clinicians can speak directly to clients about their current interaction when they are trying to use accurate empathy to establish a *working alliance* (Chapter 2):

THERAPIST: As we talk about this, does it feel like I am understanding what is most important for you here—am I getting it right?

Talking about what is going on between the therapist and client becomes even more important in working with *resistance* (Chapter 3):

THERAPIST: What's it been like for you to talk with me today? What's felt good, and what hasn't?

Working directly with the current interaction is also helpful when establishing an *internal focus* (Chapter 4):

THERAPIST: Maybe we are arm wrestling with each other a bit here today. It seems to me that you keep talking about what others are doing, and I keep asking about what you are thinking or how you responded. What do you see going on between us?

Therapists also want to work with the client's *feelings* in a way that brings the full intensity of whatever feelings the client is experiencing into the immediacy of the therapist–client relationship (Chapter 5):

THERAPIST: I can see how much this has hurt you, and how sad you are feeling right now.

Moreover, therapists intervene in the here and now by providing clients with *interpersonal feedback* about the impact their interpersonal coping style is having on others (Chapter 7):

THERAPIST: Aaron, may I have permission to share some feedback about how you come across to me at times—and maybe to others, too? I'm not sure you're aware of the effect you have on others sometimes. I know that people usually don't talk together so directly, but I want you to know I'm bringing up this sensitive issue with the good intention to be helpful. Can I share my thoughts with you?

Finally, this focus on the current interaction between the therapist and the client is central to working with *transference reactions* (Chapter 8):

THERAPIST: How do you think I am going to react to you if you do that? What am I likely to be thinking inside or feeling toward you?

The unifying theme in all of these process-oriented interventions is to *link the client's issues and concerns with others to what is currently going on between the therapist and the client.* In this way, the client's problems with others are not just talked about abstractly; they are re-experienced and potentially resolved in the real-life relationship with the therapist. In turn, this in vivo or experiential relearning with the therapist empowers clients to begin changing how they are responding to similar problems with significant others in their lives.

To illustrate, let's return to the previous client, whose basic relational theme is his repeatedly getting locked in control battles with others. To begin resolving this client's problem with others, the therapist's intention is to find an effective way (that is, not blaming or confrontational but tactful) to make this control issue overt and put it on the table for discussion. The therapist does this by wondering aloud or tentatively inquiring about this possibility, so the therapist and client together can begin to explore it directly in terms of their relationship.

THERAPIST: Does this type of control battle that you have been describing with others ever get going between us, too? Does it ever seem like you and I are jockeying over control in our relationship?

Making such a process comment and bringing the client's concerns with others into the here and now may seem a dreadful prospect to new therapists. However, by judiciously taking this risk, the therapist can bring real-life immediacy to therapist–client interactions. Problems with others are not just being talked about intellectually; instead, for the first time perhaps, they are being addressed honestly and constructively with someone *as they occur.* If the therapist waits for her own sense of good timing (for example, the therapist may say to herself, "I think this might work right now"), responds in a respectful manner or tone, and inquires tentatively and collaboratively—inviting the client's reactions as well—the potential for change is at hand.

Let's consider a range of three progressively more challenging responses to the therapist's query about control issues in their relationship.

Response 1: A Disconfirming Response Is Least Challenging to the Therapist

An example of a disconfirming response from the client might be as follows:

CLIENT: No, I don't feel like we are in a control battle—that's not going on here.

The therapist can then use this response to begin exploring what is different about their relationship.

THERAPIST: Good, I'm glad that's not a problem for us. Any ideas about what makes our relationship different?

The client's answer may provide useful information:

CLIENT: You treat me with respect; that's what's different.

The therapist can follow up on this comment by exploring the client's concerns about not being respected in other relationships. The therapist also can build on it to establish that mutually respectful relationships could be developed with some others in his life as well. Further, the therapist is modeling for this client how he can find diplomatic ways to talk with others when he feels that an unwanted control battle is coming into play.

RESPONSE 2: CLIENTS MAY AVOID THE IMMEDIACY OF THEIR CONFLICTS BY DISCOUNTING THE THERAPIST'S RELEVANCE

CLIENT: We're not in a control battle because this isn't a real relationship. You're just my therapist.

Clients make far more significant gains in treatment when this distancing defense is addressed and resolved. At this point, it is essential to clarify that the therapist and the client really are two people who have been having a relationship for some time, even though there are constraints on their relationship (for example, Therapist: Yes, I am your therapist, so in many ways our relationship is different from a social one. I'm wondering, though, what about this makes it not feel like a real relationship?). The therapist can then begin to explore the threats aroused for the client by more meaningful involvement, such as being controlled, judged, not being liked or respected, and so forth (for example, Therapist: If this relationship did feel more "real" to you, what would be different between us...or what might go wrong here that goes wrong with others?). The therapist and the client can then agree to watch for these concerns and address them if they arise in their relationship. It is usually more effective if therapists do not bring up these relational patterns in the abstract but wait until they think the client might be experiencing them right now, in their current interaction, before addressing them.

RESPONSE 3: CLIENTS MAY RESPOND TO THE THERAPIST'S QUESTION AFFIRMATIVELY

CLIENT: (*in an exasperated tone*) Of course you're in control! You insist that we stop at 10 to the hour, whether I'm finished or not. And you're subtle about it, but I see you trying to take control all the time and make things go where you want.

Many new therapists fear such an accusing response and may eschew process comments to safeguard against such criticism. Too often, student therapists take the client's comment at face value and believe that they have done something wrong and actually been too controlling if the client disapproves of them in this way. Perhaps the therapist actually has been too controlling—this possibility must always be considered and owned if necessary. In many cases, however, the client's reactions have as much or more to do with the client's own cognitive schemas than they do with the reality of how the therapist has responded. Thus, rather than posing a problem, this affirmative response from the client actually provides an important window of opportunity.

Openly acknowledging that the same conflict the client has with others is also occurring right now with the therapist creates the opportunity to explore this issue directly and begin working together to try and resolve it in their relationship. Why do we keep returning to this same point? As long as the client is feeling controlled by the therapist, the therapist is not going to succeed in helping the client change this problem with others in his life either. However, by utilizing the therapeutic relationship as a social learning lab, therapists can use their interactions with clients to begin changing this maladaptive pattern or schema in the following ways:

1. By remaining nondefensive and accepting the validity of the client's concerns whenever possible:

    THERAPIST: Yes, we do have to stop at 10 to the hour, and that is an unnatural ending for you. I can see how the control issue is brought up by those time constraints.

2. By exploring the client's perceptions further and working with the client to understand them better:

    THERAPIST: What do you think is going on for me when I try to take something we talk about in a certain direction? What does it seem like I am trying to do at those times—what might my intentions be?

3. By "differentiating" the therapist from other figures in the client's life:

    THERAPIST: Yes, I have had ideas about where we should go, but I have also been interested in following your lead, too. In fact, I may be different from some other people in your life because I genuinely liked it when you disagreed with me last week and told me what you thought. I like it when neither of us feels controlled and we can both say what we want. It makes our relationship feel more alive to me.

4. By offering to be sensitive to this concern in their relationship, expressing a willingness to handle the issue differently in the future, and inviting clients to tell the therapist right then whenever they think this conflict is occurring between them—so together they can talk it through and change it before continuing:

    THERAPIST: I don't want you to be controlled in our relationship as you have been in others. That's no good for you or anybody else. Let's try to do something about it. From now on, I will watch the clock and let you know when it's 5 minutes before we have to stop. And any time you feel like I am directing you away from where you want to go or being controlling in any way, tell me, and we'll stop right then, sort through what's happening, and change it. What else could we do to help with this? Tell me your ideas.

Adopting these approaches will work easily with some clients, whereas other clients will insist on reestablishing the same constricted relational patterns. However, *if therapists are willing to remain nondefensive and tolerate their own discomfort for a few moments*, most clients will become significantly more engaged in the therapeutic relationship and motivated to explore this and other related issues more fully. As clients repeatedly find that the expected but unwanted old scenario does not repeat with the therapist, they experience greater **interpersonal safety** than they have known before. As a result, important new issues, pathogenic beliefs, and previously threatening feelings can emerge, providing new material for exploration and clarifying the treatment focus. Routinely, many clients will make attempts during the next week to try out this new way of relating with others and test out what they have just experienced with the therapist. That is, clients often will return to the next session and report on their attempts to change these same types of patterns and respond in a different way to someone in their life. Of course, clients will have both successes and failures in their attempts to change these problematic patterns with others, and the next chapter will explore closely what therapists can do to help at this pivotal new point in treatment.

Thus, the real-life experience of change within the therapeutic relationship is a powerful relearning experience for clients and often propels clients to begin exploring new ways of responding with others in their lives where the same problems and patterns are occurring. As we have emphasized, however, intervening so directly in the here and now may break the social rules and be uncomfortable for therapists in the beginning. To help with this, we now explore several reasons for such discomfort and offer some practical guidelines to help therapists intervene effectively with the process dimension.

THERAPISTS' INITIAL RELUCTANCE TO WORK WITH THE PROCESS DIMENSION

Karen, a first-year practicum student, was seeing her first client. Everything her client talked about seemed unimportant—she just rambled on and on about nothing. Karen felt bored, and even though she knew inside that the client also felt like this wasn't going anywhere, neither one wanted to say anything. A few sessions went by like this and Karen's awkward, bad feeling just got worse. She stopped enjoying their sessions and almost began to dread them. Karen developed a knot in her stomach as she sat politely with the client and worried to herself: "Being a nice, empathetic person isn't going anywhere"; "Maybe I'm in the wrong field"; "I don't think I can do this…"

Karen's supervisor sensed her distress and tried to help. He suggested that Karen talk with the client about what she was feeling and check things out—maybe the client was feeling the same way, too? For example, the supervisor suggested that Karen reflect aloud on their interaction, and he tried to show her how by role-playing things she might say, such as:

"Sometimes I'm wondering if what you're talking about feels like what's really most important for you";

"How does it feel for you to talk with me—or others—about problems or personal things that are meaningful for you?";

OR

"What's it like for you to come each week and talk with me? What feels good and what isn't being helpful?"

Karen knew that this is what she "should" do, but there was absolutely no way she could bring herself to say things like that. She couldn't break the social rules and be so forthright; she couldn't do that with her client or with anyone—she just couldn't. Karen had never talked this way in her family, and felt torn apart about by the possibility that she might hurt her client's feelings by being so direct.

After six sessions, the client told Karen how nice and what a good listener she was, but that she wanted to stop coming. Karen continued to be nice and act socially proper by agreeing, but inside she felt devastated knowing that it could have been—should have been—so different. Her supervisor understood that this was a painful failure experience for Karen. Based on his experience with many other graduate student thera-pists, he reassured her that, in a year or two, these process comments could become natural ways of responding that felt like second nature to her. Karen was not convinced but hoped it could be true.

In this section, we explore why some beginning therapists, like Karen, may be reluctant to make process-oriented interventions, talk forthrightly with clients about their current interaction together, and explore how aspects of the clients' problems with others may be occurring in the therapeutic relationship. We will examine six reasons why therapists in training may find it difficult to work in the here and now and create immediacy, provide clients with interpersonal feedback about the impact they are having on others, and utilize process comments that focus on the current interaction between the therapist and the client. We will provide guide-lines to help therapists begin working in these initially challenging yet potent and rewarding ways.

Although the best vehicle for effecting change is often the therapeutic relation-ship, it can be anxiety arousing for new therapists to take the plunge and say, for example,

THERAPIST: Does that kind of problem ever go on here, too—you know, between us here in therapy?

OR

THERAPIST: Do you find it hard, here with me also, to let me know when you disagree with me, feel afraid to express your needs, feel angry with me, find it hard to trust me, find yourself taking care of my needs, worry about always finding the right words, are keeping things on the surface, and so forth?

Talking together so directly about issues between two people breaks the cultural norms and family rules that most therapists grew up with. As a result, and because of the immediacy that such forthright communication creates, some beginning thera-pists are reluctant to make process comments in the beginning and explore how clients may be experiencing parallel problems with the therapist that they are having with others.

A few beginning therapists find it easy to work with clients using a process approach. They have always been able to talk directly with others when problems came up in their relationships, and it makes sense to them that what tends to go wrong for clients in other relationships will also come into play with the therapist at times. For these therapists who are more comfortable with approaching inter-personal conflict, it is enlivening to make potential problems or misunderstand-ings overt and talk about them openly with the client. The process approach simply gives them the permission they may need to work with clients in this engaging way, and they welcome the intensity of relating in such a deeply personal way.

For others, however, process comments are more difficult at first. Although it makes sense intellectually to link the problems that clients are having with others to the current interaction with the therapist, they find it much harder to actually put this into practice. Because they have had so little direct clinical experience, most first- and second-year graduate students have little confidence and, understandably, are not eager to have what little self-confidence they may possess shaken by stepping outside of familiar bounds. Furthermore, even if they think that this type of intervention might be useful with a particular client, many beginning therapists struggle mightily to discern whether they are objectively observing the client's behavior or whether their perceptions simply reflect their own countertransference. For example:

THERAPIST: (*internally*) Is this just me, or does she make everybody feel this way?

Finally, even if therapists feel confident that it is indeed the client's faulty coping style and what the client tends to elicit from others as well, *they may be unsure of how best to address this reenactment and find a helpful way to make it overt—without making the client feel criticized or blamed.*

Still other therapists feel concerned that to be forthright with clients could be hurtful because it would draw out or reveal clients' pain. Others have seen directness used hurtfully—in angry confrontations or blaming personal attacks that were intended only to win arguments, put someone down, or induce shame or guilt. Therapists never want to do any of this, of course, or intrude on their clients and violate their personal boundaries in any way (Kiesler, 1996; 1988). Indeed, diminishing clients' self-esteem or sense of personal safety in these ways will impair significantly their ability to make progress in treatment. However, clinicians readily can learn to address the process dimension in respectful ways that actually reassure clients, rather than make them feel uncomfortable. Let's examine six common concerns therapists have about using the therapeutic relationship to intervene with clients' problems and suggest some solutions.

1. Uncertainty about When to Intervene. Because new therapists have not had much experience attending to the process dimension or tracking it with clients, they are often unsure when or how the client's conflicts are potentially being played out in the therapeutic relationship. As trainees gain more experience and begin to integrate all of the complex new information their clinical training presents, most will become more comfortable intervening with the process dimension. The usual sequence is, at first, therapists will find that they are beginning to recognize the process dimension and see when something important is going on between them. However, it is often weeks or months later until new therapists feel confident enough to begin speaking up and inquiring about this possibility with the client. It is something of a personal risk for therapists to suggest:

THERAPIST: I'm curious about something that might be going on between us right now, and wondering if...

Although therapists will be accurate with some of their process observations, the client will not resonate with others. Therapists have not failed in any way or made a mistake when the observation they have tentatively suggested is inaccurate; they are simply trying to understand, and clients usually appreciate these good intentions. As emphasized earlier, the therapist's aim is not to be "right."

Instead, he is trying to initiate a mutual collaboration or dialogue with the client—so together they can consider and explore what may be going on between them:

THERAPIST: (*nondefensively, in a friendly and welcoming tone*) OK, what I'm suggesting doesn't quite fit. Help me say it more accurately. What are your words for what might be going on between us here?

Nevertheless, novice therapists should not be in a hurry to make process comments or use other interventions that create immediacy until they feel ready to do so. Forcing such responses will not be good for the therapist or effective for the client. If therapists are disempowered because they do not feel in control of choosing how and when they intervene with their client, in turn, they will not be capable of empowering the client.

How can therapists discern when the client's conflicts are being reenacted between them? Let's follow a five-step sequence. First, it is helpful if the therapist has tried to identify the client's maladaptive relational patterns with others. By formulating working hypotheses about what goes wrong with others—and how those same patterns or similar themes could come about in their interaction—the therapist is better prepared to respond to what this particular client is likely to present in treatment. Thus, when therapists can anticipate the types of expectations, schema-driven distortions, and reenactments that this client is likely to enact with them, therapists do not have to "think on their feet" as much. They can better "see" what is going on between them or "hear" what the client is really saying—right now as the client is saying it, rather than "getting it" later (for example, a trainee reviewing a tape recording of his session might think to himself: "She's telling me that "*everybody*" judges her. She's probably telling me that she's worried that I'm judging her and being critical, too. Why do I get it now, when it's too late, and not right then when she's saying it to me? This always happens!").

Second, these working hypotheses will help therapists make sense of their own experience and recognize when they are starting to feel or respond as others do in the client's old scenario. For example, when clients are leading the therapist to feel bored, impatient, or overwhelmed, these working hypotheses will help the therapist consider the possibility that she is beginning to react to the client as others in the client's life often do—that is, that her feelings and reactions toward the client may be signaling that the same scenario that causes problems with others may be under way here, too.

Third, this is a good time for the therapist to consult with a supervisor and check out together what may be occurring in the therapeutic interaction. If it does seem as if a reenactment is under way, supervisees can explore any concerns they may have about making this interaction overt the next time it comes up and talking it through with the client. Looking for effective ways to explore this possibility with the client, therapists can also role-play or rehearse alternative responses with the supervisor. Again, it is not realistic for therapists in training to be able to successfully explore and try out these process-oriented interventions without active guidance from a supportive supervisor.

Fourth, with this preparation, therapists can wait to broach this possibility with the client the next time they feel this reenactment may be occurring. *It is more effective to intervene at the moment when the interaction is occurring* (immediacy) *rather than bringing it up in the abstract* (for example, at the beginning of the next session). Recall the therapist earlier who realized, after the session, that his client likely was feeling judged or criticized by him. He did not start the next session by bringing up this possibility and talking with the client about it—as he wanted to do. Instead, he listened to his supervisor's advice to wait until he thought this issue might be occurring between them again, which did indeed happen about 20 minutes into their next session, and successfully inquired about this possibility *as it was occurring*—which holds so much more meaning for the client.

Fifth, beginning therapists are encouraged to wait until they feel that this intervention (or any other) is likely to work. We need to respect our own sense of **timing** and listen to our own feeling that, "This just isn't going to work right now" versus "This might be a good time to say this." As emphasized, it is important that therapists *choose* whether to make this or any other intervention, and when to try it—compliance is just as problematic for therapists as it is for clients.

It will often take a year or two before these responses feel natural and come easily—but they will. Good advice for the novice therapist is to be patient with yourself, practice with classmates, and ask your instructors/supervisors to role-play and demonstrate these immediacy interventions.

2. Fear of Offending the Client.   Some beginning therapists are concerned that the client will feel they are being too blunt, personal, or confrontational if they ask about what may be going on between them:

THERAPIST: What's it like to talk with me about this?

Therapists may worry that the client might feel angry or hurt, not like them, or—worse yet—just stare blankly back at them in silence! Of course, process comments, like any other interventions, can be made in blunt, accusatory, intrusive, demanding, or otherwise insensitive and ineffective ways. Warmth, tact, an "inquiring" or curious presentation, and a good sense of humor can all go a long way toward making every intervention more effective.

Therapists may also be worried about trespassing on social norms and expectations. Speaking respectfully yet forthrightly about what may be going on between two people may break unspoken social rules, and many trainees are concerned that clients may be surprised or taken back by this. To help ease the transition toward this more genuine dialogue, therapists can provide **contextual remarks** that facilitate the bid for more open or authentic communication. For example, the first time the therapist addresses the process dimension with a client, the therapist can create safety for the client by offering an introductory remark that acknowledges the shift to another level of discourse:

- Can we be forthright with each other and speak directly about something?
- Let me break the social rules for a minute and ask about something that might be going on between us.
- I know that people don't usually talk together this way, but I think it would help if we could talk about...

Therapists will not be too blunt, take clients off guard, or offend clients if they respond respectfully, invite collaboration (using prompts such as "What do you think?"), and especially, *provide transition comments such as these.* These contextual remarks are very effective in helping clients understand the therapist's good intentions, as they help clients shift from the socially polite to this more straightforward approach. Rather than being threatened by this invitation for more forthright communication, most clients want this more authentic and substantive discussion. Welcoming it from the beginning, they are reassured by the therapist's willingness to get down to business and talk in plain terms about what's really wrong in their lives—something they often wanted but have not been able to do with others.

3. Therapists' Own Insecurities and Countertransference Issues.

To work with clients in the highly personal manner described here, therapists are challenged to take the risk of engaging in a genuine relationship with the client. In the interpersonal process approach, therapy is a two-way street where each participant is willing to be affected by the other. Process comments lose their impact when therapists are not willing to be emotionally present and engaged in the "real relationship" with the client (Gelso, 2002; Gelso & Samstag, 2008). For example, the client receives a contradictory or mixed message from the therapist that is highly problematic when the therapist makes a process comment, inviting a more genuine dialogue with the client, but does not follow through. This mixed message occurs when the therapist becomes defensive and distances himself from the client or is unwilling to consider the potential validity of the client's criticism or honestly explore the client's dissatisfaction with something the therapist has done. For example:

THERAPIST: What do you think as I say that to you?

CLIENT: Well, I feel like you're criticizing me a little bit there, and I guess it doesn't feel very good.

THERAPIST: (*defensively*) Well, of course I'm not being critical—geez, you really are hypersensitive to feeling criticized! You shouldn't be so sensitive. Do you overreact like this to other people, too?

In contrast, when therapists are willing to remain nondefensive and talk with the client about their participation in problems and what is occurring between them, sessions become more intense and productive. As an alternative to the previous therapist's reaction, a nondefensive therapist might respond:

THERAPIST: I'm sorry it feels like I'm critical—that sure wouldn't feel very good. I don't want to be critical in any way, and do want to give you helpful feedback in a way that's not judgmental. Thanks for being so honest with me—you're helping our work together.

Thus, therapists' willingness to be personally affected by the client and to look at their own contribution to their interpersonal process will at times be anxiety arousing for all therapists. We wouldn't want it to be otherwise—nothing ventured, nothing gained. A corrective emotional experience does not occur unless the relationship is significant and holds real meaning for *both* the client and the therapist.

By looking honestly at their own contribution to problems or misunderstandings in the relationship, therapists facilitate an egalitarian relationship that holds genuine meaning for both. Of course, this increasing mutuality will activate therapists' own personal problems or countertransference issues at times (Gelso & Hayes,

2007; Mitchell, 1993). For example, in order to respond more authentically, therapists need to be able to relinquish hierarchical control over the relationship, which will arouse anxiety for some therapists who need to be the authority, in control of what happens next, or one-up in relationships. Therapists may also be concerned that this genuine responsiveness or emotional presence with the client will lead to a loss of appropriate therapeutic boundaries and result in overinvolvement or acting out on the part of the client or therapist. However, by consulting with a supervisor and applying the guidelines on optimum interpersonal relatedness from Chapter 6, therapists will be able to recognize when their own countertransference issues are prompting them to become too close to—or too distant from—the client.

In making process comments, how do therapists know whether what is going on is due to their own personal issues or to the client's dynamics—that is, to client-induced or therapist-induced countertransference? Therapists have two useful methods of distinguishing whether the client's dynamics—eliciting maneuvers, testing behavior, transference distortions, and so forth—or the therapist's own personal issues are operating. First, new therapists need a supportive supervisor who helps them track the therapeutic process, especially in the beginning. It is unrealistic to expect trainees to be a participant-observer who can maintain a balance of emotional relatedness and objectivity without ongoing assistance. This is a normative, developmental issue in clinical training. Although new therapists should be thinking about this distinction and working hard to clarify what is the client's issue and what is their own, they should not feel badly if they are having trouble distinguishing this in the beginning. It is complex and ambiguous, so let your supervisor help you.

Second, beginning therapists can safeguard against confusing their own and clients' issues by tracking and observing, but not jumping to conclusions about the therapist–client relationship as soon as they see something significant happening. It is more effective for the therapist to generate working hypotheses first, and then wait to see whether the observation also applies to subsequent exchanges between therapist and client. *If the issue or concern is relevant for the client, it will repeat as a theme or pattern. If it does not keep coming up, then it is likely that the therapist's own issues are involved or that it is not an important concern for the client and this hypothesis should be discarded.*

Thus, therapists don't wish to be cavalier about making process comments. When in doubt, it is best to wait, gather additional information about the therapeutic process, and/or consult with a supervisor before raising the issue with the client. Remember, too, that all process comments are simply observations. They are offered to the client *tentatively*, as possibilities for joint exploration and mutual clarification, not as truth or fact.

Across varying treatment approaches, the therapist's relationship with the client may be the most significant means of effecting change (Barber et al., 2009; Najavits & Strupp, 1994; Norcross, 2002; Seligman, 1995). At the same time, one of the most likely reasons for treatment to end prematurely is that the client's interpersonal coping strategies have activated the therapists' own conflicts (countertransference). That is, the therapist becomes invested in rescuing the client who presents as a victim, the therapist becomes defensive and competitive with the client who is challenging, the therapist emotionally disengages and gives up on the client who is aloof or distancing, and so forth. In this way, therapists always need to remain open to the possibility that their own countertransference issues may have come into play, and should generate working hypotheses about how each client's conflicts could be reenacted in treatment (see Appendix B, Part 6). Likewise, beginning and experienced therapists also honor the therapeutic enterprise when they:

* make a lifelong commitment to exploring and remaining open to their own countertransference propensities,
* consult with supervisors and colleagues as an ongoing career activity, and
* seek their own personal therapy when countertransference issues persist.

Countertransference issues are most likely to create problems when clinicians disregard them. The red flag goes up when clinicians are not willing to consider their own participation or potential contribution to the conflict. In contrast, therapists who are aware that they are susceptible to countertransference and who discuss possible instances of countertransference with supervisors as they arise should feel unfettered about working with clients in a process-oriented manner.

4. Concern about Appearing Confrontational.   Some new therapists misunderstand process comments, misconstruing them as *confrontations*. In the typical scenario, these therapists are reluctant to try making a process comment because they anticipate that the client will be mad about being "confronted" in this hostile or accusatory way, and angrily will walk away and leave treatment. *There should be nothing confrontational about a process comment or any other immediacy intervention. A process comment is a tentative wondering aloud, an invitation from the therapist to the client for a dialogue about what may be going on between them right now.* If you think that the process comment you are about to make is going to make the client feel accused or blamed, or that it will "put the client on the spot," don't make it—that's not the stance we are seeking. Instead, therapists can wait for another time that feels better or, preferably, talk with the client about his reservations about sharing this observation. For example:

THERAPIST: There's something I'm thinking about right now, but I'm feeling unsure of talking about it with you. I guess I'm concerned that you might feel criticized or blamed, which is not my intention. Can we talk about this for a minute?

In a similar vein, some therapists are concerned that, if they address issues directly, the client will misconstrue their straightforwardness as an unwanted and intimidating confrontation. Again, such a confrontational stance is not our intent—most therapists do not wish to be confrontational in any way, just as most clients do not want to be "confronted." Process interventions can be presented in an aggressive, demanding, intrusive, judgmental, or confrontational manner, just as any intervention can be carried out ineffectively, but there is nothing insensitive or disrespectful about simply communicating forthrightly. Again, if therapists think these (or any other) interventions may be perceived by the client as critical, blaming, or unwanted, they should not use them. As we have just seen, therapists can either wait for what feels like a good time to try this particular intervention, **metacommunicate** about their reluctance to share observations with the client, or simply respond in other ways. We are actively seeking honest, straightforward communication that is respectful and collaborative—not confrontations.

Similarly, some therapists may worry that by responding more forthrightly, accurately reflecting the key concern or core message, or capturing the central feeling in what the client just said, they may "hurt" the client in some way. These expectations of client vulnerability are usually inaccurate, and are more likely to be related to the therapist's own countertransference issues about becoming more effective, having more impact on the client, or acting stronger. Virtually all clients will welcome the invitation for a more authentic, forthright dialogue about what's really wrong in their lives and about what's going on between the therapist and the client (and we believe that so many clients are dropping out of treatment prematurely because they don't get this straight talk and direct personal engagement with the therapist—the dialogue remains friendly, nice, and on the surface). In sum, as clients find that it is safe and helpful to speak with the therapist about what goes on between them at times, new therapists will be reassured that they can be both sensitive and direct at the same time.

Fear about hurting the client takes on more personal significance for some therapists who grew up with highly authoritarian parents. Such therapists may have been exposed to **double binding family communications** in which there was a great discrepancy between what was actually being done (for example, the child routinely may have been ignored or kept on the "outside," threatened or hit, diminished, or ridiculed) and what was being said (for example, "We are a close and loving family. There are no problems, everyone is happy"). The essential element in the double bind is the clearly understood but unspoken family rule that the child cannot acknowledge the incongruency in any way. For example, the child in such a double binding situation cannot make these contradictory messages overt by meta-communicating and saying, "You're telling me to clean my plate, but later you'll make fun of me for being fat. Stop it—you're driving me crazy!" *Children who have been intimidated and disempowered ("undone") by growing up with these maddening mixed messages are highly anxious about making them overt because further rejection, debasement, or abandonment has been threatened or implied.*

Such double binding communications occur in many dysfunctional families, and are routine in alcoholic and in physically and sexually abusive families. Further, they are especially problematic for children in emotionally abusive families where children are mistreated (for example, an angry and rejecting caregiver says in a contemptuous tone of voice, "You're a disgusting pig. Just looking at you makes me want to throw up") *but the family presents very well to others in public.* The "scars are hidden," making them much more difficult for clients to resolve because they lack **external validation** (leading bewildered clients to ask themselves, "What's real?"). Therapists who have suffered such mistreatment and had so much inefficacy engendered themselves may find it threatening (or at other times liberating) to use process comments and make overt what is occurring. An enduring legacy of such painful developmental experiences is that these therapists may be afraid (or terrified) that if they break the family rule and speak directly about what is behaviorally transpiring, something very dangerous will result—even though they cannot quite name the threat or be specific about what the catastrophe will be. For some, a pathogenic belief in their own inherent badness—or that they did not deserve to be protected—that was instilled by the mistreatment they suffered, will be confirmed. Thus, old threats can be evoked—but also resolved—by breaking double binds and no longer complying with or being ruled by unfair family rules about how family members must communicate. However, such countertransference issues are most appropriately resolved in the therapist's own counseling, rather than in the supervisory relationship.

5. Therapists' Fear of Revealing Their Own Inadequacies.

Some therapists may fear that if they bring out how the client's conflict is being replayed in the therapeutic relationship it will be revealing of their own mistakes or inability to respond effectively. For example:

THERAPIST: (*thinking to herself*) Yes, there it is, the same problem that he has with his wife is going on between us right now. It sure is getting in our way, but I don't have a clue how to bring it up. And if I tried, it'd just convince him that I really and truly don't know what I'm doing in here....

In that case, making their interpersonal process overt and trying to talk together about it may seem like the last thing in the world the therapist would want to do.

Whether the therapist chooses to address them or not, such reenactments will occur at times in most therapeutic relationships—for novice and experienced therapists alike. By making them overt, however, clinicians do not reveal their own inadequacies or mistakes, assume responsibility for causing the client's conflicts or pain because their effective responses have revealed them (a common pathogenic belief for many who enter the helping professions), or assume sole responsibility for changing them. Rather, the therapist is simply wondering aloud, in a tentative manner, about what *may* be happening in their relationship and giving clients the invitation to work together to change this unwanted pattern if the client also sees it occurring.

To address potential reenactments effectively, therapists work together with the client to *clarify the repetitive sequence of interactions* that typically unfolds in problematic scenarios with others. Continuing this collaborative effort, therapist and client can explore whether these problematic patterns that have been occurring with others ever come up in the therapeutic relationship as well. If they have, therapists can remain nondefensive and actively engage in exploring their own contributions to this shared conflict, express their willingness to change their part in the old scenario, and try to make it come out better for the client this time. Still working collaboratively, the therapist and the client search for different ways of interacting that stops or changes the old, unwanted scenario. By doing so, the client no longer feels controlled, judged or blamed, alone or unseen, and so forth—at least in this relationship—and is encouraged to begin constructing more rewarding relationships with others along these new and better lines.

When the therapist first begins to make these reenactments overt, clients may feel discouraged—which can evoke feelings of inadequacy or guilt in the therapist. Initially, the client may feel discouraged because this reenactment confirms the clients' expectations that relationships cannot be different than they have been in the past. Intellectually, the client may recognize that other ways of relating are possible. However, this possibility holds little meaning experientially when the current enactment with the therapist confirms the client's faulty schemas and follows the same, problematic scenario that she has experienced over and over. Therapists' feelings of failure will be resolved when therapists:

- realize they are already responding adequately to the client by recognizing that this is how the client's relationships have tended to go in the past;
- understand that this has been an unsolvable and painful problem that often-times has characterized the client's life;
- affirm the historical validity of the client's hopeless or discouraged feelings, based on the reality of what the client has actually experienced with others in the past; and
- help clarify how the therapist and client are changing this unwanted pattern, right now, by recognizing and working to alter it in their current interaction.

Thus, the therapist is *behaviorally demonstrating* that the relationship between the therapist and client is no longer following the same well-worn, problematic patterns and is giving clients the corrective relationship they need in order to begin changing with others. Based on this new experience with the therapist, clients may also become more aware of and receptive to instances when others do in fact respond well to them—rather than only noting confirmations of the unwanted response that is familiar, expected, and hurtful.

By thoughtful preparation—mentally formulating and revising working hypotheses, keeping written process notes, and writing case conceptualizations—new therapists will be able to resolve legitimate concerns about their own adequacy and performance. In the beginning, most therapists cannot successfully track the process dimension just by thinking on their feet—there's too much going on in the session. Instead, therapists can prepare themselves by considering in advance various hypotheses about what may be occurring in the therapeutic relationship along the process dimension. As we have already noted, therapists do not usually want to make a process comment until they have seen an issue occur several times and until they have some tentative understanding of what it may mean. When the client's coping strategy elicits anger, discouragement, anxiety, or some other strong emotion in them, therapists are encouraged to wait to respond until they better understand their own personal reactivity or countertransference reaction and can reestablish their neutrality (in other words, sort this through with your supervisor first). Finally, therapists who have prepared themselves with working hypotheses will feel more confident about making process comments. They will be less defensive about feedback the client may give them, more able to hear whatever concerns their clients' present, and better able to adjust more flexibly to their clients' needs.

As developing therapists become more experienced and confident, it will be easier for them to take the risk of "not knowing" and explore more open-endedly what may be occurring in the client–therapist relationship. When feeling confused, for example, an experienced therapist may be able to inquire simply:

THERAPIST: Hey, how did we get into this, anyway?

6. Concerns about Owning Personal Power.

A pervasive but generally unacknowledged issue in clinical training is therapists' concerns about owning their own personal power. *Many new therapists feel uncomfortable about allowing themselves to become someone who is important to clients and having a significant impact on their lives.* Especially in the beginning, it is also anxiety arousing to make strong interventions that can influence clients profoundly and to accept the responsibility that comes with exercising such personal power. As a result, *many beginning therapists overqualify their comments and water down the impact of their interventions.*

The therapist's own effectiveness can arouse anxiety for many reasons. Some clinicians were parentified or aggrandized as children. For these therapists, legitimate competencies can be readily exaggerated into unrealistic all-powerful or all-responsible grandiosities. Initially, as a child, it may have been exciting to be special and powerful vis-à-vis their parent in this way, but it soon becomes lonely, intimidating, or burdensome. As a result, these therapists may undo, or quickly retreat from, strong interventions they make or may avoid them altogether.

Some developing therapists are reluctant to have a significant impact on clients because of their own separation anxiety or separation guilt. For these therapists, competent or independent functioning in their families of origin threatened their emotional ties to parental caregivers (that is, they lacked a secure base). As the young child began to explore independently and move toward greater autonomy, caregivers may have looked sad or hurt (instilling separation guilt), withdrew physically or emotionally (instilling separation anxiety), demanded more or belittled the child's attempts at mastery (instilling feelings of inadequacy and shame), or otherwise paired anxiety with healthy competence and individuation strivings. Therapists for whom these types of developmental experiences were ongoing or characteristic may retreat from their own effective interventions, and frequently may be heard making comments like, "I don't know what to do," "I'm afraid of hurting the client," or "I'm so screwed up myself that I have no right to try to help somebody else." All therapists have their own personal problems and limitations. However, when such comments persist, it may indicate that the therapist wants to avoid the anxiety evoked by effectively addressing the client's problems—in the same way that effectively differentiating from their attachment figures and demonstrating competence and independence was (and is) anxiety evoking because it violated unspoken "rules of attachment" and left them feeling alone.

What can help trainees with these concerns? An affirming supervisory relationship is crucial if therapists are to embrace their own personal power and be as effective as they can be. The supervisor–supervisee relationship is most productive when supervisees:

- feel personally supported by the supervisor,
- receive conceptual information and practical guidelines when needed,
- obtain nonjudgmental assistance in sorting out their own conflicted reactions toward the client that have been triggered by the client's eliciting maneuvers or by the therapist's own personal issues, and
- are able to address and resolve the interpersonal conflicts (restore ruptures) that are likely to arise in the supervisor–supervisee relationship.

If the supervisee feels that the supervisor is solely directing the case, the interpersonal process between the supervisor and supervisee is problematic. In some cases, this process will be reenacting the supervisees' childhood dilemma in which they were held "responsible" for the behavior of others and for other outcomes yet never were allowed to truly voice their own opinion, develop their own ideas, and be active participants in formulating the direction of their lives—and now, as trainees, their clients' treatment. Thus, even though the supervisor is ultimately responsible for the client, the supervisor and supervisee should be consulting together collaboratively in a way that allows the supervisee to feel ownership of the treatment process—just as the supervisee collaborates actively with the client so the client feels ownership of the treatment process. (See the Motivational Interview literature for the powerful effects of using this approach (Miller & Rose, 2009; Moyers et al., 2005)). Supervisees need to hold primary responsibility for what occurs in sessions and need to feel free to act on their own ideas, while taking into account the supervisor's input. Otherwise, supervisees will not be able to experience their successes and failures as their own and will not be able to learn from them.

Ideally, the supervisor will help the supervisee evaluate the effectiveness of interventions and consider alternatives without taking away the supervisee's own initiative. When the supervisor and supervisee become stuck on a conflict in their relationship, the same issues often carry over to the supervisee–client relationship and will be reenacted there (that is, in a **"parallel process"**). In other words, the interpersonal process between the supervisor and supervisee is often reenacted or paralleled in the supervisee–client relationship. Often, the therapeutic relationship will not progress until the conflicts in the supervisory relationship are resolved. In contrast, when the supervisor and supervisee can maintain a collaborative relationship, the supervisee is better able to enact a working alliance with the client as well. It might be useful for trainees to know that certain concerns are common at various stages in the training process (Stoltenberg & McNeil, 2009). For example, early in the training process, trainees are concerned about what to say and how they should respond. Later, supervisees' identification with client issues and countertransference emerge—which often introduces challenges in the supervisor–supervisee relationship. This emergence often represents an opportunity for growth, which is facilitated by an empathic and supportive supervisor who uses reflective listening and assists supervisees in exploring the impact of their countertransference on the treatment process.

In sum, it may take several years before therapists can allow themselves to have as significant an impact on clients as possible and to fully possess their own personal power; this is a gradual developmental process. If student therapists do not progress along this feeling-of-adequacy dimension as they move through their training, they should discuss this issue with their supervisors as they go through their training program and/or seek treatment for themselves.

CLOSING

This chapter has drawn primarily from process-oriented concepts in the group therapy, family therapy, and existential psychotherapy literature, and from the counseling literature on immediacy interventions.

Therapists will find that, as in individual therapy, working with the process dimension is also the central focus in marital counseling, group therapy, and family therapy. The interpersonal patterns and relational conflicts between couples, family members, and group members similarly begin to involve the therapist. As in individual therapy, process comments that describe the current interaction with the therapist, or what is transpiring between group members or family members, can be used to identify and change these interpersonal patterns. In couples counseling, for example, the therapist may repeatedly share observations about the way in which the couple seems to be interacting or talking together:

THERAPIST: You two are arguing about who takes out the garbage, but I'm wondering if the argument is really about who has the power to make decisions or has the right to tell the other what to do. What do you two think—would it be helpful to talk about that issue more directly?

Finally, the underlying assumption in all interpersonally oriented treatment approaches is that the therapeutic relationship will come to resemble other prototypic relationships in the client's life. Clients' problems will emerge in the therapeutic relationship, especially if the therapist tracks the process dimension and attends to the reactions that clients tend to elicit from the therapist and others. Therapists do not need to artificially construct situations to recreate clients' conflicts and make them accessible for treatment. In other words, therapists can respond to clients in genuine ways. They do not need to contrive or manipulate events in order to strategically recreate clients' conflicts and thereby create "therapeutic experiences" for clients (for example, start the session late on purpose in order to make the client feel mad or unimportant, and use this experience to bring up the client's conflicts about expressing anger or readiness to feel unwanted). Such paradoxical or strategic approaches undermine the authenticity and integrity of the therapeutic relationship. Furthermore, this process often recapitulates the client's developmental history as the client is again forced to comply with others' covert control or anxiously struggle to decipher incongruent metacommunications. In contrast, if therapists are willing to work with the natural expression of the client's conflicts in the therapeutic relationship, they can offer the client a meaningful, real-life experience of trust, empowerment, and change.

SUGGESTED READINGS

1. Drawn from the film *Ordinary People* (Academy Award–winning director, Robert Redford, 1980), a case conceptualization of Conrad and the Gared family is provided in Chapter 9 of the Student Workbook.
2. For an excellent discussion of the role of the therapist–client alliance as critical to change in therapy, and the importance of empathic attunement in this process, see Angus, L. and Kagan, F., (2007), "Empathic Relational Bonds and Personal Agency in Psychotherapy: Implications for Psychotherapy Supervision, Practice, and Research." *Psychotherapy: Theory, Research, Practice, Training* 14, 371–77. This important article highlights how "therapeutic empathy" facilitates clients' agency and engagement, and provides examples of how these authors actively work on developing this skill with students in their training program.
3. Several helpful training texts are available to teach new therapists how to use immediacy interventions effectively. Practical guidelines to help therapists learn how to metacommunicate and when to intervene with the process dimension

are found in Chapter 12 of *Helping Skills* (Hill, 2009); Chapter 15 of *Handbook of Interpersonal Psychotherapy* (Anchin & Kiesler, 1982); Chapters 3 and 5 of Cashdan's (1988) *Object Relations Therapy*; and Chapter 4 of *Negotiating the Therapeutic Alliance* (Safran & Muran, 2000).

4. Useful illustrations of how therapists can integrate other theoretical modalities with the interpersonal process approach are provided by P. Wachtel in *Psychoanalysis, Behavior Therapy, and the Relational World* (1997) and *Relational Theory and the Practice of Psychotherapy* (2008).

CHAPTER 10 | WORKING–THROUGH AND TERMINATION

CONCEPTUAL OVERVIEW

Working–through and *termination* are two distinct phases of treatment. As clients continue to change dysfunctional patterns with the therapist and find that familiar but problematic scenarios do not reoccur in this relationship, they begin to generalize this experience of change beyond the therapy setting. Based on the experience that emotional needs can be met and conflicts can be resolved in the therapeutic relationship, clients begin to explore how they can make the same types of changes in other relationships as well. This working-through phase of treatment is an exciting period of growth and change *as clients try out with others the emotional relearning that has occurred with the therapist.* As the therapist helps clients resolve with others the conflicts that they have already been able to resolve with the therapist, treatment evolves to a natural close. The termination phase provides an opportunity to reexperience and resolve old conflicts, internalize the helping relationship the client has had with the therapist, and successfully terminate treatment. Although termination is considered an ending, it is in reality a transitional phase to the client's next stage of development. It ends the treatment phase where therapist and client have worked together to address the client's presenting problems; it also signals the next stage of life where the client feels more contained and better able to manage her life with more balance and flexibility. This also includes knowing and feeling comfortable with asking for help, should that become needed.

WORKING–THROUGH

THE COURSE OF CLIENT CHANGE: AN OVERVIEW

New therapists usually lack a conceptual overview of how client change comes about. Often, because they have not worked with many clients or may not have experienced their own successful therapy, student therapists lack a sense of the ordered sequence in which change typically occurs. In the previous chapter, we reviewed the sequence of therapist activities over the course of treatment. We now begin this section by extending this framework for conceptualizing the course of therapy. To get an overview of the change process, we will describe when and how clients tend to resolve their presenting problems and adopt new, more adaptive responses with others. As you will note, clients often change in a relatively predictable sequence of steps, and this unfolding course of change is reviewed from the beginning of treatment to the final working-through and termination phases.

For some clients, change begins as soon as they decide to enter treatment. These clients are in the committed, or action, stage of the change process (Prochaska & Norcross, 2006). They recognize that a problem exists and are no longer going to deny or avoid their problem. Further, they are now ready to devote time and energy to addressing the problem(s). These clients, by entering treatment, are also acknowledging that they cannot resolve their problems alone and need help. Some clients recognize that their resolution to seek help is a healthy step forward, and they feel good about making this decision. Sadly, others may experience it as a failure, as evidence of their inadequacy, disloyalty to their family, or as a violation of their cultural values and beliefs. When clients can allow themselves to feel good about the decision to seek help, or when the therapist can help clients reframe their need for help so they can have more self-empathy, some initial relief of symptoms—such as anxiety or depression—may result. This internal commitment to seek help is a crucial first step.

Some changes in feelings and behavior may also occur during clients' initial sessions with the therapist. Clients are reassured when the therapist:

- invites them to express their concerns directly and fully;
- listens intently with respect and empathy for their distress;
- is able to enter their subjective worldview and grasp the core meaning that these concerns hold for them;
- works collaboratively with the client on a treatment plan that yields hope;
- helps them feel connected to a caring "other" (a basic human need and facilitator of healthy emotional development; see Townsend & McWhirter, 2005); and
- demonstrates a practical ability to help solve at least some small part of the client's immediate problem or symptom (for example, by providing practical information about discipline/child rearing, teaching relaxation techniques for anxiety, identifying thought processes or behavior patterns that are maladaptive, role-playing new behaviors that will help with an upcoming situation, and so forth).

Case Example. Ella came to the initial session saying that she had come because she wanted to know how to help her depressed cousin who was isolating himself from others. As the intake progressed, Ella repeated "he doesn't really know what a hard time really means." Taking the cue, the therapist gently asked, "What has been hard at times for you, Ella?" She began to tear and explained that she and her older sister had been repeatedly ridiculed and "shoved around" by their stepfather. Neither she nor her sister had ever told anyone or even talked to each other about it except to acknowledge to each other minimally what was happening at the time. As soon as they were able to work and could afford an apartment, they moved out of the family home.

THERAPIST: You told no one....Did your mother or father know you were being mistreated?

ELLA: I don't know...*(tears become sobs)*...I think maybe...

THERAPIST You're not quite sure....Do you think they didn't know?

ELLA: My dad wasn't in the picture by then. He was gone. I didn't tell my mom, though, even though I think she knew. I wanted to protect her from really knowing.

THERAPIST You didn't tell her because you wanted to protect her...you didn't want her to *really* know?

ELLA: Yes, I wanted her to be safe...safe from really knowing...keep her protected, you know.

THERAPIST You wanted to protect her from having to deal with all of this, and maybe that left you feeling alone and needing someone to protect you and keep you safe? Ella, this is difficult to think about....I'm glad you are bringing me in on this— you've been alone with it for so long. What's been the hardest part?

ELLA: She knew, I know my mother knew. She had to hear it, and see the bruises we had. I just didn't want to make her feel badly because if I was to say, "Ma, can't you see what he's doing to us?' what could she do? Besides, if I did say anything to her, she'd have to choose, us or him.

THERAPIST You're saying she didn't want to hear about this physical abuse because it would disrupt her marriage, and that you think she might have chosen him over you and your sister? Am I getting that right?

ELLA: Yeah, I think she ignored what was going on for us just to keep her life easier. That's probably been the hardest part. Even now, my sister won't talk about any of it. He was one of those ragers, you know. It's probably why both of us still feel scared all the time....

As clients find that the therapist can understand their experience and respond compassionately to them, a working alliance is established that also reduces anxiety and depression. Such validation and support do not resolve most clients' conflicts but do engender hope and help to relieve their distress. Finding this safe haven in the therapeutic relationship, clients are no longer alone in their problems—they have the benevolent ally they need who can see and understand what's wrong. As we have emphasized, the essence of a secure attachment is that the child is secure in the expectation that when she is distressed, the caregiver will see or register the distress and try to help her solve her problem (thus, a secure attachment has nothing to do with common misconceptions about warmth, friendliness, being nice, or "bonding" through common interests). The key, which is also our role as therapists, is to provide clients the sense of safety that they are no longer facing their problems alone—as with Ella above. In the therapeutic relationship, needs, concerns, and fears can be shared, deeply understood, and empathically responded to. This then would be followed with helping clients discern who, in their environment, is also safe to share their concerns with—and who isn't. This process of providing clients with a safe haven is accompanied by actively listening for and encouraging their own voice—to clarify and claim what they like, believe in, and choose. The therapist's task goes *beyond* empathic attunement and also includes respectfully challenging clients' faulty schemas and beliefs about how relationships must and always will be (Bernier & Dozier, 2002; Bowlby, 1988; Dozier & Tyrrell, 1998). Here the client is encouraged and supported in developing a **differentiated** sense of self. This does not mean disconnection from others; rather, it refers to connection to a coherent, reflective, and integrated inner voice that is the foundation for self-efficacy and self-agency. Taken together, these interventions form the bedrock of therapeutic effectiveness for therapists working within most theoretical orientations and treatment modalities.

In terms of the sequence of change, once the client experiences the therapist as a "go-to" person when in need, the next phase where substantial changes are observed is when the therapist succeeds in focusing clients inward on their own thoughts, feelings, and response patterns and away from their preoccupation with the problematic behavior of others. When clients explore their own internal and interpersonal reactions, they often recognize how their behavior contributes to an interpersonal conflict and may be able to change their participation in it. Adopting an internal focus for change is also an important way to identify maladaptive schemas and reframe how clients think about their problems (Levensen, 1998). Redefining the conflict with the other person as, in part, an internal problem usually reveals a wider array of new and more adaptive alternatives that clients can begin to try out with others.

Next, as the therapist focuses clients inward on their own experience, the conflicted emotions that accompany most clients' problems will emerge. The pace of change accelerates as clients begin to express each sequential feeling in their affective constellations. Clients have the opportunity to resolve their problems when they stop defending against their core conflicted emotions, feel the safety to begin sharing them, and receive a more affirming response from the therapist than they have come to expect from others. Far-reaching changes are set in motion as clients begin to integrate their affective constellations through the holding environment provided by the therapist's acceptance and understanding.

As clients' conflicted emotions emerge, the therapist is also better able to conceptualize clients' relational schemas and the pathogenic beliefs that accompany them (Levenson & Strupp, 1997). The therapist then focuses treatment along these conceptual lines and clarifies how clients are reexperiencing the same relational patterns, faulty assumptions, and affective themes throughout the various issues and concerns they present. Some further change may occur as the therapist helps clients identify when each of these three patterns occur in their current interaction, and begin to recognize when and how they are occurring with others (Beitman & Soth, 2006). This new awareness often contributes to behavior change, and it facilitates the next and most significant point of change.

As described in the previous chapter, the same conflicts that originally led clients to seek treatment, and that they have been discussing with the therapist, will usually be reenacted in the therapeutic relationship, especially along the process dimension—that is, the way in which the therapist and client interact. This reenactment in the therapeutic relationship provides both an opportunity for therapy to fail—by reenacting the client's conflict—and an opportunity for therapy to succeed—by resolving the conflict that has emerged in the therapeutic relationship. *Treatment reaches a critical juncture when clients feel that the therapist is responding in the same problematic manner as significant others have done in the past.* This usually occurs in one of three ways: transference, client-induced countertransference, or therapist-induced countertransference.

With his delightful sense of humor, the marital therapist Harville Hendrix (2001) refers to this as "the 3 P's." Clients *pick* others who keep presenting them with old patterns and problems, they *provoke* others to respond in unwanted but expected ways, and they *perceive* others as responding in the same problematic ways—even when they haven't! However this reenactment comes about, most clients cannot make significant progress in resolving their conflicts with others until this replay with the therapist is resolved. Only when the therapist and client achieve such resolution does the client have a *real-life experience* of change. This experience of change, as opposed to insight, cognitive reframing, or challenging faulty assumptions, is often the pivotal step that shapes whether enduring alteration in the client's characteristic ways of relating, thinking, and behaving can occur. One episode of such in vivo or experiential relearning will usually not be sufficient, however, as clients repeatedly need to experience that their maladaptive relational patterns have a different (reparative) outcome with the therapist compared to the hurtful responses they have had with others.

When the client experiences the therapeutic relationship as different and finds that problematic expectations are not confirmed, the door opens to change with others in two ways. First, *the therapist begins in earnest to assist the client to actively transfer the client's here-and-now relearning to other relationships*:

THERAPIST: Good, it's getting clear between us that I'm really not judging or criticizing you. How would it be if we started sorting this through better with others in your life?

CLIENT: Yeah, that would help a lot.

THERAPIST: Who consistently does this to you? And who in your life doesn't?

CLIENT: (*chuckles*) The list of doers and non-doers. Well, my dad is definitely at the top of that list of doers, and I'm thankful to say that my boyfriend is a non-doer.

Second, under their own initiative, *many clients will come back the next session, following a corrective emotional experience with the therapist, and relate successes and failures in their attempts to relate in this new way with others in their life*:

CLIENT: I tried speaking up with my girlfriend this week, you know, like I did with you last time. In a way it felt good, but in a way it didn't. When she said...

At this point in the change process, the therapist often acts as a "coach" to help clients deal with the positive and negative responses of others to their new ways of relating. Through behavioral rehearsal or role-playing, and direct guidance or teaching, the therapist helps clients transfer the new ways of interacting that have occurred with the therapist to others in their lives.

With this overview in mind, let's now enter the working-through phase of therapy more closely, and see how the changes that occur with the therapist can be generalized to others.

THE WORKING–THROUGH PROCESS

When clients change, this usually begins in the therapeutic relationship. Some clients, especially those who generally function well, can successfully adopt new behavior with just the therapist's encouragement and advice. When the new behavior is integrally linked to their core conflict, however, most clients will need to practice this new response in the therapeutic relationship. Thus, *therapists will need to actively encourage clients to try out new ways of responding with them in the therapy setting* and reassure their clients that they intend to respond in affirming ways, rather than in the problematic ways that clients have grown to expect from others. It is crucial that therapists convey verbally *and* enact behaviorally their support of their clients' new behavioral repertoire. Therapists can anticipate that their clients will soon test the therapist and try out these anxiety-arousing new responses (for example, by being more assertive or disagreeing with the therapist). If the therapist passes the clients' test—responds affirmingly to the new behavior, rather than unwittingly repeats unwanted but familiar responses—clients have the real-life experience that change can occur.

This corrective emotional experience is the linchpin in clients' reworking of their inflexible interpersonal coping strategies, rigid schemas, and dysfunctional internal working models. As previously observed, one corrective episode with the therapist is not sufficient to effect change. Most clients will need to reenact this and other reparative relational themes over and over again in the therapeutic relationship. As a rule of thumb, the more clients have been hurt, the more often they will need to reexperience a new and safer response that doesn't fit the old schema or template. However, once clients have seen that at least one relationship can be different, the therapist can actively engage clients in generalizing this new experience of change beyond the treatment setting. Thus, **working-through** is not about gaining insight or exploring the past—it is about linking the changes that have occurred with the therapist to changes with significant others in their current lives. Typically—but by no means always—clients try out new ways of responding with others in the following progression.

Change often occurs first in the client's relationship with the therapist—especially if the therapist is working actively with "you and me" and asking about and sorting through what's going on this relationship. Clients then change with acquaintances they do not know well or with others who are not especially important to them—that is, where the stakes are not so high if things go poorly. This is often followed by change with supportive others who are important to the client, such as caring friends, teachers, or mentors. Per client response specificity, some clients will become interested in trying to change old response patterns with the developmental figures with whom the conflicts originally arose, such as caregivers and important family members. In contrast, other clients will want to explore potential changes in their internal reactions and behavioral responses to primary others with whom the conflict is currently being lived out, such as spouses or children, and subsequently transition to formative issues with developmental figures. These final two arenas are the most challenging because the consequences have so much greater import. Although all of this will overlap or co-occur to some extent for many clients, as a rule of thumb, clients generally will try to make changes first in the interpersonal sphere that feels safest or most likely to succeed.

During the working-through stage of therapy, some clients will rapidly assimilate the new ways of responding and readily apply them throughout their lives. Other clients, who have more pervasive conflicts or who have been traumatized or deprived more severely, will work through their problems more slowly. For example, adult clients with a Fearful attachment style and who have experienced overt rejection from a parent (such as being cast in the family role of a "bad child"), grew up in families where familial rules prohibited talking about or making overt mistreatment that was occurring (often by merely cueing the child with a certain look or eyebrow raise), or have been deeply "un-done" by pervasive invalidation of their experience (and been left disturbingly unsure of the validity of their own thoughts, feelings, and perceptions), will need to confront and address the same fears and expectations repeatedly. Each time, the therapist's aim is to clarify how, here again, the same relational patterns, affective themes, or faulty beliefs are being played out in this situation that just occurred with someone in the client's life this week, and help to clients find a better way to respond to this particular manifestation or variant of their core conflict. In this regard, the core conflict will be repeatedly expressed in four areas of client functioning:

- In the current interaction or interpersonal process that is transpiring with the therapist;
- When reviewing crisis events in the client's life or the crisis events that originally prompted the client to seek treatment;
- In developmental relationships with family members;
- In current relationships with friends and significant others in which the client's emotional problems are being activated.

For many clients, their attachment style and attachment history will have an impact on this working-through phase. For example, clients with secure attachments are better able to risk exploring their issues in more depth, address and sort through real conflict with the therapist, and better use the therapist as a secure base from whom to effectively "launch" (Mallinckrodt et al., 2005; Romano et al., 2008). That is, they are able to be emotionally close to their therapist and are also able to transfer what they have learned in their relationship with the therapist to relationships outside the therapy setting. In contrast, clients who have preoccupied attachment styles, while able to go to their therapists to express their distress and needs, are anxious that the therapist will not be consistently available and that their needs will not be met. They talk about their problems in diffuse, overwhelming terms. This diffuse and superficial manner, coupled with their inability to trust others, makes it difficult for these clients to accept and internalize the therapists' benevolence toward them and so they have difficulty developing a sustained, coherent, and compassionate sense of self. Thus, working-through, which requires the flexibility to become more **reflective** and consider differing perspectives, exploring new ways of being with others (which includes relating based on real rather than magnified needs), and discerning more selectively with whom to disclose and trust, will be a significant challenge for these clients. Unlike clients who have a preoccupied attachment style, those who have a dismissing style are likely to have difficulty both recognizing and disclosing their needs, and engaging emotionally with the therapist. As we have seen, these clients are especially apprehensive about intimacy and vulnerability. The relational message they convey, overtly and nonverbally, is that they need no one and can manage on their own. One of the primary tasks here is to help the client consider the possibility that "being-in-relationship" does not inevitably hold the threat of intrusive control from others. The other task is to help the dismissive client see that disclosing needs and asking for help sometimes (for example, when one is sick) are separate from "weakness." Many clients fail to see that realistically assessing strengths and needs and asking for help when needed is a healthy and natural process. Thus, the best way to help both preoccupied and dismissive clients become more capable of sustained intimate relationships is to find that they can have a meaningful relationship with their therapist—one that is safe or reparative because it isn't overwhelming or distant this time. Only then, when they have had the real-life experience that the unwanted responses they expected did not occur in this relationship, can the client begin to translate this reparative experience to relationships outside the therapy room where they often have been hurt in the past in these familiar ways. This resolving process, termed **earned security**, is a healthy sign of emotional well-being and it facilitates improved functioning in many ways (Farber & Metzger, 2009; Mallinckrodt et al., 2005; Townsend & McWhirter, 2005).

In terms of the treatment process, many therapists find that the working-through phase is one of the most rewarding phases of treatment. The therapist has the opportunity to take pleasure in clients' newly obtained mastery and celebrate with them as they successfully adopt new responses in progressively more challenging situations. During this stage, as change occurs, the therapist can also become more actively involved in answering direct questions and in providing suggestions and information to help clients enact new behavior or solve problems. For example:

THERAPIST: Maybe you could try saying something like this to her…

Therapists also can provide more interpersonal feedback about how the client comes across to the therapist and others (Brammer & MacDonald, 1996):

THERAPIST: Hey, that wasn't that "small voice" we've been talking about. You sounded strong and clear as you said that to me—it held real conviction. It's great to see you like this. Have you been acting stronger like this with others, too?

Therapists often find that cognitive interventions and behavioral procedures for rehearsing alternative responses are especially helpful during this phase. Once the interpersonal process is enacting a corrective experience, the client becomes more responsive to other types of intervention, such as assertiveness training and parenting education, self-monitoring and self-instructional training, role-playing and other modeling techniques, and educational inputs or readings. The nature of the therapeutic relationship also changes during this period. As clients improve through the working-through period, they will increasingly perceive the therapist in more realistic terms. As this occurs—that is, once the transference projections and eliciting maneuvers have been jointly identified and are being resolved—the therapist can disclose more personal information to the client and enjoy further mutuality with the client.

However, the therapist is still prepared to respond to the reenactments, faulty beliefs, faulty coping patterns, and resistance that clients will continue to present.

As clients live out a new and different relationship with their therapist, a relationship that is resolving rather than reenacting, they will find that new ways of relating with others in their lives is possible. This is a critical juncture in treatment, and *the therapist's task is to actively help clients anticipate and negotiate the successes and failures that are likely to follow.* As rigid schemas change and expand, clients need to be informed that, although some individuals will respond positively to the clients' new changes, others will not. It certainly will be exciting when others respond affirmingly to clients' changes and new behavior. However, therapists must also help clients discern who is likely to respond poorly (for example, when a client stops an old caretaking role and begins to say no sometimes). *Therapists need to help clients anticipate realistically how it will feel, and how they are likely to react, if the other person responds in the old, unwanted way to their new response.* This is especially important in primary relationships with spouses and parents, where it can be so deeply discouraging when clients, yet again, receive the same unwanted but familiar responses. To illustrate:

> Wendy, a 35-year old mother of two, had been molested by her older stepbrother
> (father's son from a previous marriage) throughout her childhood. After being
> in treatment for a year, she felt able for the first time to tell her husband about it.
> He responded well and she was then able to talk about this to her closest friend.
> Feeling she might be able to broach this with her family of origin, she chose to
> talk first to her other (biological) brother before talking to her parents. To her
> surprise, her brother's response was negative and Susan suffered a setback. Susan
> and her therapist wondered what they had failed to "filter" or discern in deciding the
> time to talk to her brother had come. They decided to invite him to a therapy session
> and he was willing to attend. As they processed the subject, Susan's brother realized
> that being told by Susan about the molestation had been greatly disturbing for
> him because his young daughters often visited her stepbrother's family. He
> acknowledged that he had suspected Susan's abuse but had not really wanted to
> believe it as true—so he had recast it in his mind as "sexual exploration." He
> had not thought much about it in recent years as his stepbrother had become
> such a highly respected professional in the community. Registering his profound
> (but not uncommon) denial of his daughters' vulnerability, he now was able to
> support Susan. They planned to talk to their families together and jointly file
> a CPS report because their stepbrother had contact with their own and other
> children.

Therapists are encouraged to be forthright and help clients anticipate some of the difficulties and potential unwanted responses they might receive in attempting to effect change in their interactions with others. This is especially critical when there are complicated family issues.

THERAPIST: It's great that your husband has been able to listen to you better, and take your concerns more seriously now that you are expressing them more directly. But I'm wondering how this will go with your mother when she visits this weekend. What if you speak up with her, as you have been doing with others, and she keeps turning everything back around to her needs and what she wants? Let's think this through before you see her.

In sum, most clients will have experiences with others that both affirm the new ways of being and other experiences that painfully reenact the old relational patterns or expectations. Thus, we must examine what therapists can do to help clients anticipate these disappointments, which are an inevitable part of the change process, and how therapists can prepare clients to respond more effectively to such unwanted responses than they have been able to do in the past.

As clients find that they can safely respond in new ways with the therapist, and then make successful changes with certain supportive individuals in their current lives, their expectations of change in some other relationships may become unrealistically high. For example, suppose an adult survivor of childhood abuse has been deeply understood and validated by an effective therapist. Following this success with the therapist, the client risks disclosing this shame-laden secret to her husband, and then to her best friend, and they both respond affirmingly as well. Next, the client's lifelong wish that the abusing parent can now acknowledge the reality of what occurred long ago is evoked, or that the nonabusing parent, who could not be protective at the time, will now hear or believe the client. With expectations buoyed by successes with others, this client may address parents or other family members, *only to encounter the same invalidation, scapegoating, and threats of ostracization that she received decades ago.* Or, some clients may be profoundly discouraged to find that, unlike the therapist and some others in their lives, their marital partner cannot change or respond positively to their healthy new behavior (for example, being more independent, affectionate, assertive, or limit setting). In this way, it is important for therapists to help clients examine their expectations of others *before* they try out new responses. In addition, the therapist's goal is to prepare clients by anticipating (1) how others are likely to respond to their new behavior, and (2) how the clients are likely to respond (what they have said and done in the past) if they receive the same types of unwanted responses they have received before.

Thus, in order to manage these disappointments that are an inevitable part of the change process, therapists can prepare clients in the following ways:

1. By helping clients *realistically anticipate* how each new person is likely to respond to their changes. For example, if the client is a married woman, what will her husband probably say and do when she acts more assertively with him for the first time?

    CLIENT: He's likely to smile at me like I'm a cute child or something, and then just change the topic as if I didn't even say anything;

    OR

    CLIENT: My mother is likely to say, "Even though you are 50, you are *still* my baby";

    OR

    CLIENT: My family is likely to say, "You've gotten so selfish and bossy since you got that degree and started working."

2. By helping clients spell out in detail how they are likely to feel, and what they are likely to say and do, if they receive a version of the old unwanted response to their healthier new behavior.

CLIENT: I think I'd feel ashamed. You know, feel stupid for trying to stand up for myself, and just give up and withdraw inside—like I've always done before;

OR

CLIENT: I'd feel anxious and keep trying to convince them to see and understand how I feel but, you know, then I'll just give up and go get a glass of wine—or maybe two or three.

3. By role-playing and providing new, more adaptive responses that, after rehearsing in therapy, clients can use at this discouraging moment, instead of repeating what they have usually done in such circumstances in the past. Recalling the concept of "change from the inside out," the key concept is trying to help clients change their own response in problematic interactions, rather than focusing on the often futile attempt to get others to respond in new or better ways.

CLIENT: I'm trying to help our marriage and tell you something important right now, but you're not taking me seriously. I don't like feeling dismissed like this. Are you willing to take this more seriously or should we just stop this conversation now?

Realistically, clients will continue to have experiences with others that reenact hurtful old scenarios and confirm their pathogenic beliefs. *However, by helping clients walk through these three steps, the therapist can help them find ways of changing their own internal and interpersonal responses—even when others are unable to change.* For example, clients can learn that they are not to blame for, or deserving of, the way they were mistreated—even though family members still cannot affirm them. Or, the married woman in the prior example learns that she can still advocate for herself. Maybe her husband will not be able to change, but she does not have to go along with or comply with her husband's dismissal in the same painful way she has in the past, and she can establish new relationships with others where her limits or opinions can be respected—even though her spouse or other family members cannot do this. In this process, clients are empowered to learn that they can change their own internal and interpersonal responses, even when significant others in their lives are not supportive.

Thus, the familiar but unwanted reactions that clients sometimes receive as they try out new ways of responding with others can be deeply disappointing, yet they also provide an opportunity to further work through and resolve the conflict. Through each setback with a significant other, the therapist and client both gain further awareness of the client's historical response pattern, deeper access to the painful feelings and faulty assumptions that accompany them, and continue exploring new responses that clients might try out the next time they are in that situation. More specifically, the therapist might explore each successive step with the client:

*   What were you thinking and feeling inside when he dismissed you like that?
*   Tell me what you did when he said that to you.
*   What would you like to be able to say to him instead the next time this comes up?
*   Would you like to role-play that together and try it out to see how it might go?

In this way, therapists can help clients recognize what they say and do to contribute to these prototypical problematic interactions. With this awareness, clients benefit significantly from role-playing or practicing new and more empowering responses that change their role in the relational pattern, even when others cannot change—as often occurs. This process will also facilitate clients' ability to establish new, more affirming relationships with others that accept or even welcome these important changes. In other words, clients' resolutions do not rest on changes in the parent, spouse, or others, as most clients inaccurately believe when they enter treatment, but on changing how they respond to themselves and others in current interactions.

It is not always easy for therapists to provide a corrective emotional experience, however. Although clients are improving during the working-through period, they will still be struggling with their core conflicts. Eliciting behavior, testing behavior, and transference distortions that narrowly slot the therapist into the old relational template or role will still have to be negotiated. Especially when clients are distressed or feel vulnerable, therapists need to work through the distortions that stem from the old schemas—for example, misperceptions that the therapist "doesn't really like seeing me," "isn't strong enough to handle me," "feels burdened by my problems," "is judging me," "will try to control me," "will get tired of trying to helping me," "is someone I should be very cautious about being vulnerable with," and so forth.

In addition, most clients will become discouraged at times by the repetitious working-through process. Seeing that the same old conflict is confronting them again, clients may feel that "nothing has changed," that it is futile to keep trying, and that they should consider dropping out of treatment. Therapists can make no guarantees of change, of course, and cannot assume responsibility for the clients' motivation to continue. At these critical junctures, however, therapists want to actively reach out and extend themselves to their clients. Only as therapists come to grasp the profound influence of early maladaptive schemas in shaping the subjective experience of "reality" can they be empathic with clients' discouragement. Although clients may realize it intellectually, experientially it really does seem to them as if this is the only way that relationships can be (Client: "Why do I *always* have to be the responsible one who takes care of everything!").

The client needs the therapist to hold steadfast that there can be another way— some relationships can be different some of the time. Therapists can point out possible instances where such other ways have occurred and convey that they remain committed to working with the client, even though they appreciate how frustrated or discouraged the client is feeling right now. It is therapists' resolve to effect change in the relationship with the client, and their personal commitment to helping this particular client change with others, that pulls the client through these expectable crisis points in the working-through phase. And it is, very crucially, the therapist's ability to continue interacting with the client in ways that offer the client a corrective emotional experience that will provide hope that some relationships can be different. In providing this reparative experience, the therapist provides the client with a safe person or a secure base that will help them reorganize how they see themselves and others, and meaningfully integrate their thoughts and feelings into a more cohesive self (Elliott et al., 2004; Pos et al., 2003).

As an illustration of the working-through phase, we now consider a case example in which the client changes first in her relationship with the therapist and then, with the therapist's assistance, is able to extend that change to other relationships in her life. The episode in which client change first emerges—the *critical incident*— may be an overt, interpersonal conflict with the therapist or a compelling, transference-laden misunderstanding. More likely, however, as in this example, it is a subtle aspect of the interaction with the therapist that could easily go unnoticed unless the therapist, informed by working hypotheses, has been prepared to look for this relational theme or pattern.

Our case example concerns Tracy, a moving-toward client who habitually responded to others in a compliant, pleasing manner. She usually went along with others, acting as if her wishes and feelings didn't matter. Tracy said that she didn't believe that others would be very interested in listening to her or doing what she wanted to do—she long had felt that either she had to "go along" or be alone. Her therapist linked Tracy's compliant behavior to her dysthymia, especially her presenting symptom of crying spells, and went on to explore Tracy's low self-esteem and feelings of worthlessness. In addition, the therapist actively encouraged Tracy to respond differently in their relationship. The therapist behaviorally demonstrated that she cared about what Tracy had to say and invited her to express her own opinions in counseling—even if that meant disagreeing with the therapist. During their next session, Tracy tested to see whether the therapist meant what she said and risked this significant new behavior in the therapeutic relationship:

TRACY: You're right. I guess I never have felt very good about myself.

THERAPIST: From other things you've said, it seems to me that your mother just wasn't there for you as a child. You didn't get the support you needed from her. Maybe we should explore that further.

TRACY: No, I don't really want to talk about her. I remember her more fondly. My father is the one who was pretty harsh.

THERAPIST: (*exclaiming happily*) You just said no, you don't want to talk about your mother and that I was wrong about her!

TRACY: I'm sorry; sure, we can talk about her. What do you want to know?

THERAPIST: Forget your mother; you just disagreed with me! You just told me you didn't want to do what I wanted to do and suggested what you thought would be better. That's great!

TRACY: What?

THERAPIST: You just did what we have been talking about. You expressed your own opinion, said what you thought, and disagreed with me. I am so happy for you!

TRACY: Aren't you mad at me? Didn't that hurt your feelings?

THERAPIST: Oh, no, we can disagree and still be close. That was a brave thing you just did.

TRACY: So…I guess what you're saying is that maybe it is OK for me to say what I think sometimes?

THERAPIST: Sure it is. I care about what you think. I want to know what you feel, and I want to do things your way, too.

TRACY: (*tearing*) But you're safe. It's easier to do that with you—you're not like other people.

THERAPIST: Yes, I am "safe," and it is easier to do things like that with me than with other people. But you have changed to be able to do that with me; you've grown. You couldn't do that before, you know.

TRACY: Yeah, that's right.

THERAPIST: And if you can value yourself enough to express what you think in here with me, then you can begin to do that out there with other people as well.

TRACY: Do you really think so?

THERAPIST: Yes, I'm sure you can. If you can do it with me, you can do it with some of them.

TRACY: Oh, I would love to be able to do this with my boyfriend.

THERAPIST: Tell me how this usually plays out with your boyfriend.

TRACY: (*describes the usual scenario*)

THERAPIST: OK, what do you want to be able to say to him instead?

TRACY: (*describes an alternative scenario*)

THERAPIST: Would it be helpful to you if we role-play that together? You can be him— and say the things that he's likely to say that will be hard for you to deal with. And I'll take your part and role-play ways you might want to respond to him. Then we can talk through together what might work for you and what doesn't—and where the problems might develop if you decide to try some of that with him.

Why is the therapist so excited about such a seemingly small change? Is this single manifestation of a new behavior all that is necessary for Tracy to resolve her problems? No, but this highly significant event brings her to the working-through phase of treatment. Within the safe orbit of the therapeutic relationship, Tracy was able to try out a new behavior that taps into the most significant issue in her life and arouses intense anxiety. For Tracy, that is, adopting such a seemingly insignificant self-assertive behavior arouses lifelong feelings of worthlessness. And, at the same time, she is discarding the primary means she has developed to protect herself from these painful feelings—that is, to "move toward" and go along with others. In order to resolve her painful conflict, Tracy will have to repeatedly experience the same sequence:

1. Her old relational pattern is activated—she has to "go along" and act as if everything is OK even when things are painfully wrong.
2. She tries out a new and more adaptive response with the therapist, which is to assert her own preference.
3. She receives a different and more satisfying response from the therapist—the therapist affirms Tracy's initiative.
4. She generalizes the new behavior to others beyond the therapy setting by considering how she typically behaves with significant others in her life and, specifically, what she might want to say and do instead the next time she is in this situation with her boyfriend. This repetitive reworking of the new behavior and the old conflict is typical of the working-through phase.

When Tracy returned for her next session, she made no mention of this incident with the therapist and said nothing about what went on with her boyfriend. Recognizing this as resistance, the therapist waited for the opportunity to acknowledge what had occurred between them the week before and to reassure Tracy that she was supportive of her new assertive response.

THERAPIST: We haven't talked yet today about the important new way that you responded to me last week. Did you have any thoughts about that during the week?

TRACY: No, not really.

THERAPIST: OK. I just wanted to say again how happy I was for you. It felt very good to me to see you express your own opinion and be able to disagree with me.

Evidently, this reassurance must have given Tracy the permission she needed. The following week, she was able to adopt the same type of assertive behavior with her boyfriend for the first time. This successful experience, in turn, encouraged her to confront the same issue in another more threatening relationship—with her father.

TRACY: (*beaming*) Guess what? I did it! I've been dying to tell you all week!

THERAPIST: (*enthusiastically*) Ha! What did you do?

TRACY: I did what we talked about with my boyfriend. He was talking about something he thought we should do, and I disagreed with him and told him what I thought we should do instead. It felt great, and I don't think he minded, either.

THERAPIST: Good for you! You're on your way.

TRACY: I can't believe how easy it was. (*pause*) I wish it could be that easy with my father. He's always putting me down when I say something.

THERAPIST: Maybe you're ready to start changing your relationship with your father as well. How do you respond to him when he puts you down?

TRACY: I get real quiet when he does that. I feel like crying, but I don't. I wish I could just tell him that I don't like it and I want him to stop.

THERAPIST: Yeah, speaking up at the time he does that and setting limits with him would certainly change your relationship. It would be saying to him—and to yourself—that you value who you are and what you have to say and you're not going to let him hurt you like that anymore.

TRACY: But I could never do that. I'd like to, but I just couldn't.

THERAPIST: You haven't been able to do that in the past, but you have been changing in here with me, and now with your boyfriend, too. So maybe you can do something different with your father as well. What holds you back from setting limits with him? What are you most afraid of?

TRACY: He wouldn't take me seriously; he'd just laugh at me.

THERAPIST: How would that make you feel?

TRACY: Worthless. (*begins crying*)

THERAPIST: So that's where that awful feeling comes from.

TRACY: Yeah. (*long pause*) It makes me mad, too.

THERAPIST: Yes, that makes absolute sense. It hurts you very much when he diminishes you like that, and you have every right to be angry.

TRACY: It's not fair! I don't want to let him do that anymore. It's not good for me.

THERAPIST: That's right, it's not good for you, and you don't deserve it. Right now, you are making the transition from feeling worthless to feeling worthy—that you do matter. Now that you are seeing this so clearly, maybe you don't have to go along with it the way you used to.

TRACY: But what could I do?

THERAPIST: What would you like to do the next time he does that?

TRACY: Well, all right, maybe I can just tell him to stop it—tell him that I don't want to be treated like that anymore.

THERAPIST: Yes. That would be a strong and appropriate response on your part. But before you try it, let's explore further how that conversation might play out. Maybe we can identify the trouble spots and role-play some responses to help you through them.

In ever-widening circles such as these, clients confront and work through the same relational patterns in the different spheres of their lives.

Notice that the critical incident in this case was very subtle: When the therapist suggested that Tracy talk about her mother, Tracy said she didn't want to. Therapists could easily miss such slight shifts in behavior unless they have formulated working hypotheses about how the client's conflicts with others are likely to be reenacted in the therapeutic relationship. If Tracy's therapist hadn't anticipated that Tracy would tend to comply with her, and if the therapist hadn't been alert for any budding sign of self-direction or assertiveness from Tracy, this particular opportunity for change would most likely have been lost. That is, the therapist might have responded by focusing on the content (by saying, for example, "But it's important to look at your relationship with your mother because…"), rather than on the process (Tracy's self-assertion). In that case, Tracy would have complied readily with the therapist and dutifully talked about her relationship with her mother. Then, the interpersonal process between Tracy and her therapist would be repeating her problematic pattern, and little change with others in her everyday life would result.

When working with a client who seems unable to change, the therapist may eventually grow discouraged and become critical toward, or disengaged from, the client. In those circumstances, the therapist should consider whether the therapeutic interaction is subtly reenacting the client's problematic interpersonal patterns. That would explain why the client is unable to change, even though the therapist and client continue to talk about adopting more assertive behaviors with others, the therapist sincerely supports and encourages this, and the client learns useful skills that should facilitate the new behavior. Such subtle reenactments along the process dimension are expectable. If therapists utilize process notes (Appendix A) and generate working hypotheses (Appendix B), they can get an idea of how each particular client's conflicts are likely to be expressed or played out in the therapeutic relationship. *By anticipating what to listen and watch for, therapists will be able to recognize such reenactments—in the moment as they are occurring—*and be prepared to respond more effectively to them.

Again, as clients are trying to generalize change to relationships outside the therapy setting, they will regularly receive the familiar but unwanted responses from some in their lives that confirm faulty beliefs and evoke familiar unwanted scenarios. As we will see later, for example, Tracy's father did not respond well to her more assertive voice, as her boyfriend did. Therapists can use this deeply disappointing experience as an opportunity to help clients explore and discern which relationships in their life can be different and better and which cannot change.

Going from Current Problems, Through Family-of-Origin Work, and on to "The Dream"

In the working-through phase, the therapeutic action is primarily in the present. That is, many of the significant events that take place in treatment are current interactions with the therapist that disconfirm pathogenic beliefs and maladaptive schemas. As these reparative interactions occur, they also lead to internal reworkings of clients' schemas as new behavior comes into play (Akhtar, 2007). These cognitive changes are necessary to help clients assimilate and maintain these new and more effective ways of relating after treatment ends. The process of experiencing a different response from the therapist, along with new ways of thinking about self and others, contributes meaningfully to **emotional regulation**—clients no longer have to reactively avoid others, impulsively hurt themselves by cutting, fighting, or drinking, or persistently (even angrily) demand attention, and so forth. Following these changes in clients' relationship with the therapist, the next step is to generalize these corrective experiences and expand clients' interpersonal range beyond the therapy setting as they now try out these new ways of thinking, feeling, and behaving with others. As we have seen, clients will have both successes and disappointments. Some individuals will welcome the client's changes, whereas others will insist on continuing along well-worn, problematic lines.

As clients explore the potential for change in current relationships and come to terms more realistically with the possibilities and limits of these relationships, two new subphases in the working-through process may emerge. First, clients, on their own, often start to look back in time in order to better understand the formative experiences that originally shaped the problems they now are resolving. To elucidate this process, we will examine therapeutic guidelines to help with this "family-of-origin work." Second, as clients develop a more realistic narrative that helps them better make sense of their life experiences and how they have become who they are, they also begin to look ahead. That is, they begin to think more about what they want their life to be in the future and to reformulate life plans to better fit the person they are becoming. This subphase, often ushered in by the emergence of "the Dream," will be explored second.

Family-of-Origin Work.   It can be liberating for clients to explore the familial interactions and developmental experiences that shaped their current conflicts. As they gain an understanding of the family rules, roles, and childhood dilemmas that shaped how they learned to cope and adapt, they become more accepting of the choices, compromises, and "best effort" adaptations they have had to fashion in their lives (Russell & Fosha, 2008). Therapists can also, during this time, *help clients feel more empathy for:*

* themselves and what they have endured or suffered,
* how they have chosen to cope, and
* the personal limitations or difficult life circumstances that led their caregivers to respond in the problematic ways they did.

Clearly, this developmental perspective can be profoundly enriching. In most cases, however, it is not productive for therapists to *lead* clients back in time, to make historical interpretations, or try to make links between current relational patterns and formative family relationships. Although these developmental interpretations and historical connections may indeed be highly accurate, few clients will find them helpful with their current problems. Per client response specificity, some clients may be able to utilize historical connections—and therapists can assess whether they are likely to be helpful for this particular client. However, many clients will not find these developmental links relevant to their current problems—especially early in treatment, even though they may be quite accurate. For example:

THERAPIST: I keep hearing the same theme: that what you are complaining about with your spouse seems in some ways to be similar to what used to occur with your mother.

CLIENT: My mother? I don't think my mother has anything to do with this. She was great, and my wife's driving me crazy. Besides, she's been dead for almost 10 years, so I'm really not sure why we're talking about my mother?

In this way, many clients will not be able to use this type of historical connection to make progress in treatment. Although the developmental suggestion certainly may be apt, often it is too far from their current experience to hold meaning for them. Instead, a *mutual exploration* of familial and developmental experiences can teach therapists (and clients) a great deal about the faulty schemas and coping strategies that clients originally learned, how these are shaping problems in current relationships, and the corrective experiences that clients need to find now in treatment with the therapist. Thus, this developmental understanding may not help clients much with their current problems, but it does offer therapists a great deal of help in generating working hypotheses about the themes and patterns that are likely to go on between them in their interpersonal process. Toward this different purpose, for example, the therapist might ask:

THERAPIST: What was it like to be a child growing up in your family?

OR

THERAPIST: Tell me about your parents' marriage. Can you bring it to life for me?

The therapeutic dialogue might go something like this:

THERAPIST: Tell me how your parents responded, what they would say and do, when you were successful or felt proud of an accomplishment?

CLIENT: I'm not sure what you mean?

THERAPIST: I'm wondering if this depression didn't start right after you got that promotion and pay raise? That possibility is what's leading me to ask, What was the look on your mother's face when you showed her that spelling award years ago?

CLIENT: *(pensively)* Well, I don't think it was a very happy face. I don't know why, but she almost seemed sort of sad or hurt.

THERAPIST: Uh-huh, your success seemed to hurt her in some way, almost as if she might have felt it as a loss. That sure would make it hard for a girl to feel good about her success experiences. How did your father respond when you told him about winning?

CLIENT: There was nothing subtle about that—he was mad and told me to stop bragging in front of my brother.

To recap, the developmental information gained from this type of mutual exploration, as opposed to historical interpretations, will inform therapists about the formative experiences that shaped this client's cognitive schemas and expectations. This provides therapists with invaluable guidelines to begin formulating the kinds of therapeutic responses that are likely to reenact problematic relational patterns for this client, and that are likely to be reparative or resolving for this client. For example, the client just cited will need to find that her therapist takes unambiguous pleasure in her success experiences and personal strengths. Early in treatment, however, most clients will not be able to make meaningful bridges between current and past relationships from this developmental exploration. When the therapist highlights seemingly obvious connections, the relational patterns may not come alive for clients until they first have experienced a corrective response to these old relational patterns with the therapist. *Following this corrective experience, many clients will spontaneously lead the therapist back to the developmental experiences that originally shaped some of their problems.* These client-initiated explorations, which usually occur immediately after clients have a corrective emotional experience with the therapist or have successfully adopted a new response with others, continue throughout the working-through period. In this sequence, the explorations are not sterile interpretations or abstract concepts. Instead, they are enlivening for clients, feel salient and informative, and readily contribute to productive change with others.

Let's pause for a moment and highlight the important sequence of change that is being emphasized here. In contrast to many other counseling theories, the interpersonal process approach does not suggest that behavior change leads to insight or, conversely, that insight leads to behavior change. Although both of these change processes occur, a different mechanism of change is proposed here. *Both meaningful insight and sustainable behavior change follow or result from clients' new or reparative experience with the therapist* (Levenson, 2003; Wachtel, 1987).

As clients make their own meaningful links between formative and current relationships, they often ask the therapist what they should do about problems and patterns that are continuing in their current relationships with family members. We now examine two broad guidelines to help clients with this family-of-origin work. First, recall the "internal focus for change" described in Chapter 4. We will again see how clients can learn to change how *they* respond in current problematic interactions with family members—rather than engage in futile attempts to try to get parents or others to be different and change how they have always responded.

Second, we look at the concept of "grief work," in which clients mourn and come to terms with what they have missed developmentally. Working in both of these areas results in an interpersonal and internal resolution of family dynamics.

Clients who struggle with more pervasive or long-standing problems usually have **family-of-origin work** to do. Many clients do not wish to discuss problems from the past with caregivers or family members, and therapists should not press clients to do so. However, other clients will want to acknowledge or address past conflicts with parents and other family members—as a way to stop participating in them or to better resolve them. It is deeply gratifying—and a powerful impetus for change—when such rapprochements succeed. As we have been emphasizing, however, such current resolutions of historical problems often do not occur. Sometimes caregivers have grown and changed over the years and can talk more openly or nondefensively for the first time about hurtful interactions that once occurred. However, in many cases, and especially with more serious problems or abuse, the client will often receive the same invalidating, guilt-inducing, or threatening responses as they did years ago in childhood. This is especially painful when the client's motivation for talking with the attachment figure is the unrecognized wish to get the parental approval, affirmation, or protection that the client has always longed to receive. If the therapist has not realistically prepared the client for the possibility that this disappointment may well occur, the client may despair and feel hopeless about change occurring in any relationship. Oftentimes, *the client will be breaking unspoken but strongly held family rules against acknowledging that any family problems or that a certain specific conflict with a parent ever existed* and, in many cases, punitive homeostatic mechanisms will be set in motion. For example, when Tracy confronted her father about always putting her down, he again disparaged her—just as he had for the past 20 years. As if that doesn't make change difficult enough, look what happens next in the family system. Pressuring Tracy to just "go along" with this mistreatment, *her mother threatened to disown her for being "disrespectful" and "ungrateful," and the next day her depressed and obese sister telephoned from another city and tried to make her feel guilty for "hurting Daddy" and "stirring up trouble."* No, for clients like Tracy, change is not easy and problems are not simple.

As we see here, the problem is not simply the parent's mistreating the child. If that occurs but the mistreatment can be "named" or spoken about in the family, the child has a problem but it is much easier to resolve (that is, it's not the content—what happened, but the process—how others responded to it). Instead, far more serious consequences—such as the self-hatred, bulimia, and dysthymia that Tracy suffered—result when either of the following occur: (1) the parent and extended family system make the adult offspring again feel responsible for, or deserving of, the parent's original and current mistreatment; or (2) the child and his experience are invalidated—as if everyone seems to be in agreement that nothing significant or problematic occurred (mystification).

Clients usually feel hopeless and blame themselves when parental caregivers and the broader family system cannot change. Although these clients may go on to act helpless in other current relationships, in actuality they are not. Their resolution does not rest on the parent/family changing but on how they change and respond in current interactions with living parents or, equally important, in their ongoing internal relationship with deceased caregivers. In order for clients to make enduring changes in deep-seated relational conflicts, they can change their own responses in these prototypical interactions. For example, Tracy's father was too limited to be able to hear her concern and talk with her about the problem. However, Tracy stopped going along with her father's disparagement of her. Instead, she began setting limits with her father and bravely metacommunicated with him:

TRACY: I keep asking you to stop it when you put me down, and you laugh at me and keep doing it. I can't make you stop, but I don't have to pretend it's OK anymore. It hurts, and I don't like it.

Although her father stayed in the same disparaging mode and again made fun of her, things did change profoundly for Tracy when she changed how she responded to him. As she was able to sustain this stronger stance toward him and the family system over the next few months, she felt less insecure and intimidated with them and others in her life than she had ever been. Her long-standing feelings of worthlessness and depression significantly improved, and her intermittent battles with bulimia largely disappeared. These far-reaching changes occurred for Tracy even though her father—as well as her mother and sister, who pressured her to retreat and "go along" again—could not change.

In this way, clients make progress by changing how they respond in current interactions with significant others in their lives. Clients also change by doing **grief work**—coming to terms with the feelings left in them by hurtful or emotionally depriving parent–child interactions. Clients do not need to discuss with caregivers problems in historical relationships. However, they do need to:

1. stop disavowing and acknowledge to themselves what was legitimately wrong;
2. stop their own participation in any ongoing mistreatment; and
3. grieve for the support, protection, or validation they wanted but have missed.

That is, they have to grieve the loss of the "wished for" parent. Many adult clients struggle mightily with highly ambivalent relationships with their painfully inconsistent caregivers. Oftentimes it is especially important to help them become aware that sometimes the unrealistic feeling of hope for their idealized parent can emerge when they glimpse the brief episodes of kindness in their caregiver that regularly occur, yet the parent cannot sustain this benevolent responsiveness and readily returns to their usual stance of being largely self-absorbed or self-centered, rejecting or contemptuous, and so forth. Helping clients be realistic in their appraisals, providing support in their grief, and changing their own responses to their caregivers' typical behavior helps clients sustain emotional stability in the face of hurtful parental behavior. Thus, regardless of whether formative attachment figures can change—and whether they are living or deceased—clients can resolve these problems through their own internal work.

Exploring the potential for better relationships with family members is often important, but it may or may not lead to improved relations. Often, aging caregivers and other family members cannot change, so *therapists need to prepare adult clients that when they introduce healthier new behavior into their family systems it will often be met with significant resistance.* Even if the family system has improved, however, grown offspring will still function better in current relationships when they can more realistically acknowledge what was wrong (and what was right) in the past and mourn what they missed developmentally. The same model applies again: The therapist provides a supportive holding environment that allows clients to come to terms with the sadness, anger, shame, and other feelings that were too threatening or unacceptable to contain before. However, to succeed in this work, therapists are encouraged to remember that *clients will be safe to discuss the "bad" or problematic part of the caregiver (or spouse) only after being reassured that therapists also appreciate and affirm the "good" or well-intended parts of these relationships as well.* The therapist's goal here is to resolve or integrate the ambivalence, without resorting to the splitting defenses or triangulation that the client has relied on in the past.

In another way, the therapist must also help clients come to terms with both the good news and the bad news about their current situations. The good news is that, through grief work, clients can disconfirm the pathogenic belief that they were in some way to blame for their abandonment, exploitation, rejection, parentification, and so forth. The bad news is that clients' developmental needs—such as the need for secure attachment ties—were not adequately met then, and these original developmental needs cannot now be met in current adult relationships (for example, with a spouse, friend, or therapist). Only after such losses or deprivation can be acknowledged as real, that is, mourned rather than disavowed, can clients go on for the first time to get appropriate and realistic adult versions of their emotional needs met in current relationships. As long as developmental needs are operating unacknowledged, clients are impaired in their ability to get legitimate adult needs met in current relationships. Finally, once clients have accepted both the "good news" and the "bad news" associated with their caregivers, they are then free to accept some of those same characteristics in themselves, and relinquish their own rigidity and perfectionism.

Two countertransference propensities can prevent therapists from helping clients achieve resolution of such family-of-origin work. First, because of their own splitting defenses, therapists sometimes do not have the breadth or internal security to help clients accept both what was good and what was problematic in their development. Therapists' countertransference propensities lead them to make one of two errors. On the one hand, *some therapists characteristically want to downplay or minimize the extent and continuing influence of painful developmental experiences*:

THERAPIST: That was then. You need to forget about what happened in the past. Let's try to do something about the problems you're having now. Besides, forgiveness is such a liberation.

On the other hand, in contrast to this denial, other therapists only want to blame the caregivers, make them "bad," and ignore the strengths and positive contributions that also were present in the relationship:

THERAPIST: *(judgmental tone)* Your father was sick to do that! What's wrong with him? I can't imagine someone wanting anything to do with him after that.

Instead, therapists need to affirm the full reality of what went wrong in formative relations, while appreciating the need to maintain whatever ties can be preserved. Except in extreme circumstances, therapists want to discourage clients from breaking off all contact with conflicted caregivers. When a client cuts off all communication with family members, splitting defenses are usually operating and lasting change is less likely. In such cases, the therapist will often be idealized and clients will be compelled to recreate the split-off, bad side of their conflicts in other relationships or to make themselves "bad." Therapists will be far more effective—and clients will be better able to stay with this difficult work—when therapists can acknowledge realistically what was wrong without demonizing parental figures and making them "all bad":

EFFECTIVE THERAPIST: I can see how much it hurt you when he did that. He did go way over the top sometimes. Those times must have been really tough for you. I wonder what was going on for him when he lost it like that—you know, what led him to such a desperate place?

VERSUS

INEFFECTIVE THERAPIST: I can't believe he did that to you. What a destructive SOB. He's hopelessly narcissistic!

The second countertransference propensity that interferes with family-of-origin work is that therapists often want to bypass the work of mourning. *As clients become able to acknowledge realistically to themselves the ways in which they were hurt developmentally or missed what they needed, painful feelings of loss are evoked.* Therapists may find it difficult to help clients come to terms with those. Most clients do not want to end the hope—or fantasy—of getting what they missed; some may even become angry when therapists acknowledge these limitations. Therapists' reluctance to address these issues will be reinforced if, as routinely occurs, such material arouses therapists' own unfinished family-of-origin and grief work. However, if the therapist tries to bypass the work of mourning and moves too quickly to a problem-solving approach, clients will not be able to act on or sustain these practical suggestions—even though they may be useful ideas. Only after original losses have been acknowledged and integrated (that is, grieved for in a supportive holding environment with the therapist) can clients open up their adult emotional needs and feel sustained by new relationships in a way they have not been able to risk before. As clients improve in these significant ways, they become able to leave the past behind, to turn the page and begin looking ahead to a different and better future.

The Dream.   Clients make behavioral changes and feel better as they work through their problems. As they improve, the focus of therapy moves away from developmental conflicts and, to some extent, beyond current interpersonal problems. In a very positive transition, the therapeutic dialogue often moves toward future plans that reflect life-enhancing aspirations and goals for the client. Referencing the literature on the psychology of hope, for example, therapists can help clients explore how they would like to be—"imagining possible selves" and clarifying what they might like to become (Russell & Fosha, 2008; Snyder, 1994).

Similarly, therapists can work in an existential tradition with issues regarding personal choice and explore the question: How can clients create more meaning in their lives, live more authentically, and make the most of the time and relationships they have? This freedom to explore what they really want—what matters most to them in life, develop their own personal voice more fully, and risk making choices and life changes based on self-discovery, often follows experiencing the therapist as a secure base and learning that change does not lead to catastrophe. Rather, this process is liberating and allows for more authentic and meaningful living. This exciting transition from solving problems to leading a fuller and more meaningful life is often signaled by the emergence of **the Dream**.

In his classic books, *The Seasons of a Man's Life* and *The Seasons of a Woman's Life*, Levinson (1978, 2000) describes the profound influence of the Dream in shaping the structure of adult life and the course of personality development. As Levinson points out, the Dream does not refer to casual waking or sleeping dreams but, in the largest sense, to the kind of life that one wants to lead. At first, the Dream may be poorly articulated and tenuously connected to reality, but it holds imagined possibilities of self in the adult world that generate vitality. It is the central issue in Martin Luther King's historic "I Have a Dream" speech or in Delmore Schwartz's story "In Dreams Begin Responsibilities."

The Dream has roots in the grandiose and unrealistic hero fantasies of adolescence, but it is more than that. In early adulthood, the Dream still has the quality of a vision—of who and how we would like to be in the world. In this way, the Dream is inspiring and sustaining for the individual, even though it may be mundane to others. For example, individuals have Dreams to be a good mother or responsible husband, a respected community leader, an ethical attorney, a skilled craftsperson, a successful but honest businessperson, an artist or a spiritual leader. *If the individual is to have purpose and a sense of being alive, occupational and marital choices need to incorporate some aspects of the Dream.* When the Dream has been abandoned or set aside, life will not be infused with vitality and meaning, even though the person may be successful.

Up until the working-through phase of therapy, therapists are usually responding to the despair of broken dreams, the disillusionment of unfulfilling dreams, and the cynicism of abandoned dreams. In most cases, the life-infusing Dream that Levinson has articulated has not yet been addressed in treatment. In the working-through phase, however, clients are resolving their problems and emerging as healthier individuals. At this point, therapists can help clients to better articulate and renew their Dream and to find ways of incorporating aspects of their Dream in their everyday lives. To achieve this, therapists first need to be able to differentiate the Dream from the broken dreams that contribute to clients' presenting symptoms. These broken dreams usually reflect failed attempts to rise above their conflicts and become special, as discussed in Chapter 7.

In the process of working-through, therapists repeatedly help clients recognize their compensatory strivings to be special and rise above their conflicts by pleasing others (moving toward), achieving success and power (moving against), or becoming safely aloof and cynically superior (moving away). Therapists help clients progressively relinquish these inflexible coping styles and adopt a wider interpersonal range. In doing so, clients stop blocking their generic conflict, which presents the opportunity for clients' core conflicts to be expressed and resolved within the thera-peutic relationship—and then generalized to others. However, therapists can antici-pate that when clients relinquish these attempts to rise above their conflicts they may experience a sense of failure or loss of self-esteem. Thus, one important compo-nent of helping clients resolve their problems is to replace these defensive, grandiose strivings with clients' own attainable, yet sustaining, Dream. That is, clients need to be actively encouraged as they enter this stage to explore and formulate their *own* personal dreams based on their own interests. *Many clients have never had the op-portunity in their families of origin to know themselves, to clarify what their genuine interests and passions might be, or to choose their own goals and pursue their own ambitions.* This phase of therapy can become enlivening with the support of thera-pists who are not, in the way the client may have experienced in her original family, going to direct her toward their own agenda for her, discourage or disparage her efforts, be uninterested in or unresponsive to whatever is most enlivening or meaning-ful to her, expect too much of her, and so forth. Here, instead, the "Dream" is her dream—motivating, empowering, and attainable.

Arthur Miller's (1949) Pulitzer Prize–winning play, *Death of a Salesman*, poi-gnantly illustrates how the neurotic striving to rise above conflicts inevitably fails, how it can lock clients in their generic conflicts for a lifetime, and how it stifles the Dream. The protagonist, Willy Loman, has adopted a moving-toward interper-sonal style to cope with his profound feelings of inadequacy and shame. With his smile and his freshly shined shoes, Willy pathetically strives not just to be liked but to be "well liked." In a brilliant exposition of multigenerational family roles and relationships, Miller shows how Willy's grandiose strivings to rise above his shame-based sense of self by being well liked are also acted out through his oldest son, Biff. Willy exaggerates Biff's accomplishments as "magnificent," excuses or ignores his shortcomings, and never responds to the reality of his son's actual feel-ings, needs, or wishes.

Near the end of the play, both the father's and the son's defensive strivings to rise above fail. Willy is fired from his job—a crushing humiliation that shatters his myth of being well liked and his lifelong coping strategy of winning approval. That same day, Biff fails to secure an important business opportunity that his father has encouraged. With this setback, Biff realizes that he has been living out his father's grandiose expectations that he would become a fabulously successful entrepreneur. In the closing scene, however, father and son respond very differently to their re-spective crises.

Biff is able to relinquish the unrealistic script of super-achievement, popularity, and wealth that his father has demanded of him. In the last scene, Biff begs his fa-ther to release him from this grandiose role, to "speak the truth" in the family for the first time, and allow him to drop the false appearances and preoccupation with success and allow him to be just a decent, regular person—no longer fulfilling the familial role of "Hero"—and to let that be good enough. However, Willy cannot relinquish his own rising-above defense and, therefore, cannot grant Biff's plea to be let "off the hook" and stop striving to be superior and special. Instead, Willy alternately rejects Biff contemptuously and then idealizes him as "magnificent."

Biff finally recognizes the family role and myth that he and his father have been playing out all of their lives. By addressing it with his father, Biff is able to change his part in this symptomatic scenario. Although this honest communication does not lead to an interpersonal resolution with Willy, it does lead to an internal resolution for Biff. Soon afterward, Biff decides to leave home. He clarifies that he knows better now who he is and what he wants to do with his life. He is going to pursue his own Dream: to leave the crowded, noisy city, move to a more rural setting, and work outdoors with his hands where he can "feel the sun on my back and smell the grass." Whereas Biff could relinquish the neurotic strivings to be special and allow his own attainable Dream to emerge, Willy could not.

In one last attempt to change their dysfunctional relationship and find a more authentic connection, Biff reaches out to put his arms around his father. Horrified, Willy recoils in fear and disgust. Although his lifelong obsession has been to win the approval and respect of others, he forcefully rejects this genuine affection from his son. If Biff is no longer willing to fulfill the role of the superachieving Hero that Willy demands in order to compensate for his own shame-based feelings of inadequacy, he can't embrace him. To accept the love that Biff was offering, and let him off the hook and out of the Hero role, would reveal Willy's own generic conflict—his unmet needs to be respected and wanted, his profound sense of being unlovable, and his intense feelings of inadequacy and shame. With no understanding of what these disavowed feelings are about, and no supportive relationship to help him contain this flooding, these unacceptable feelings would be overwhelming. Thus, Willy is now trapped in his conflict: He is no longer able to maintain his own (or Biff's) defensive strivings to rise above, and he is unable to let his true feelings and concerns emerge. Desperately bound in that unresolvable conflict, Willy's tragic solution is to end his own life.

As clients' neurotic strivings to be special and rise above their conflicts fail, their generic conflicts are revealed. Without a holding environment to help them contain their feelings, some individuals—like Willy—feel hopeless, and a few may see suicide as the only way out. In contrast, if the therapist can provide a more satisfying response to the conflict and disconfirm what the client has found before and come to expect, the client can begin to come to terms with these feelings in a better way and change. As clients begin to improve during the working-through phase, the Dream will often emerge.

To help clients change, therapists can support and encourage the Dream in many ways. First, the therapist is listening actively for the issues that genuinely enliven clients, engage their intrinsic interests, and bring them pleasure. By attending closely to what clients really want to do and inquiring about what feels right for them, therapists can help clients discover and articulate their Dream. More specifically, the therapist's intention is to ask questions that will help clients clarify what they like, find interesting, or want in order to make a better life for themselves. For example, clients may respond:

- I want to stop losing my temper—I want my children's respect back.
- I want to make peace with my father and bury the hatchet.
- I don't want to keep driving to this same dead-end job. I want to finish school and get my credential.
- I want to be less social and spend more time with my family and a few true friends.

For other clients, their Dream is completely undeveloped and they cannot articulate what they want in any meaningful way. Oftentimes, their caregivers didn't track their experience or keep their "mind" in mind very well, and didn't register what they were feeling or join them in their interests. These clients, with a less developed sense of self, are unaware not only of what they want to do in life but, more basically, even of what they like or dislike. For clients like these, the therapist's task is to help them begin to attend to and discover their feelings, interests, values, and ultimately formulate and pursue their Dream by repeatedly:

- inviting clients to attend to what they are experiencing at different moments;
- orienting clients to listen to themselves and simply be mindfully aware of what they want to do in different situations—even if they cannot act on their wishes (for example, to realize "I'm not enjoying this right now—I want to leave");
- being interested in and entering into clients' subjective experience;
- calling attention to, and trying to participate supportively in, whatever seems to interest or hold meaning for clients;
- acknowledging and taking overt pleasure in what clients do well; and
- encouraging clients to act on their own feelings or preferences when possible (for example, even something as simple as recognizing, "I'm sleepy—I'm going to take a nap").

Although many clients have a much more developed Self than this, some will need help to begin attending to their own internal experience in these very basic ways. It is rewarding for the therapist to help clients **differentiate a self** by responding—perhaps for the first time in their lives—to their own feelings and interests. Clinicians working within many theoretical orientations have emphasized this process, such as the concept of "mirroring" in self psychology (Kohut, 1977), "experiencing" in the client-centered tradition (Carkhuff, 1999), or "attuned responsiveness" in attachment theory (Bowlby, 1988).

As clients begin to experience their inner life more fully, and the therapist responds affirmingly, their Dream will start to emerge. At this point, the therapist can do several things to help clients bring aspects of their Dream into their current lives. For example, the therapist can provide a "practicing sphere" in which clients can discuss and explore various possibilities of their Dream without having to become committed prematurely to any action. The therapist can also help clients tailor their Dream to reality and find ways to express aspects of their Dream in their everyday lives. Although most clients probably can't quit their day job and become, say, full-time performers, a client who loves music can sing in the choir, take piano lessons, put together a garage band, get a degree in music, and so forth. The therapist can also help clients identify training or education routes that will facilitate realization of their Dream, and give clients permission to develop close relationships with others who can act as mentors in their chosen fields and pursuits.

Finally, clients will confront the same core conflicts in exploring their Dream that they have experienced in other aspects of their lives. Thus, the therapist will also have to help clients work through variations of their old conflicts that are aroused by pursuing their Dream. For example, the therapist will have to help clients differentiate between their Dream and their interpersonal coping style and defensive strivings to rise above. This might include strivings to feel secure by being loved by everyone, to gain a sense of personal adequacy by becoming wealthy and powerful, or to escape criticism or rejection by being safely aloof and superior. Progress usually will not be simple or straightforward, however. For many clients, *actively pursuing what they really want to do, or successfully attaining what they want in life, commonly threatens attachment ties and arouses both separation anxiety and separation guilt—leading clients to retreat from progress or undo their success in treatment.* However, as clients successively work through these conflicts and find that they do have a right to their own life, they can follow through on realistic plans for attaining aspects of their Dream. With this success, treatment evolves to a natural conclusion.

TERMINATION

CLIENT ENDS TREATMENT EARLY

Premature termination has long been recognized as a serious problem in psychotherapy (Garfield, 1994; Reis & Brown, 2006). However, there is no consensus in the field on the definition of premature termination (Dew & Brown, 2005) or real understanding of why it occurs so frequently. Does premature termination refer to dropout after the first session or dropout anytime prior to the client meeting his/her treatment goals? Bergin and Garfield (1994) have suggested "premature termination" be considered whenever an individual has attended a session and then discontinues on his own by failing to come to future scheduled appointments. According to published data, disturbingly, approximately 30 percent of clients drop out after completing only an intake and 40 to 60 percent before experiencing sustainable benefits, and that on average, clients attend only three to six treatment sessions (Callahan & Hynan, 2005; Garfield, 1994; Hansen et al., 2002; Wiezbick & Pekarik, 1993). Why are we failing to help so many clients? A substantial focus of the early dropout literature has been on demographic and lifestyle characteristics of those who terminate early. These include people who come from deprived backgrounds, those of minority status, individuals who have a history of substance abuse, those with a history of involvement with the criminal justice system, and so forth (see O'Brien et al., 2009). While these features are important, the real issue for therapists is understanding what clients are actually experiencing in treatment; in particular, what clients are not getting or finding in their therapy that leads to such extensive dissatisfaction and disengagement (Muran et al., 2005; Safran et al., 2005; Swift & Callahan, 2008). Research suggests that one important factor implicated in early termination of treatment results from clients' unmet expectations—that is, their expectations of what they would gain from being in treatment doesn't match what they actually experience. Many therapists have interpreted this to mean they need to engage in further "role induction;" that is, giving client more explanations about what therapy entails. We believe, however, that role induction is but a small and "structural" feature of therapy and not the major reason clients disengage early. We believe that it is the therapy process itself and the client's *experience*, especially in that important first session, that is key to whether clients will continue in treatment. As clinical supervisors and training clinic directors, we find that beginning students struggle with knowing or understanding their own role; they consider one stance to be that of functioning in a guidance or directive mode (here they would tell the client, from a more "expert" perspective, how to structure their day, give them homework, advise them on decisions, and so forth). The other role they often take is a passive, "neutral" stance, where they are simply in the same room with the client, convey very little of themselves to the client, avoid questions the client asks as much as possible, and function as nondirectively as possible. Their hope here is that the client will somehow know they are trying to be fully accepting, nonjudgmental, and are not trying to direct the client's life or tell them what to do. The problem with each of these stances, however, is that clients come to therapy because they have real problems and they want to have an authentic, active, and engaged relationship with a person who is truly present—who clearly wants to be with them and enter their experience with genuine interest and concern, demonstrates some real understanding about what's going on for them and what's really wrong, and is able to offer them something that is meaningful or helpful right from the beginning. They want the experience of being heard and seen and want to feel that their therapist is trying to understand *their* personal issue or problem and join with them in clarifying what it is they need. The key here is that *before they leave the session, the therapist needs to give the client the experience that there is someone here who is worth all the trouble and expense of coming to talk to.* Thus, in this first session in particular, but carrying through in subsequent sessions, the therapist needs to:

- Give clients a sense of hope that they can get the help they need in this setting by "hearing" what is most pressing, painful or important for them right now;
- Help clients clarify what it is that is troubling them, that is, begin figuring out what's really wrong and what they can begin doing now to help with this issue;
- Work together to discern what they need most, today, before they leave each session;
- Provide hope that by working together it is possible to find ways to make the future be different and see that it does not have to repeat the past; nor do they have to face their challenges alone;
- Actively engage with clients and provide them with the sense that you are going to be a highly involved collaborator with them on this journey—actively sharing ideas, observations, alternatives, and feedback, but without making decisions for them and telling them "what to do" or how to lead their lives; and especially,
- Talk forthrightly together about what is going on between you in your relationship—How is it to come here today and ask me for help? What does and doesn't feel good about talking to me? What are we doing in here that is helping you and what could I do differently to be more helpful next time? What are your thoughts about coming back and talking with me again next week? What do you think I might be thinking about you right now after you've just taken the risk to share all of that with me? Was there anything I said or did today that bothered you or didn't quite feel comfortable? And so forth.

We find that when student therapists (1) stretch their personal comfort zone and actively engage with clients in these forthright and direct ways right from the beginning; and (2) are able to "get" what is most important to the client and provide empathic understanding, few clients drop out prematurely.

In this vein, Sue & Zane (2009) write eloquently about the importance of offering the client a "Gift"—this needs to be something that demonstrates behaviorally that you, as a therapist, are someone who has the ability to be helpful and make a difference (that is, demonstrate or achieve credibility). This capacity to offer a **Gift**, and your behavioral communication that you are going to be a present and actively engaged helper, will have your client leaving the session with a sense of hope; it also increases the client's likelihood of returning. This Gift can range from:

- simply giving the client the experience of feeling affirmed, such as understanding and validating the burdensome impact on the client of growing up as a parentified child, or the lifelong anxiety instilled in a child by lying in bed at night and listening to your parents fight;
- letting certain clients know that you understand that some children were not protected, got hit or molested, and that you are a safe person to talk with about those issues and that many others have not wanted to hear;
- providing him with some basic educational information about sex education or sexual dysfunction so this client no longer feels so different or alone;
- role playing how to approach an upcoming situation or person that is particularly troubling;
- providing an overwhelmed parent with positive parenting strategies or other ways to regain control and discipline a child, and so forth.

A very important component of all sessions, but the initial session in particular, is that clients leave the session feeling a sense of connection, collaboration, and hope that they will get the help they need, and not feeling that, primarily, *they are merely answering questions for the therapist so she can complete the agency's intake form or make a provisional diagnosis for a new patient chart*. Find a balance: make real contact with the client, provide a Gift if you can, and then tell your client that you need to shift gears and ask some different questions in order to complete some needed paperwork.

INEFFECTIVE THERAPIST: I'd like to talk about anything you want to talk about and follow your lead. I don't want to give you any answers. I'm going to help you by listening and not by telling you what to do.

EFFECTIVE THERAPIST: Welcome. I'm glad you're here. We spoke briefly on the phone, but let's get right to work on what's troubling you. Why don't you tell me more about what's wrong and bringing you in, and then we can start sorting things through together.

You want your client to have the feeling that "this is a person who has something to offer me that other people in my life can't offer me right now, and I can benefit from what is being offered." In terms of role induction, it is useful to communicate to clients that improvement or recovery from treatment takes longer than most expect (for example, 50 percent of clients improve after eight sessions and most need 13 to 18 sessions; see Swift & Callahan, 2008). Clients may also benefit from knowing that therapy may evoke painful feelings or difficult memories but that you will be with them through all the challenging moments of this journey. For therapists working within the interpersonal-process model, however, *role induction for the client emphasizes education about the therapeutic relationship and*

re-defining social norms to include straight talk about what might be going on be-
tween the client and the therapist. To illustrate, toward the end of the initial intake
appointment, the intake worker can say:

THERAPIST: Bob, here's the best way that you can help the therapist help you. We can
   speak up and be much more direct or forthright here in therapy than we can be in
   other settings. So, if the therapist isn't being helpful to you, or is doing something
   you don't like—such as being too quiet, or you don't know what they're thinking,
   or you aren't getting enough feedback—speak up and tell the therapist. Don't be
   nice and go along, or get mad and go away, bring it up and let the therapist know
   what isn't working for you. How does that sound—is that something you could do?

CLIENT: Well, I don't usually talk that that way to others, but I can see that it might be
   helpful. But what if I say something I don't like or isn't working for me, and the
   therapist doesn't change?

THERAPIST: If you are telling your therapist something you don't like or would prefer
   instead, and the therapist doesn't want to hear your feedback, talk it through with
   you, and try to respond to it in some way that makes sense to both of you, it's
   probably not a good match and I'd recommend that you ask to be transferred to a
   different therapist.

To recap, stemming the premature terminations that pervade our field necessi-
tates that in the initial sessions therapists go beyond being "nice" or "passively"
present. Therapists must be actively engaged, assume a consistent **mentalizing perspective** in which they have the client's "mind in mind," *so that what is most distressing or urgent for this client is carefully discerned and clearly articulated.*
Following this, the therapist will need to respond in the corrective, reparative ways
we have been discussing. If this is accomplished, the client and therapist will have
co-created a productive therapeutic space where termination is most likely to occur
only after meaningful and enduring treatment goals have been met.

Client and Therapist Talk About Therapy Ending

Termination is an important phase of therapy and needs to be negotiated thought-
fully. Ending the therapeutic relationship will often be of great significance to
clients. For some, *this may be the first positive ending of a relationship they have
ever experienced. How it is managed is so important that it often influences how
well clients will be able to manage future separations, endings, and losses in their
lives.* If carefully planned, it provides an opportunity to convey to clients that they
have been valued, allows therapists and clients time to talk together about areas
of growth and anticipate where clients' future challenges may emerge, influences
clients' ability to enter into relationships with others in the future, and provides a
healthy model for ending relationships (Many, 2009; Rappleyea et al., 2009).
Most beginning therapists greatly underestimate what an important experience a
successful termination can be for clients. It holds the potential to diminish or even
undo the changes that have come about in treatment, or to help clients consolidate
and further their gains—and better be able to maintain these changes after treatment has ended. Therefore, the therapist's intention is to make the most of this significant opportunity for further change.

Early in their training, therapists often ask, "How will I know when it's time to end therapy?" It is time to end when clients have achieved significant relief from their symptoms, can respond more flexibly in current situations rather than slotting diverse experiences into narrow, predisposed categories, and have begun to take steps toward promising new directions in their lives. Therapists know that clients are ready to terminate when they receive converging reports of client change from three different sources:

- from clients—when clients report that they consistently feel better and can respond in more adaptive ways to previously problematic situations;
- from their own observations—when clients can consistently respond to the therapist in new ways that expand their old coping styles and do not reenact their maladaptive relational patterns; and
- from significant others in the client's life—when significant others in clients' lives give them feedback that they are different or make comments such as, "You never used to do that before."

With this convergence of perceptions that important changes have occurred, it is time for treatment to end. Before going on to discuss what therapists need to do to make terminations successful, we must distinguish between two types of endings: natural and unnatural.

The termination sequence just described is a **natural ending** because the client's work is finished. One of the therapist's primary goals in these natural endings is to affirm both sides of clients' feelings about ending—wanting to go but also wanting to be supported. Thus, *therapists are trying to provide a secure base during the termination phase* by taking unambiguous pleasure in clients' independence and actively supporting their movement out "on their own." However, a secure base also means providing continuing support as clients differentiate and emancipate, so therapists can let clients know that they will accept their request for help or contact in the future if the need arises. Clients' old schemas, transference distortions, and faulty expectations *will often be reactivated* toward the therapist as they begin to discuss ending. To correct these transference distortions, therapists need to clarify overtly that they will not be disappointed, burdened, troubled, and so forth (as many clients will inaccurately expect) if clients want to send a letter to stay in touch, to share important life events such as marriages and births, or to recontact the therapist for help if they have problems in the future.

The key issue here is that *so many clients have not had permission to be both separate and related at the same time.* For example, growing up, becoming more independent, and differentiating from their caregiver by pursuing their own interests, goals, and relationships often meant losing their emotional support and attachment ties. Or, staying connected and emotionally supported too often meant remaining dependent or not having their own voice, views, or feelings distinct from those of the people they were with. Thus, the termination phase provides these clients with an important opportunity to further resolve the separateness–relatedness dialectic, and significantly enhance their capacity both for greater autonomy and intimacy in the future.

Therapists offer many clients a far reaching reparative experience when clients are assured of the therapist's support for both sides of their feelings. That is, many clients benefit greatly from finding that the therapist takes pleasure in their success and welcomes their independence yet is still available and interested in their well-being if the client needs to recontact them (or another therapist) in the future. With this support for both sides of the separateness–relatedness dialectic, clients can further internalize the therapist and keep all that he has offered, solidify their own sense of self and personal efficacy, and successfully end the relationship.

Unfortunately, most graduate student therapists do not have the satisfaction of seeing many cases evolve to a natural close. Instead, therapy ends before the client is finished for a number of reasons. As discussed above, sometimes the client initiates the termination and simply does not return to the next scheduled appointment. At other times, therapy ends because the graduate student must move on to another placement, the school year ends, or the training clinic provides only time-limited services. Sometimes treatment ends when the therapist and client have made substantial gains; at other times, the treatment process may be only in midstream. In any case, significant therapeutic gains can still be made during the termination phase. To accomplish this more difficult task, however, it is essential that the therapist and client address and work through the complex issues that an unnatural or externally imposed ending arouses. Because graduate student therapists often have to initiate terminations before clients are ready, we will focus on **unnatural endings** in the discussion that follows. *It is necessary to examine unnatural or imposed endings closely because they routinely evoke angry, blaming, and distancing reactions in the client that, in turn, activate guilt, defensiveness, and other countertransference reactions in the therapist that can undo significant gains that have been made.*

It is often difficult for both the therapist and the client to end their relationship. This occurs because the ending stirs up emotional reactions that are far more intense and significant than either the therapist or the client had anticipated: for example, evoking loved ones who have died; divorces where a child loses their relationship with a parent in the aftermath of the breakup; important relationships that ruptured and could not be repaired; emotional deprivation and other lasting disappointments with, and unremitting longing for more from, a parent or caretaker; multiple family relocations; attachment figures who became emotionally unavailable because they became alcohol or drug dependent; parental anxiety and awkward disengagement as a child enters puberty; and so forth. As a result, *therapists and clients regularly **collude** to deny the reality of the impending ending and to avoid the difficult feelings it arouses for both of them.* When therapists and clients stumble in this way and do not address the ending squarely, their interpersonal process will often reenact clients' core conflicts and prevent them from resolving their problems as fully as they could have. Clearly, this becomes especially important with clients who have a history of painful and unresolved interpersonal losses, which so many clients bring to treatment. In particular, clients who received little validation or support for previous interpersonal losses now need a new and reparative type of ending with the therapist that differs from the unacknowledged, unpredictable, unplanned, and unpaced losses in their pasts (Many, 2009). Thus, *the single most important guideline for negotiating a successful termination is to unambiguously acknowledge and sensitively address the reality of the ending.* Even with clients who are ready to terminate treatment, old issues of loss and abandonment are unexpectedly but commonly evoked as termination approaches.

EFFECTIVE THERAPIST: We have three more sessions left before the school year ends and we have to stop. Let's talk about what this ending is going to mean for both of us.

The therapist and client can then discuss all of the client's positive and negative reactions about the ending, *especially the client's emotional reactions toward the therapist*. Although the therapist and client may both want to avoid this topic, the therapist cannot let that happen. Once the therapist and client mutually decide that the client is ready to terminate, or that the therapist must terminate because of outside constraints, they need to establish the specific date for the final session and then explore the client's reactions to ending their relationship. For example:

THERAPIST: Our last session will be in three weeks, on Thursday, December 9, at 4:00. What are your thoughts about that?

Because the therapist and client need time to work through their ending, the final session should be scheduled at least two or more weeks hence. If the client announces, out of the blue, that he has decided to terminate today and this is his last session, the therapist should actively encourage the client to return just one more time in order to sort through what this ending might mean for the client and to have a better opportunity to say good-by to each other. We always want to support the client's individuation and independence but this type of abrupt and unilateral ending is often acting out and avoiding issues.

In some cases, it will be difficult for the therapist to set a specific termination date and discuss the ending because this arouses the therapist's own conflicts over the ending—or more commonly, endings in general. Setting a clear termination date will also often bring to the fore the client's presenting symptoms and core conflicts again. For some clients, it may be necessary to rework aspects of these issues once more. That is, it is not uncommon for some clients to temporarily retreat from the changes they have made when termination becomes a reality, or to miss the next session (in particular, therapists may need to share with certain clients that they are concerned that they might "forget" the last session and will want to talk through this unwanted possibility with them).

For all of these reasons, *therapists need to acknowledge the termination forthrightly*, count down the final sessions, and repeatedly invite clients to discuss their reactions to the termination:

- After today, we will have only three sessions left. How is it for you to hear me say that?
- We have only two sessions left now, and I think it's important that we talk together about our ending. What comes up for you when I acknowledge that soon we are going to have to stop working together?
- Next week will be our last session. We've accomplished some things together, but problems remain and our work is unfinished. What do you find yourself thinking about our relationship and the time we have spent together?
- This is the last time we are going to be able to meet. I'm feeling sad about that, and I'm wondering what kind of feelings you're having.

This type of **countdown** precludes any ambiguity, bargaining, or denial that clients may have about the ending. For example, it will prevent the client from asking, at the end of the last session:

CLIENT: So, will I be seeing you next week?

It also keeps the therapist from acting out and avoiding the separation as well—for example, by saying at the end of the last session:

INEFFECTIVE THERAPIST: Oh, I see our time is up. Well, I guess that's it.

When therapists do not address the termination directly they are often acting out their own separation anxieties or separation guilt. They may have their own unresolved feelings from unwanted and painful endings in their own lives that were not responded to supportively and they had to negotiate on their own—which is problematic because they dovetail so often with their clients' own unresolved feelings over past painful endings that were not dealt with. Other types of countertransference–transference dovetail during the termination phase as well, such as clients who were parentified and now are avoiding this topic because they sense it is difficult for the therapist to address the termination. In sum, unresolved conflicts over ending relationships may be the most common countertransference issue that therapists bring to treatment. When this occurs, therapists can reassure themselves that this is a common and understandable problem to have, *but they also need to do something about it by seeking consultation from a supervisor or colleague.*

Because both clients' transference distortions and therapists' countertransference reactions are especially likely to occur around unnatural terminations, and at the end of time-limited treatments, we need to explore these issues further. Both therapists and clients often have experienced painful endings with significant others as "just happening to them." In many cases, they:

- were not prepared in advance for the separation,
- did not understand when or why this particular ending was occurring, or
- were not able to participate in the leave-taking by discussing it with the departing person or by saying good-bye.

Such problematic endings have left many therapists and clients feeling powerless in regard to some of the most important experiences in their lives. In contrast, the approach suggested here enhances clients' efficacy by allowing them to be active, informed participants in the ending. Clients know when the ending is occurring (experience some predictability), have the safety to express their feelings (whether appreciative, disappointed, or angry), receive a nondefensive response, and have input on future challenges. Why is this so important? Especially with unnatural endings, they often activate early maladaptive schemas and leave the client feeling that this relationship has just failed in the same way that others have in the past. Let's explore this expectable problem more fully.

As the old saying goes, children are supposed to grow up and put their feet under their own table, but it's hard to get up from the table when you're still hungry. The emotional deprivation that many clients bring to therapy is activated when treatment ends prematurely or before they feel ready to stop. Routinely, *clients' anger and disappointment about what they have wanted but missed in formative relationships—even their sense of abandonment or betrayal—may be directed toward the therapist,* no matter how clearly the treatment parameters were explained at the beginning of treatment. Clients' guilt-inducing accusations can be very difficult to hear for new and experienced therapists alike:

- You're just like my mother. You're leaving me, too.
- Our relationship never meant much to me, anyway.
- Nothing's really changed for me in therapy. This was kind of just a waste of time.

The success of treatment in these cases depends in part on *therapists' ability to remain nondefensive and tolerate clients' protests*—rather than feel guilty, become defensive, and make ineffective attempts to talk clients out of their feelings:

INEFFECTIVE THERAPIST: But you *have* changed. You're not as depressed as you used to be, and now you can do _____, and before you couldn't do that.

Instead of these defensive, futile attempts, therapists need to begin by acknowledging clients' feelings:

EFFECTIVE THERAPIST: You're really angry at me as our relationship is ending, and I respect those important feelings. Right now, it just seems like nothing meaningful has occurred and that nothing has gotten better for you.

By accepting clients' angry protests in these ways—without agreeing that their factual content is true—therapists give clients a reparative response that doesn't match their schemas or previous experiences. In turn, this affirmation of their current frustration or disappointment may allow clients to connect their strong emotional reactions to the developmental figures who originally disappointed or left them. Thus, *what therapists often find hardest about unnatural endings—and about time-limited treatment—is managing their own guilt or defensiveness in the face of clients' anger over the ending.* However, by drawing on the guidelines outlined here, the therapist can help clients see how this ending is in fact different in some important ways from other unwanted endings in their past—even though it feels as if the same unwanted scenario is happening to them again. Therapists can help make the termination of therapy different from past problematic endings and, in doing so, help resolve this **unfinished grief work** from the past in the following ways:

- talking with clients about the ending in advance;
- inviting clients to share their angry, disappointed, or sad feelings and being able to accept those feelings without becoming defensive;
- talking with clients about the meaning the termination holds for them, and clarifying how it is similar to and different from other important endings in their lives;
- validating clients' experience by acknowledging the ways that this ending may be evoking other painful or unwanted endings with others;
- discussing the meaning this relationship has held for the therapist and sharing some of the therapist's own feelings about ending this relationship; and
- ensuring that therapists and clients can say good-bye to each other.

In other previous endings that have been problematic, clients were not able to have such experiences with the departing person. Although most clients will have difficulty recognizing these differences initially, the therapist offers clients an opportunity to resolve long-standing problems by helping them differentiate this type of mutually acknowledged ending from the incomplete or unsatisfying endings with which they have had to cope in the past. In this way, successfully managing the termination also becomes the prototype for learning how to cope effectively with future endings and losses. One of the illuminating pioneers in short-term treatment modalities, Mann (1973) writes informatively about these issues and the strong feelings of loss that are activated for so many clients during termination.

In the face of clients' withdrawal or blaming accusations that their termination is "just like" prior losses and disappointments, therapists, too, may lose sight of the real differences that exist between this termination and clients' problematic past endings. In that case, therapists may accept the clients' blame, feel guilty, and avoid the ending—or invalidate the clients' sad, disappointed, or angry feelings by trying to talk clients out of them. In that case, paradoxically, therapists make the clients' accusations come true: They metaphorically reenact important aspects of the clients' past conflicted endings. Therapists do better by acknowledging the similarity of the clients' feelings at termination and at past separations, and then exploring with clients what they can do together to make this termination different and better from such problematic endings in the past.

During the termination period, clients often have a range of sad, hopeful, grateful, and other feelings toward the therapist. As therapists track clients' reactions to the ending, they will often see how clients' core conflicts are simultaneously being aroused in two contexts. First, clients' primary feelings about termination (such as guilt over feeling better and no longer needing the therapist; feeling rejected, abandoned, or let down by the therapist; and so forth) may match the clients' feelings about the formative relationship in which the core conflict originally arose. Second, clients' reaction toward termination may also match the feelings that were aroused in the crisis situation that originally brought them to treatment. By responding affirmingly to the clients' feelings, and clarifying their connection to these two sources when appropriate, the therapist is helping them further resolve their core conflicts.

Finally, to help with all of this challenging material, Marx and Gelso (1987) emphasize three main steps in effective terminations:

1. looking back and reviewing what has changed;
2. looking ahead and making realistic plans for coping with the problems that are likely to come up; and
3. saying good-bye.

Exploring effective terminations further, a useful intervention for the termination phase is to go through a **Review-Predict-Practice sequence** with all clients. Therapists review progress, accomplishments, successful changes and transitions, and unfinished issues with clients. Especially important, therapists help clients predict the challenging events or anxiety-arousing situations that are likely to activate old wounds or bring up their dysfunctional coping patterns in the future. To help clients maintain the changes they have achieved, it is essential to anticipate these potentially problematic situations or relational scenarios. Therapists want to identify these activating circumstances or triggers as specifically as possible. As illustrated with Tracy, earlier in this chapter, the therapist might ask:

THERAPIST: What is your dad likely to say if you bring this up? How is that going to make you feel, and what are you likely to do at that moment?

Finally, practice in dealing with these **triggering issues** or familiar patterns also prepares clients to succeed on their own. Therapists best can provide this by role-playing effective and ineffective responses to the threatening or unwanted interactions that clients are likely to face in the future.

To sum up, the therapist's goal is to address the client's feelings about ending the relationship and help the client work through them. This is difficult to do, however, when both the therapist and the client may wish to avoid the ending—as so often occurs. The therapist and client often have shared much in their work together and, hopefully, the relationship has grown to hold real meaning for both of them. To avoid the conflicts that are activated by the termination, clients—and therapists, too—often deny or try to diminish the importance of the ending. As previously noted, some clients may try to devalue the significance of the relationship, others may diminish the importance of the work that has been accomplished, and others will become symptomatic again and anxiously communicate that they are still too troubled to terminate and make it on their own. A few clients will ask the therapist if they can continue the relationship by becoming friends (Client at last session: Wanna meet at Starbucks next week?). These common reactions during the termination phase keep clients from facing current and past losses, preclude clients from receiving a new and more reparative response from the therapist, and keep clients from being able to internalize the therapist and carry forth with others the gains they have made in treatment.

ENDING THE RELATIONSHIP

Although some clients will reexperience their presenting problems at the time of termination, many will not. Both clients and therapists still need to talk about the ending, however. For example, therapists can talk with the client about the different ways in which they have seen the client change over the course of treatment. Therapists can also acknowledge the limitations of their work together and the unfinished issues that clients will need to continue to work with on their own or perhaps with another therapist at some point in the future. Fearing that they are being disloyal, many clients will need explicit permission from the therapist to work with another therapist if they have problems in the future. The therapist needs to actively reassure these clients that the therapist would be happy to know that they are taking good care of themselves and are getting the help they need. Recollections of close moments, awkward misunderstandings, risks ventured, and humorous incidents also can be shared. This also is a time when therapists can share some of their own feelings about the client including the significance of this particular relationship for the therapist.

Finally, in natural and unnatural endings, we have seen that termination brings the therapist and client back to the separateness–relatedness dialectic. In the case of natural endings, the therapist's intention is to give clients permission to leave.

Clients need to know with absolutely no uncertainty that the therapist enjoys their success, is pleased by their independence, and takes pleasure in seeing them become committed to new relationships and activities. That is, the therapist wants to help the client celebrate and feel supported as he transitions out of treatment into a new phase of life. The client:

- having connected with the therapist, can now feel more competent or confident about relating to others ("I have had a connected relationship here and can continue to engage in connecting with others");
- having struggled with emotional or behavioral dysregulation and needed help with this, can now self-regulate with more balance or flexibility ("I feel better able to manage my life," "I can think more compassionately about myself and others," "I can be prudent rather than impulsive in my behavior," "I no longer view the world in all or none terms," and so forth); and
- having arrived at an understanding that change is also an opportunity, can view this "loss" as signaling the transition to the next stage in life (in much the same way leaving home to attend college consists of both "loss" and "success"—so that feeling nervous or sad, as well as feeling excited and delighted, both make sense).

Per client response specificity, some clients will find it helpful to be reassured that they can have future contact with the therapist if the need arises; others will not need or want this. Knowing that the therapist is available if needed but also supports their decision to leave, clients are able to keep and apply the help the therapist has provided. This type of natural termination supports clients' own strivings for individuation in a supportive relational context, and it empowers clients to further claim their own authentic voice.

A thoughtful termination is extremely helpful in clients' ability to trust in the possibility that future relationships can be nurturing and stable. The model presented by the therapist is important—it conveys that stable, emotionally nurturing, and attuned relationships do exist, and that these can end in caring, manageable, and affirming ways. Significantly, addressing termination overtly is also an opportunity to observe mutually the clients' "developmental progression" and how much more connected she now is to her own voice and authentic self or how much more confident she may now be about engaging with others.

To recap, many clients appreciate the affirmation that they are quite ready to transition to the next phase on their own, yet they welcome knowing they can return should the need arise. However, some clinicians, and graduate student therapists in particular, often cannot be available in this way because endings with their clients are of necessity unnatural. In these cases, the finality of the ending must be honestly acknowledged. Therapists, with the help of a supportive supervisor, want to anticipate their own personal reactions to ending with this particular client, and be prepared to manage their own countertransference reactions effectively. In particular, many new therapists will need assistance from their supervisors in order to respond acceptingly, rather than defensively, to the client's angry, disappointed, dismissing, and other defensive or self-protective reactions. The unfinished business that remains also needs to be realistically acknowledged, and the therapist may need to help the client find a new therapist to finish the work that was begun.

When the therapist can do this, clients will be able to accept the limitations of the relationship without having to reject the therapist or undo the work that was accomplished. Although treatment is often incomplete when the relationship comes to an end, when clients can live out with others what they have experienced with the therapist, and therapists can appreciate what they have learned from and given to the client, both will have been enriched by this relationship.

CLOSING

This book has followed the course of therapy from beginning to end. Just as clients still have unresolved problems when treatment ends, many questions about therapy still remain unanswered for therapists-in-training. Challenging personal issues and complex interpersonal processes have been introduced here, and therapists will need further reading, supervision, and experience to help with the myriad exceptions that occur in every therapeutic relationship. However, the reader has learned essential guidelines for understanding the interaction taking place in therapeutic relationships, and for using the process dimension to help clients change. The conceptual framework presented here will help new therapists work more effectively with their clients and, hopefully, better integrate training from a variety of theoretical perspectives.

The theme of this text is that enduring change occurs within the context of significant relationships. Therapists working within cognitive and behavioral approaches, interpersonal and psychodynamic frameworks, and family systems and other theoretical traditions can utilize the interpersonal process approach to make their own work more effective. Accordingly, therapists are encouraged to use themselves and the relationships they provide to help clients change.

SUGGESTED READINGS

1. In the Student Workbook at the end of Chapter 4, readers were encouraged to complete the self-assessment scale, "Interpersonal Process Skills: Assessing Competencies." Having completed this text, students may wish to assess themselves again on these basic clinical skills. In addition, readers are encouraged to read the case study of Dorothy (a parentified child), from the *Wizard of Oz*, in Chapter 10 of the Student Workbook.

2. Student therapists who have found helpful the ideas presented here are also likely to enjoy *Helping Skills: Facilitating Exploration, Insight, and Change* (Hill, 2009), and for more experienced trainees, *Interpersonal Reconstructive Therapy* (Benjamin, 2003). Additionally, *Between Therapist and Client* (Kahn, 1997), *Relational Concepts in Psychoanalysis* (Mitchell, 1988), and *Time-limited Dynamic Therapy* (Levenson, 1995) provide informative presentations of interpersonally oriented therapy.

3. Helpful information about identifying clients' core conflicts and how they can be reactivated during the termination phase of treatment can be found in Mann's (1973) classic primer, *Time-Limited Psychotherapy*, and in Chapter 8 ("Termination") of Binder's (2004) *Key Competencies in Brief Dynamic Psychotherapy: Clinical Practice Beyond the Manual.*

4. Much useful information about successful terminations can be found in Wachtel (2002), "Termination of Therapy: An Effort at Integration," *Journal of Psychotherapy Integration* 12 (3), 373–83. Also, M. M. Many (2009), "Termination as a Therapeutic Intervention When Treating Children Who Have Experienced Multiple Losses." *Infant Mental Health Journal* 30 (1), 23–39. This article provides informative ways to conceptualize and intervene with traumatized children and their caregivers. She highlights the importance of providing these families with "a new model for loss" (p. 23).

5. The Dream is an important psychological dimension in people's lives that has received too little attention. The reader is encouraged to read Daniel Levinson's (1978) groundbreaking book, *The Seasons of a Man's Life*; see especially pages 91–109. *The Seasons of a Woman's Life* (2000), published soon after Levinson's death, tracks developmental stages and transitions in women's lives. The Dream, as it relates to women, is given special attention on pages 238–39.

PROCESS NOTES

Client_____ Therapist_____ Session Number_____ Date_____

1. **Issues presented.** Key concerns expressed by client or noted by therapist (e.g., what is the client distressed about today?).

   _____
   _____
   _____

2. **Themes.** Repetitive patterns or issues observed (are there concerns that the client presents recurrently, such as feeling diminished, unappreciated, taken advantage of, etc.?).

   _____
   _____
   _____

3. **Relationship and Resistance.** Did the client make any overt statements or covert references (embedded messages) about you or your interaction together (e.g., difficulty trusting, not feeling heard or understood, and so forth)? Did the client express any ambivalence about attending sessions or concerns about any aspect of being in treatment (e.g., question if treatment is helpful, if she is being disloyal to family or spouse by talking to you, etc.)?

   _____
   _____
   _____

4. **Interpersonal Process.** What can you learn about your client's problematic interactions with others from your own personal reactions toward the client (e.g., do you feel bored, frustrated, protective, etc.)? Are there ways in which your interaction (and your internal response to the client) is similar to what the

client elicits in others or in some way parallels what tends to go wrong with others in the client's life?

_____

_____

_____

5. **Transference and Countertransference.** Who are you in this client's life? That is, are there characteristic ways in which the client tends to distort or misperceive you and others? Does this client respond to you as he does to significant others—or differently? Are there ways in which your own personal history or current life circumstances are activated by this client and influence how you respond to him?

_____

_____

_____

6. **Evaluating Interventions.** What were your primary interventions (e.g., empathic listening, reframing, role-playing, etc.)? How did the client respond to your interventions—what was effective and ineffective? Looking ahead, what might you want to do differently next time?

_____

_____

_____

7. **Treatment Focus.** What do you see as the key issues or concerns that are central to this client's problems? What do you want to focus on and explore further?

_____

_____

_____

8. **Critical incidents.** Identify potential legal, ethical, or reporting concerns for consultation with supervisor (e.g., child, elder, and dependent abuse; danger to self and others; etc.)

_____

_____

_____

GUIDELINES FOR TREATMENT PLANNING

Although writing a case formulation is often challenging, completing one is important because it helps therapists provide more focused treatments and helps them respond to treatment planning requests from HMOs and other third-party insurers.

1. **Formulation of the problem(s).** Summarize the client's presenting problem(s) and suggest an initial diagnosis. Why is the client seeking treatment now? What has the client done to cope in the past and what has helped with these problems? How does the client tend to feel about herself (for instance, worthless, superior, burdened), and what does this feeling tend to evoke in you and others? What organizing themes or patterns characterize the client's most important experiences with others?

2. **Treatment focus.** What has treatment focused on to date? Identify the maladaptive patterns (including style of relating to others, pathogenic beliefs, cognitive schemas, and emotional conflicts) that recur for the client and contribute to his presenting problems. For example, thinking of the client's internal working models, ask yourself the following questions: (1) What does the client want from others (e.g., to be cared for, protected, given support to pursue her own interests, etc.)? (2) What does the client expect from others (e.g., to be criticized, idealized, abandoned, etc.)? (3) What is the client's experience of self in relationship to others (e.g., a sense of being unimportant or disconnected, burdensome, exploited, etc.)? (4) What conflicted feelings are typical for this client (e.g., fear, shame, guilt, anger, etc.)? (5) What interpersonal coping strategies are characteristic for this client (e.g., to comply and go along, to withdraw and convey no need for others, to dominate and try to control, and so on)? (6) What do these coping strategies tend to elicit from the therapist and others (e.g., disinterest, advice-giving)? (7) What stressors, crises, and social supports does the client have? Use your responses to these questions to identify two or three primary treatment focuses that clarify what's wrong and needs to change for this client.

3. **Developmental context.** How did the client's problems originally come about? Identify developmental experiences (individual, familial, cultural), crises or ongoing stressors, and social supports that shaped the client's subjective

worldview and current conflicts. What were the client's experiences in the family? For example, how did they nurture (was love and support conditional?), discipline (were families able to set and follow through with limits?), establish generational boundaries and roles (was the child parentified and scripted to meet adult needs?), engage in abuse or neglect, and so forth? How have familial and sociocultural factors (e.g., ethnicity, religion, economic opportunities, gender, sexual orientation, physical disability, etc.) affected the client and shaped her development? In other words, how did the client's early maladaptive schemas and interpersonal coping strategies develop?

4. **Therapeutic process**. Describe the quality of the working alliance and how you and the client have interacted together. In what ways is your interaction together parallel to her experiences with others (i.e., reenacting), and how is it different (corrective or reparative)? What automatic thoughts and pathogenic beliefs, distressing emotional states, and maladaptive relational patterns are evident in this client's life, and how have these been expressed in the therapeutic relationship and addressed in your work together? In what ways does the client engage in "tests" with you to confirm or disconfirm past hurtful experiences that have led to her current problems? When the client is distressed, how does he tend to misperceive and react to you? Have you been able to address potential problems and misunderstandings between you and restore ruptures in the working alliance?

5. **Goals and interventions**. As you focus on the two to three core treatment issues identified in item 2 above, indicate how you plan to address these. Articulate as clearly as possible the specific experiences (affective, cognitive, behavioral) that this client needs in order to meet the treatment goals identified. That is, what reparative experiences does this client need to live out with the therapist, and what might be needed to help this client translate new experiences with you to relationships outside the therapy setting? How would you evaluate the appropriateness of the goals set and interventions proposed given the client's personal, familial, and cultural context? Consider also whether referrals might be appropriate (i.e., group counseling, possible need for medication, parenting education, and so forth). In other words, based on where you want to go with this client, how do you plan to get there?

6. **Impediments to change**. What client characteristics are most likely to impede your ability to sustain a strong working alliance with this client? If the client terminates prematurely or treatment is not successful, anticipate the factors that could be contributing. Are there any significant personal, familial, or cultural factors that may make it difficult for this client to acknowledge having problems or to ask for help? Are you able to talk with the client regularly, and in depth, about the way you interact and work together, her reactions toward you and the treatment process, and what you could do to make treatment more helpful? Are you able to help the client identify interpersonal styles that are no longer adaptive, and help her become more flexible in her current coping strategies? If not, what factors are contributing to the impasse? Most importantly, suggest how the client's maladaptive relational patterns with others could be subtly reenacted with you in the therapeutic relationship, especially when the client's problems or interpersonal style activates your own countertransference issues.

REFERENCES

Ackerman, S., & Hilsenroth, M. (2003). A review of therapist characteristics and techniques positively impacting the therapeutic alliance, *Clinical Psychology Review*, 23, 1–33.

Akhtar, S. (2007). Diversity without fanfare: Some reflections on contemporary psychoanalytic technique. *Psychoanalytic Inquiry* 27: 690–704.

Alexander, F., & French, T. M. (1946). *Psychoanalytic therapy*. New York: Ronald Press.

Allen, J., & Fonagy, P. (2006). *Handbook of mentalization-based treatment*. Sussex, Eng.: Wiley.

Allen, J., Fonagy, P., & Bateman, A. (2008). *Mentalizing in clinical practice*. Arlington, VA: American Psychiatric Publishing.

American Psychological Association (2003). Guidelines on multicultural education, training, research, practice and organizational change for psychologists. *American Psychologist* 58: 377–402.

American Psychological Association Presidential Task Force on Evidence-Based Practice, (2003). Evidence-based practice in psychology. *American Psychologist* 61: 271–85.

Amrhein, P., Miller, W., Yahne, C., & Palmer, M. (2003). Client commitment language during motivational interviewing predicts drug use outcomes. *Journal of Consulting and Clinical Psychology* 71: 862–78.

Anchin, J., & Kiesler, D. (1982). *Handbook of interpersonal psychotherapy*. New York: Pergamon Press.

Anderson, C. M., & Steward, S. (1983). *Mastering resistance: A practical guide to family therapy*. New York: Guilford Press.

Anderson, T., Ogles, B. M., Patterson, C. L., & Lambert, M. J. (2009). Therapist effects: Facilitative interpersonal skills as a predictor of therapist success. *Journal of Consulting and Clinical Psychology* 65: 755–68.

Andrews, J. (1991). *The active self in psychotherapy: An interpretation of therapeutic styles*. New York: Gardner.

Andrusyna, T. P., Luborsky, L., Pham, T., & Tang, T. Z. (2006). The mechanisms of sudden gains in supportive-expressive psychotherapy for depression. *Psychotherapy Research* 16: 526–35.

Angus, L., & Hardtke, K. (2006). Insight and story change in brief experiential therapy for depression: An intensive narrative process analysis. In L. Castonguay & C. Hill (Eds.), *Insight and psychotherapy*. (pp. 187–207). Washington, DC: American Psychological Association.

Angus, L., & Kagan, F. (2007). Empathic relational bonds and personal agency in psychotherapy: Implications for psychotherapy supervision, practice, and research. *Psychotherapy: Theory, Research, Practice, Training* 4: 371–77.

Angus, L., Lewin, J., Bouffard, B., & Rotondi-Trevisan, D. (2004). "What's the story?" Working with narrative in experiential psychotherapy. In L. Angus & J. McLeod (Eds.), *The handbook of narrative and psychotherapy: Practice, theory and research*. (pp. 87–101). Thousand Oaks, CA: Sage.

Archineiga, G. M., & Anderson, T. C. (2008). Toward a fuller conception of machismo: Development of a traditional machismo and

455

caballerismo scale. *Journal of Counseling Psychology* 55 (1): 19–33.

Aron, L. (1996). *A meeting of minds: Mutuality in psychoanalysis.* Hillsdale, NJ: Analytic Press.

———. (1999). Clinical choices and the relational matrix. *Psychoanalytic Dialogue* 9: 1–29.

Bachelor, A. (1995). Clients' perception of the therapeutic alliance: A qualitative analysis. *Journal of Counseling Psychology* 42: 323–27.

Balcom, D. (1991). Shame and violence: Considerations in couples' treatment. *Journal of Independent Social Work* 5: 165–81.

Balcom, D., Lee, R., & Tager, J. (1995). The systemic treatment of shame in couples. *Journal of Marital and Family Therapy* 21 (1): 55–65.

Baldwin, S., Wampold, B., & Imel, Z. (2007). Untangling the alliance-outcome correlation: Exploring the relative importance of therapist and patient variability in the alliance. *Journal of Consulting and Clinical Psychology* 75 (6): 842–52.

Bandura, A. (1977). Self-efficacy: Toward a unifying theory of behavioral change. *Psychological Review* 84 (2): 191–215.

———. (1982). Self-efficacy mechanisms in human agency. *American Psychologist* 37: 122–47.

———. (1997). *Self-efficacy: The exercise of control.* New York: Freeman.

———. (2006). Toward a psychology of human agency. *Perspectives on Human Science* 1: 164–80.

Barber, J. P. (2007). Issues and findings in investigating predictors and moderators of psychotherapy outcome: Introduction to the specialization. *Psychotherapy Research* 17: 131–36.

Barber, J. P., Luborsky, L., Gallop, R., Crits-Christoph, P. (2001). Therapeutic alliance as a predictor of outcome and retention in the national institute on drug abuse collaborative cocaine treatment study. *Journal of Consulting and Clinical Psychology* 69: 119–24.

Barber, J., Connolly, M., Crits-Christoph, P., Gladis, L. (2009). Alliance predicts patients' outcome beyond in-treatment change in symptoms. *Personality Disorders: Research, Theory and Treatment* 1: 80–89.

Barkham, M., & Shapiro, D. (1986). Counselor verbal response modes and experienced empathy. *Journal of Counseling Psychology* 33 (1): 3–10.

Bartholomew, K., & Horowitz, L. M. (1991). Attachment styles among young adults: A test of a four-category model. *Journal of Personality and Social Psychology* 61: 226–44.

Bateson, G. (1972). *Steps to an ecology of mind.* New York: Dutton.

Baumeister, R. F., & Scher, S. J. (1988). Self-defeating behavior patterns among national individuals: Review and analysis of common self-destructive tendencies. *Psychological Bulletin* 104: 3–22.

Baumrind, D. (1967). Child care practices anteceding three patterns of preschool behavior. *Genetic Psychology Monographs* 75: 43–88.

———. (1971). Current patterns of parental authority. *Developmental Psychology Monograph* 4 (1, pt. 2).

———. (1983). Familial antecedents of social competence in young children. *Psychological Bulletin* 94 (1): 132–42.

———. (1991). The influences of parenting style on adolescent competence and substance use. *Journal of Early Adolescence* 11: 56–95.

Beavers, W. R. (1982). Healthy, midrange and severely dysfunctional families. In F. Walsh (Ed.), *Normal family processes* (pp. 45–66). New York: Guilford Press.

Beavers, W. R., & Hampson, R. B., (1993). Measuring family competence: The Beavers Systems Model. In F. Walsh (Ed.), *Normal family processes* (2nd ed.). New York: Guilford Press.

Beck, A. (1967). *Depression: Causes and treatment.* Philadelphia: University of Pennsylvania Press.

———. (1976). *Cognitive therapy and the emotional disorders.* New York: International Universities Press.

———. (1995). *Cognitive therapy: Basics and beyond.* New York: Guilford Press.

Beck, A., Freeman, A., & Davis, D., (2003). *Cognitive therapy of personality disorders.* New York: Guilford Press.

Beckham, E. E. (1992). Predicting patient dropout in psychotherapy. *Psychotherapy: Theory, Research, Practice, Training* 29: 177–182.

Beebe, B., Knoblauch, S., Rustin, J., Sorter, D., Jacobs, T. J., & Polly, R. (2005). *Forms of intersubjectivity in infant research and adult treatment.* New York: Other Press.

Beebe, B., Jaffe, J., & Lachman, F. (2004). A dynamic systems view of communication. In J. Auerbach, K. Levy, & C. Schaffer (Eds.), *Relatedness, self definition and mental representation: Essays in honor of Sidney Blatt.* New York: Routledge.

Beebe, B., & Lachman, F. (2002). *Infant research and adult treatment: Co-constructions interactions.* Hillsdale, NJ: Analytic Press.

Beitman B. D., & Soth, A. M. (2006). Activation of self-observation: A core process among the psychotherapies. *Journal of Psychotherapy Integration* 16: 383–397.

Bender, D. (2005). The therapeutic alliance in the treatment of personality disorders. *Journal of Practical Psychiatry and Behavioral Health* 11 (2): 73–87.

Bender, H. L., Allen, J. P., McElhaney, K. B., Antonishak, J. (2007). Use of harsh physical discipline and developmental outcomes in adolescence. *Development and Psychopathology* 19: 227–42.

Benjamin, J. (2009). A relational psychoanalysis perspective on the necessity of acknowledging failures and containing features of the intersubjective relationship (the shared third). *The International Journal of Psychoanalysis* 9 (3): 441–450.

Benjamin, L. (1987a). Use of the SASB dimensional model to develop treatment plans for personality disorders: I. Narcissism. *Journal of Personality Disorders* 1: 43–70.

———. (1987b). Commentary on the inner experience of the borderline self-mutilator. *Journal of Personality Disorders* 1: 334–39.

———. (1993). Every psychopathology is a gift of love. *Psychotherapy Research* 3: 1–24.

———. (2003). *Interpersonal diagnosis and treatment of personality disorders* (2nd ed.). New York: Guilford Press.

———. (2003). *Interpersonal reconstructive therapy*. New York: Guilford Press. Bergin, A. E. (1997).

———. Neglect of the therapist and the human dimensions of change: A commentary. *Clinical Psychology: Science and Practice* 4: 83–89.

Bergin, A., & Garfield, S. (Eds.). (1994). *Handbook of psychotherapy and behavior change* (4th ed.). New York: Wiley.

Berk, L. E. (2005). Why parenting matters. In S. Olfman (Ed.), *Childhood lost: How American culture is failing our kids*. Westport, CT: Praeger.

Berk, M., Berk, L., & Castle, D. (2004). A collaborative approach to the treatment alliance in bipolar disorder. *Bipolar Disorder* 6: 504–18.

Berlin, L., Zeanah, C., & Lieberman, A. (2005). Prevention and intervention programs for supporting early attachment security. In P. Shaver & J. Cassidy (Eds.), *Handbook of attachment theory and research* (2nd ed.). New York: Guilford Press.

Bernier, A. T., & Dozier, M. (2002). The client-counselor match and the connective emotional experience: Evidence from interpersonal and attachment research. *Psychotherapy: Theory, Research, Practice, Training* 31 (1): 32–43.

Beutler, L. E. (1997). The psychotherapist as a neglected variable in psychotherapy: An illustration by reference to the role of therapist experience and training. *Clinical Psychology: Science and Practice* 4: 44–52.

Beutler, L. E., Moleiro, C., & Talebi, H. (2002). Resistance in psychotherapy: What conclusions are supported by research? *Journal of Clinical Psychology* 58: 207–17.

Binder, J. L. (2004). *Key competencies in brief dynamic psychotherapy*. New York: Guilford Press.

Binder, J., & Strupp, H. (1997). "Negative process": A recurrently discovered and underestimated facet of therapeutic process and outcome in the individual psychotherapy of adults. *Clinical Psychology: Science and Practice* 4: 121–39.

Bischoff, M. M., & Tracey, T. J. (1995). Client resistance as predicted by therapist behavior: A study of sequential dependence. *Journal of Counseling Psychology* 42: 487–95.

Bjorgvinsson, T., & Hart, J. (2006). Cognitive behavioral therapy promotes mentalizing. In J. G. Allen & P. Fonagy (Eds.), *Handbook of mentalization-based treatment*. Sussex: Wiley.

Blatt, S. J., Sanislow, C. A., Zuroff, D., & Pilkonis, P. (1996). Characteristics of the effective therapist: Further analysis of the data from the NIMH TDCRP. *Journal of Consulting and Clinical Psychology* 64: 1276–84.

Blatt, S. J., & Zuroff, D. C. (2005). Empirical evaluation of the assumptions underlying evidence-based treatment in mental health. *Clinical Psychology Review* 25: 459–86.

Bloom, M., Fischer, J., & Orme, J. G. (2003). *Evaluating practice: Guidelines for the accountable professional*, (5th ed.). Englewood Cliffs, NJ: Prentice-Hall.

Bohart, A., Elliott, R., & Greensberg, L. (2002). In J. C. Norcross (Ed.), *Psychotherapy relationships that work* (pp. 89–108). New York: Oxford University Press.

Bordin, E. S. (1979). The generalizability of the psychoanalytic concept of the working alliance. *Psychotherapy* 16: 252–60.

———. (1994). Theory and research on the therapeutic working alliance: New directions. In A. O. Horvath & L. S. Greenberg (Eds.), *The working alliance* (pp. 13–37). New York: Wiley.

Boszormenyi-Nagy, I., & Spark, G. (1973). *Invisible loyalties: Reciprocity in intergenerational family therapy*. New York: Harper & Row.

Boszormenyi-Nagy, I., Grunebaum, J., & Ulrich, D. (1991). Contextual therapy. In A. S. Gurman & D. P. Knishern (Eds.), *Handbook of family therapy* (vol. 51). New York: Bruner/Mazel.

Bowen, M. (1966). The use of family theory in clinical practice. *Comprehensive Psychiatry* 7: 345–76.

———. (1978). *Family therapy in clinical practice*. New York: Aronson.

Bowlby, J. (1973). *Attachment and loss*. Vol. 2, *Separation: Anxiety and anger*. New York: Basic Books.

———. (1988). *A secure base*. New York: Basic Books.

Brammer, L. M., & MacDonald, G. (1996). *The helping relationship: Process and skills* (6th ed.). Boston: Allyn & Bacon.

Brennan, K., Clark, C., & Shaver, P. (1998). Self-report measurements of adult attachment: An integrative overview. In J. A. Sampson & W. F. Rholes (Eds.), *Attachment theory and close relationships* (pp. 46–76). New York: Guilford Press.

Breuer, J., & Freud, S. (1955). Studies on hysteria. *Standard Edition* (vol. 2, pp. 1–319). London: Hogarth Press. (Original work published 1893–95).

Bridges, M. (2006). Activating the corrective emotional experience. *Journal of Clinical Psychology* 62 (5): 551–68.

Brisch, K. H. (2002). *Treating attachment disorders: From theory to therapy.* New York: Guilford Press.

Bromberg, P. (2006). *Awakening the dreamer.* Mahwah, NJ: Analytic Press.

Brogan, M., Prochaska, J. O., Prochaska, J. M. (1999). Predicting termination and continuation status in psychotherapy using the transtheoretical model. *Psychotherapy* 36: 105–13.

Burns, D., & Auerbach A. (1996). Therapeutic empathy in cognitive-behavioral therapy: Does it really make a difference? In P. Salkovskis (Ed.), *Frontiers of Cognitive Therapy* (pp. 135–64). New York: Guilford Press.

Callahan, J., & Hynan, M. (2005). Models of psychotherapy outcome: Are they applicable in training clinics? *Psychological Services*, 2, 65–69.

Callahan, J. L., Swift, J. K., & Hynan, M. T. (2006). Test of the phase model of psychotherapy in a training clinic. *Psychological Services* 3: 129–36.

Callahan, P. E. (2000). Indexing resistance in short-term dynamic psychotherapy: Change in breaks in eye contact during anxiety. *Psychotherapy Research* 10: 87–99.

Cardemil, E. V., & Battle, C. L. (2003). Guess who's coming to therapy? Getting comfortable with having conversations about race and ethnicity in psychotherapy. *Professional Psychology: Research and Practice* 34: 278–86.

Carkhuff, R. R. (1999). *The art of helping in the 21st century* (8th ed.). Amherst, MA: Human Resource Development Press.

Cartwright, D. (2004). Anticipatory interpretations: Addressing "cautionary tales" and the problematic premature termination. *Bulletin of the Menninger Clinic* 68 (2): 95–114.

Cashdan, S. (1988). *Object Relations Therapy: Using the relationship.* New York: Norton.

Cassidy, J., & Shaver, P. R. (2008). *Handbook of attachment: Theory, research, and clinical applications* (2nd ed.). New York: Guilford Press.

Castonguay, L., & Beutler, L. (Eds.). (2006). *Principles of therapeutic change that work.* New York: Oxford University Press.

Catley, D., Harris, K. J., Mayo, M. S., & Hall, S. (2006). Adherence to principles of motivational interviewing and client within-session behavior. *Behavioral and Cognitive Psychotherapy* 34: 43–56.

Cervone, D. (2000). Thinking about self-efficacy. *Behavior modification* 24: 30–56.

Cervone, D., Artistico, D., & Berry, J. (2006). Self-efficacy and adult development. In C. Hoare (Ed.), *Handbook of adult development and learning.* New York: Oxford University Press.

Colin, V. L. (1996). *Human attachment.* New York: McGraw-Hill.

Comas-Diaz, L. (2006). Cultural variation in the therapeutic relationship, In C. D. Goodheart, A. E. Kazdin, & R. J. Steinberg (Eds.). *Evidence-based psychotherapy: Where practice and research meet.* Washington, DC: American Psychological Association.

Connolly, M. B., Crits-Christoph, P., Shappell, S., & Barber, J. P. (1999). Relation of transference interpretations to outcome in the early sessions of brief supportive-expressive psychotherapy. *Psychotherapy Research* 9: 485–95.

Constantine, M. (Ed.). (2007). *Clinical practice with people of color: A guide to becoming culturally competent.* New York: Teachers College Press.

Constantine, M., & Sue, D. W. (2005). Strategies for building multicultural competence in mental health and education settings. Hoboken, NJ: Wiley.

Cooper, A. M. (Ed.). (2006). *Contemporary psychoanalysis in America: Leading analysts present their work.* Washington, DC: American Psychiatric Publishing.

Cooper, G., Hoffman, K., Powell, B., & Marvin, R. (2005). *Enhancing early attachments.* New York: Guilford Press.

Cowan, C., & Cowan, P. (2005). Two central roles for couple relationships: Breaking negative intergenerational patterns and enhancing children's adaptation. *Sexual and Relationship Therapy*, 20 (3): 275–88.

Crits-Christoph, P. & Barber, J. P. (Eds.). (1991). *Handbook of short-term dynamic psychotherapy.* New York: Basic Books.

Crits-Christoph, P., & Gibbons, B. B. (2002). Relational interpretations. In J. C. Norcross (Ed.), *Psychotherapy relationships that work: Therapist contributions and responsiveness to patients* (pp. 285–300). New York: Oxford University Press.

Crits-Christoph, P., Siqueland, L., Blain, J., & Frank, A. (1999). Psychosocial treatments for cocaine dependence: National Institute on Drug Abuse Collaborative Cocaine Treatment Study. *Archives of General Psychiatry* 56: 493–502.

Curtis, J. M. (1981) Indications and contradictions in the use of therapist's self disclosure. *Psychological Reports* 49: 499–507.

Curtis, J. T., & Silberschatz, G. (1997). Plan formulation method. In T. D. Eells (Ed.), *Handbook of psychotherapy case formulation* (pp. 116–36). New York: Guilford Press.

Curtis, J. T., Silberschatz, G., Sampson, H., Weiss, J., & Rosenberg, S. E. (1988). Developing reliable psychodynamic case formulations: An illustration of the plan diagnosis method. *Psychotherapy* 25: 256–65.

Daly, K., & Mallinckrodt, B. (2009). Experienced therapists' approach to psychotherapy for adults with attachment avoidance or attachment anxiety. *Journal of Counseling Psychology* 56 (4): 549–63.

Davanloo, H. (Ed.). (1980). *Short-term dynamic psychotherapy*. New York: Aronson.

deRoten, Y., Drapeau, M., Stigler, M., & Despland, J. N. (2004). Yet another look at the CCRT: The relation between core conflict relational themes and defense functioning. *Psychotherapy Research* 14: 252–60.

Dew, S. E., & Bickman, L. (2005). Client expectancies about therapy. *Mental Health Services Research* 17: 21–33.

DiCaccavo, A. (2006). Working with parentification: Implications for clients and counseling psychologists. *Psychology and Psychotherapy: Theory, Research and Practice* 79: 469–78.

Diener, M. J., Hilsenroth, M. J., & Weinberger, J. (2007). Therapist affect focus and patient outcomes in psychodynamic psychotherapy: A meta-analysis. *American Journal of Psychiatry* 164: 936–41.

Dollard, J., & Miller, N. (1950). Personality and psychotherapy: An analysis in terms of learning, thinking, and culture. New York: McGraw-Hill.

Domenech-Rodriguez, M., Donovick, M., & Crowley, S. (2009). Parenting styles in a cultural context: Observations of protective parenting in first-generation Latinos. *Family Process* 48 (2): 195–210.

Dozier, M., Cue, K. L., & Bernett, L. (1994). Clinicians as caregivers: Role of attachment organization in treatment. *Journal of Counseling and Clinical Psychology* 62: 793–800.

Dozier, M., & Tyrrell, C. (1998). The role of attachment in therapeutic relationships. In J. A. Simpsons and W. S. Rholes (Eds.), *Attachment theory and close relationships* (pp. 221–28). New York: Guilford Press.

Draguns, J. (2007). Psychotherapeutic and related interventions for global psychology. In M. J. Stevens (Ed.), *Toward a global psychology: Theory, research, intervention, and pedagogy* (pp. 233–66). Mahwah, NJ: Erlbaum.

Dutton, D. G. (1995). *The batterer: A psychological profile*. New York: Basic Books.

Eagle, M., & Wolitzky, D. (2009). Adult Psychotherapy from the Perspectives of Attachment Theory and Psychoanalysis, in J. Obegi & E. Berant (Eds.), *Attachment Theory and Research in Clinical Work with Adults*, NY: Guilford.

Ecker, B., & Hulley, L. (1996). *Depth-oriented brief therapy*. San Francisco: Jossey-Bass.

Egan, G. (2002). *The skilled helper* (7th ed.). Pacific Grove, CA: Brooks/Cole.

Elliott, R., Greenberg, L. S., & Lietaer, G. (2003). Research on experiential psychotherapies. In M. Lambert, A. Bergin, & S. Garfield (Eds.), *Handbook of psychotherapy and behavior change* (5th ed.). New York: Wiley.

Elliott, R., Shapiro, D. A., Firth-Cozens, J., Stiles, W. B., Hardy, G. E., Llewelyn, S. P., & Margison, F. R. (1994). Comprehensive process analysis of insight events in cognitive-behavioral and psychodynamic-interpersonal psychotherapies. *Journal of Counseling Psychology* 41: 449–63.

Elliott, R., Watson, J., Goldman, R., & Greenberg, L. (2004). *Learning emotion-focused therapy: The process experiential approach to change*. Washington, DC: American Psychological Association.

Ellis, A. (1999). *How to make yourself happy and remarkably less disturbable*. San Luis Obispo, CA: Impact Publishers.

Ellis, A., & Greiger, R. (1996). *What is rational-emotive therapy?* Baltimore: John Hopkins University Press.

Engel, D., & Arkowitz, H. (2006). *Ambivalence in psychotherapy: Facilitating readiness to change*. New York: Guilford Press.

Engel, L., & Ferguson, T. (1990). *Hidden guilt*. New York: Pocket Books.

Erikson, E. (1968). *Childhood and society* (2nd ed.). New York: Norton.

Eron, L. D., & Huesmann, L. R. (1990). The stability of aggressive behavior: Even unto the third generation. In M. Lewis & S. M. Miller (Eds.), *Handbook of developmental psychopathology* (pp. 147–55). New York: Plenum.

Falender, C., & Shafransky, E. (2004). *Clinical supervision: A competency-based approach*. New York: Guilford Press.

Faoud, N., & Arredondo, P. (2007). *Becoming culturally oriented: Practical advice for psychologists and educators*. Washington, DC: American Psychological Association.

Farber, B. A. (2006). *Self-disclosure in psychotherapy*. New York: Guilford Press.

Farber, B. A., Berano, K. C., & Capobianco, J. A. (2004). Clients' perceptions of the process and consequences of self-disclosure in psychotherapy. *Journal of Counseling Psychology* 51: 340–46.

Farber, B., & Geller, J. (1994). Gender and representation in psychotherapy. *Psychotherapy* 31: 318–26.

Farber, B. A., Khurgin-Bott, R., & Feldman, S. (2009). The benefits and risks of patient self-disclosure in the psychotherapy of women with a history of childhood sexual abuse. *Psychotherapy: Theory, Research, Practice, Training* 46 (1): 52–67.

Farber, B. A., & Metzger, T. A. (2009). The therapist as secure base. In J. H. Obegi & E. Berant (Eds.), *Attachment theory and research in clinical work*. New York: Guilford Press.

Farley, J. (1979, January). Family separation-individuation tolerance: A developmental conceptualization of the nuclear family. *Journal of Marital and Family Therapy*, 61–67.

Field, T. (1996). Attachment and separation in young children. *Annual Review of Psychology* 47: 541–61.

Finkelhor, D. (1990). Early & long-term effects of child sexual abuse: An update. *Professional Psychology: Research and Practice*, 21 (5), 325–330.

Fiori, K., Consedine, N., & Magai, C. (2009). Late life attachment in context: Patterns of relating among men and women from seven ethnic groups. *Journal of Cross Cultural Gerontology* 24 (2): 121–41.

Fitzpatrick, M. R., Stalikas, A., & Iwakabe, S. (2001). Examining counselor interventions and client progress in the context of the therapeutic alliance. *Psychotherapy: Theory, Research, Practice, Training* 38: 160–70.

Fonagy, P. (2000). Attachment and borderline personality disorder. *Journal of the American Psychoanalytic Association* 48 (4): 1129–47.

Fonagy, P. (2002). *What works for whom?: A critical review of treatments for children and adolescents*. New York: Guilford Press.

Fonagy, P., Gergeley, G., Jurist, E. J., & Target, M. (2002). *Affect regulation, mentalization, and the development of the self*. New York: Other Press.

Fonagy, P., Leigh, T., Steel, M., & Kennedy, R. (1996). The relation of attachment status, psychiatric classifications, and response to psychotherapy. *Journal of Consulting and Clinical Psychology* 64: 22–31.

Fonagy, P., & Target, M. (1997). Attachment and reflective function: Their role in self-organization. *Developmental Psychopathology* 9: 679–700.

————. (2002). Early intervention and the development of self-regulation. *Psychoanalytic Inquiry* 22: 307–35.

————. (2006). The mentalization focused approach to self pathology. *Journal of Personality Disorders* 20 (6): 544–76.

Fragoso, J. M., & Kashubeck, S. (2000). Machismo, gender role conflict, and mental health in Mexican American men. *Psychology of Men and Masculinity* 1 (2): 87–97.

Frailberg, S., Adelson, E., & Shapiro, V. (1975). Ghosts in the nursery: A psychoanalytical approach to the problems of impaired mother–infant relationships. *Journal of the American Academy of Child Psychiatry* 14: 387–421.

Friedman, R. C., Garrison, W. B., Bucci, W., & Gorman, B. S. (2005). Factors effecting change in private psychotherapy patients of senior psychoanalysts: An effectiveness study. *Journal of the American Academy of Psychoanalysis and Dynamic Psychiatry* 33: 583–610.

Fromm, E. (1982). *Escape from freedom*. New York: Avon.

Fromm-Reichmann, F. (1960). *Principles of intensive psychotherapy*. Chicago: University of Chicago Press.

Fuendeling, J. M. (1998). Affect regulation as a stylistic process within adult attachment. *Journal of Social and Personal Relationships* 15: 291–322.

Gabbard, G. O. (2007). Unconscious enactments in psychotherapy. *Psychiatric Annals* 37: 268–75.

Galassi, J. P., & Bruch, M. A. (1992). Counseling with social interaction problems: Assertion and social anxiety. In S. D. Brown & R. W. Lent (Eds.), *Handbook of counseling psychology* (pp. 753–91). New York: Wiley.

Garfield, S. L. (1994). Research on client variables in psychotherapy. In A. E. Bergin & S. L. Garfield (Eds.), *Handbook of psychotherapy and change* (4th ed., pp. 190–228). New York: Wiley.

Garfield, S. L. (1997). The therapist as a neglected variable in psychotherapy research. *Clinical Psychology: Science and Practice* 4: 40–43.

Gaston, L. (1990). The concept of alliance and its role in psychotherapy. In A. E. Bergin & S. L. Garfield (Eds.), *Handbook of psychotherapy and behavior change* (4th ed., pp. 190–228). New York: Wiley.

Geller, J. D. (2005). Style and its contribution to a patient-specific model of therapeutic technique. *Psychotherapy: Theory, Research, Practice, Training* 21: 469–82.

Gellner, P. (2006). The concept of objective countertransference and its role in a two-person psychology. *American Journal of Psychoanalysis* 66: 25–42.

Gelso, C. J. (2002). The real relationship: The "something more" of psychotherapy. *Journal of Contemporary Psychology* 32 (1): 35–40.

Gelso, C. J., & Carter, J. (1994). Components of the psychotherapy relationship: Their interaction and unfolding during treatment. *Journal of Counseling Psychology* 41 (3): 296–306.

Gelso, C. J., & Hayes, H. (1998). *The psychotherapy relationship: Theory, research, and practice*. New York: Wiley.

————. (2007). *Countertransference and the therapist's inner experience: Perils and possibilities*. Mahwah, NJ: Erlbaum.

Gelso, C. J., Latts, M. C., Gomez, M. J., & Fassinger, R. E. (2002). Countertransference management and therapy outcome: An initial evaluation. *Journal of Clinical Psychology* 58: 861–67.

Gelso, C. J., & Samstag, L. (2008). A tripartite model of the therapeutic relationship in psychotherapy. In S. Brown & R. Lent (Eds.), *Handbook of counseling psychology* (4th ed., pp. 267–83). New York: Wiley.

Gendlin, E. T. (1996). *Focusing-oriented psychotherapy: A manual of the experiential method*. New York: Guilford Press.

———. (1998). *Focusing-oriented psychotherapy*. New York: Guilford Press.

Gilbert, P. (2009). Developing a compassion based approach in cognitive behavioral therapy. In G. Gimos (Ed.), *Cognitive behavioral therapy: A guide for the practicing clinician* (pp. 205–20). New York: Routledge.

Gilbert, P., & Procter, S. (2006). Compassionate mind training for people with high shame and self criticism. *Clinical Psychology and Psychotherapy* 13: 353–76.

Gill, M. (1994). *Psychoanalysis in transition: A personal view*. Hillsdale, NJ: Analytic Press.

Gill, M., & Muslin, H. (1976). Early interpretations of transference. *Journal of the American Psychoanalytic Association* 24: 779–94.

Gilligan, C. (1982). *In a different voice: Psychological theory and women's development*. Cambridge, MA: Harvard University Press.

Gilligan, J. (2009). Sex, gender, & violence: Estela Weldon's contribution to our understanding of the psychopathology of violence, *British Journal of Psychotherapy*, 25(2), 239–256.

Ginot, E. (2001). The holding environment and intersubjectivity. *The Psychoanalytic Quarterly* 70 (2): 417–46.

Goldenberg, I., & Goldenberg, H. (2008). *Family therapy: An overview* (7th ed.). Pacific Grove, CA: Brooks/Cole.

Goldfried, M. R. (1980). Toward the delineation of therapeutic change principles. *American Psychologist* 35 (11): 991–99.

———. (2004). Integrating interactively oriented brief psychotherapy. *Journal of Psychotherapy Integration* 14 (1): 93–105.

———. (2006). Cognitive-affect-relational-behavior therapy. In G. Stricker & J. Gold (Eds.), *A casebook of psychotherapy integration*. Washington, DC: American Psychological Association.

Goldfried, M. R., Raue, P. J., & Gastonguay, L. G. (1998). The therapeutic focus in significant sessions of master therapists: A comparison of cognitive-behavioral and psychodynamic-interpersonal interventions. *Journal of Consulting and Clinical Psychology* 66: 803–10.

Gollwitzer, P. M., & Schaal, B. (1998). Metacognition in action: The importance of implementation intentions. *Personality and Social Psychology Review* 2: 124–36.

Gottman, J., & DeClair, J. (2001). *The Relationship Cure*. New York: Three Rivers Press.

Gottman, J., & Gottman, S. (2004). *Principles for making your marriage work*. New York: Crown.

Gottman, J., & Siluer, N. (1999). *The seven principles for making marriage work*. New York: Wiley.

Greenberg, J., & Mitchell, S. (1983). *Object relations in psychoanalytic theory*. Cambridge, MA: Harvard University Press.

Greenberg, L. S. (2002). *Emotion-focused therapy: Coaching clients to work through their feelings*. Washington, D.C.: American Psychological Association.

Greenberg, L. S., & Goldman, R. N. (2008). *Emotion-focused couples therapy: The dynamics of emotion, love, and power*. Washington, DC: American Psychological Association.

Greenberg, L. S., & Paivio, S. (2003). *Working with emotions in psychotherapy*. New York: Guilford Press.

Greenberg, L. S., Rice, L. N., & Elliott, R. (1993). *Facilitating emotional change: The moment-by-moment process*. New York: Guilford Press.

Greenberg, L. S., & Safran, J. D. (1987). *Emotion in Psychotherapy*. New York: Guilford Press.

Greenson, R. (1965). The working alliance and the transference neurosis. *Psychoanalytic Quarterly* 34: 155–81.

Greenson, R. (1967). *The technique and practice of psychoanalysis*. Vol. 1. New York: International Universities Press.

Groth, M. (2008). Authenticity in existential analysis. *Existential Analysis* 19 (1): 81–101.

Guame, J., Gmel, G., & Daeppen, J. B. (2008). Brief alcohol interventions: Do counselors' and patients' communication characteristics predict change? *Alcohol and Alcoholism* 43: 62–69.

Guerin, P., Fogarty, T., Fay, L., & Kautto, J. (1996). *Working with relationship triangles: The one-two-three of psychotherapy*. New York: Guilford Press.

Gurman, A. S., & Messer, S. B. (2003). *Essential psychotherapies* (2nd ed.). New York: Guilford Press.

Gutheil, T., & Brodsky, A. (2008). *Preventing boundary violations in clinical practice*. New York: Guilford Press.

Guthrie, E., Moorey, J., Margison, F., & Barter, H. (1999). Cost effectiveness of brief psychodynamic-interpersonal therapy in high utilizes of psychiatric services. *Archives of General Psychiatry* 56: 519–26.

Guy, J. (1987). *The personal life of the psychotherapist*. New York: Wiley.

Haley, J. (1967). Toward a theory of pathological systems. In G. H. Zuk & I. Boszormenyi-Nagy (Eds.), *Family therapy and disturbed families.* Palo Alto, CA: Science and Behavior Books.

————. (1980). *Leaving home: Therapy with disturbed young people.* New York: McGraw-Hill.

————. (1996). *Learning and teaching therapy.* New York: Guilford Press.

Hammond, R. T., & Nichols, M. P. (2008). How collaborative is structural family therapy? *Family Journal* 16: 118–24.

Hansen, N. B., Lambert, M. J., & Forman, E. M. (2002). The psychotherapy dose-response effect and its implications for treatment delivery services. *Clinical Psychology: Science and Practice* 9: 329–43

Hanson, R., & Spratt, E. (2002). Reactive attachment disorder: What we know about the disorder and implications for treatment. *Child Maltreatment* 5: 137–45.

Hardy, G., & Shapiro, D. (1987). Therapist verbal response mode in prescriptive vs. exploratory psychotherapy. *British Journal of Clinical Psychology* 24: 235–45.

Hartman, E. (1978, October). Using eco-maps and genograms in family therapy. *Social Casework*, 464–76.

Hatcher, R., & Barends, A. (2006). How a return to theory could help alliance research. *Psychotherapy: Theory, Research, Practice, Training* 43: 292–99.

Hayes, J. A., & Gelso, C. J. (2001). Clinical implications of research on countertransference: Science informing practice. *Journal of Clinical Psychology* 57: 1041–51.

Hendricks, M. N. (2002). Focusing-oriented/experiential psychotherapy. In D. J. Cain & J. Seeman (Eds.), *Humanistic psychotherapies: Handbook of research and practice* (pp. 221–52). Washington, DC: American Psychological Association.

Hendrix, H. (2001). *Getting the love you want: A guide for couples.* New York: Owl Books.

Henry, W., Schacht, T., & Strupp, H. (1990). Patient and therapist introject, interpersonal process, and differential psychotherapy outcome. *Journal of Consulting and Clinical Psychology* 58: 768–74.

Henry, W. & Strupp, H. (1994). The therapeutic alliance as interpersonal process. In A. O. Horvath & S. Greenberg (Eds.) *The Working Alliance: Theory, Research, and Practice*, Oxford, England: Wiley & Sons, pp. 51–84.

Henry, W., Strupp, H., Schacht, T., & Gaston, L. (1994). Psychodynamic approaches. In A. E. Bergin & S. L. Garfield (Eds.), *Handbook of psychotherapy and behavior change* (4th ed., pp. 467–508). New York: Wiley.

Hesse, E., & Main, M. (1999). Second generation effects of unresolved trauma in non-maltreating parents: Dissociated, frightened and threatening parental behavior. *Psychoanalytic Inquiry* 19: 481–540.

Hill, C. (2009). *Helping Skills: Facilitating exploration, insight, and action.* (3rd ed.). Washington, DC: American Psychological Association.

Hill, C. E., & O'Brien, K. M. (1999). *Helping skills: Facilitating exploration, insight, and action.* Washington, DC: American Psychological Association.

Hill, C. E., & O'Grady, K. (1985). List of therapist intentions illustrated in a case study and with therapists of varying theoretical orientations. *Journal of Counseling Psychology* 32 (1): 3–22.

Hill, C. E., Kellems, I., Kolchakian, M., & Wonnel, T. (2003). The therapist experience of being the target of hostile versus suspected-unasserted client anger: Factors associated with resolution. *Psychotherapy Research* 13: 475–91.

Hill, C. E., & Knox, S. (2002). Self-disclosure. In J. C. Norcross (Ed.), *Psychotherapy relationships that work: Therapist contribution and responsiveness to patient needs* (pp. 249–59). Oxford, Eng.: Oxford University Press.

Hill, C. E., & Knox, S. (2009). Processing the therapeutic relationship. *Psychotherapy Research*.

Hill, C. E., Knox, S., Hess, S., & Crook-Lyon, R. (2007). The attainment of insight in the Hill dream model: A single case study. In L. Castonguay & C. E. Hill, *Insight in psychotherapy* (pp. 207–30). Washington, DC: American Psychological Association.

Hill, C. E., Thompson, B. J., Cogar, M. M., & Denman, D. W. (1993). Beneath the surface of long-term therapy: Client and therapist report of their own and each other's covert processes. *Journal of Counseling Psychology* 40: 278–88.

Hill, C. E., Thompson, B. J., & Corbett, M. M. (1992). The impact of therapist ability to perceive displayed and hidden client reactions on immediate outcome in first sessions of brief therapy. *Psychotherapy Research* 2: 143–55.

Hill, C. E., Sim, W., Spangler, P., Stahl, J., Sullivan, C., & Teyber, E. (2008). Therapist immediacy in brief psychotherapy: Case study II. *Psychotherapy Theory, Research, Practice, Training* 45 (3): 298–315.

Hilsenroth, M. J. (2007). A programmatic study of short-term psychodynamic psychotherapy: Assessment, process, outcome, and training. *Psychotherapy Research* 17: 31–45.

Hintikka, U., Laukkanen, E., Marttunen, M., & Lehtonen, J. (2006). Good working alliance and psychotherapy are associated with positive changes in cognitive performance among adolescent psychiatric inpatients. *Bulletin of the Menninger Clinic* 70: 316–35.

Hobson, R. F. (1984). *Forms of feeling: The heart of psychotherapy.* London: Tavistock.

Hoffman, K. T., Marvin, R. S., Cooper, G., & Powell, B. (2006). Changing toddlers' and preschoolers' classifications: The circle of security intervention. *Journal of Consulting and Clinical Psychology* 74 (6): 1017–26.

Horner, A. J. (2005). *Dealing with resistance in psychotherapy.* Lantham, MD: Aronson.

Horney, K. (1966). *Our inner conflicts.* New York: Norton.

———. (1970). *Neurosis and human growth.* New York: Norton.

Horowitz, M. (1976). *Stress response syndromes.* Northvale, NJ: Jason Aronson.

Horowitz, M., Marmar, C., Krupnick, J., Wilner, N., Kaltreider, N., & Wallerstein, R. (1984). *Personality styles and brief psychotherapy.* New York: Basic Books.

Horowitz, M., Rosenberg, S. E., & Bartholomew, K. (1993). Interpersonal problems, attachment styles, and outcome in brief dynamic psychotherapy. *Journal of Consulting and Clinical Psychology* 61: 549–60.

Horvath, A. O. (2000). The therapeutic relationship: From transference to alliance. *Journal of Clinical Psychology* 56: 163–73.

Horvath, A. (2006). The alliance in context: Accomplishments, challenges, and future directions, *Psychotherapy: Theory, Research, and Practice,* 43(3), 258–263.

Horvath, A. O., & Bedi, R. P. (2002). The alliance. In J. C. Norcross (Ed.), *Psychotherapy relationships that work: Therapist contributions and responsiveness to patients* (pp. 37–69). New York: Oxford University Press.

Horvath, A. O., & Breensberg, L. S. (1989). Development and validation of the Working Alliance Inventory. *Journal of Counseling Psychology* 36: 223–33.

Horvath, A., Gaston, L., & Luborsky, L. (1993). The therapeutic alliance and its measures. In N. E. Miles, L. Luborsky, J. P. Barber, & J. P. Docherty (Eds.), *Psychodynamic treatment research* (pp. 247–73). New York: Basic Books.

Horvath, A. O., & Greenberg, L. S. (1994b). *The working alliance: Theory, research, and practice.* New York: Wiley.

Huntsinger, C., & Jose, P. (2009). Relations among parental acceptance and control and children's social adjustment in Chinese American and European American families. *Journal of Family Psychology* 23 (3): 321–30.

Imel, Z., & Wampold, B. (2008). The importance of treatment and the science of common factors in psychotherapy. In S. D. Brown, & R. W. Lent (Eds.), *Handbook of counseling psychology* (4th ed.). Hoboken, NJ: Wiley.

Ivey, A. E., & Ivey, M. B. (1999). *Intentional interviewing and counseling* (4th ed.). Pacific Grove, CA: Brooks/Cole.

Jacobsen, B. (2007). Authenticity and our basic existential dilemmas. *Existential Analysis* 18 (2): 289–97.

James, R. K., & Guilliland, B. (2000). *Crisis intervention strategies* (4th ed.). Monterey, CA: Wadsworth.

Johnson, B., Taylor, E., D'elia, J., Tzanetos, T., Rhodes. R., & Geller, J. D. (1995). The emotional consequence of therapeutic misunderstandings. *Psychotherapy Bulletin* 30: 139–49.

Jouhnson, S. M. (2009). Attachment theory and emotionally focused therapy for individuals and couples: Perfect partners. In J. H. Obegi & E. Berant (Eds.), *Attachment theory and research in clinical work with adults.* New York: Guilford Press.

Jones, A. S., & Gelso, C. J. (1988). Differential effects of style of interpretation: Another look. *Journal of Counseling Psychology* 35: 363–69.

Jones, J. M. (1995). *Affects as process: An inquiry into the centrality of affect in psychological life.* Hillsdale, NJ: Analytic Press.

Jordan, J., Kaplan, A., Miller, J., Striver, P., & Surrey, J. (1991). *Women's growth in connection* (pp. 126–84). New York: Guilford Press.

Joyce, A., Piper, W., Ogrodniczuk, J., & Klein, R. (2007). *Termination in psychotherapy: A psychodynamic model of processes and outcomes.* New York: Guilford Press.

Kafka, F. (1966). *Letter to his father.* New York: Schocken Books.

Kahn, M. (1997). *Between therapist and client.* New York: Freeman.

Karen, R. (1992, February). Shame. *Atlantic Monthly,* 40–70.

———. (1998). *Becoming attached.* New York: Oxford University Press.

Kasper, L., Hill, C. E., & Kivligham, D. (2008). Therapist immediacy in brief psychotherapy: Case study I. *Psychotherapy: Theory, Research, Practice, Training,* 45, 281–287.

Kaufman, G. (1989). *The psychology of shame.* New York: Springer.

Kelly, A. E. (1998). Clients' secret keeping in outpatient therapy. *Journal of Counseling Psychology* 45: 50–57.

Kerr, M. E., & Bowen, M. (1988). *Family evaluation: An approach based on Bowen theory.* New York: Norton.

Kelly, G. (1963). *The psychology of personal constructs.* New York: Norton.

Kiesler, D. J. (1966). Some myths of psychotherapy research and the search for a paradigm. *Psychological Bulletin* 65: 110–36.

———. (1988). *Therapeutic metacommunication: Therapist impact* disclosure as feedback in psychotherapy. Palo Alto, CA: Consulting Psychologists Press.

————. (1996). *Contemporary interpersonal theory and research: Personality, psychopathology, and psychotherapy.* New York: Wiley.

Kiesler, D., & Van Denburg, T. (1993). Therapeutic impact disclosure: A last taboo in psychoanalytic theory and practice. *Clinical Psychology & Psychotherapy* 1 (1): 3–13.

Kiesler, D., & Watkins, L., (1989). Interpersonal complementary and the therapeutic alliance: A study of relationship in psychotherapy. *Psychotherapy* 26 (2): 183–94.

Kirsh, B., & Tate, E. (2006). Developing a comprehensive understanding of the working alliance in community mental health. *Qualitative Health Research* 16: 1054–74.

Kitzmann, K. (2000). Effects of marital conflict on subsequent triadic family interacting and parenting. *Developmental Psychology* 36: 3–13.

Klerman, G. L., Weissman, M., Rounsaville, B., & Chevron, E. (1984). *Interpersonal psychotherapy of depression.* New York: Basic Books.

Kobak, R., Cassidy, J., & Lyons-Ruth, K. (2006). Attachment stress and psychopathology: A developmental pathways model. *Developmental Psychopathology* 2 (1): 357–88.

Kohut, H. (1971). *The analysis of the self.* New York: International Universities Press.

————. (1977). *The restoration of the self.* New York: International Universities Press.

Kring, A., & Sloan, D. (2009). *Emotion regulation and psychopathology.* New York: Guilford Press.

Krishnakumar, A., & Buehler, C. (2000). Interparental conflict and parenting behaviors: A meta-analytic review. *Family Relations* 49: 25–44.

Lafferty, P., Beutler, L. F., & Crago, M. (1991). Differences between more and less effective psychotherapists: A study of select therapist variables. *Journal of Consulting and Clinical Psychology* 57: 76–80.

Laing, R. D., & Esterson, A. (1970). *Sanity, madness and the family.* Middlesex, Eng.: Penguin.

Lambert, M., & Barley, D. (2002). Research on the therapeutic relationship and psychotherapy outcome. In J. Norcross (Ed.), *Psychotherapy relationships that work: Therapist contributions and responsiveness to patients* (pp. 17–22). New York: Oxford University Press.

Lambert, M., & Ogles, B. (2004). The efficacy and effectiveness of psychotherapy. In M. J. Lambert (Ed.), *Bergin and Garfield's handbook of psychotherapy and behavior change* (5th ed., pp. 139–93). New York: Wiley.

Lazarus, A. (1973). Multimodal behavior therapy: Treating the BASIC I.D. *Journal of Nervous and Mental Disease* 156: 404–11.

————. (1993). Tailoring the therapeutic relationship, or being an authentic chameleon. *Psychotherapy* 30: 404–7.

Lazarus, A. O. (1989). *The practice of multi-modal therapy.* Baltimore: Johns Hopkins University Press.

Lazarus, A., Lazarus, C. & Fey, A. (1993). *Don't believe it for a minute!: Forty toxic ideas that are driving you crazy.* San Luis Obispo, CA: Impact.

Levant, R. F. (2005). *Report of the 2005 presidential task force on evidence based practice.* Washington, DC: American Psychological Association.

Levenson, E. (1982). Language and healing. In S. Slip (Ed.), *Curative factors in dynamic psychotherapy.* New York: McGraw-Hill.

————. (1995). *Time-limited dynamic therapy.* New York: Basic Books.

————. (1998). *Time-limited dynamic psychotherapy: Making every session count.* Video and viewer's manual. San Francisco: LIFT.

————. (2003). Time-limited psychotherapy: An integrationist perspective. *Journal of Psychotherapy Integration* 13 (3, pt. 4): 300–33.

————. (2004). Time-limited dynamic psychotherapy: Formulation and intervention. In D. Mantosh (Ed.), *The art and science of brief psychotherapies: A practitioner's guide* (pp. 21–91). Washington, D.C.: American Psychiatric Publishing, Inc.

Levenson, H., & Strupp. H. (1997). Cyclical maladaptive patterns in time-limited dynamic psychotherapy. In T. D. Ells (Ed.), *Handbook of psychotherapy case formulation* (pp. 84–115). New York: Guilford Press.

Levinson, D. (1978). *The seasons of a man's life.* New York: Ballantine.

————. (2000). *The seasons of a woman's life.* New York: Knopf.

Levy, K. N., Clarkin, J. F., Yeomans, F. E., Scott, L. N., Wasserman, R. H., & Kernberg, O. F. (2006). The mechanisms of change in the treatment of borderline personality disorder with transference-focused psychotherapy. *Journal of Clinical Psychology* 62: 481–501.

Levy, K. N., Meehan, K. B., Weber, M., & Reynoso, J. (2005). Attachment and borderline personality disorder: Implications for psychotherapy. *Psychopathology* 38: 64–74.

Lewis, H. (1971). *Shame and guilt in neurosis.* New York: International Universities Press.

Lewis, M. (1995). *Shame: The exposed self*, NY: The Free Press.

Lieberman, A., & Van Horn, P. (2008). *Psychotherapy with infants and young children*. New York: Guilford Press.

Ligero, D. P., & Gelso, C. J. (2002). Countertransference, attachment, and the working alliance: The therapist's contribution. *Psychotherapy* 39: 3–11.

Linehan, M. (1992). *Cognitive-behavioral treatment of borderline personality disorder*. New York: Guilford Press.

Linehan, M. (1997). Validation and psychotherapy. In A. L. Bohart and L. S. Greenberg (Eds.) *Empathy reconsidered: New directions* (pp. 353–392), Washington, DC: American Psychological Association.

Liu, W. (2005). The working alliance, therapy ruptures and impasses, and counseling competence: Implications for counselor training and education. In R. T. Carter (Ed.), *Handbook of racial-cultural psychology and counseling* (vol. 2, pp. 148–67). Hoboken, NJ: Wiley.

Lohr, M., Olatunji, B., Baumeister, R., & Bushman, B. (2007). The psychology of anger venting and empirically supported alternatives that do no harm. *The Scientific Review of Mental Health Practice* 5: 53–64.

Lothane, Z. (2006). Reciprocal free association: Listening with the third ear as an instrument in psychoanalysis. *Psychoanalytic Psychology* 23: 711–27.

Luborsky, L. (1976). Helping alliances in psychotherapy. In J. L. Clanghorn (Ed.), *Successful psychotherapy*. New York: Brunner/Mazel.

———. (1984). *Principles of psychoanalytic psychotherapy: A manual for supportive-expressive treatment*. New York: Basic Books.

Luborsky, L., & Crits-Christoph, P. (1990). *Understanding transference: The CCRT method*. New York: Basic Books.

Luborsky, L., & DeRubeis, R. (1984). The use of psychotherapy treatment manuals: A small revolution in psychotherapy research style. *Clinical Psychology Review* 4 (5): 602–11.

Luborsky, L., & Marks, D. (1991). Short-term supportive-expressive psychoanalytic psychotherapy. In P. Crits-Cristoph & J. Barber (Eds.), *Handbook of short-term dynamic psychotherapy* (pp. 110–36). New York: Basic Books.

Luborsky, L., McLellan, A. T., Diguer, L., Woody, G., & Seligman, D. A. (1997). The psychotherapist matters: Comparison of outcomes across twenty-two therapists and seven patient samples. *Clinical Psychology* 4: 53–65.

Luborsky, L., McLellan, A. T., Woody, G. E., O'Brien, C. P., & Auerbach, A. (1985). Therapist success and its determinants. *Archives of General Psychiatry* 42, (6): 602–611.

Luborsky, L., et al. (1986). The nonspecific hypothesis of therapeutic effectiveness: A current assessment. *American Journal of Orthopsychiatry* 56: 501–12.

———. (1999) The researcher's own therapy allegiances: A "wild card" in comparisons of treatment efficacy. *Clinical Psychology: Science and Practice* 6: 95–106.

Lyons-Ruth, K. (1999). The two-person unconscious: Intersubjective dialogue, enactive relational representation, and the emergence of new forms of relational organization. *Psychoanalytic Inquiry* 19: 576–617.

Mahalik, J. R. (1994). Development of the Client Resistance Scale. *Journal of Counseling Psychology* 41: 58–68.

Malan, D. H. (1976). *The frontier of brief psychotherapy: An example of the convergence of research and clinical practice*. New York: Plenum.

Malik, M., Beutler, L., Alimohamed, S., & Gallagher-Thompson. (2003). Are all cognitive therapies alike? A comparison of cognitive and noncognitive therapy process and implications for the application of empirically supported treatments. *Journal of Counseling and Clinical Psychology* 71: 150–58.

Mallinckrodt, B. (2000). Attachment, social competencies, social support, and interpersonal process in psychotherapy. *Psychotherapy Research* 10: 239–66.

Mallinckrodt, B., Gantt, D., & Coble, H. (1995). Attachment patterns in the psychotherapy relationship: Development of the client attachment to therapist scale. *Journal of Counseling Psychology* 42: 307–17.

Mallinckrodt, B., Porter, M. J., & Kivlingham, D. M. (2005). Client attachment to therapist, depth of in-session exploration, and object relations in brief psychotherapy. *Psychotherapy: Theory, Practice, Training* 42: 85–100.

Mann, J. (1973). *Time-limited psychotherapy*. Cambridge, MA: Harvard University Press.

Mann, J., & Goldman, R. (1982). *A casebook in time-limited psychotherapy*. New York: McGraw-Hill.

Mantsios, G. (1998). Class in America: Myths and realities. In P. A. Rothenberg (Ed.), *Race, class, and gender in the United States: An integrated study* (4th ed., pp. 202–14). New York: St. Martin's Press.

Many, M. M. (2009). Termination as a therapeutic intervention when treating children who have experienced multiple losses. *Infant Mental Health Journal* 30 (1): 23–39.

Marmarosh, C., Gelso, C., Markin, R., & Majors, R. (2009). The real relationship in psychotherapy: Relationships to adult attachments, working alliances, transferences, and therapy outcomes. *Journal of Counseling Psychology* 56 (3): 37–50.

Martin, A., Buchheim, A., Berger, U., & Straus, B. (2007). The impact of attachment organization on potential countertransference reactions. *Psychotherapy Research* 17: 46–58.

Martin, D. J., Garske, J. P., & Davis K. D. (2000). Relation of therapeutic alliance with outcome and other variables: A meta-analytic review. *Journal of Counseling and Clinical Psychology* 68: 428–50.

Marvin, R., Cooper, G., Hoffman, K., & Powell, B. (2002). The Circle of Security project: Attachment-based intervention with caregiver-pre-school child dyads. *Attachment and Human Development* 4 (1): 107–24.

Marx, S. A., & Gelso, C. J. (1987). Termination of individual counseling a university counseling center. *Journal of Counseling Psychology* 34: 3–9.

Mash, E., & Hunsley, J. (1993). Assessment considerations in the identification of failing psychotherapy: Bringing the negatives out of the darkroom. *Psychological Assessment 5*: 292–301.

Masterson, J. (1972). *Treatment of the borderline adolescent: A developmental approach.* New York: Wiley.

———. (1976). *Psychotherapy of the borderline adult: A developmental approach.* New York: Brunner/Mazel.

Matsakis, A. (1998). *Managing client anger: What to do when a client is angry at you.* Oakland: New Harlinger Publications.

May, R. (1977). *The meaning of anxiety.* New York: Norton.

McCarthy, P. (1982). Differential effects of counselor self-referent responses and counselor status. *Journal of Counseling Psychology* 29: 125–31.

McClure, F., & Teyber, E. (2003). *Casebook in child and adolescent treatment: Cultural and familial contexts.* Pacific Grove, CA: Wadsworth.

McGoldrick, M., Gerson, R., & Shellenberger, S. (1999). *Genograms: Assessment and intervention.* New York: Norton.

McGoldrick, M., Gerson, R., & Petry, S. (2008). *Genograms: Assessments and interventions* (3rd ed.). New York: Norton.

McGoldrick, M., & Hardy, K. (2008). *Re-visioning family therapy: Race, culture and gender in clinical practice* (2nd ed). New York: Guildford Press.

McWhirter, E. H. (1996). *Counseling for empowerment.* Alexandria, VA: American Counseling Association.

McWilliams, N. (1999). *Psychoanalytic case formulation.* New York: Guilford Press.

Meissner, W. W. (2007). Therapeutic alliance: Theme and variations. *Psychoanalytic Psychology* 24: 231–54.

Mikulincer, M., & Shaver, P. (2003). The attachment behavioral system in adulthood: Activation, psychodynamics, and interpersonal processes. In M. P. Zanna (Ed.), *Advances in experimental social psychology* (vol. 35, pp. 53–152). New York: Academic Press.

———. (2007). *Attachment in adulthood.* New York: Guilford Press.

———. (2008). Adult attachment and object relations. In J. Cassidy & P. R. Shaver (Eds.), *Handbook of attachment.* New York: Guilford Press.

Mikulincer, M., Shaver, P., & Pereg, D. (2003). Attachment theory and affect regulation: The dynamics, development, and consequences of attachment-related strategies. *Motivation and Emotion* 27: 77–102.

Miller, A. (1949). *Death of a salesman.* New York: Penguin.

———. (1984). *Prisoners of childhood.* New York: Basic Books.

Miller, W. R., Benefield, R. G., & Tonigan, J. S. (1993). Enhancing motivation for change in problem with drinking: A controlled comparison of two therapist styles. *Journal of Counseling and Clinical Psychology* 61: 433–61.

Miller, W., & Rollnick, S. (2002). *Motivational interviewing: Preparing people for change.* New York: Guilford Press.

———. (2002). *Motivational interviewing* (2nd ed.). New York: Guilford Press.

Miller, W., & Rose, G. (2009). Toward a theory of motivational interviewing. *American Psychological Association* 64 (6): 527–37.

Millon, T. (1981). *Disorders of personality: DSM111, Axis 11.* New York: Wiley.

Millon, T. (1999). Reflections on psychosynergy: A model for intergrating science, theory, classification, assessment, and therapy. *Journal of Personality Assessment,* 72 (3), 437–456.

Millon, T., & Grossman, S. (2007). *Overcoming resistant personality disorders: A personalized psychotherapy approach.* Hoboken. NJ: Wiley.

Mills, J., Bauer, G., & Miars, R. (1989). Use of transference in short-term dynamic psychotherapy. *Psychotherapy* 26: 338–43.

Minuchin, S. (1974). *Families and family therapy.* Cambridge, MA: Harvard University Press.

———. (1984). *Family kaleidoscope.* Cambridge, MA: Harvard University Press.

Minuchin, S., Lee, W. Y., & Simon, G. (1996). *Mastering family therapy: Journeys of growth and transformation.* New York: Wiley.

Minuchin, S., & Nichols, M. P. (1998). Structural family therapy. In F. M. Datilio (Ed.), *Case studies in couple and family therapy: Systemic and cognitive perspectives*. New York: Guilford Press.

Minuchin, S., Nichols, M. P., & Lee, W. (2007). *Assessing couples and families: From symptom to system*. Boston: Allyn and Bacon.

Mitchell, S. (1988). *Relational concepts in psychoanalysis*. Cambridge, MA: Harvard University Press.

————. (1993). *Hope and dread in psychoanalysis*. New York: Basic Books.

————. (2000). *Rationality: From attachment to intersubjectivity*. Hillsdale, NJ: Analytic Press.

Mitchell, S., & Black, M. (1995). *Freud and beyond*. New York: Basic Books.

Moyers, T. B., Miller, W. R., & Hendrickson, S. M. L. (2005). How does motivational interviewing work? Therapist interpersonal skill predicts client involvement within motivational interviewing sessions. *Journal of Consulting and Clinical Psychology* 73: 590–98.

Moyers, T. B., & Martin, T. (2006). Therapist influence on client language during motivational interviewing sessions: Support for a potential causal mechanism. *Journal of Substance Abuse Treatment* 30: 245–51.

Moyers, T., Martin, T., Christopher, P., Houck, J., Tonigan, J., Amrhein, P. (2007). Client language as a mediator of motivationak interviewing efficacy: Where is the evidence? *Alcoholism: Clinical and Experimental Research* 31(Suppl. 3), 40–47.

Mueller, W. J., & Aniskiewics, A. S. (1986). *Psychotherapeutic intervention in hysterical disorders*. Northvale, NJ: Aronson.

Mueller, W., & Kell, B. (1972). *Coping with conflict: Supervising counselors and psychotherapists*. New York: Appleton-Century-Crofts.

Muran, J. C., Safran, J. D., Gorman, B. S., Samstag, L. W., Eubanks-Carter, C., & Winston, A. (2009). The relationship of early alliance ruptures and their resolution to process and outcome in three time-limited psychotherapies for personality disorders. *Psychotherapy: Theory, Research, Practice, Training* 46 (2): 233–48.

Muran, J. C., Safran, J. D., Samstag, L. W., & Winston, A. (2005). A comparative treatment study of personality disorders. *Psychotherapy: Theory, Research, Practice, Training* 42: 512–31.

Murray, C., & Waller, G. (2002). Reported sexual abuse and bulimic psychopathology among non-clinical women: The mediating role of shame. *International Journal of Eating Disorders* 32: 186–91.

Nathanson, D. L. (Ed.). (1987). *The many faces of shame* (pp. 246–70). New York: Guilford Press.

Najavits, L., & Strupp, H. (1994). Differences in the effectiveness of psychodynamic therapists: A process-outcome study. *Psychotherapy Research*, 31: 114–23.

Neimeyer, R. (2009). *Constructivist Psychotherapy: Distinctive Features*, NY: Routledge/Taylor & Francis Group.

Newman, C. (1998). Therapeutic and supervisory relationships in cognitive-behavioral therapies: similarities and differences. *Journal of Cognitive Psychotherapy*, 12 (2), 95–108.

Norcross, J. C. (2002). *Psychotherapy relationships that work: Therapist contributions and responsiveness to patients*. New York: Oxford University Press.

Norcross, J. C., Beutler, L. E., & Levant, R. L. (2006). *Evidence-based practices in mental health: Debate and dialogue on the fundamental questions* (pp. 299–307). Washington, DC: American Psychological Association.

Obegi, J., & Berant, E. (2009) (Eds.). *Attachment theory and research in clinical work with adults*. New York: Guilford Press.

O'Brien, A., Fahmy, R, & Singh, S. P. (2009). Disengagement from mental health services: A literature review. *Social Psychiatry & Psychiatric Epidemiology* 44 (7): 558–68.

Ogden, T. H. (2002). New reading of the origins of object relations theory. *International Journal of Psychoanalysis* 83: 767–82.

Ogrodniczuk, J. S., Joyce, A. S., & Piper, W. T. (2005). Strategies for reducing patient-inpatient premature termination of psychotherapy. *Harvard Review of Psychiatry* 13 (2): 57–70.

Olson, D. H., McCubbin, H. I., Barnes, H., Larsen, A., Muxen, M., & Wilson, M. (1983). *Families: What makes them work*. Newbury Park, CA: Sage.

Orlinsky, D. E., Ronnestad, M. H., & Willutzki, U. (2004). Process and outcome in psychotherapy. In M. J. Lambert (Ed.), *Bergin and Garfield's handbook of psychotherapy and behavior change* (5th ed., pp. 307–89). Hoboken, NJ: Wiley.

Parish, M., & Eagle, M. N. (2003). Attachment to the therapist. *Psychoanalytic Psychology* 20: 271–86.

Parker, S., & Thomas, R. (2009). Psychological differences in shame vs. guilt: Implications for mental health counselors. *Journal of Mental Health Counseling* 31: 213–24.

Passons, W. (1975). *Gestalt approaches in counseling*. New York: Holt, Rinehart & Winston.

Pederson, P., Crethar, H., & Carlson, J. (2008). *Inclusive cultural empathy: Making relationships central in counseling and psychotherapy*. Washington, DC: American Psychological Association.

Pedersen, P., Draguns, J., Lonner, W., & Trimble, J. (Eds.). (2008). *Counseling Across Cultures* (6th ed.). Thousand Oaks, CA: Sage.

Persons, J., Curtis, J., & Silberschatz, G. (1991). Psychodynamic and cognitive-behavioral formulations of a single case. *Psychotherapy: Theory, Research, Practice, Training* 28: 4.

Persons, J. B., & Silberschatz, G. (1998). Are results of randomized controlled trials useful to psychotherapists? *Journal of Consulting and Clinical Psychology* 66: 126–35.

Pesale, F. P., & Hilsenroth, M. T. (2009). Patient and therapist perspectives on session depth in relation to technique during psychodynamic psychotherapy. *Psychotherapy: Theory, Research, Practice, Training* 46: 390–96.

Peterson, D., Friedman, S., Geshmay, S., & Hill, C. E. (1998). Client perspectives on impasses. Manuscript in preparation, University of Maryland.

Pezzarossa, B., Della Rosa, A., & Rubino, A. I. (2002). Self-defeating personality and memories of parents' child-rearing behavior. *Psychological Reports* 91: 436–38.

Pinderhughes, H. (1989). *Understanding race, ethnicity and power*. New York: Free Press.

Pipher, M. (1994). *Reviving Ophelia*. New York: Ballantine.

Poorman, P. *Microskills*. Boston: Pearson, 2003.

Pos, A. E., Greenberg, L. S., Goldman, R. N., & Korman, L. M. (2003). Emotional processing during experiential treatment of depression. *Journal of Consulting and Clinical Psychology* 71: 1007–16.

Prochaska, J., DiClemente, C., & Norcross, J. (1992). In search of how people change: Applications to addictive behaviors. *American Psychologist* 47: 1102–14.

Prochaska, J., & Norcross, J. (2002). Stages of change. In J. C. Norcross (Ed.)., *Psychotherapy relationships that work*. New York: Oxford University Press.

———. (2006). *Systems of psychotherapy: A transtheoretical analysis* (6th ed.). Pacific Grove, CA: Brooks/Cole.

Prochaska, J., Norcross, J., & DiClemente, C. (2005). Stages of change: Perspective guidelines. In G. P. Koocher, J. C. Norcross, & S. S. Hill (Eds.), *Psychologists' desk reference* (2nd ed., pp. 226–31). New York: Oxford University Press.

Rappleyea, D., Harris, S. M., White, M., & Simon, K. (2009). Termination: Legal and ethical considerations for marriage and family therapists. *American Journal of Family Therapy* 37: 12–27.

Regan, A. M., & Hill, C. E. (1992). Investigation of what clients and counselors do not say in brief therapy. *Journal of Counseling Psychology* 39: 168–74.

Reik, T. (1948). *Listening with the third ear*. New York: Farrar, Straus & Giroux.

Reis, B. F., & Brown, L. G. (2006). Preventing therapy dropout in the real world: The clinical utility of video tape preparation and client estimate of treatment duration. *Professional Psychology: Research and Practice* 37: 311–16.

Renik, O. (1998). The role of countertransference enactment in a successful clinical psychoanalysis. In S. J. Ellman & M. Moskowitz (Eds.), *Enactment: Toward a new approach to the therapeutic relationship* (pp. 111–28). Northvale, NJ: Aronson.

Renik, O. (1999). Playing one's cards face up in analysis. *Psychoanalytic Quarterly* 68: 521–39.

Rennie, D. L. (1992). Qualitative analysis of the client's experience of psychotherapy: The unfolding of reflexivity. In S. G. Toukmanian & D. L. Rennie (Eds.), *Psychotherapy process research: Paradigmatic and narrative approaches* (pp. 211–233). Newbury Park, CA: Sage.

Rhodes, R. H., Thompson, B. J., & Elliott, R. (1994). Client retrospective recall of resolved and unresolved misunderstanding events. *Journal of Counseling Psychology* 41: 473–483.

Rizq, R. (2009). Teaching and transformation: A psychoanalytic perspective on psychotherapeutic training. *British Journal of Psychotherapy* 25 (3): 363–80.

Robbins, S. B., & Jolkovski, M. P. (1987). Managing countertransference feelings: An international model using awareness of feeling and theoretical framework. *Journal of Counseling Psychology* 34: 276–82.

Robitschek, C., & McCarthy, P. (1991). Prevalence of counselor self-reference in the therapeutic dyad. *Journal of Counseling & Development* 69: 218–21.

Rocklin R., & Levitt, D. (1987). Those who broke the cycle: Therapy with non-abusive adults who were physically abused as children. *Psychotherapy* 24 (4): 769–78.

Rogers, C. (1951). *Client-centered therapy*. Boston: Houghton Mifflin.

———. (1959). A theory of therapy, personality and interpersonal relationships as developed in the client-centered framework. In S. Koch (Ed.), *Psychology: A study of science*. Vol. 3, *Formulations of the person and the social context* (pp. 184–256). New York: McGraw-Hill.

———. (1975). Empathy: An unappreciated way of being. *Counseling Psychologist* 21: 95–103.

———. (1980). *A way of being*. Boston: Houghton Mifflin.

Rollnick, S., & Miller, W. R. (1995). What is motivational interviewing? *Behavioral and Cognitive Psychotherapy* 23: 325–34.

Rollnick, S., Miller, W. R., & Butler, C. C. (2008). *Motivational interviewing in health care.* New York: Guilford Press.

Romano, V., Fitzpatrick, M., & Janzen, J. (2008). The secure-base hypothesis: Attachment, attachment to counselor, and session exploration in psychotherapy, global. *Journal of Counseling Psychology* 55 (4): 495–504.

Roth, A., & Fonagy, P. (2005). *What works for whom? A critical review of psychotherapy research* (2nd ed.). New York: Guilford Press.

Rubak, S., Sandback, A., Lauritzen, T., & Christensen, B. (2005). Motivational interviewing: A systematic review and meta-analysis. *British Journal of General Practice* 55: 305–12.

Rubino, A. I., Pezzarossa, B., Della Rosa, A., & Siracusano, A. (2004). Self-defeating personality and memories of parent's child-rearing behavior : A replication. *Psychological Reports* 94: 733–35.

Rubino, G., Barker, C., Roth, T., & Faeron, D. (2000). Therapist empathy and depth of interpretation in response to potential alliance ruptures: The role of therapist and patient attachment styles. *Psychotherapy Research* 10: 408–20.

Russell, E., & Fosha, D. (2008). Transformational affects and core state in AEDP: The emergence and consolidation of joy, hope, gratitude, and confidence in (the solid goodness of) the self. *Journal of Psychotherapy Integration* 18 (2): 167–90.

Safran, J., & Muran, J. (1995). Resolving therapeutic alliance ruptures: Diversity and integration. *Psychotherapy in Practice* 1: 81–91.

————. (2000). *Negotiating the therapeutic alliance: A relational treatment guide.* New York: Guilford Press.

————. (2003). *Negotiating the therapeutic alliance: A relational treatment guide.* New York: Guilford Press.

Safran, J., & Muran, J. (Eds.). (1998). *The therapeutic alliance in brief psychotherapy.* Washington: DC: American Psychological Association.

Safran, J., Muran, J., & Proskurov, B. (2009). Alliance, negotiation, and rupture resolution. *Handbook of evidence-based psychodynamic psychotherapy.* New York: Humana Press.

Safran, J., Muran, J. C., Samstag, I., & Stevens, C. (2002). Repairing alliance ruptures. In J. C. Norcross (Ed.), *Psychotherapy relationships that work: Therapist contributions and responsiveness to patients* (pp. 235–54). London: Oxford University Press.

Safran, J. D., Muran, J. C., Samstag, L. W., & Winston, A. (2005). A comparative treatment study of potential treatment failures. *Psychotherapy: Theory, Research, Practice, Training* 42: 532–45.

Safran, J., & Segal, Z. (1990). *Interpersonal process in cognitive therapy.* New York: Basic Books.

Samoilov, A., & Goldfried, M. R. (2000). Role of emotion in cognitive-behavior therapy. *Clinical Psychology: Science and Practice* 7: 373–85.

Sampson, H. (2005). Treatment by attitudes. In G. Silberschatz (Ed.), *Transformative relationships: The control-mastery theory of psychotherapy* (pp. 111–120). New York: Routledge.

Samstag, L., Muran, J., & Safran, J. (2004). Defining and identifying alliance ruptures. In D. P. Charman (Ed.). *Core Processes in Brief Psychodynamic Psychotherapy: Advancing Effective Practice*, Mahwah, NJ: Lawrence Erlbaum Publishers. Pp. 187–214.

Samstag, L., Muran, J. C., Wachtel, P., & Slade, A. (2008). Evaluating negative process: A comparison of working alliance, interpersonal behavior, and narrative coherency among three psychotherapy outcome conditions. *American Journal of Psychotherapy* 62 (2): 165–94.

Satir, V. (1967). *Conjoint family therapy.* Palo Alto, CA: Science and Behavior Books.

Satir, V., & Bitter, J. R. (2000). The therapist and family therapy: Satir's human validation model. In A. Horne & J. L. Passmore (Eds.), *Family counseling and therapy* (3rd ed.). Pacific Grove, CA: Wadsworth-Brooks/Cole.

Sauer, E. M., Lopez, F. G., & Gormley, B. (2003). Respective contributions of therapist and client adult attachment orientations to the development of the working alliance: A preliminary growth modeling study. *Psychotherapy Research* 13: 371–82.

Scharff, J., & Scharff, D. (1997). Object relation couple therapy. *American Journal of Psychotherapy* 51: 141–73.

Schatzman, M. (1973). *Soul murder: Persecution in the family.* New York: Random House.

Scheff, T. J. (2000). *Bloody Revenge: Emotions, Nationalism, and War.* Lincoln, NE: iUniverse.com, Inc.

Schill, T., & Williams, D. (1993). Attachment histories for people with characteristics of self-defeating personality. *Psychological Reports* 73: 1232–34.

Schore, A. N. (2003). *Affect regulation and disorders of the self.* New York: Norton.

Seligman, M. (1975). *Helplessness.* New York: Freeman.

————. (1995). The effectiveness of psychotherapy: The Consumer Reports study. *American Psychologist* 50: 965–74.

————. (2000). Clinical implication of attachment theory. *Journal of the American Psychoanalytic Association* 48 (4): 1189–96.

Seligman, M., Rashid, T., & Parks, A. (2006). Positive psychotherapy. *American Psychologist* 61: 774–88.

Shapiro, D. (1989). *Psychotherapy of neurotic character*. New York: Basic Books.

Shaver, J., & Cassidy, P. R. (Eds.). (2008). *Handbook of attachment: Theory, research, & clinical applications* (2nd ed.). New York: Guilford Press.

Shaver, P. R., & Mikulincer, M. (2002). Attachment-related psychodynamics. *Attachment and Human Development* 4: 133–61.

Sheff, T. J. (1995). Shame and related emotions. (Special Issue). *American Behavioral Scientist* 38 (8).

Siegel, D. J. (2007). *The mindful brain: Reflection and attunement in the cultivation of well-being*. New York: Norton.

Sifneos, P. (1987). *Short–term dynamic psychotherapy: Evaluation and techniques* (2nd ed.). New York: Plenum.

Silberschatz, G. (Ed.). (2005). *The control-mastery theory: An integrated cognitive-psychodynamic relational theory*. New York: Routledge.

Silberschatz, G., & Curtis, J. T. (1993). Measuring the therapist's impact on the patient's therapeutic progress. *Journal of Consulting and Clinical Psychology* 61: 403–11.

Simon, J. L. (2002). Analysis of the relationship between shame, guilt, and empathy in intimate relationship violence (abstract). *Dissertation Abstracts International* 63 (6): 3026.

Slade, A. (1999). Attachment theory and research: Implications for the theory and practice of individual psychotherapy with adults. In J. Cassidy & P. R. Shaver (Eds.), *Handbook of attachment: Theory, research, and clinical applications* (pp. 575–94). New York: Guilford Press.

———. (2000). The development and organization of attachment: Implications for psychoanalysis. *Journal of the American Psychoanalytic Association* 48: 1147–74.

———. (2004). The move from categories to process: Attachment phenomena and clinical evaluation. *Infant Mental Health Journal* 25 (4): 269–83.

———. (2008). The implications of attachment theory and research for adult psychotherapy: Research and clinical perspective. In J. Cassidy and P. R. Shaver (Eds.), *Handbook of attachment: Theory, research, and clinical practice* (2nd ed.). New York: Guildford Press.

Slicker, E. K., & Thornberry, I. (2002). Older adolescent well-being and authoritative parenting. *Adolescent and Family Health* 3: 9–19.

Smith, A. E., Msetfi, R. M., & Golding, L. (2010). Client self rated adult attachment patterns and the therapeutic alliance: A systematic review. *Clinical Psychology Review* 30: 326–37.

Smith, J. (2006). Form and forming a focus in brief dynamic therapy. *Psychodynamic Practice* 12 (3): 261–79.

Snyder, C. R., (1994). *The psychology of hope: You can get there from here*. New York: Free Press.

Solomon, M. (1973). A developmental, conceptual premise for family therapy. *Family Process* 12 (2): 179–88.

Speight, S., Myers, L., Cox, C., & Highlen, P. (1991). A redefinition of multicultural counseling. *Journal of Counseling Development* 70 (1): 29–36.

Spiegel, D., & Alpert, J. L. (2000). The relationship between shame and rage: Conceptualizing the violence at Columbine High School. *Journal for the Psychoanalysis of Culture* 5 (2): 237–46.

Springmann, R. (1986). Countertransference: Clarifications in supervision. *Contemporary Psychoanalysis* 22: 252–77.

Sroufe, L. (2005). Attachment and development: A prospective, longitudinal study from birth to adulthood. *Attachment & Human Development* 7 (4): 349–67.

Sroufe, L., Egeland, B., Carlson, E., & Collins, W. (2005). *The development of the person: The Minnesota study of risk and adaptation from birth to adulthood*. New York: Guilford Press.

Steinbeck, J. (1952). *East of Eden*. New York: Viking Press.

Stern, S. (1994). Needed relationships and repeated relationships: An integrated relational perspective. *Psychoanalytic Dialogues* 4: 317–46.

Stierlin, H. (1972). *Separating parents and adolescents*. New York: Quadrangle.

Stolorow, R. D., Atwood, G. E., & Brandchaft, B. (1994). *The intersubjective perspective*. New York: Jason Aronson.

Stolorow, R. D., Brandchaft, B., & Atwood, G. E. (2000). *Psychoanalytic treatment: An intersubjective approach*. Mahwah, NJ: Analytic Press.

Stoltenberg, C. D., & McNeill, B. W. (2009). *IDM supervision: An integrated developmental model for supervising counselors and therapists* (3rd Ed.). San Fransisco: Jossey-Bass.

Strupp, H. (1980a). Success and failure in time-limited psychotherapy. *Archives of General Psychiatry* 37: 595–613, 708–16, 831–41, 947–54.

———. (1980b). Success and failure in time-limited psychotherapy: A systematic comparison of two cases. *Archives of General Psychiatry* 37: 595–603.

————. (1993). The Vanderbilt Psychotherapy Studies: Synopsis. *Journal of Consulting and Clinical Psychology* 61 (3): 431–33.

————. (1995). The psychotherapist's skills revisited. *Clinical Psychology* 2: 70–74.

Strupp, H., & Binder, J. (1984). *Psychotherapy in a new key: A guide to time-limited dynamic psychotherapy.* New York: Basic Books.

Strupp, H., & Hadley, S. (1979). Specific versus nonspecific factors in psychotherapy: A controlled study of outcome. *Archives of General Psychiatry* 36: 1125–36.

Strupp, H. H. (1995). The psychotherapist's skills revisited. *Clinical Psychology* 2: 70–74.

Sturge-Apple, M., Davies, P., & Cummings, E. (2006). Hostility and withdrawal in marital conflict: Effects on parental emotional unavailability and inconsistent discipline. *Journal of Family Psychology* 20: 227–38.

Sue, D., & Sue, H. (1998). *Counseling the culturally different* (3rd ed.). New York: Wiley.

Sue, D. W., & Sue, D. (2008). *Counseling the culturally different: Theory and practice* (5th ed.). Hoboken, NJ: Wiley.

Sue, S., & Zane, N. (1987). The role of culture and cultural techniques in psychotherapy: A critique and reformulation. *American Psychologist* 42: 37–45.

————. (2009). The role of culture and cultural techniques in psychotherapy: A critique and reformulation. *Asian American Journal of Psychology* 1: 3–14.

Sullivan, H. S. (1953). *The interpersonal theory of psychiatry.* New York: Norton.

————. (1968). *The interpersonal theory of psychiatry.* New York: Norton.

————. (1970). *The psychiatric interview.* New York: Norton.

Swift, J., & Callahan, J. (2008). A delay discounting measure of great experiences and the effectiveness of psychotherapy. *Professional Psychology: Research and Practice* 39 (6): 581–88.

Tangney, J. P. (1998). Are shame and guilt related to distinct self discrepancies? A test of the 1999 Higgins Hypothesis. *Journal of Personality and Social Psychology* 75 (1): 256–68.

Tangney, J., & Dearing, R. (2002). *Shame and guilt.* New York: Guilford Press.

Teyber, E. (1981). Structural family relations: A review. *Family Therapy* 1: 39–48.

————. (1983). Effects of the parental coalition on adolescent emancipation from the family. *Journal of Marital and Family Therapy* 9: 89–99.

————. (2001). *Helping children cope with divorce.* San Francisco: Jossey-Bass.

Teyber, E., & McClure, F. (2000). Therapist variables. In C. Snyder & R. Ingram (Eds.), *Handbook of psychological change;*

Psychotherapy processes and practices for the 21st century. New York: Wiley.

Thomaes, S., Stegge, H., & Olthof, T. (2007). Externalizing shame responses in children: The role of fragile positive self-esteem. *British Journal of Developmental Psychology* 25: 559–77.

Thompson, R. A. (2008). Early attachment and later development: Familiar questions, new answers. In J. Cassidy & P. R. Shaver (Eds.), *Handbook of attachment,* New York: Guilford Press.

Tomkins, S. (1962–1992). *Affect, imagery, consciousness* (4 vols.). New York: Springer.

Tomkins, S. (1967). Shame, In D. Nathanson (Ed.), *The many faces of Shame,* NY: Guilford, pp. 133–161.

Townsend, K., & McWhirter, B. (2005). Connectedness: A review of the literature with implications for counseling, assessment, and research. *Journal of Counseling & Development* 83: 191–201.

Tracey, J., Robins, R., & Tagney, J. (Eds.). (2007). *The self-conscious emotions: Theory and research.* New York: Guilford Press.

Tronick, E. (1998). Dyadically expanded states of consciousness and the process of therapeutic change. *Infant Mental Health Journal* 19 (3): 290–99.

Truax, C. B., & Carkhuff, R. R. (1967). *Toward effective counseling and psychotherapy.* Chicago: Aldine.

Tyron, G., & Kane, A. (1993). Relationship of working alliance to mutual and unilateral termination. *Journal of Counseling Psychology* 40 (1): 33–36.

Tyrone, G. S., & Winograd, G. (2002). Goal consensus and collaboration. In J. C. Norcross (Ed.), *Psychotherapy relationships that work.* New York: Oxford University Press.

Varghese, A., & Jenkins, S. (2009). Parental overprotection, cultural value conflict, and psychological adaptation among Asian Indian women in America. *Sex Roles* 61 (3): 235–51.

Vasquez, M. (2007). Cultural difference and the therapeutic alliance: An evidence-based analysis. *American Psychologist* 62 (8): 878–85.

Velasquez, R. J., Arellano, L. M., & McNeill, B. W. (2004). *The handbook of chicana/o psychology of mental health.* Mahwah, NJ: Erlbaum.

Vera R., Marilyn F., & Jennifer J. (2008). The secure-base hypothesis: Global attachment, attachment to counselor, and session exploration in psychotherapy. *Journal of Counseling Psychology* 55 (4): 495–504.

Villanueva, M., Tonigan, J. S., & Miller, W. R. (2007). Responses of Native American clients to three treatment methods for alcohol dependence. *Journal of Ethnicity in Substance Abuse*, 6 (12), 41–48.

Wachtel, P. (1982). *Psychoanalysis and behavior therapy: Toward an integration.* New York: Basic Books.

———. (1987). *Action and insight.* New York: Guilford Press.

———. (1993). *Therapeutic communication: Principles and effective practice.* New York: Guilford Press.

———. (1997). *Psychoanalysis, behavior therapy, and the relational world.* Washington, DC: American Psychological Association.

———. (2002). Termination of therapy: An effort at integration. *Journal of Psychotherapy Integration* 12 (3): 373–83.

———. (2008). *Relational theory and the practice of psychotherapy.* New York: Guilford Press.

Walborn, F. (1996). *Process variables: Four common elements of counseling and psychotherapy.* Pacific Grove, CA: Brooks/Cole.

Wallin, D. (2007). *Attachment in psychotherapy.* New York: Guilford Press.

Wampold, B. E. (2001). *The great psychotherapy debate: Models, methods, and findings.* Mahwah, NJ: Erlbaum.

———. (2006). Not a scintilla of evidence to support empirically supported treatments as more effective than other treatments. In J. C. Norcross, L. E. Beutler, & R. F. Levant (Eds.), *Evidence-based practices in mental health: Debate and dialogue on the fundamental questions* (pp. 299–307). Washington, DC: American Psychological Association.

Wampold, B. E., Imel, Z., Bhati, K., & Johnson, M. (2006). Insight as a common factor. In L. G. Castonguay & C. E. Hill (Eds.), *Insight in psychotherapy* (pp. 119–40). Washington, DC: American Psychological Association.

Wampold, B. E., Mondin, G. W., Moody, M., Stich, F., Benson, K., & Ahn, H. (1997). A meta-analysis of outcome studies comparing bona fide psychotherapies: Empirically "all must have prizes." *Psychological Bulletin* 122: 203–15.

Watson, J. C., & Greenberg, L. (2000). Alliance ruptures and repairs in experimental therapy. *Journal of Clinical Psychology* 56: 175–86.

Wei, M., Heppner, P. P., & Mallinckrodt, B. (2003). Perceived coping as a mediator between attachment and psychological distress: A structural equation modeling approach. *Journal of Counseling Psychology* 50: 438–47.

Wei, M., Shaffer, P. A., Young, S. K., & Zakalik, R. A. (2005). Adult attachment, shame, depression, and loneliness: The mediation role of basic psychological needs satisfaction. *Journal of Counseling Psychology* 52: 591–601.

Wei, M., & Ku, T. (2007). Testing a conceptual model of working through self-defeating patterns. *Journal of Counseling Psychology* 53 (3): 295–305.

Weiner, I. B., & Bornstein, R. F. (2009). *Principles of psychotherapy: Promoting evidence-based psychodynamic practice* (3rd ed.). Hoboken, NJ: Wiley.

Weinfield, N. S., Sroufe, L. A., Egeland, B., & Carlson, E. (2008). Individual differences in infant-caregiver attachment: Conceptual and empirical aspects of security. In J. Cassidy & P. R. Shaver (Eds.), *Handbook of attachment.* New York: Guilford Press.

Weiss, J. (1993). *How psychotherapy works.* New York: Guilford Press.

———. (2005). Safety. In G. Silberschatz (Ed.). *Transformative relationships: The control-mastery theory of psychotherapy* (pp. 31–43). New York: Routledge.

Weiss, J., & Sampson, H. (1986). *The psychoanalytic process: Theory, clinical observation and empirical research.* New York: Guilford Press.

Wells, M., Bruss, K. V., & Katrin, S. (1998). Abuse and addiction: Expressions of a wounded self and internalized shame. *Psychology: A Journal of Human Behavior* 35: 11–14.

Wells, M., & Jones, R. (2000). Childhood parentification and shame-proneness: a preliminary study. The *American Journal of Family Therapy* 28: 19–27.

Wenar, C., & Kerig, P. (2006). *Psychopathology from infancy through adolescence: A developmental approach* (5th ed.). New York: Random House.

Westin, D., Morrison, K., & Thompson-Brenner, H. (2004). The empirical status of empirically supported psychotherapies. *Journal of Consulting and Clinical Psychology* 69: 875–99.

Westin, D., Novotny, C. M., & Thompson-Brenner, H. (2004). The empirical status of empirically supported psychotherapies: Assumptions, findings, and reporting in controlled clinical trials. *Psychological Bulletin* 130: 631–63.

Westra, H. (2004). Managing resistance in cognitive behavioral therapy: The application of motivational interviewing in mixed anxiety and depression. *Cognitive Behavioral Therapy* 33: 161–75.

Wheelis, A. (1974). *How people change.* New York: Harper & Row.

Whelton, W. T., Paulson, B., & Marusiak, C. W. (2007). Self-criticism and the therapeutic relationship. *Counseling Psychology Quarterly* 20: 135–48.

White, B. (2007). Working with adult survivors of sexual and physical abuse. In T. Ronen & A. Freeman (Eds.), *Cognitive behavioral therapy in clinical social work practice.* New York: Springer.

White, W. L., & Miller, W. R. (2007). The use of confrontation in addiction treatment: History, science, and time for change. *Counselor* 8 (4): 12–30.

Wierzbicki, M., & Pekarik, G. (1993). A meta-analysis of psychotherapy dropout. *Professional Psychology: Research and Practice* 24: 190–95.

Wiger, D. (1999). *The psychotherapy documentation primer.* New York: Wiley.

Williams, D. C., & Levitt, H. M. (2007). Principles for facilitating agency in psychotherapy. *Psychotherapy and Research* 17: 66–82.

Williams, E. N., Judge, A., Hil, C. E., & Hoffman, M. A. (1997). Experience of novice therapists in prepracticum: Trainees', clients', and supervisees' perceptions of therapists' personal reactions and management strategies. *Journal of Counseling Psychology* 44: 390–99.

Williams, W. N., & Fauth, J. (2005). A psychotherapy process study of therapist in session self-awareness. *Psychotherapy Research* 15: 374–81.

Winnicott, D. W. (1965). Ego distortion in terms of true and false self (pp. 140–152). In D. W. Winnicott (1965), *The maturational process and the facilitating environment.* New York: International Universities Press.

Woodhouse, S. S., Schlosser, L. Z., Crook, R. E., & Ligiero, D. P. (2003). Client attachment to therapist: Relations to transference and client recollections of parental caregiving. *Journal of Counseling Psychology* 50: 395–408.

Woodward, B. (1984). *Wired.* New York: Simon & Schuster.

Yalom, I. (1975). *The theory and practice of group psychotherapy* (2nd ed.). New York: Basic Books.

———. (1981). *Existential psychotherapy.* New York: Basic Books.

———. (2003). *The gift of therapy: An open letter to a new generation of therapists and their patients.* New York: Perennial Currents.

Young, J. E. (1999). *Cognitive therapy for personality disorders: A schema-focused approach* (3rd ed). Sarasota: Professional Resource Press.

Young, J. E., Klosko, H. S., & Weishaar, M. E. (2003). *Schema therapy: A practitioner's guide.* New York: Guilford Press.

Young, J. E., Weinberger, A. D., & Beck, A. T. (2001). Cognitive therapy for depression. In D. H. Barlow (Ed.), *Clinical handbook of psychological disorders* (3rd ed.). New York: Guilford Press.

Zaki, J., Bolger, N., & Ochner, K. (2008). It takes two: The interpersonal nature of empathic accuracy. *Psychological Science* 19: 399–404.

Zeidner, M., & Endler, N. (Eds.). (1996). *Handbook of coping: Theory, research, and application.* New York: Wiley.

Zuckerman, A., & Mitchell, G. L. (2004). Psychology interns' perspectives on the forced termination of psychotherapy. *Clinical Supervisor* 23: 55–70.

Name Index

Subject Index

481